# Plant Genetics and Biotechnology

# Plant Genetics and Biotechnology

Editor: Edgar Crombie

R CALLISTO
REFERENCE

www.callistoreference.com

**Callisto Reference,**
118-35 Queens Blvd., Suite 400,
Forest Hills, NY 11375, USA

Visit us on the World Wide Web at:
www.callistoreference.com

ISBN: 978-1-64116-072-8 (Hardback)

**Cataloging-in-Publication Data**

Plant genetics and biotechnology / edited by Edgar Crombie.
    p. cm.
Includes bibliographical references and index.
ISBN 978-1-64116-072-8
1. Plant genetics. 2. Plant biotechnology. 3. Crops--Genetics. 4. Agricultural biotechnology.
I. Crombie, Edgar.
QK981 .P53 2019
581.35--dc21

# Table of Contents

# Preface

The world is advancing at a fast pace like never before. Therefore, the need is to keep up with the latest developments. This book was an idea that came to fruition when the specialists in the area realized the need to coordinate together and document essential themes in the subject. That's when I was requested to be the editor. Editing this book has been an honour as it brings together diverse authors researching on different streams of the field. The book collates essential materials contributed by veterans in the area which can be utilized by students and researchers alike.

Plant biotechnology is the science of developing methods of genetic modification in plants to achieve certain desired characteristics. Genetic modification is the altering of plants' genetic traits by manipulating genes. In recent years, plant breeders have developed innovative methods of genetic modification for increased crop yield. The techniques related to this field have applications across a number of scientific fields such as plant physiology, botany, agronomy, plant biochemistry, plant pathology, etc. This book presents upcoming concepts and theories related to the fields of plant biotechnology and genetics. It strives to provide a fair idea about this discipline and to help develop a better understanding of the latest advances within this field. Those in search of information to further their knowledge will be greatly assisted by this book.

Each chapter is a sole-standing publication that reflects each author's interpretation. Thus, the book displays a multi-facetted picture of our current understanding of application, resources and aspects of the field. I would like to thank the contributors of this book and my family for their endless support.

**Editor**

# A modular toolbox for gRNA–Cas9 genome engineering in plants based on the GoldenBraid standard

Marta Vazquez-Vilar[1], Joan Miquel Bernabé-Orts[1], Asun Fernandez-del-Carmen[1], Pello Ziarsolo[2], Jose Blanca[2], Antonio Granell[1] and Diego Orzaez[1*]

## Abstract

**Background:** The efficiency, versatility and multiplexing capacity of RNA-guided genome engineering using the CRISPR/Cas9 technology enables a variety of applications in plants, ranging from gene editing to the construction of transcriptional gene circuits, many of which depend on the technical ability to compose and transfer complex synthetic instructions into the plant cell. The engineering principles of standardization and modularity applied to DNA cloning are impacting plant genetic engineering, by increasing multigene assembly efficiency and by fostering the exchange of well-defined physical DNA parts with precise functional information.

**Results:** Here we describe the adaptation of the RNA-guided Cas9 system to GoldenBraid (GB), a modular DNA construction framework being increasingly used in Plant Synthetic Biology. In this work, the genetic elements required for CRISPRs-based editing and transcriptional regulation were adapted to GB, and a workflow for gRNAs construction was designed and optimized. New software tools specific for CRISPRs assembly were created and incorporated to the public GB resources site.

**Conclusions:** The functionality and the efficiency of gRNA–Cas9 GB tools were demonstrated in *Nicotiana benthamiana* using transient expression assays both for gene targeted mutations and for transcriptional regulation. The availability of gRNA–Cas9 GB toolbox will facilitate the application of CRISPR/Cas9 technology to plant genome engineering.

**Keywords:** Plant gene editing, Plant gene activation, Plant gene repression, CRISPR/Cas9, gRNAs, Multigenic assemblies, GoldenBraid, Luciferase/renilla assay

## Background

Since its discovery, the clustered regularly interspaced short palindromic repeats (CRISPR)-Cas immune bacterial system has rapidly become a powerful technology for genome editing in many organisms. This system is based on a guide RNA (gRNA) that directs the *Streptococcus pyogenes* Cas9 nuclease to its target site. The application of the RNA-guided Cas9 technology is being widely exploited by the scientific community in cell cultures [1], animals [2, 3] or plants [4, 5].

On the plant field, RNA-guided genome engineering via Cas9 has been employed in diverse approaches, from single and/or multiple gene knock-outs [6–8] to targeted insertions of donor sequences [9] or even targeted transcriptional regulation through the fusion of transcriptional activation or repressor domains to an inactivated Cas9 [10]. A remarkable feature of gRNA–Cas9 is that facilitates targeting multiple sequences simultaneously. While similar technologies such as the ZFNs (zinc finger nucleases) [11] or the TAL effectors [12] require recoding of a new protein for each target sequence, with the gRNA–Cas9 a change of 20 nts in the guide RNA is

*Correspondence: dorzaez@ibmcp.upv.es
[1] Instituto de Biología Molecular y Celular de Plantas (IBMCP), Consejo Superior de Investigaciones Científicas, Universidad Politécnica de Valencia, Camino de Vera s/n, 46022 Valencia, Spain
Full list of author information is available at the end of the article

enough, paving the way for multiplex editing and design of complex regulatory circuits among other engineering possibilities [13].

The direct transfection of Cas9 and guide RNAs into plant protoplasts followed by plant regeneration from single-cell has been shown effective for genome editing in rice and tobacco, however the efficiency remained relatively low, and besides, whole plant regeneration from protoplasts is not currently feasible for many crop species [14]. A successful alternative for plants is the use of Agrobacterium mediated T-DNA transformation, followed by callus induction and organogenic plant regeneration (or floral dip transformation in the case of Arabidopsis). In this case, T-DNA-delivered gRNA–Cas9, besides acting transiently during callus formation, can also integrate in the genome and continue its activity in somatic tissues [4]. To exploit the full potential of the T-DNA strategy it is important to expand the ability to combine different gRNAs together with Cas9 within a single T-DNA, as it has been demonstrated that all-in-one plasmid approaches significantly increase editing efficiency [15].

Modular cloning methods are being increasingly adopted by the plant research community as they greatly facilitate the combinatorial assembly of pre-made DNA elements into multigene constructs [16, 17]. GoldenBraid is a modular cloning standard that makes use of the Type IIS restriction enzyme BsaI for the assembly of basic, so-called "level 0" DNA elements (promoters, coding regions, terminators, etc.) into transcriptional units (TUs), and then incorporates a second enzyme, BsmBI, to build higher level structures using a double-loop iterative strategy [18]. Level 0 parts are flanked by 4 nucleotides overhangs, the sequence of which determines the relative position of each part in the transcriptional unit. To be usable in GB cloning, all level 0 parts need to be previously adapted with the incorporation of flanking BsaI recognition sites, the addition of flanking 4 bp standard barcodes, and the removal of internal BsmBI and BsaI sites. The whole process of adaptation to the standard is often referred to as "domestication". Once domesticated, GB parts can be efficiently combined to create large multigenic constructs within binary destination plasmids ready to be used in Agrobacterium-mediated plant transformation. A key feature of GB is that all constructs can be reused in new combinations following the same cloning scheme, fostering the exchange of genetic elements. Interestingly, GB part reusability enables the unequivocal association of physical parts with experimental information, as no further modifications (i.e. subcloning, re-assembly or PCR re-amplification) are required to incorporate a GB part into different genetic modules. The GB webpage (https://gbcloning.upv.es/) offers a set of online tools for 'in silico' multigenic assemblies and a

database for the collection and exchange of GB standard parts [19]. Although Type IIS cloning methods have been employed for multi-gene assemblies with a wide range of applications in several organisms [20, 21], the GB framework is specially designed for plants since the GB destination plasmids are two sets of binary vectors (one based on pGreen and a second one based on pCambia) and all the GB standard parts including promoters and terminators are suitable for plant biotechnology.

The GB cloning strategy is especially suited for the construction of vectors incorporating Cas9 together with multiple guide RNAs in the same T-DNA. Here, we report the implementation of a GB-adapted gRNA–Cas9 toolbox for plants, which includes the domestication of gRNA/Cas9 elements, the definition of a CRISPR cloning workflow and incorporation of new online tools for building CRISPR-based genome engineering constructs in binary vectors.

## Results
### GB-adapted cloning strategy for CRISPR/Cas9 plant constructs

To facilitate the assembly of CRISPR/Cas9 constructs and the delivery of multiple guide RNAs in the same T-DNA, we designed the CRISPR cloning workflow depicted on Fig. 1a. As a first step, twenty nucleotides sequences designed against a specific genomic target can be incorporated to the GoldenBraid scheme using the 'GB CRISPR domesticator' tool available at https://gbcloning.upv.es/do/crispr/. This tool generates a new target-specific GB element (D-Target/M-Target, syntax structure B3c–B4–B5c or B3c–B4–B5d), which can be used immediately or stored in the database for future assemblies. The D/M-Target comprises two partially complementary oligonucleotides yielding a double-stranded DNA fragment flanked by four nucleotides overhangs. In a next step, the D/M-Target is combined with a PolIII promoter (currently, Arabidopsis U6-26 and U6-1 and rice U3 promoters are available in the GB collection) and with the scaffold RNA in a cyclic digestion/ligation Golden Gate reaction [22] to build the complete gRNA expression cassette. This step is assisted by the 'CRISPR Assembler' tool available at https://gbcloning.upv.es/tools/crisprsassembler.

The conditions for gRNA assembly were optimized by checking three key parameters, namely primer concentration, primer dilution buffer and annealing conditions in a total of 12 combinations. The resulting assemblies were then transformed into E. coli and the efficiency assessed by the number of colonies obtained (Fig. 1b, c). Two colonies of each of the 12 assembly reactions were selected for restriction analysis resulting in a 100 % of positive clones (see Additional File 1: Figure S1). Primer dilution was found the main factor affecting reaction efficiency, with

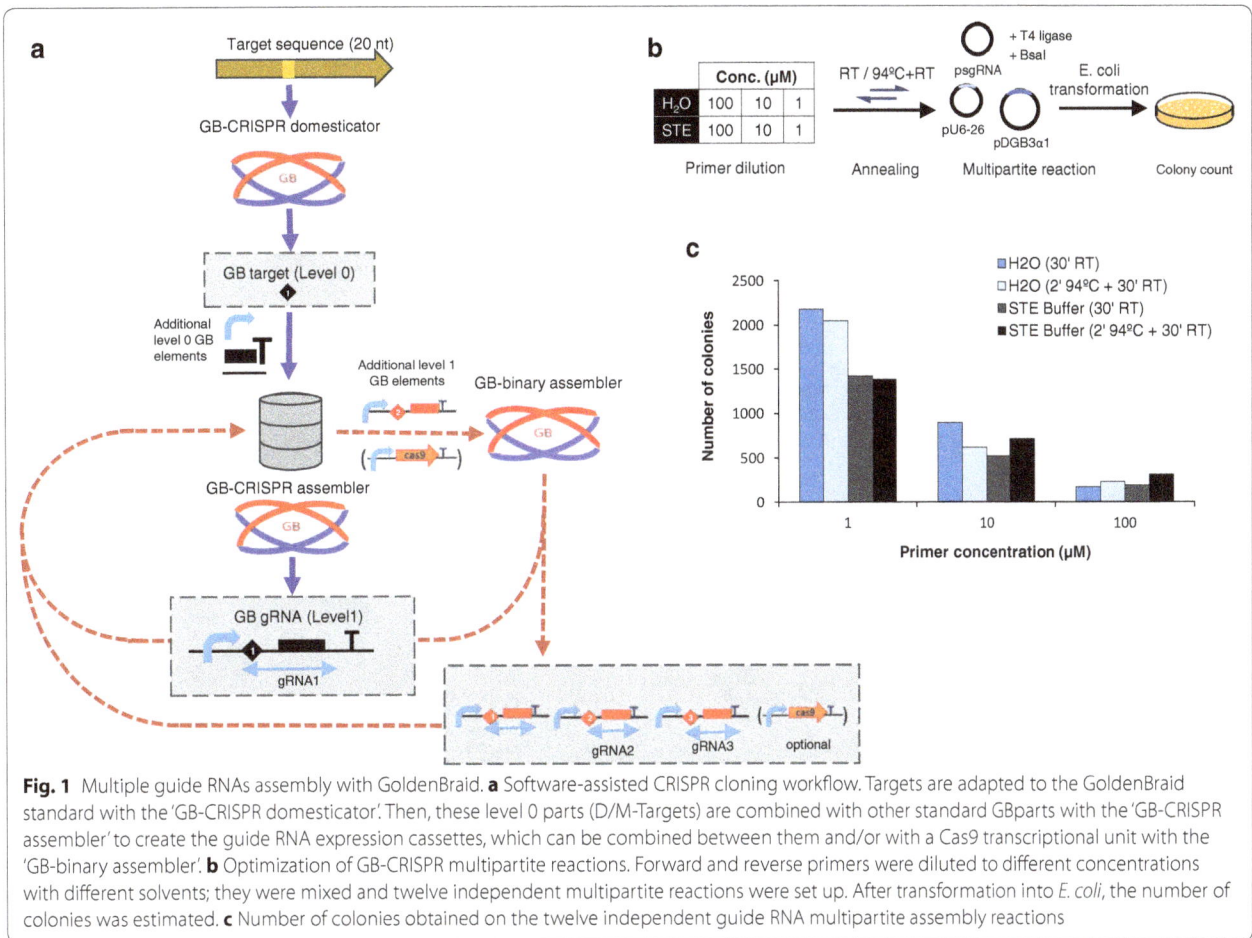

**Fig. 1** Multiple guide RNAs assembly with GoldenBraid. **a** Software-assisted CRISPR cloning workflow. Targets are adapted to the GoldenBraid standard with the 'GB-CRISPR domesticator'. Then, these level 0 parts (D/M-Targets) are combined with other standard GBparts with the 'GB-CRISPR assembler' to create the guide RNA expression cassettes, which can be combined between them and/or with a Cas9 transcriptional unit with the 'GB-binary assembler'. **b** Optimization of GB-CRISPR multipartite reactions. Forward and reverse primers were diluted to different concentrations with different solvents; they were mixed and twelve independent multipartite reactions were set up. After transformation into *E. coli*, the number of colonies was estimated. **c** Number of colonies obtained on the twelve independent guide RNA multipartite assembly reactions

best results obtained at low primer concentrations. Only minor effects were observed associated to buffer or denaturing condition (Fig. 1b, c). Accordingly, recommended conditions for CRISPR assembly in multipartite GB reactions were set at 1 µM primer concentration in water with a 30 min annealing step performed at room temperature.

Following the GB workflow, every gRNA expression cassette assembled in GB compatible vectors can be combined with each other and/or with a Cas9-encoding transcriptional unit (Fig. 1a) with the 'GB Binary Assembler' web tool (https://gbcloning.upv.es/do/bipartite/). GB binary reactions were highly efficient as previously described Sarrion-Perdigones et al. [23] and accurate since white colonies analyzed resulted in 100 % correct assemblies in most cases (see Additional file 1: Figure S1; Additional file 2: Table S3). The current GB-adapted gRNA–Cas9 toolbox incorporates seven different Cas9-encoding TUs which have been designed for gene editing, gene activation and gene repression projects. All Cas9 TUs described in this paper were created by combining only protein-coding GBparts, leaving constitutive plant expression elements invariant. The assembly of

inducible and/or tissue-specific expression of Cas9 is also possible using other standard parts from the collection.

### Transient expression of GB-adapted Cas9 TUs provides efficient targeted mutagenesis in *N. benthamiana* leaves

To experimentally validate the different GB modules for gRNA–Cas9-mediated gene mutation, we tested them in *N. benthamiana* by targeting the endogenous xylosyltransferase (XT) gene. A BLAST search on the *N. benthamiana* genome with the GenBank accession ABU48858, resulted in scaffolds Niben101Scf04205Ctg025 and Niben101Scf04551Ctg021 corresponding to predicted cDNAs Niben101Scf04205g03008 (XT1) and Niben101Scf04551g02001 (XT2) respectively. We decided to target the two of them using a specific guide RNA for each one. The 20-bp target sequences for each guide RNAs were designed with the CRIPSR-P online tool [24], imposing the requirement for a G at the 5′ end of the sequence and minimizing off-targeting. An extra criterion for selection was the presence of a restriction site overlapping the Cas9 cleavage site to facilitate the detection of the mutations. The selected targets are depicted on Fig. 2a.

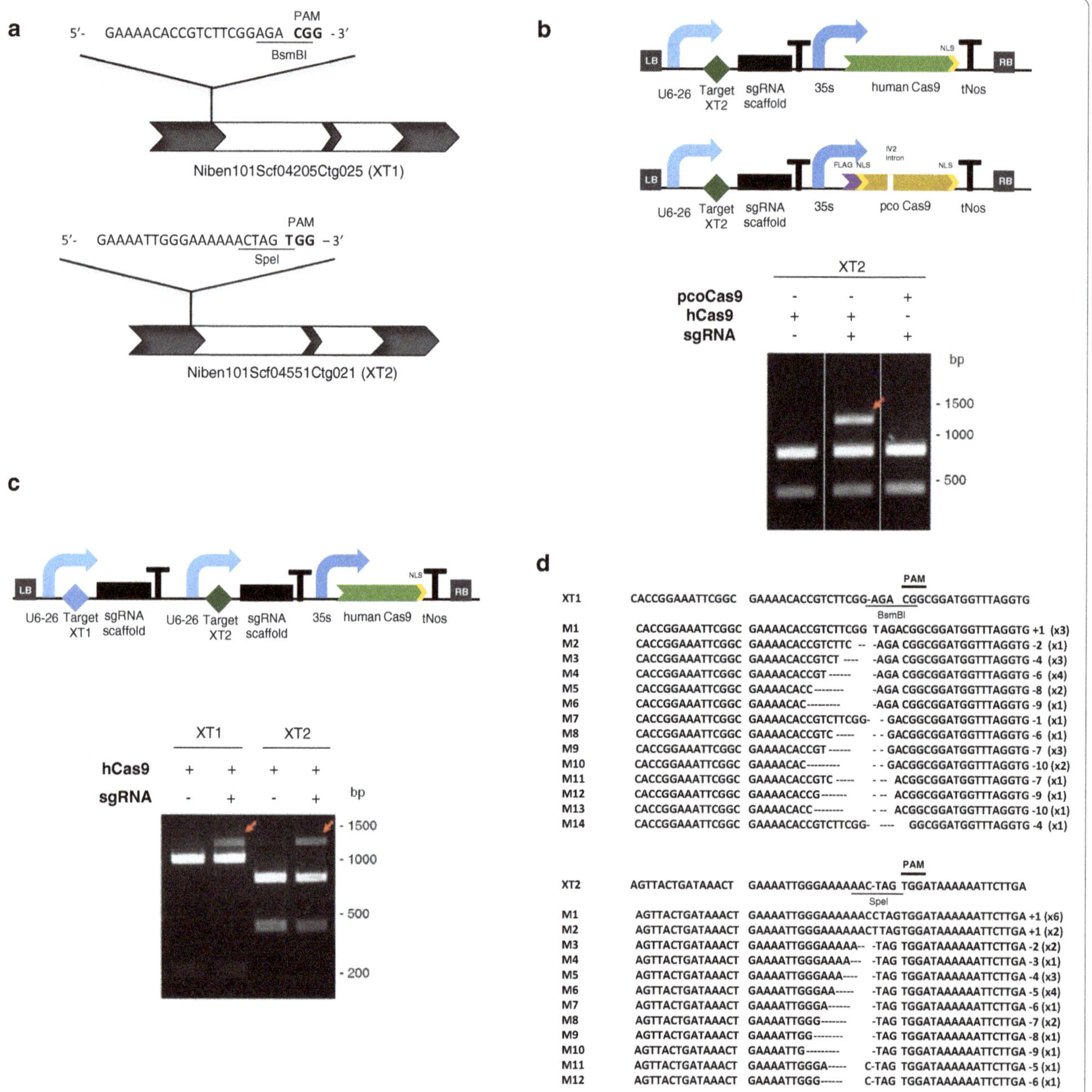

**Fig. 2** Targeted mutagenesis using the CRISPR/Cas9 system in transient expression in *N. benthamiana*. **a** Schematic representation of the structure of Niben101Scf04205Ctg025 (XT1) and Niben101Scf04551Ctg021 (XT2) (exons in *grey*, introns in *white*) with the sequences of the target sites. Diagnostic restriction sites are underlined and the PAM sequence is shown in *bold*. **b** Comparison of the mutation efficiency of hCas9 and pcoCas9 targeting the XT2. *Red arrow* shows SpeI resistant PCR fragments only visible on the gRNA and hCas9 combination. **c** PCR/RE assay to detect simultaneous targeted mutations on XT1 and XT2. *Red arrows* show BsmBI and SpeI resistant PCR fragments amplified from *N. benthamiana* genomic DNA. **d** Alignment of XT1 and XT2 sequences obtained from different clones of uncleaved bands (see **c**). XT1 target site appears in blue and XT2 target site in green. *Red letters* and *dashes indicate* insertions and deletions respectively

GB-based gene targeting constructs carrying human-optimized (h) [25] and plant-optimized (pco) [26] Cas9 variants directed to the single target of XT2 were transferred to Agrobacterium and infiltrated into *N. benthamiana* leaves. To test the mutation efficiency, genomic DNA was extracted from leaves, the targeted region amplified by PCR and the presence of mutated fragments estimated based on the elimination of the internal SpeI restriction enzyme (RE) site. The mutation efficiency for the hCas9 was estimated as 11 % based on the intensity of the undigested band (Fig. 2b Lanes 2 and 3) relative to the undigested DNA present on the negative control

(Fig. 2b Lane 1). For pcoCas9 mutation efficiency was below detection levels as it was not possible to visualize the undigested band on the agarose gel.

According to these results we assembled both gRNAs targeting XT1 and XT2 together with the hCas9 TU in a single T-DNA and transiently expressed them in *N. benthamiana* leaves. hCas9-induced mutations were detected as above with the restriction enzyme site loss method using BsmBI for XT1 and SpeI for XT2 (Fig. 2c). The gRNA-guided Cas9 activity resulted in part of the DNA being resistant to RE digestion (see undigested band in Lanes 2 and 4) that was not detected when only hCas9 was expressed (Lanes 1 and 3). To corroborate the presence of mutations on the undigested PCR products, the undigested amplicons were cloned and individual clones were sequenced. The most prevalent mutations observed for XT1 were deletions of less than 10 nucleotides, while for XT2 a 32 % of the mutated clones had single nucleotide insertions (C or T) (Fig. 2d). Mutation rates of 17 % (XT1) and of 14.5 % (XT2) were observed for the new construct. Since 29 % (XT1) and 32 % (XT2) of the clones showed the wild type sequence, we included this correction factor to obtain a more accurate estimation of the mutation rate. As result, we obtained a mutation rate of 12.1 % for XT1 and a mutation rate of 9.9 % for XT2, consistent with the 11 % obtained for the same target when a single gRNA was used. The differences in the mutation efficiencies observed in both targets could be due to a GC content of 30 % for target XT2 in contrast to a 50 % GC content of target XT1.

## GB-adapted dCas9 variants modulate transcriptional activity *in N. benthamiana* transient assays

The modularity of GoldenBraid assembly facilitates the design of Cas9 variants with novel functions as e.g. transcriptional activators, repressors, chromatin remodeling factors, etc., by incorporating additional coding modules as translational fusions to an inactive (dead) version of Cas9 (dCas9). To validate this option we built and tested a number of GB-based transcriptional regulators which were targeted to a nopaline synthase promoter (pNOS) fused to a luciferase reporter.

Making use of level 0 standard genetic parts, we assembled five different transcriptional units (TUs) expressing either the dCas9 (D10A H840A) alone or C-terminus chimeric versions of it fused either to an activator (VP64 or EDLL) or a repressor (SRDX and BRD) (Additional file 1: Figure S2). These five chimeric transcriptional regulators were tested in combination with five gRNAs directed against different regions of pNOS on both sense and antisense strands (Fig. 3a). Changes in the transcriptional activity in these construct were estimated with the luciferase/renilla system using a reporter construct (REP)

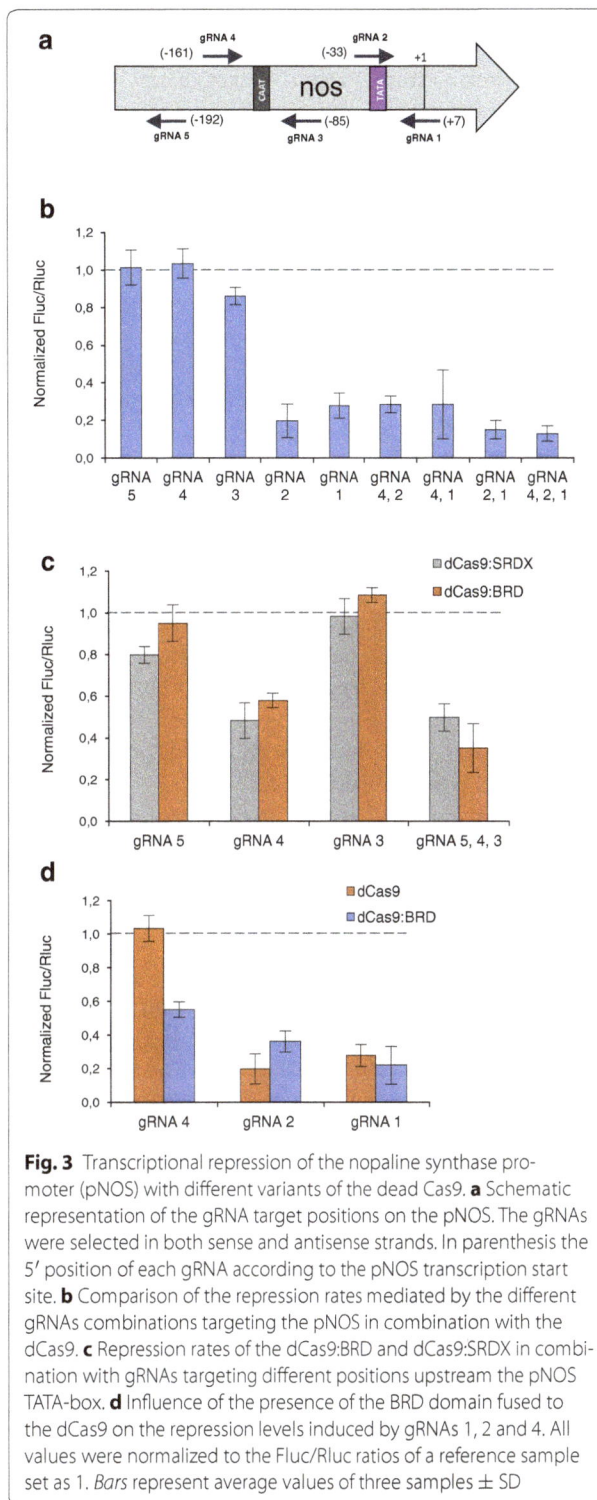

**Fig. 3** Transcriptional repression of the nopaline synthase promoter (pNOS) with different variants of the dead Cas9. **a** Schematic representation of the gRNA target positions on the pNOS. The gRNAs were selected in both sense and antisense strands. In parenthesis the 5′ position of each gRNA according to the pNOS transcription start site. **b** Comparison of the repression rates mediated by the different gRNAs combinations targeting the pNOS in combination with the dCas9. **c** Repression rates of the dCas9:BRD and dCas9:SRDX in combination with gRNAs targeting different positions upstream the pNOS TATA-box. **d** Influence of the presence of the BRD domain fused to the dCas9 on the repression levels induced by gRNAs 1, 2 and 4. All values were normalized to the Fluc/Rluc ratios of a reference sample set as 1. *Bars* represent average values of three samples ± SD

that included the firefly luciferase (Fluc) driven by the pNOS and the renilla luciferase (Rluc) driven by the 35S promoter as an internal reference. Transient co-transformations of REP with Cas9 and gRNA constructs were performed in order to test the ability of GB-built dCas9 chimeras to modulate transcription.

Since previous studies reported that dCas9 itself could act as a transcriptional repressor [27], we first tested the repressor activity of the non-chimeric dCas9 TU. All five gRNAs targeting pNOS induced variable repression rates depending on their position (Fig. 3b). The Fluc/Rluc ratios decreased as the position of the gRNA gets closer to the Transcription Start Site (TSS) whereas no repression was detected neither for gRNA4 (positions −161 to −142) nor for gRNA5 (positions −211 to −192). Co-expression of the two most effective gRNAs, gRNA 1 and 2, showed a nearly additive effect. However, the addition of a further gRNA, such as gRNA4, to one or both of them did not change the repression level.

Next, the dCas9 fusions to the BRD and the SRDX repressor domains were tested in combination with gRNAs 3, 4 and 5, all three designed to bind upstream the TATA-box. Figure 3c shows that only gRNA4, the gRNA designed on the sense strand, was capable of producing a significant repression on the transcriptional activity. A slight decrease in the Fluc/Rluc ratio was observed when gRNA4 was combined with the two additional gRNAs. The repression levels found with the dCas9:BRD and dCas9:SRDX were similar (Fig. 3c).

To determine whether the presence of the repressor domain modified the effect of the dCas9 itself, we compared the transcriptional activity obtained for the gRNAs 1, 2 and 4 in presence of the dCas9 with the ones obtained with the dCas9:BRD (Fig. 3d). While in the case of the gRNA4 only dCas9:BRD had an effect on the reduction of the transcriptional activity, for the gRNAs overlapping the TATA-box and the TSS, both dCas9 and dCas9:BRD achieved similar repression levels.

Next, we decided to test whether the dCas9 fused to an activator domain could increase the transcriptional activity on the same reporter construct. The results showed that dCas9:VP64 and dCas9:EDLL raised the reporter levels in combination with gRNA4, while in combination with gRNA5 only a small induction rate was detected and no induction was observed with gRNA3, corroborating the functionality observed for the same gRNAs with dCas9:SRDX and dCas9:BRD (Fig. 4a). Using both the dCas9:VP64 and the dCas9:EDLL variants in combination with 3× multiplexed gRNAs (gRNA 3, 4 and 5), the pNOS transcriptional activity was doubled.

These results demonstrated that it is possible to modulate the transcriptional activity driven by the pNOS using one or more gRNAs in combination with different chimeric versions of the dCas9. The maximum induction rate, calculated with the values of the best reported repression and activation Fluc/Rluc ratios, was 6.5× (Fig. 4b).

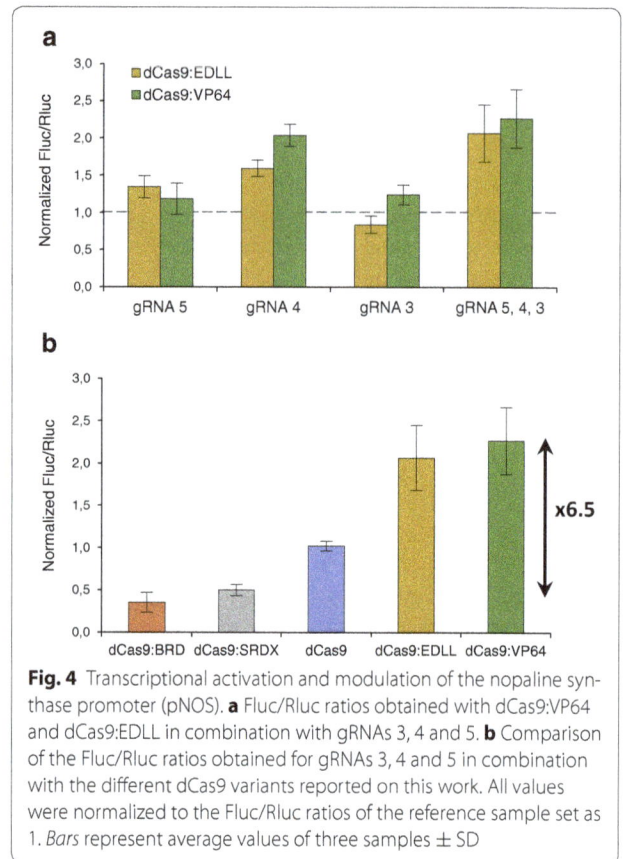

**Fig. 4** Transcriptional activation and modulation of the nopaline synthase promoter (pNOS). **a** Fluc/Rluc ratios obtained with dCas9:VP64 and dCas9:EDLL in combination with gRNAs 3, 4 and 5. **b** Comparison of the Fluc/Rluc ratios obtained for gRNAs 3, 4 and 5 in combination with the different dCas9 variants reported on this work. All values were normalized to the Fluc/Rluc ratios of the reference sample set as 1. *Bars* represent average values of three samples ± SD

## Second-dimension multiplexing using GoldenBraid

To further increase the gRNA multiplexing capacity we decided to incorporate a polycistronic strategy to the GB pipeline. This strategy, which has been validated in rice [28], allows the simultaneous expression in a single transcript of multiple gRNAs, which are later processed by the endogenous tRNA ribonucleases P and Z to produce the individual gRNAs. To adapt the general GB cloning system to the polycistronic strategy we incorporated single tRNA–gRNA oligomers as level 0 GBparts, which are then multipartitely assembled on level 1 to create polycistronic tRNA–gRNAs (Fig. 5a). To avoid using PCR reactions during the construction of each tRNA–gRNA oligomer, we designed new level −1 plasmids containing both the tRNA and the gRNA flanked by BsmBI restriction sites. The BsmBI assembly of level −1 plasmids with the D-target primers heteroduplex results in level 0 GB-oligomers. In turn, these level 0 elements are combined together with the level 0 PolIII promoter to create a level 1 polycistronic tRNA–gRNA in a software-assisted step available at https://gbcloning.upv.es/do/multipartite/free/. We validated the assembly efficiency of the 2-D multiplexing schema by assembling a level 2 construct targeting simultaneously *N. benthamiana*

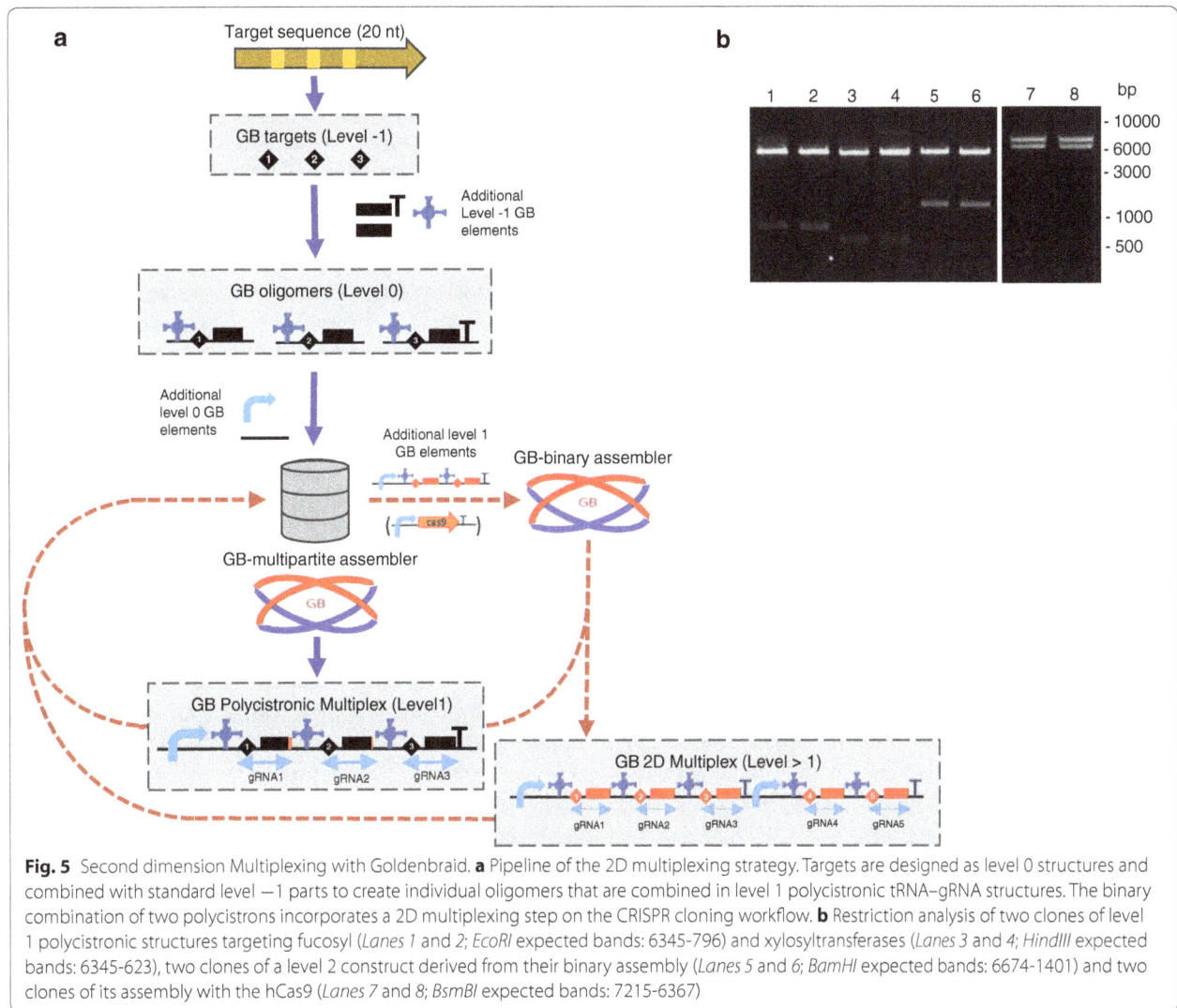

**Fig. 5** Second dimension Multiplexing with Goldenbraid. **a** Pipeline of the 2D multiplexing strategy. Targets are designed as level 0 structures and combined with standard level −1 parts to create individual oligomers that are combined in level 1 polycistronic tRNA–gRNA structures. The binary combination of two polycistrons incorporates a 2D multiplexing step on the CRISPR cloning workflow. **b** Restriction analysis of two clones of level 1 polycistronic structures targeting fucosyl (*Lanes 1* and *2*; *EcoRI* expected bands: 6345-796) and xylosyltransferases (*Lanes 3* and *4*; *HindIII* expected bands: 6345-623), two clones of a level 2 construct derived from their binary assembly (*Lanes 5* and *6*; *BamHI* expected bands: 6674-1401) and two clones of its assembly with the hCas9 (*Lanes 7* and *8*; *BsmBI* expected bands: 7215-6367)

fucosyl and xylosyltransferase genes. As the two gRNAs targeting XTs have been previously tested in this work, we used the same targets (Additional file 2: Table S2) for the assembly of a polycistronic tRNA–gRNA combining two GBoligomers. Since the number of genes encoding fucosyltransferases in the *N. benthamiana* genome is very high, we decided in this example to target only five of them using a combination of three gRNAs (Additional file 2: Table S2), one of them targeting three genes and the remaining two gRNAs targeting a single gene. After assembling firstly all five level 0 oligomers and subsequently the two level 1 polycistronic structures, they were combined together in a GB binary reaction (Fig. 5b) to generate a single binary plasmid containing all five gRNAs targeting a total of seven genes encoding fucosyl and xylosyltransferases. All the assembly steps resulted in 100 % accuracy rates (at least 4 white colonies analysed in

each step) demonstrating the efficiency of the proposed scheme for 2D multiplexing. The whole process took just nine working days, and in three extra days the Cas9 was added to the assembly.

## Discussion

The adoption of standard rules and modular design has promoted the expansion of many engineering disciplines from mechanics to electronics and is likely to have an impact in genome engineering as well. Modular cloning methods based on TypeIIS restriction enzymes such as Golden Gate [22], MoClo [29] and GoldenBraid [23], greatly facilitate the construction of large multigene assemblies enabling the concurrent delivery of multiple pieces of genetic information into the cell. Moreover, Type IIS cloning systems are especially well suited for the definition of standard assembly rules. Very recently, a common DNA assembly syntax for TypeIIS cloning has been agreed by 26 different Plant laboratories and research groups worldwide, constituting one of the first examples of a Bioengineering Standard adopted by the Scientific Community [16]. We have introduced the necessary modifications in GB to make the gRNA–Cas9 toolbox fully compliant with the new standard.

The first step towards GB adaptation for gene targeted mutation consisted in the design of a GB-compatible assembly scheme that facilitates both gRNA multiplexing and Cas9 modification. We decided to build both gRNAs and Cas9 transcriptional units as level 1 structures to maximize their exchangeability while preserving the combinatorial potential. In the GB system, level 1 constructs grow only binarely, which poses a certain limitation in terms of cloning speed. Other systems growing multipartitely using Golden Gate assembly have been proposed for mammalian and plant systems, however this is at the cost of flexibility and reusability of the constructs [30–32]. Conversely, level 1 GB constructs are exchangeable, offering the possibility to reuse efficient gRNA constructs in new editing or regulatory combinations. Furthermore, this initial decision proved to be most adequate with the incorporation of polycistronic tRNA–gRNA constructs at level 1, which provides a new combinatorial dimension for multiplexing, and makes possible to hierarchically combine gRNAs using different assembly levels. Hence, in our 2D editing example we grouped homologous functions (either xylosyl or fucosyltransferases) in level 1, and later combined them in level 2 in a binary assembly step. Similarly hierarchical assembly approaches can be used to build increasingly complex gRNA–Cas9-based transcriptional regulatory circuits in few days.

The assembly and functional validation of several gRNA–Cas9 constructs provides evidence of the efficiency of the process and the functionality of the elements that were incorporated to the GB toolkit. GB is based on Golden Gate typeIIS cloning which is an extremely efficient multipartite assembly method when parts are conveniently cloned within an entry plasmid. Whether the same high efficiency is maintained when one of the parts is made of two partially overlapping 23–25 mer oligonucleotides encoding the target sequence remained to be tested. Counterintuitively, the efficiency of the reaction was shown to be significantly higher when low concentrations of oligonucleotides (nM range) were employed in the reaction mix. Also, it is worth to notice that in the proposed GB gRNA building scheme, the only variable input specific for each new construct are the two 25 mer oligonucleotides; all the remaining building elements are invariant and stored in the GB collection, a feature that significantly reduces gene synthesis costs for building gRNA–Cas9 constructs for plants.

The first functional characterization of the new GB targeted mutagenesis tools was the quantification of Cas9 nuclease activity in a *N. benthamiana* transient expression method [26, 33]. As shown, efficiencies up to 12 % were observed using a human codon optimized Cas9 (hCas9) directed against two independent targets. In our hands hCas9 performed better than plant-optimized pcoCas9 in *N. benthamiana* transient assays, although it remains to be seen if the same differences are observed in other experimental systems. The mutation rate observed here with the hCas9 is consistent with those described when hCas9 and gRNAs were assembled in the same T-DNA [34] and much higher than the rates obtained by [34] and [33] when the same were co-delivered in different plasmids by in *trans* co-transformation. The reported efficiency for the plant-optimized pcoCas9 when co-expressed with the gRNA on the same vector was substantially lower (4.8 %) [26]. Therefore it is possible that our detection system based on the presence of an undigested band was not sensitive enough to detect this mutation rate.

The ability of GB-adapted gRNA/Cas9 elements to conduct RNA-guided transcriptional regulation was assessed by using the pNOS fused to luciferase as a reporter system. We observed that, by directing a nuclease-inactivated Cas9 to promoter regions around the transcription origin of the reporter gene, expression levels were severely reduced. These results were in line with previous reports showing an intrinsic repressor activity of a dCas9 without further modifications [10, 27]; however in our experimental conditions dCas9 intrinsic repression was almost completely abolished when paired to gRNAs targeting distal regions upstream of the −100 position. In the same upstream regions, however, the translational fusion of dCas9 with specific transcription modulating protein domains efficiently conducted the downregulation (BRD, SRDX) or upregulation (VP64, EDLL) of the reporter activity respectively. It was also observed that, by targeting several gRNAs towards the same promoter, the activation/repression effect was increased, highlighting the convenience of multiplex targeting to achieve efficient transcriptional regulation. Altogether, the range of transcriptional activities that we were able to modulate using current GB gRNA–Cas9 tools was relatively modest, approximately seven times from the strongest repressor to the strongest activator. Further optimization of the system (e.g. improved fusion linkers, optimization of fusion sites, etc.) will be necessary to increase this efficiency. Nevertheless it should be noticed that, given that in the *N. benthamiana* agroinfiltration system several T-DNA copies of the reporter gene are co-delivered simultaneously in each cell there is probably a high demand for dCas9 fusions to achieve substantial activation/repression. In future experiments the quantification of the effect of dCas9 fusions on single copy genes stably integrated in the plant genome will be investigated.

Very recently, new gRNA–Cas9 toolkits for targeted mutagenesis or transcriptional regulation have been reported including animal [35, 36] and plant-dedicated [31, 32, 37] systems, although none of them involve a standardized strategy. Interestingly, the toolbox reported by Lowder et al. incorporates gRNA–Cas9 elements for targeted mutagenesis and transcriptional regulation using a combination of type IIs and gateway recombination for multiplex assembly. In comparison, the GB toolbox showed here present a number of distinctive features. First, the GB toolbox includes a number of software tools that generate standardised protocols in each gRNA–Cas9 assembly step. The implementation of assembly software tools not only serves to facilitate construct-making for non-trained users, but most importantly, it turns GB into a self-contained, fully traceable assembly system, where all elements generated with GB software tools, now including also gRNA/Cas9 elements, are perfectly catalogued and their genealogy documented. Second, the modularity of GB facilitates combinatorial arrangements as e.g. between pre-set gRNA arrays and different Cas9 versions and enables the exchange of pre-made combinations. Finally, the GB cloning loop enables endless assembly of both monocistronic and polycistronic tRNA–gRNA expression cassettes, enhancing the multiplexing capacity of the system.

## Conclusions

A modular gRNA–Cas9 toolbox conforming to the GoldenBraid standard for Plant Synthetic Biology was developed and functionally validated. The GB-gRNA/Cas9 toolbox, comprising an adapted cloning pipeline, domesticated gRNA/Cas9 elements and a dedicated software tool, was shown to facilitate all-in-one-T-DNA cloning and gRNA multiplexing. The GB-adapted gRNA/Cas9 elements combined among them and/or with other GB elements were shown effective in targeting reporter genes for mutagenesis, transcriptional activation and transcriptional repression in *N. benthamiana* transient assays. The GB adaptation enhances CRISPRs/Cas9 technology with traceability, exchangeability and improved combinatorial and multiplexing capacity.

## Methods

### GBparts construction

GBparts used in this work were created following the domestication strategy described in [18]. For parts GB0575, GB1001 and GB1079, PCR amplifications with the primers obtained at https://gbcloning.upv.es/do/

domestication/were performed using the Phusion High-Fidelity DNA polymerase (Thermo Scientific). For level 0 parts GB0273, GB0645, GB1175, GB1185, GB1186, GB1187 and for level −1 parts GB1205, GB1206, GB1207 double-stranded DNA was synthesized using IDT gBlocks® Gene Fragments. GB1041 was amplified from GB0575 to incorporate the D10A and H840A mutations. For level 0 parts, 40 ng of the PCR products or gBlocks® were cloned into the pUPD with a *BsmBI* restriction–ligation reaction. Level −1 parts were cloned into the pVD1 (GB0101) with a *BsaI* restriction–ligation reaction following the same protocol. A list of the level −1 and level 0 parts is provided in the Additional file 2: Table S3; their nucleotide sequences can be searched at https://gbcloning.upv.es/search/features/with their corresponding ID numbers. All level −1 and level 0 GB parts were validated by restriction enzyme (RE) analysis and confirmed by sequencing.

### Guide RNA assembly on level 0 and level 1

Assembly optimization reactions were performed as follows: primers gRNA_XT2_F/gRNA_XT2_R were resuspended in water and STE buffer (10 mM Tris pH 8.0, 50 mM NaCl, 1 mM EDTA) to final concentrations of 100, 10 and 1 µM. Equal volumes of forward and reverse primers were mixed. The mixture was split into two different tubes and one of them was incubated at 94 °C for 2 min prior to a 30 min incubation at room temperature while the other was directly incubated at room temperature for 30 min. The BsaI restriction–ligation reactions were set up in 10 µl with 1 µl of primers mix, 75 ng of GB1001 (U626 promoter), 75 ng of GB0645 (scaffold RNA) and 75 ng of pDGB3α1 destination vector. One microliter of the reaction was transformed into *E. coli* TOP10 electrocompetent cells and the number of white colonies growing on agar plates counted.

The selected conditions for the gRNA assemblies were dilution in water, incubation at room temperature for 30 min and set the restriction–ligation reaction with a final primer concentration of 0.1 µM. For gRNA assemblies on level 1, two complementary primers designed at http://www.gbcloning.upv.es/do/crispr/and listed on Additional file 2: Table S2, were included in a *BsaI* restriction–ligation reaction following the selected conditions. For the assembly of guide RNAs on level 0, the primers listed on Additional file 2: Table S2 were included in a *BsmBI* restriction–ligation reaction following the selected conditions together with the pUPD2 and 75 ng of the corresponding level −1 tRNA-scaffold plasmid depending on the desired position of each target on the level 1 assembly. All level 1 gRNA constructs were validated by RE-analysis, analyzed by sequencing and confirmed correct.

### Cloning in α and Ω-level destination vectors

Multipartite BsaI restriction–ligation reactions from level 0 parts and binary *BsaI* or *BsmBI* restriction–ligation reactions were performed as described in [18] to obtain all the level ≥1 assemblies. A list with all the TUs and modules used in this work is provided on the Additional file 2: Table S3. All level ≥1 were validated by restriction enzyme (RE) analysis. Furthermore, partial sequencing was carried out to check part's boundaries. The sequences of all level ≥1 constructs can be found entering their IDs (displayed at Additional file 2: Table S3) at https://gbcloning.upv.es/search/features/.

### *Nicotiana benthamiana* agroinfiltration

For transient expression, plasmids were transferred to *Agrobacterium tumefaciens* strain GV3101 by electroporation. *N. benthamiana* plants were grown for 5 to 6 weeks before agroinfiltration in a growing chamber compliant with European legislation. Growing conditions were 24 °C (light)/20 °C (darkness) with a 16-h-light/8-h-dark photoperiod. Agroinfiltration was carried out with overnight-grown bacterial cultures. The cultures were pelleted and resuspended on agroinfiltration solution (10 mM MES, pH 5.6, 10 mM $MgCl_2$, and 200 μM acetosyringone) to an optical density of 0.2 at 600 nm. After incubation for 2 h at room temperature on a horizontal rolling mixer, the bacterial suspensions were mixed in equal volumes. The silencing suppressor P19 was included in all the assays; in the same T-DNA for the transcriptional regulation experiments and co-delivered in an independent T-DNA for the targeted mutagenesis assays. Agroinfiltrations were carried out through the abaxial surface of the three youngest leaves of each plant with a 1 ml needle-free syringe.

### Genomic DNA extraction and PCR/restriction enzyme assay

Samples for genomic DNA extraction were collected from 5 days post infiltrated leaves. For genomic DNA extraction, 50 mg of tissue powder coming from a pool of three leaves were ground in 500 μl of DNA extraction buffer (200 mM TrisHCl-pH 7.5, 250 mM NaCl, 25 mM EDTA, 0.5 % SDS). The plant extract was mixed gently and it was spin at 14,000×g for 3 min. The supernatant was transferred to a new tube and an equal volume of isopropanol was added for DNA precipitation. The supernatant was removed after centrifugation (5 min at 14,000×g) and the DNA was washed twice with 70 % ethanol. The pellet was dried for half an hour and it was dissolved with 100 μl of elution buffer (10 mM TrisHCl-pH 8, 1 mM EDTA).

DNA amplicons covering the XT1 and XT2 target sites were obtained by PCR of genomic DNA using the Phusion High-Fidelity DNA polymerase (Thermo Scientific) and two pairs of gene specific primers: XT1_F/XT1_R for XT1 and XT2_F/XT2 _R for XT2 (Additional file 2: Table S1). The resulting PCR products were purified with the QIAquick PCR purification kit (QIAGEN) following the manufacturer's protocol and restriction reactions were set up with 500 ng of purified DNA and the corresponding restriction enzyme; BsmBI (Fermentas) for XT1 and SpeI (Fermentas) for XT2. Band intensities were estimated using the 'Benchling Gels' (https://benchling.com) tool.

### Gel band purification and *BsaI*-cloning

PCR products resistant to *BsmBI* and *SpeI* digestion were purified from a 1 % agarose gel with the QIAEX II Gel Extraction Kit following the manufacturer's protocol. For sequence analysis, the purified PCR products were subsequently amplified with XT12BsaI_F/XT12BsaI_R primers (Additional file 2: Table S1) to incorporate BsaI sites for improving cloning efficiency. Finally, they were cloned into the pDGB3α1 with a *BsaI* restriction–ligation reaction and individual clones were sequenced.

### Luciferase/Renilla activity determination

Samples of leaves coinfiltrated with the REP (GB1116), different activator/repressor TUs (GB1172 and GB1188 to GB1191) and the independent or combined gRNAs targeting the pNOS were collected at 4 days post infiltration. For the determination of the luciferase/renilla activity one disc per leaf (d = 0.8 cm, approximately 18–19 mg) was excised, homogenized and extracted with 150 μl of 'Passive Lysis Buffer', followed by 15 min of centrifugation (14,000×g) at 4 °C. Then, the supernatant was diluted 2:3 in Passive Lysis Buffer resulting in the working plant extract. Fluc and Rluc activities were determined following the Dual-Glo® Luciferase Assay System (Promega) manufacturer's protocol with minor modifications: 10 μl of working plant extract, 40 μl of LARII and 40 μl of Stop&Glo Reagent were used. Measurements were made using a GloMax 96 Microplate Luminometer (Promega) with a 2-s delay and a 10-s measurement. Fluc/Rluc ratios were determined as the mean value of three samples coming from three independent agroinfiltrated leaves of the same plant and were normalized to the Fluc/Rluc ratio obtained for a reference sample including the REP (GB1116) co-infiltrated with an unrelated gRNA (GB1221) and the corresponding activator/repressor TU.

### Additional files

**Additional file 1: Figure S1.** Cloning efficiency of representative Level ≥1 GBelements. **Figure S2.** Schema of the dead Cas9 transcriptional units tested on the repression and activation experiments.

**Additional file 2: Table S1.** Primers used for the amplification of the *N. benthamiana* xylosyltransferases XT1 (Niben101Scf04205Ctg025) and XT2 (Niben101Scf04551Ctg021) regions. **Table S2.** List of forward and reverse primers used to construct the targets. **Table S3.** List of GBelements generated in this work.

## Abbreviations

pNOS: nopaline synthase promoter; gRNA: guideRNA; GB: GoldenBraid; TU: transcriptional unit; XT: xylosyltransferase; Fluc: firefly luciferase; Rluc: renilla luciferase.

## Authors' contributions

DO, MV-V, JMB-O and AF-D-C conceived and designed the experiments. MV-V and JMB-O performed the experimental work. PZ and JB developed the software tools. DO and MV-V drafted the manuscript. DO, AG and AF-D-C discussed and revised the manuscript. All authors read and approved the final manuscript.

## Author details

[1] Instituto de Biología Molecular y Celular de Plantas (IBMCP), Consejo Superior de Investigaciones Científicas, Universidad Politécnica de Valencia, Camino de Vera s/n, 46022 Valencia, Spain. [2] Centro de Conservación y Mejora de la Agrodiversidad Valenciana (COMAV), Universidad Politécnica de Valencia, Camino de Vera s/n, 46022 Valencia, Spain.

## Acknowledgements

This work has been funded by Grant BIO2013-42193-R from Plan Nacional I + D of the Spanish Ministry of Economy and Competitiveness. Vazquez-Vilar M. is a recipient of a Junta de Ampliación de Estudios fellowship. Bernabé-Orts J.M. is a recipient of a FPI fellowship. We want to thank Nicola J. Patron and Mark Youles for kindly providing humanCas9 and U6-26 clones. We also want to thank Eugenio Gómez for providing *Arabidopsis thaliana* genomic DNA and Concha Domingo for providing rice genomic DNA. We also want to thank the COST Action FA1006 for the support in the development of the software tools.

## Competing interests

The authors declare that they have no competing interests.

## References

1. Ran FA, Hsu PD, Wright J, Agarwala V, Scott DA, Zhang F. Genome engineering using the CRISPR-Cas9 system. Nat Protoc. 2013;8(11):2281–308. doi:10.1038/nprot.2013.143.
2. Yang X. Applications of CRISPR-Cas9 mediated genome engineering. Mil Med Res. 2015;2:11. doi:10.1186/s40779-015-0038-1.
3. Wang H, Yang H, Shivalila CS, Dawlaty MM, Cheng AW, Zhang F, et al. One-step generation of mice carrying mutations in multiple genes by CRISPR/Cas-mediated genome engineering. Cell. 2013;153(4):910–8. doi:10.1016/j.cell.2013.04.025.
4. Bortesi L, Fischer R. The CRISPR/Cas9 system for plant genome editing and beyond. Biotechnol Adv. 2015;33(1):41–52. doi:10.1016/j.biotechadv.2014.12.006.
5. Belhaj K, Chaparro-Garcia A, Kamoun S, Patron NJ, Nekrasov V. Editing plant genomes with CRISPR/Cas9. Curr Opin Biotechnol. 2015;32:76–84. doi:10.1016/j.copbio.2014.11.007.
6. Shan Q, Wang Y, Li J, Zhang Y, Chen K, Liang Z, et al. Targeted genome modification of crop plants using a CRISPR-Cas system. Nat Biotechnol. 2013;31(8):686–8. doi:10.1038/nbt.2650.
7. Gao J, Wang G, Ma S, Xie X, Wu X, Zhang X, et al. CRISPR/Cas9-mediated targeted mutagenesis in *Nicotiana tabacum*. Plant Mol Biol. 2015;87(1–2):99–110. doi:10.1007/s11103-014-0263-0.
8. Fauser F, Schiml S, Puchta H. Both CRISPR/Cas-based nucleases and nickases can be used efficiently for genome engineering in *Arabidopsis thaliana*. Plant J. 2014;79(2):348–59. doi:10.1111/tpj.12554.
9. Schiml S, Fauser F, Puchta H. The CRISPR/Cas system can be used as nuclease for in planta gene targeting and as paired nickases for directed mutagenesis in Arabidopsis resulting in heritable progeny. Plant J. 2014;80(6):1139–50. doi:10.1111/tpj.12704.
10. Piatek A, Ali Z, Baazim H, Li L, Abulfaraj A, Al-Shareef S, et al. RNA-guided transcriptional regulation in planta via synthetic dCas9-based transcription factors. Plant Biotechnol J. 2015;13(4):578–89. doi:10.1111/pbi.12284.
11. Beerli RR, Barbas CF 3rd. Engineering polydactyl zinc-finger transcription factors. Nat Biotechnol. 2002;20(2):135–41. doi:10.1038/nbt0202-135.
12. Bogdanove AJ, Voytas DF. TAL effectors: customizable proteins for DNA targeting. Science. 2011;333(6051):1843–6. doi:10.1126/science.1204094.
13. Nielsen AA, Voigt CA. Multi-input CRISPR/Cas genetic circuits that interface host regulatory networks. Mol Syst Biol. 2014;10:763. doi:10.15252/msb.20145735.
14. Eeckhaut T, Lakshmanan PS, Deryckere D, Van Bockstaele E, Van Huylenbroeck J. Progress in plant protoplast research. Planta. 2013. doi:10.1007/s00425-013-1936-7.
15. Mikami M, Toki S, Endo M. Comparison of CRISPR/Cas9 expression constructs for efficient targeted mutagenesis in rice. Plant Mol Biol. 2015. doi:10.1007/s11103-015-0342-x.
16. Patron NJ, Orzaez D, Marillonnet S, Warzecha H, Matthewman C, Youles M, et al. Standards for plant synthetic biology: a common syntax for exchange of DNA parts. New Phytol. 2015. doi:10.1111/nph.13532.
17. Liu W, Stewart CN Jr. Plant synthetic biology. Trends Plant Sci. 2015;20(5):309–17. doi:10.1016/j.tplants.2015.02.004.
18. Sarrion-Perdigones A, Vazquez-Vilar M, Palaci J, Castelijns B, Forment J, Ziarsolo P, et al. GoldenBraid 2.0: a comprehensive DNA assembly framework for plant synthetic biology. Plant Physiol. 2013;162(3):1618–31. doi:10.1104/pp.113.217661.
19. Vazquez-Vilar M, Sarrion-Perdigones A, Ziarsolo P, Blanca J, Granell A, Orzaez D. Software-assisted stacking of gene modules using GoldenBraid 2.0 DNA-assembly framework. Methods Mol Biol. 2015;1284:399–420. doi:10.1007/978-1-4939-2444-8_20.
20. Duportet X, Wroblewska L, Guye P, Li Y, Eyquem J, Rieders J, et al. A platform for rapid prototyping of synthetic gene networks in mammalian cells. Nucleic Acids Res. 2014;42(21):13440–51. doi:10.1093/nar/gku1082.
21. Guo Y, Dong J, Zhou T, Auxillos J, Li T, Zhang W, et al. YeastFab: the design and construction of standard biological parts for metabolic engineering in *Saccharomyces cerevisiae*. Nucleic Acids Res. 2015;43(13):e88. doi:10.1093/nar/gkv464.
22. Engler C, Gruetzner R, Kandzia R, Marillonnet S. Golden gate shuffling: a one-pot DNA shuffling method based on type IIs restriction enzymes. PLoS ONE. 2009;4(5):e5553. doi:10.1371/journal.pone.0005553.
23. Sarrion-Perdigones A, Falconi EE, Zandalinas SI, Juarez P, Fernandez-del-Carmen A, Granell A, et al. GoldenBraid: an iterative cloning system for standardized assembly of reusable genetic modules. PLoS ONE. 2011;6(7):e21622. doi:10.1371/journal.pone.0021622.
24. Lei Y, Lu L, Liu HY, Li S, Xing F, Chen LL. CRISPR-P: a web tool for synthetic single-guide RNA design of CRISPR-system in plants. Mol Plant. 2014;7(9):1494–6. doi:10.1093/mp/ssu044.
25. Mali P, Yang L, Esvelt KM, Aach J, Guell M, DiCarlo JE, et al. RNA-guided human genome engineering via Cas9. Science. 2013;339(6121):823–6. doi:10.1126/science.1232033.
26. Li JF, Norville JE, Aach J, McCormack M, Zhang D, Bush J, et al. Multiplex and homologous recombination-mediated genome editing in *Arabidopsis* and *Nicotiana benthamiana* using guide RNA and Cas9. Nat Biotechnol. 2013;31(8):688–91. doi:10.1038/nbt.2654.
27. Bikard D, Jiang W, Samai P, Hochschild A, Zhang F, Marraffini LA. Programmable repression and activation of bacterial gene expression using an engineered CRISPR-Cas system. Nucleic Acids Res. 2013;41(15):7429–37. doi:10.1093/nar/gkt520.
28. Xie K, Minkenberg B, Yang Y. Boosting CRISPR/Cas9 multiplex editing capability with the endogenous tRNA-processing system. Proc Natl Acad Sci USA. 2015;112(11):3570–5. doi:10.1073/pnas.1420294112.
29. Weber E, Engler C, Gruetzner R, Werner S, Marillonnet S. A modular cloning system for standardized assembly of multigene constructs. PLoS ONE. 2011;6(2):e16765. doi:10.1371/journal.pone.0016765.
30. Sakuma T, Nishikawa A, Kume S, Chayama K, Yamamoto T. Multiplex genome engineering in human cells using all-in-one CRISPR/Cas9 vector system. Sci Rep. 2014;4:5400. doi:10.1038/srep05400.
31. Ma X, Zhang Q, Zhu Q, Liu W, Chen Y, Qiu R, et al. A robust CRISPR/Cas9 system for convenient, high-efficiency multiplex genome editing in monocot and dicot plants. Mol Plant. 2015. doi:10.1016/j.molp.2015.04.007.
32. Lowder LG, Zhang D, Baltes NJ, Paul JW 3rd, Tang X, Zheng X, et al. A CRISPR/Cas9 toolbox for multiplexed plant genome editing and transcriptional regulation. Plant Physiol. 2015;169(2):971–85. doi:10.1104/pp.15.00636.

33. Nekrasov V, Staskawicz B, Weigel D, Jones JD, Kamoun S. Targeted mutagenesis in the model plant *Nicotiana benthamiana* using Cas9 RNA-guided endonuclease. Nat Biotechnol. 2013;31(8):691–3. doi:10.1038/nbt.2655.

34. Upadhyay SK, Kumar J, Alok A, Tuli R. RNA-guided genome editing for target gene mutations in wheat. G3. 2013;3(12):2233–8. doi:10.1534/g3.113.008847.

35. Senis E, Fatouros C, Grosse S, Wiedtke E, Niopek D, Mueller AK, et al. CRISPR/Cas9-mediated genome engineering: an adeno-associated viral (AAV) vector toolbox. Biotechnol J. 2014;9(11):1402–12. doi:10.1002/biot.201400046.

36. Port F, Chen HM, Lee T, Bullock SL. Optimized CRISPR/Cas tools for efficient germline and somatic genome engineering in Drosophila. Proc Natl Acad Sci USA. 2014;111(29):E2967–76. doi:10.1073/pnas.1405500111.

37. Xing HL, Dong L, Wang ZP, Zhang HY, Han CY, Liu B, et al. A CRISPR/Cas9 toolkit for multiplex genome editing in plants. BMC Plant Biol. 2014;14:327. doi:10.1186/s12870-014-0327-y.

# 2

# In situ hybridization for the detection of rust fungi in paraffin embedded plant tissue sections

Mitchell A. Ellison[1], Michael B. McMahon[2], Morris R. Bonde[2], Cristi L. Palmer[3] and Douglas G. Luster[2]* ⓘ

## Abstract

**Background:** Rust fungi are obligate pathogens with multiple life stages often including different spore types and multiple plant hosts. While individual rust pathogens are often associated with specific plants, a wide range of plant species are infected with rust fungi. To study the interactions between these important pathogenic fungi and their host plants, one must be able to differentiate fungal tissue from plant tissue. This can be accomplished using the In situ hybridization (ISH) protocol described here.

**Results:** To validate reproducibility using the ISH protocol, samples of *Chrysanthemum* × *morifolium* infected with *Puccinia horiana*, *Gladiolus* × *hortulanus* infected with *Uromyces transversalis* and *Glycine max* infected with *Phakopsora pachyrhizi* were tested alongside uninfected leaf tissue samples. The results of these tests show that this technique clearly distinguishes between rust pathogens and their respective host plant tissues.

**Conclusions:** This ISH protocol is applicable to rust fungi and potentially other plant pathogenic fungi as well. It has been shown here that this protocol can be applied to pathogens from different genera of rust fungi with no background staining of plant tissue. We encourage the use of this protocol for the study of plant pathogenic fungi in paraffin embedded sections of host plant tissue.

**Keywords:** Basidiomycota, Pucciniomycotina, Rust fungus, In situ hybridization, *Puccinia horiana*, *Uromyces transversalis*, *Phakopsora pachyrhizi*, *Chrysanthemum* × *morifolium*, *Gladiolus* × *hortulanus*, *Glycine max*

## Background

Rust fungi (Basidiomycota, Pucciniomycotina) are obligate parasites that infect many species of vascular plants [1, 2]. Recent studies in this laboratory have focused on Chrysanthemum white rust, caused by *Puccinia horiana*, Gladiolus rust, caused by *Uromyces transversalis* and Asian soybean rust, caused by *Phakopsora pachyrhizi*. [3, 4]. Studies on the interactions between these pathogenic fungi and plants would benefit from approaches that allow visualization of the pathogen within host plant tissue, including the use of in situ hybridization (ISH) technology.

ISH was first used to localize specific DNA sequences on chromosomes using probes labeled with radioisotopes [5, 6]. The technique was later used for the detection of viral particles and high copy number mRNA in cultured cells and sectioned tissue making it useful for localizing gene expression patterns [6–10]. Non-radioactive methods were also developed that employed digoxygenin or biotin-conjugated nucleotides allowing for detection with antibody and streptavidin conjugates [9–11]. The development of non-radioactive methods eventually gave rise to fluorescent ISH (FISH), which employs various fluorescent-labeling techniques to produce fluorescence at the site of hybridization [12–14]. Non-radioactive ISH methods have been used to accomplish such tasks as chromosome mapping [14–16], gene expression localization [6, 8, 9, 11], and pathogen detection [17–26]. Chromogenic ISH (CISH) is an alternative to FISH that

*Correspondence: doug.luster@ars.usda.gov
[2] USDA-ARS Foreign Disease-Weed Science Research Unit, Ft. Detrick, MD, USA
Full list of author information is available at the end of the article

has become popular in diagnostic laboratories studying human pathogens [19–24].

Recently ISH has been used to identify microorganisms by targeting rRNA [18, 21, 24, 27–32]. The abundance of rRNA in the cell offers ample target for probes to bind to allowing for clear visualization of microorganisms within the sample being assayed. This technique has been used to identify and characterize prokaryotic organisms [27, 29, 32] and has been used for the detection of fungi in cultures [24, 31], plant [17, 18, 20, 30] and animal tissue [19, 21–23, 25].

The aim of this study was to develop a basic ISH protocol that plant pathologists can use for the detection of rust pathogens in paraffin embedded plant tissue. Here we report the development of an optimized protocol tested on three genera of rust fungi from three plant species. The results of this investigation demonstrate the utility of ISH as a tool for visualizing the infection of plant tissue by rust fungi.

## Methods

### Generation of infected leaf material

Leaves of *Chrysanthemum × morifolium* infected with *Puccinia horiana* isolate PA-11 [33], *Gladiolus × hortulanus* infected with *Uromyces transversalis* isolate CA-07 [34], and *Glycine max* infected with *Phakopsora pachyrhizi* isolate Taiwan 72-1 [35] were generated for experiments as described. These plant pathogens are regulated under the Plant Protection Act of 2000 and inoculations were conducted in a BSL-3 Plant Pathogen Containment Facility at Ft. Detrick MD under conditions specified in valid USDA APHIS PPQ 526 permits.

### DNA extraction and sequencing for ISH probe design

Genomic DNA was extracted from 50 mg of fungal basidiospores of *Puccinia horiana* using a hexadecyltrimethylammonium bromide (CTAB) extraction protocol beginning with 1 min of homogenization in 500 µL of CTAB extraction buffer (1 % CTAB, 0.7 M NaCl, 100 mM Tris (pH 7.5), 10 mM EDTA, 0.3 mg/mL proteinase K). Homogenized samples were incubated at 65 °C for 30 min, placed on ice for 2 min, and extracted with 500 µL of chloroform: isoamyl alcohol (24:1) by 10 s of vortexing followed by centrifugation at $14,000 \times g$ for 10 min. A volume of 300 µL of aqueous phase liquid was collected from each sample, combined with an equal volume of CTAB extraction buffer, re-extracted with 500 µL of chloroform: isoamyl alcohol (24:1), vortexed, and centrifuged at $14,000 \times g$ for 10 min. An equal volume of isopropanol was added to 400 µL of aqueous phase extract, which was gently mixed, and centrifuged at $14,000 \times g$ for

15 min. Following removal of isopropanol, DNA pellets were washed with 500 µL of 70 % ethanol, and centrifuged at $14,000 \times g$ for 20 min at 4 °C. Once 70 % ethanol was removed and nucleic acid pellets were allowed to dry a volume of 50 µL of TE buffer (10 mM Tris, pH 8.0, 1 mM EDTA) containing 1 mg/mL RNase A was added to each sample. Final concentrations of extracted DNA were determined using a Nanodrop 2000 (Thermo Fisher Scientific Inc, Waltham, MA).

Basidiospore DNA was amplified using primers for 18S rDNA (Table 2) in conventional PCR reactions and products were verified by gel electrophoresis. Post-amplification products were purified using ExoSAP-IT reagent (USB Corporation) before sequencing with BigDye Terminator version 3.1 cycle sequencing kit (ABI, Foster, CA). BigDye reaction products were purified using a DyeEx® 2.0 Spin Kit (Qiagen), and analyzed on an ABI 3130XL sequencer (Applied Biosystems).

### Tissue sample collection and fixation

Rectangular leaf tissue samples measuring 4 cm by 1 cm were cut from infected leaves using a sterile razor blade and placed into a 50 mL conical tube containing 30 mL of FAE fixative (2 % formaldehyde, 5 % acetic acid, 60 % ethanol) (see supplementary protocol). After 48 h of incubation at 4 °C FAE fixative was removed from the sample tubes and samples were washed with 70 % ethanol for 5 min before incubation at room temperature in 70 % ethanol for one week. After fixation in FAE fixative and clearing in 70 % ethanol samples were shipped to American Histolabs (Rockville, MD) for RNase-free preparation, including paraffin embedding, sectioning, mounting on positively charged microscope slides, and deparaffinization. All microscope slide samples were stored at −80 °C prior to pre-hybridization.

### Prehybridization

All steps listed in the pre-hybridization protocol were carried out under ribonuclease (RNase) free conditions (see supplementary protocol). Slide mounted tissue samples were rehydrated by incubation in 100 % ethanol, 50 % ethanol, and diethylpyrocarbonate (DEPC)-treated water for 3 min each. Following rehydration samples were treated with 0.2 M HCl for 20 min, washed for 2 min in DEPC-treated water, and incubated at 70 °C for 20 min in 2× SSPE (0.3 M NaCl, 2 mM EDTA, 20 mM $NaH_2PO_4$, pH 7.4), before digestion with 10ug/mL proteinase K in proteinase K buffer (20 mM Tris–HCl, pH 7.0, 2 mM $CaCl_2$). Digestion was stopped by washing with 2× SSPE for 5 min at room temperature, and tissue was treated with 0.1 M TEA (0.1 M triethanolamine-Cl, pH 8.0) containing

0.5 % acetic anhydride (Sigma, St. Louis, MO) (v/v) for 10 min preceding treatment with biotin and streptavidin blocking solutions. Blocking of endogenous biotin is achieved by treating samples with 1 mL of 1× Blocking Reagent (Roche, Indianapolis, IN) containing 4 drops/mL of streptavidin blocker (Vector Laboratories, Burlingame, CA) for 15 min, and washing for 5 min in 2× SSPE before applying 1 mL of 1× Blocking Reagent containing 4 drops/mL of biotin blocker (Vector Laboratories, Burlingame, CA) for 15 min. After blocking, slides were washed in 2× SSPE for 5 min and dehydrated by incubation with DEPC-treated water, 50 % ethanol, and 100 % ethanol consecutively for 3 min each. Samples were allowed to air dry while hybridization mixture was prepared.

### Hybridization

Hybridization mixture (0.3 M NaCl, 20 mM Tris (pH 7.5), EDTA 2 mM, 500 µg/mL tRNA, 500 µg/mL poly(A) RNA, 1× Denhardt's Solution [36], 10 % deionized formamide, 9 ng/µL DNA probe) was prepared and heated to 85–95 °C for 2 min before being placed into ice for 2 min. After the hybridization mixture was prepared, a volume of 100 µL was added to each sample slide, a coverslip was applied, and sample slides were placed into a plastic container lined with paper towels moistened with 4X SSPE. Sample slides were incubated overnight at 42 °C in one pint polypropylene snap-lid plastic containers in a H9270 Dual-Chamber Hybridization Oven (Thermo Fisher Scientific, Waltham, MA).

### Post hybridization, staining, and permanent mounting

Slides were removed from the 42 °C hybridization oven and dipped in 2× SSPE to aid in coverslip removal. Once coverslips were removed slides were washed in 2× SSPE for 30 min at room temperature, 1× SSPE for 30 min at 52 °C, and blocked for 30 min at room temperature using 1× Blocking Reagent. Samples were treated with Streptavidin-HRP at a concentration of 1:2500 (Invitrogen Carlsbad, CA.) in 1× PBS for 1 h at room temperature, and washed three times in 1× PBS, prior to application of ImmPACT™ VIP Peroxidase Substrate Kit (Vector Laboratories) following manufacturer's instructions. To rinse and dehydrate tissue samples slides were placed consecutively into DEPC-treated water for 5 min, and 50 % ethanol and 100 % ethanol for 3 min each. Slides were allowed to dry completely before application of Permaslip Mounting Medium and Liquid Coverslip Solution (American MasterTech, Lodi, CA). Immediately following the addition of mounting media, a new coverslip was applied to permanently mount and preserve the tissue sections.

### Microscopy

All images were captured using a Nikon DS-Fi1 camera coupled with a Nikon Eclipse 80i microscope and NIS Elements imaging software (Nikon Inc., Melville, NY). The images reported here were produced through differential interference contrast (DIC) microscopy using a Nikon D-DA DIC light filter. A Nikon CFI Plan Achromat DL 10× objective lens was used to magnify samples to 100×.

## Results and discussion

A review of the literature showed that very few studies had been conducted using ISH to target ribosomal RNA in a fungal pathogen during active infection [20, 30], several studies had targeted fungal 18S rRNA for ISH [18–24, 26, 30], and a few applied biotin-streptavidin as a reporter system [19, 37]. We believed that the field of plant pathology could benefit from the development of a non-radioactive ISH method designed for detecting rust fungi in host plant tissue. Therefore, a baseline protocol (Fig. 1) was developed from consensus information drawn from the current literature for further testing and refinement.

### Sequencing for probe design

Target sequences were required for development of probes to detect rust fungi in order to test the protocol. ISH probes for the targeted species were developed through a process of database searching, sequencing, and sequence alignment. The National Center for Biotechnology Information (NCBI) Taxonomy Browser website was used to search GenBank for 18S rRNA sequences from many fungal genera. Sequences (Table 1) were aligned in a CLUSTALW multiple sequence alignment using Biology Workbench 3.2 (http://workbench.sdsc.edu/). Sequencing primers were selected from highly conserved regions of the 18S sequence for the purpose of sequencing a variable region of the 18S gDNA of two rust fungi of interest to our laboratory; *Puccinia horiana* and *Uromyces transversalis*. These primers (Table 2) amplify a 596 bp fragment of the 18S rRNA gene. The newly obtained sequences were aligned to each of the sequences in Table 1 by pair wise local alignments, using NCBI Basic Local Alignment Search Tool (BLAST) (http://blast.ncbi.nlm.nih.gov/Blast.cgi). These alignments showed high variability when comparing rusts to non-rust fungi, but variation within rust fungi was low. From these alignments two 60 bp regions were chosen, and upon examining the properties of their sequences using OligoCalc (http://www.basic.northwestern.edu/biotools/oligocalc.html), a single region was selected.

## Fixation and Slide Preparation

~ 2 weeks

- Sample Collection
- Fix
- Dehydrate
- Embed

- Section
- Mount
- De-wax
- De-paraffinize

## Pre Hybridization

~ 3.5 hours

- Rehydrate
- 0.2M HCl
- 2X SSPE at 70 °C
- Proteinase K digest at 65 °C
- 2X SSPE Stop digest

- Acetic Anhydride in TEA
- 2X SSPE Wash
- Streptavidin Blocking
- Biotin Blocking
- Dehydrate

## Hybridization

~ 16 hours

- Prepare Hybridization Mix
- Heat Mix to 85-95 °C
- Rapidly Cool Mix

- Apply to Slides
- Place slides in moist chamber
- Place at 42 °C overnight

## Post Hybridization

~ 1 hour

- Float off cover slips
- 2X SSPE wash at room temperature

- 1X SSPE wash at 52 °C

## Detection

~ 2 hour

- Block
- Streptavidin-HRP in 1X PBS
- 1X PBS wash
- Detect with HRP substrate

- Wash
- Dehydrate
- Dry
- Preserve

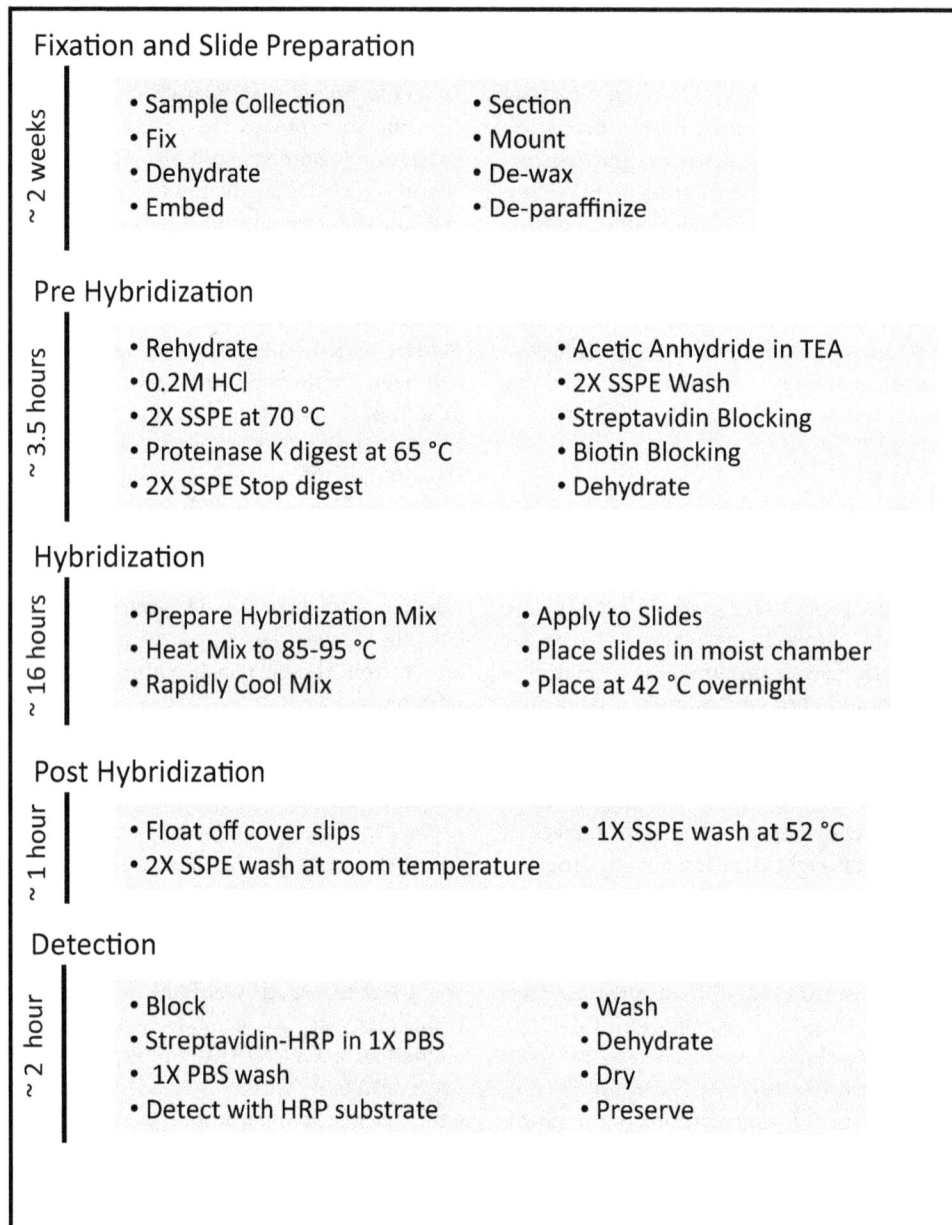

**Fig. 1** Simplified list of steps included in each phase of the *ISH* protocol with estimated length of time spent processing samples through each phase written vertically on the left hand edge

Probes were purchased from Integrated DNA Technologies (IDT) with biotin conjugated to their 5' end for use ISH (probe sequences are listed in Table 2). The probes were designed to be exact matches to the 18S rRNA for their respective pathogen (Table 3). With precisely match probes, it was possible to refine the basic ISH protocol making modifications along the way based upon experimental results.

### Development of an optimized ISH protocol

We selected a few key steps in the ISH protocol to optimize the method, using *P. horiana*-infected

**Table 1 Accession numbers of sequences used in CLUSTALW and BLAST alignments**

| Species | Accession |
| --- | --- |
| *Puccinia poarum* | GenBank:DQ831029 |
| *Aspergillus sojae* | GenBank:D63696 |
| *Aspergillus versicolor* | GenBank:AB008411 |
| *Eurotium herbariorum* | GenBank:AB008402 |
| *Cladosporium cladosporioides* | GenBank:AY251093 |
| *Fusarium culmorum* | GenBank:AF548073 |
| *Gremmeniella abietina* | GenBank:AF548074 |
| *Monographella nivalis* | GenBank:AF064049 |
| *Paecilomyces lilacinus* | GenBank:AB103380 |
| *Penicillium brevicompactum* | GenBank:AF548083 |
| *Rhizopus azygosporus* | GenBank:AB250156 |
| *Stachybotrys sp.* | GenBank:DQ069246 |
| *Trichoderma harzianum* | GenBank:AF548100 |
| *Alternaria botrytis* | GenBank:AF548105 |
| *Saccharomyces cerevisiae* | GenBank:Z75578 |
| *Wallemia sebi* | GenBank:AF548108 |
| *Phakopsora pachyrhizi* | GenBank:DQ354536 |

chrysanthemum leaves as the test material. Steps tested were proteinase K concentration, presumably affecting accessibility of the target RNA within the tissue, formamide concentration, which controls the strength of the hybridization process, final wash temperature, which controls the stringency of hybridization, and probe concentration. The protocol was tested both with and without a proteinase K digestion (for 20 min with 10 μg/mL of proteinase K at 65 °C) and the results illustrated the necessity of this step, with no signal observed in the undigested samples (see Fig. 2a). Additionally, four proteinase K concentrations (10, 20, 40, and 80 μg/mL) were tested to determine an optimum concentration. A concentration of 10 μg/mL proteinase K worked best to allow probe access without disturbing tissue morphology (see Fig. 2b). Next four concentrations of formamide (10, 30, 50, and 70 %) were tested along with two wash temperatures (42 and 52 °C carried out in $1\times$ SSPE for 30 min). Minimal differences were observed between wash temperatures, but formamide concentrations showed large differences, with lower concentrations resulting in increased signal (see Fig. 3). The results of this experiment led to the selection of two formamide concentrations for testing (50 and 10 %) with healthy plant control samples and sense probe (which should not hybridize to the target RNA sequence). Negligible background was observed in the sense-probe-treated infected tissue with a 42 °C wash and no background was observed in healthy plant samples treated with either probe. The optimal signal was obtained with 10 % formamide in the hybridization mix and the least background was observed with a 52 °C wash (see Fig. 4). Lastly probe concentration was varied (1, 3, and 9 ng/μL), and increased signal was observed when using 9 ng/

**Table 2 List of sequencing primers and ISH probes used in this study**

| Primer/probe | Sequence |
| --- | --- |
| Primer 1 *Puccinia* 18S Forward (30 bp) | 5′ CAATTGGAGGGCAAGTCTGGTGCCAGCAGC 3′ |
| Primer 2 *Puccinia* 18S Reverse (30 bp) | 5′ TGGACCTGGTGAGTTTCCCCGTGTTGAGTC 3′ |
| *Puccinia horiana* 18S Anti-Sense | 5′/biotin/AAGTTCACCAAGAGGTAAGCCTCCAACAA ATCAGTACACACCAAAAGGCAGACCAACTGC 3′ |
| *Puccinia horiana* 18S Sense | 5′/biotin/GCAGTTGGTCTGCCTTTTGGTGTGTACT GATTTGTTGGAGGCTTACCTCTTGGTGAACTT 3′ |
| *Uromyces transversalis* 18S Anti-Sense | 5′/biotin/AAGTTCACCAAGAGGTAAGCCTCCAACA AATCAGTACACACCAAAAGGCGGACCAACTGC 3′ |
| *Uromyces transversalis* 18S Sense | 5′/biotin/GCAGTTGGTCCGCCTTTTGGTGTGTACT GATTTGTTGGAGGCTTACCTCTTGGTGAACTT 3′ |
| *Phakopsora pachyrhizi* 18S Anti-Sense | 5′/biotin/AGGTTCACCAAGAGGTAAGCCTCCAAC AAATCAGTACACACCAAATGGCGGACCAACTGC 3′ |
| *Phakopsora pachyrhizi* 18S Sense | 5′/biotin/GCAGTTGGTCCGCCATTTGGTGTGTACT GATTTGTTGGAGGCTTACCTCTTGGTGAACCT 3′ |

**Table 3 CLUSTALW alignment of anti-sense probes**

| Species | Sequence |
| --- | --- |
| *Uromyces transversalis* | AAGTTCACCAAGAGGTAAGCCTCCAACAAATCAGTACACACCAAAAGGCGGACCAACTGC |
| *Puccinia horiana* | AAGTTCACCAAGAGGTAAGCCTCCAACAAATCAGTACACACCAAAAGGCAGACCAACTGC |
| *Phakopsora pachyrhizi* | AGGTTCACCAAGAGGTAAGCCTCCAACAAATCAGTACACACCAAATGGCGGACCAACTGC |
| | * ************************************* *** ********** |

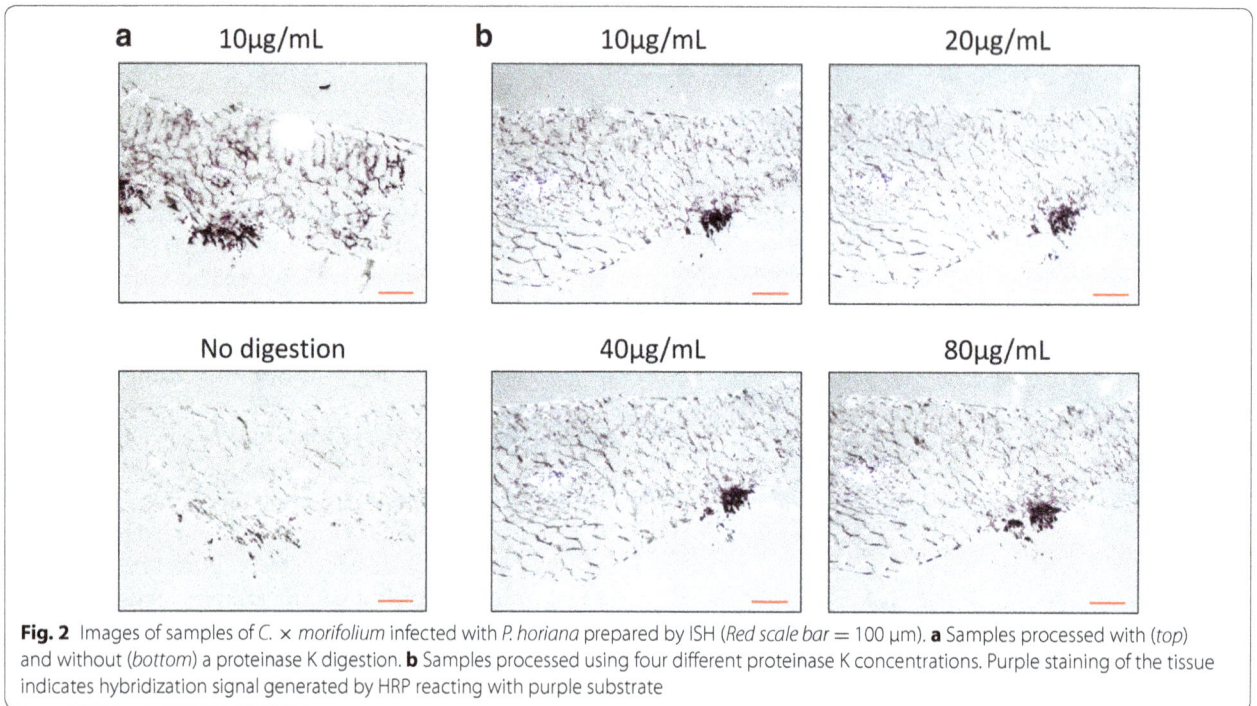

**Fig. 2** Images of samples of *C. × morifolium* infected with *P. horiana* prepared by ISH (*Red scale bar* = 100 μm). **a** Samples processed with (*top*) and without (*bottom*) a proteinase K digestion. **b** Samples processed using four different proteinase K concentrations. Purple staining of the tissue indicates hybridization signal generated by HRP reacting with purple substrate

μL with no additional background. These empirically determined conditions discovered as the result of these experiments were incorporated into a final optimized protocol. The protocol had, at this point, only been tested using sections of *Chrysanthemum × morifolium* infected with *Puccinia horiana* and further testing was required to determine the generality and reproducibility of the protocol.

### Application of the protocol to rust fungi

In order to validate the applicability and reproducibility of the refined protocol, samples were taken from two other research subjects in our laboratory; *Gladiolus × hortulanus* infected with *U. transversalis* and *Glycine max* infected with *P. pachyrhizi*. Slides containing samples of both pathogens were treated with both sense and anti-sense probes using the optimized protocol. These experiments demonstrated that the refined ISH protocol is effective on species within the genera Puccinia (Figs. 2, 3, 4) Uromyces and Phakopsora (Figs. 5, 6). When applied to *Glycine max* infected with *P. pachyrhizi*, the signal was weak compared to the other two species (Fig. 6). *U. transversalis* samples showed strong signal in all slides prepared with the anti-sense probe and no signal in the infected tissue prepared with the sense probe. Healthy plant tissue processed with both sense and anti-sense probes showed no signal, demonstrating further that this technique is preferentially staining pathogen tissue and not the host plant. The strength of signal obtained from *U. transversalis* samples was equivalent to signal observed in *P. horiana* samples. *P. pachyrhizi* showed weak signal in one out of five slides prepared with anti-sense probe and the other four appear the same as slides treated with the sense probe. The fact that some signal was observed indicates that this technique requires further optimization for *P.*

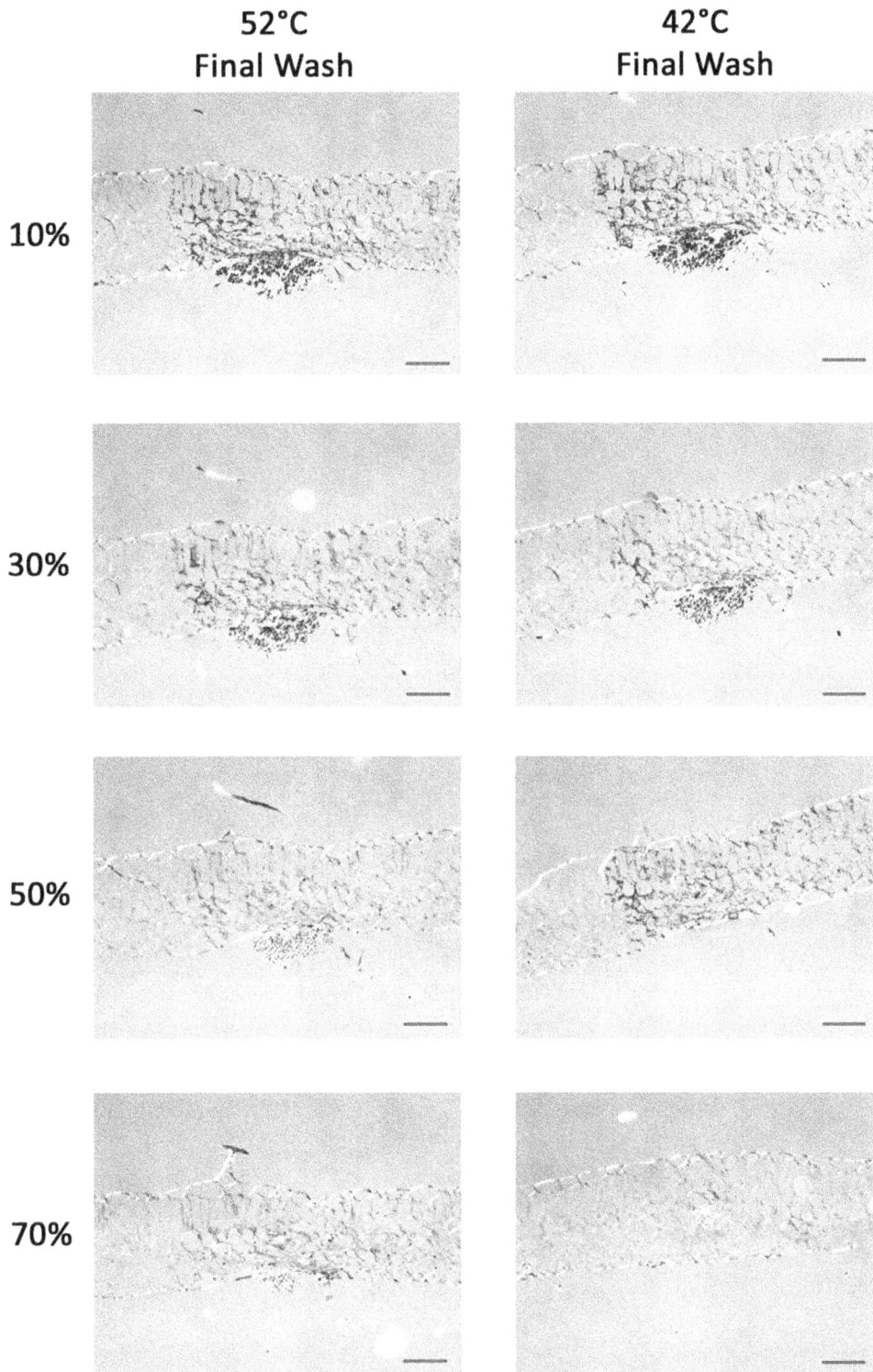

**Fig. 3** Images of *C.* × *morifolium* samples infected with *P. horiana* and prepared by ISH (*Red scale bar* = 100 µm). Photos are organized so that *columns* represent samples treated at a given final wash temperature (indicated by *column names*) and *rows* indicate samples treated with varying formamide concentrations (indicated by *row names*) in the hybridization mix for overnight hybridization. Purple staining of the tissue indicates hybridization signal generated by HRP reacting with purple substrate

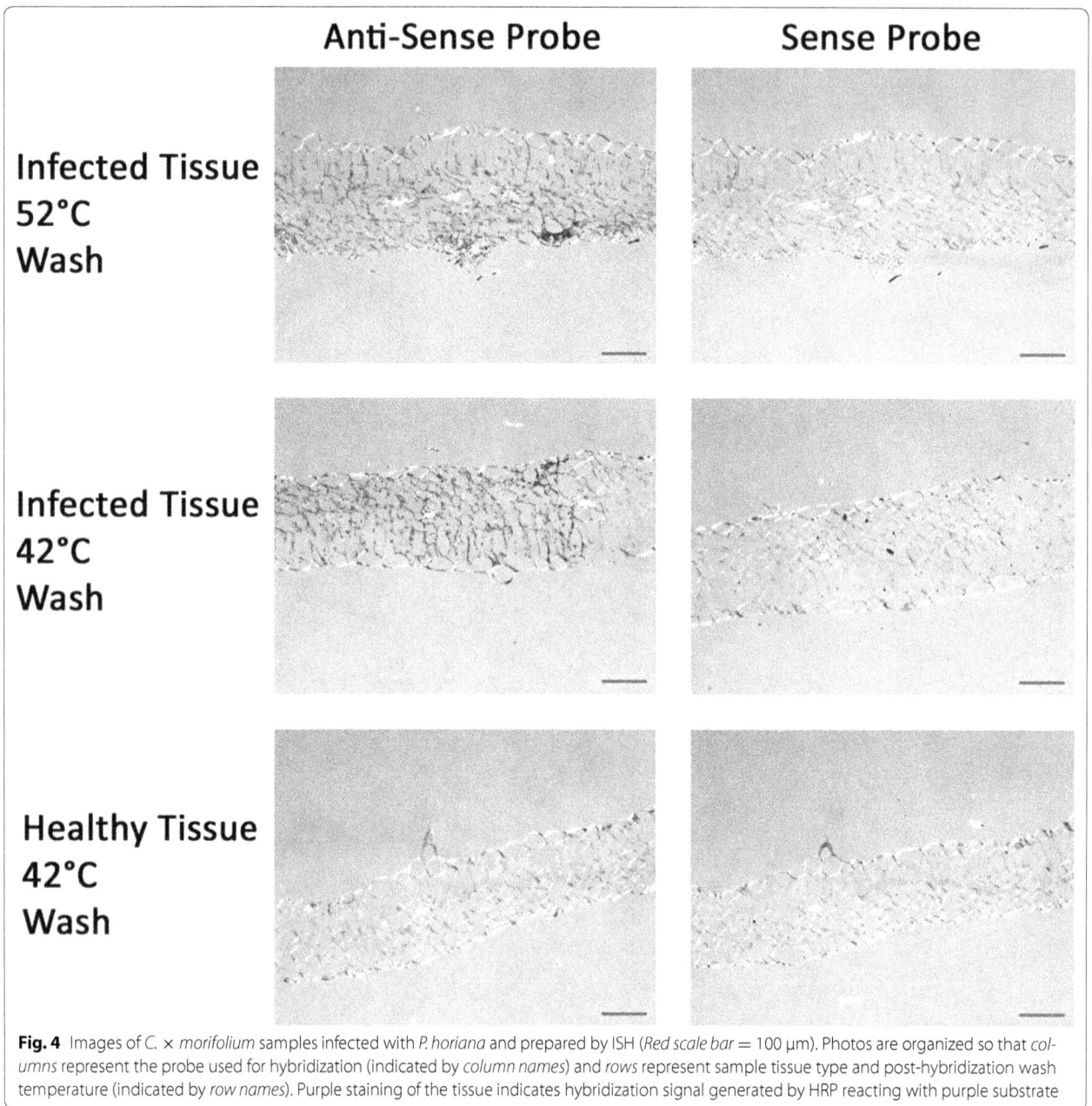

**Fig. 4** Images of *C. × morifolium* samples infected with *P. horiana* and prepared by ISH (*Red scale bar* = 100 μm). Photos are organized so that *columns* represent the probe used for hybridization (indicated by *column names*) and *rows* represent sample tissue type and post-hybridization wash temperature (indicated by *row names*). Purple staining of the tissue indicates hybridization signal generated by HRP reacting with purple substrate

*pachyrhizi.* In particular the alteration of the high temperature steps could potentially increase signal strength by reducing sample loss.

The images presented here are at relatively low magnification, and resolution is therefore at the tissue level. Further magnification of images did not resolve at the

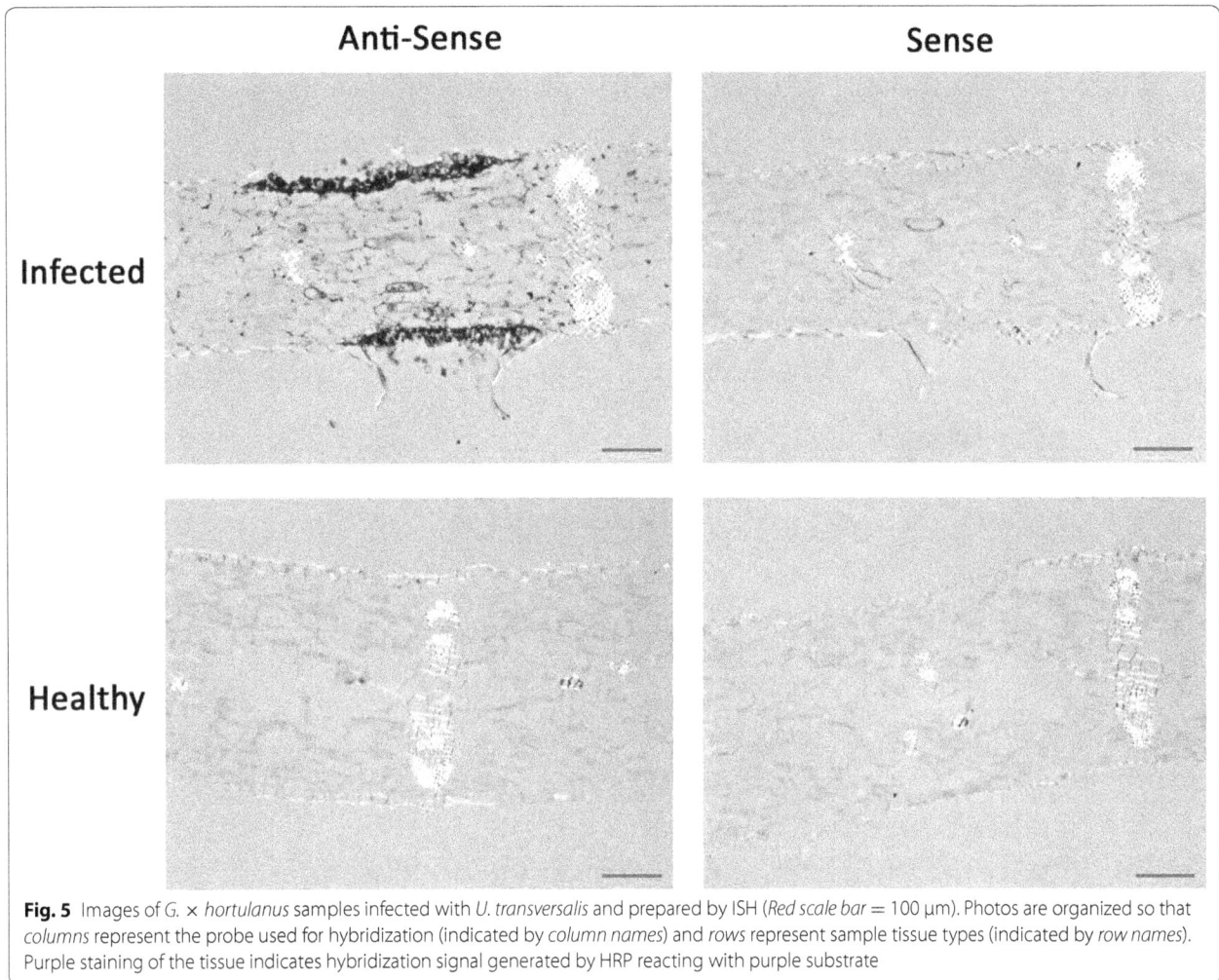

**Fig. 5** Images of *G. × hortulanus* samples infected with *U. transversalis* and prepared by ISH (*Red scale bar* = 100 μm). Photos are organized so that *columns* represent the probe used for hybridization (indicated by *column names*) and *rows* represent sample tissue types (indicated by *row names*). Purple staining of the tissue indicates hybridization signal generated by HRP reacting with purple substrate

intracellular level (not shown). Use of a fluorescent label rather than precipitable stain may result in higher-level resolution.

## Conclusion

We present here a generalized ISH protocol for localization of rust fungi in paraffin embedded sections of host tissues. This technique provides plant pathologists with a tool to study the morphology of rust fungi within the plants they infect, which may aid in the elucidation of the life cycles of these plant pathogenic fungi and determination of fungal growth patterns in host tissue. We have demonstrated that this protocol can be applied to pathogens from different genera of rust fungi residing in different plant hosts with little to no non-specific background staining of plant tissue. Our protocol is easy to apply and

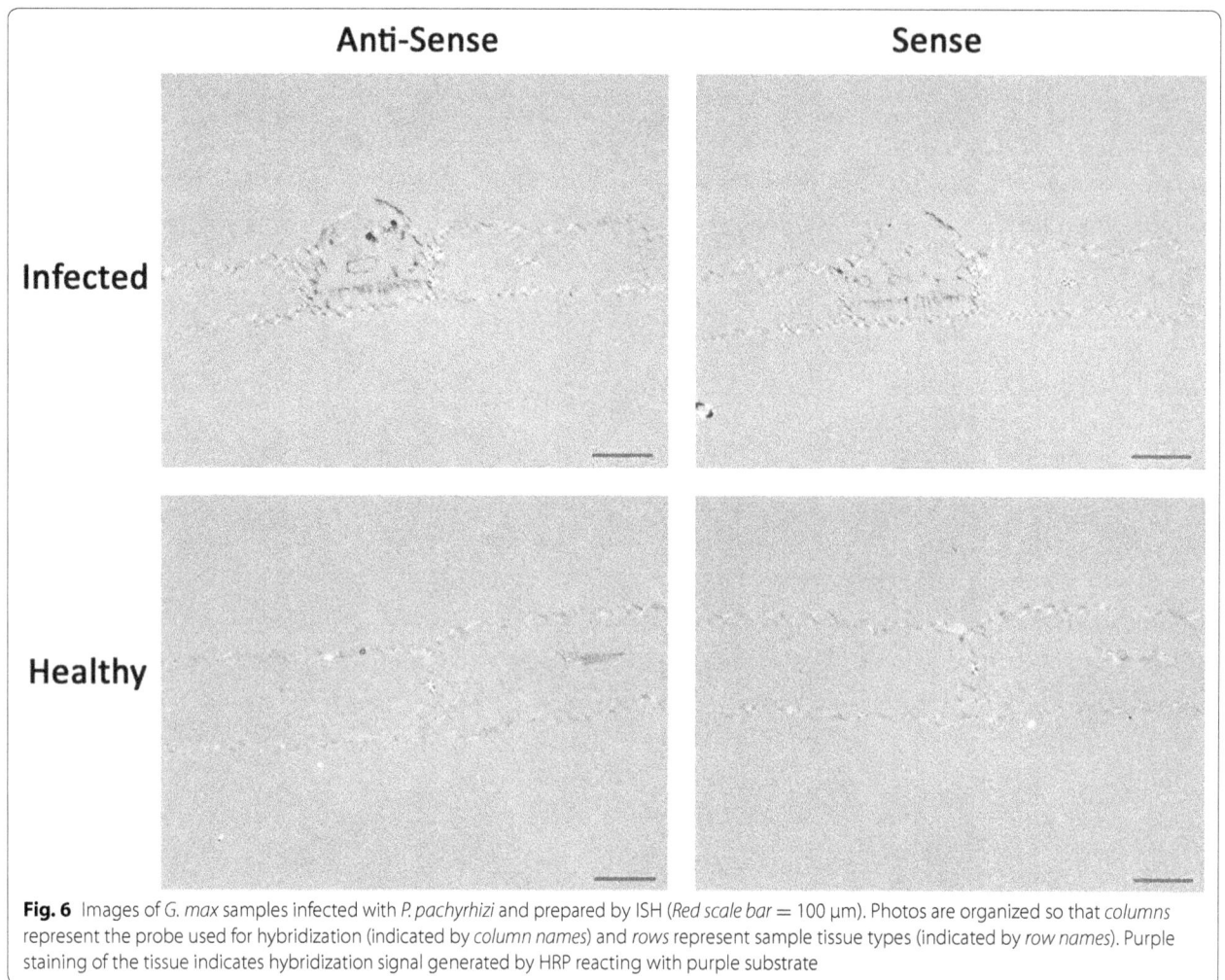

**Fig. 6** Images of *G. max* samples infected with *P. pachyrhizi* and prepared by ISH (*Red scale bar* = 100 μm). Photos are organized so that *columns* represent the probe used for hybridization (indicated by *column names*) and *rows* represent sample tissue types (indicated by *row names*). Purple staining of the tissue indicates hybridization signal generated by HRP reacting with purple substrate

modify if necessary. The generic protocol thus serves as a starting point that may be modified to suit other plant pathogen systems of interest.

### Authors' contributions
MAE co-conceived the project, adapted methods, conducted the experiments, tested protocols and wrote the manuscript, MBM co-conceived the project, provided guidance, improvements and reagents, and edited the manuscript, MRB and CLP edited the manuscript, DGL co-conceived the project and edited the manuscript. All authors read and approved the final manuscript.

### Author details
[1] Department of Biology, University of Pittsburgh School of Medicine, Pittsburgh, PA, USA. [2] USDA-ARS Foreign Disease-Weed Science Research Unit, Ft. Detrick, MD, USA. [3] IR-4 Project, Rutgers University, Princeton, NJ, USA.

### Acknowledgements
The authors thank Dr. Kerry Pedley and Amy Ruck for providing *U. transversalis* spore DNA and *P. pachyrhizi* infected soybean samples, Dr. Reid Frederick for advice and healthy soybean tissue, and Susan Nester for providing *P. horiana*—infected Chrysanthemum tissue and *U. transversalis*—infected Gladiolus tissue. MAE wishes to thank Dr. Oney P. Smith, Hood College, for many stimulating ideas and discussions.

### Competing interests
The authors declare that they have no competing interests.

### Commercial endorsement disclaimer
The use of trade, firm, or corporation names on this page is for the information and convenience of the reader. Such use does not constitute an official endorsement of approval by the USDA Agricultural Research Service, NAL or BIC of any product or service to the exclusion of others that may be suitable.

### Equal opportunity statement
USDA is an equal opportunity provider and employer.

### Funding
This work was supported by USDA, ARS and by USDA APHIS Farm Bill 10007 research grants coordinated by author Dr. Cristi L. Palmer, IR-4 Project, Rutgers University.

### References
1. Dean R, Van Kan JA, Pretorius ZA, Hammond-Kosack KE, Di Pietro A, Spanu PD, Rudd JJ, Dickman M, Kahmann R, Ellis J, et al. The Top 10 fungal pathogens in molecular plant pathology. Mol Plant Pathol. 2012;13:414–30.
2. Yamaoka Y. Recent outbreaks of rust diseases and the importance of basic biological research for controlling rusts. J Gen Plant Pathol. 2014;80:375–88.

3. Bromfield KR. Soybean Rust (Monograph). St. Paul: APS Press; 1984.
4. Schneider RW, Hollier CA, Whitam HK. First report of soybean rust caused by *Phakopsora pachyrhizi* in the continental United States. Plant Dis. 2005;89:774.
5. Gall JG, Pardue ML. Nucleic acid hybridization in cytological preparations. Methods Enzymol. 1971;21:470–80.
6. Zeller R, Rogers M, Haramis AG. *In situ* hybridization to cellular RNA. Curr Protoc Mol Biol. 2001:14.13. 11–14.13. 16.
7. Brahic M, Haase AT. Detection of viral sequences of low reiteration frequency by *in situ* hybridization. Proc Natl Acad Sci USA. 1978;75:6125–9.
8. Haramis A, Carrasco A. Whole-mount *in situ* hybridization and detection of RNAs in vertebrate embryos and isolated organs. Curr Protoc Mol Biol. 1996:14.19.
9. Lawrence JB, Singer RH. Quantitative analysis of *in situ* hybridization methods for the detection of actin gene expression. Nucleic Acids Res. 1985;13:1777–99.
10. Singer R, Lawrence JB, Villnave C. Optimization of *in situ* hybridization using isotopic and nonisotopic detection methods. Biotechniques. 1986;4:230–50.
11. Singer RH, Ward DC. Actin gene expression visualized in chicken muscle tissue culture by using *in situ* hybridization with a biotinated nucleotide analog. Proc Natl Acad Sci USA. 1982;79:7331–5.
12. Landegent J, De Wal NJI, Baan R, Hoeijmakers J, Van der Ploeg M. 2-Acetylaminofluorene-modified probes for the indirect hybridocytochemical detection of specific nucleic acid sequences. Exp Cell Res. 1984;153:61–72.
13. Pinkel D, Landegent J, Collins C, Fuscoe J, Segraves R, Lucas J, Gray J. Fluorescence *in situ* hybridization with human chromosome-specific libraries: detection of trisomy 21 and translocations of chromosome 4. Proc Natl Acad Sci USA. 1988;85:9138–42.
14. Pinkel D, Straume T, Gray J. Cytogenetic analysis using quantitative, high-sensitivity, fluorescence hybridization. Proc Natl Acad Sci USA. 1986;83:2934–8.
15. Brandriff B, Gordon L, Fertitta A, Olsen A, Christensen M, Ashworth L, Nelson D, Carrano A, Mohrenweiser H. Human chromosome 19p: a fluorescence *in situ* hybridization map with genomic distance estimates for 79 intervals spanning 20 Mb. Genomics. 1994;23:582–91.
16. Gordon L, Bergmann A, Christensen M, Danganan L, Lee D, Ashworth L, Nelson D, Olsen A, Mohrenweiser H, Carrano A. A 30-Mb metric fluorescence *in situ* hybridization map of human chromosome 19q. Genomics. 1995;30:187–92.
17. Assmus B, Hutzler P, Kirchhof G, Amann R, Lawrence JR, Hartmann A. *In situ* localization of azospirillum brasilense in the rhizosphere of wheat with fluorescently labeled, rRNA-targeted oligonucleotide probes and scanning confocal laser microscopy. Appl Environ Microbiol. 1995;61:1013–9.
18. Glöckner FO, Amann R, Alfreider A, Pernthaler J, Psenner R, Trebesius K, Schleifer K-H. An *in situ* hybridization protocol for detection and identification of planktonic bacteria. Syst Appl Microbiol. 1996;19:403–6.
19. Li AY, Crone M, Adams PJ, Fenwick SG, Hardy GE, Williams N. The microscopic examination of *Phytophthora* in plant tissues using fluorescent *in situ* hybridization. J Phytopathol. 2014;162:747–57.
20. Montone KT, Guarner J. *In situ* hybridization for rRNA sequences in anatomic pathology specimens, applications for fungal pathogen detection: a review. Adv Anat Pathol. 2013;20:168–74.
21. Montone KT, LiVolsi VA, Lanza DC, Kennedy DW, Palmer J, Chiu AG, Feldman MD, Loevner LA, Nachamkin I. *In situ* hybridization for specific fungal organisms in acute invasive fungal rhinosinusitis. Am J Clin Pathol. 2011;135:190–9.
22. Schröder S, Hain M, Sterflinger K. Colorimetric *in situ* hybridization (CISH) with digoxigenin-labeled oligonucleotide probes in autofluorescent hyphomycetes. Int Microbiol. 2010;3:183–6.
23. Shinozaki M, Okubo Y, Sasai D, Nakayama H, Ishiwatari T, Murayama S, Tochigi N, Wakayama M, Nemoto T, Shibuya K. Development and evaluation of nucleic acid-based techniques for an auxiliary diagnosis of invasive fungal infections in formalin-fixed and paraffin-embedded (FFPE) tissues. Med Mycol. 2012;53:241–5.
24. Sterflinger K, Hain M. *In situ* hybridization with rRNA targeted probes as a new tool for the detection of black yeasts and meristematic fungi. Stud Mycol. 1999;43:23–30.
25. Tanaka E. Specific *in situ* visualization of the pathogenic endophytic fungus *Aciculosporium take*, the cause of witches' broom in bamboo. Appl Environ Microbiol. 2009;75:4829–34.
26. Vági P, Knapp DG, Kósa A, Seress D, Horváth ÁN, Kovács GM. Simultaneous specific *in planta* visualization of root-colonizing fungi using fluorescence *in situ* hybridization (FISH). Mycorrhiza. 2014;24:259–66.
27. Amann R, Fuchs BM, Behrens S. The identification of microorganisms by fluorescence *in situ* hybridisation. Curr Opin Biotechnol. 2001;12:231–6.
28. Bloch B. Biotinylated probes for *in situ* hybridization histochemistry: use for mRNA detection. J Histochem Cytochem. 1993;41:1751–4.
29. Brigati DJ, Myerson D, Leary JJ, Spalholz B, Travis SZ, Fong CK, Hsiung G, Ward DC. Detection of viral genomes in cultured cells and paraffin-embedded tissue sections using biotin-labeled hybridization probes. Virology. 1983;126:32–50.
30. Moter A, Göbel UB. Fluorescence *in situ* hybridization (FISH) for direct visualization of microorganisms. J Microbiol Methods. 2000;41:85–112.
31. Wagner M, Horn M, Daims H. Fluorescence *in situ* hybridisation for the identification and characterisation of prokaryotes. Curr Opin Microbiol. 2003;6:302–9.
32. Zachgo S. *In situ* hybridization. In: Gilmartin PM, Bowler C, editors. Molecular plant biology. 2nd ed. New York: Oxford University Press; 2002. p. 41–63.
33. Bonde MR, Palmer CL, Luster DG, Nester SE, Revell JM, Berner DK. Sporulation capacity and longevity of *Puccinia horiana* teliospores in infected chrysanthemum leaves. Plant Health Prog. 2013. doi:10.1094/PHP-2013-0823-01-RS.
34. Bonde MR, Nester SE, Luster DG, Palmer CL. Longevity of *Uromyces transversalis*, causal agent of gladiolus rust, under various environmental conditions. Plant Health Prog. 2015. doi:10.1094/PHP-RS-14-0036.
35. Lamour KH, Finley L, Snover-Clift KL, Stack JP, Pierzynski J, Hammerschmidt R, Jacobs JL, Byrne JM, Harmon PF, Vitoreli AM, et al. Early detection of Asian soybean rust using PCR. Plant Health Prog. 2006. doi:10.1094/PHP-2006-0524-01-RS.
36. Denhardt DT. A membrane-filter technique for the detection of complementary DNA. Biochem Biophys Res Commun. 1966;23:641–6.
37. McManus J, Cason JE. Carbohydrate histochemistry studied by acetylation techniques: I. Periodic acid methods. J Exp Med. 1950;91:651.

# 3

# Protocol: transient expression system for functional genomics in the tropical tree *Theobroma cacao* L.

Andrew S. Fister[1], Zi Shi[2], Yufan Zhang[3], Emily E. Helliwell[4], Siela N. Maximova[1,5] and Mark J. Guiltinan[1,5*] ●

## Abstract

**Background:** *Theobroma cacao* L., the source of cocoa, is a crop of significant economic value around the world. To facilitate the study of gene function in cacao we have developed a rapid *Agrobacterium*-mediated transient genetic transformation protocol. Here we present a detailed methodology for our transformation assay, as well as an assay for inoculation of cacao leaves with pathogens.

**Results:** *Agrobacterium tumefaciens* cultures are induced then vacuum-infiltrated into cacao leaves. Transformation success can be gauged 48 h after infiltration by observation of green fluorescent protein and by qRT-PCR. We clarify the characteristics of cacao leaf stages and demonstrate that our strategy efficiently transforms leaves of developmental stage C. The transformation protocol has high efficacy in stage C leaves of four of eight tested genotypes. We also present the functional analysis of cacao chitinase overexpression using the transient transformation system, which resulted in decreased pathogen biomass and lesion size after infection with *Phytophthora tropicalis*.

**Conclusions:** Leaves expressing transgenes of interest can be used in subsequent functional genetic assays such as pathogen bioassay, metabolic analysis, gene expression analysis etc. This transformation protocol can be carried out in 1 day, and the transgenes expressing leaf tissue can be maintained in petri dishes for 5–7 days, allowing sufficient time for performance of additional downstream gene functional analysis. Application of these methods greatly increases the rapidity with which candidate genes with roles in defense can be tested.

## Background

*Theobroma cacao* L., the source of cocoa, is a tree crop of great international economic importance and the center of the multi-billion-dollar chocolate industry. While the tree is native to the Amazon basin [1], approximately 70 % of cocoa is now produced in West Africa, with the remainder coming from South America and Southeast Asia [2, 3]. Each year the crop suffers significant losses to a variety of fungal, oomycete, and viral diseases [4], resulting in significant financial loss for cacao farmers and nations exporting cocoa. Cacao research has benefited from the recent publication of the genome sequences of two genotypes [5, 6]. Availability of this data increases the speed with which putatively important cacao genes can be functionally characterized, which could lead to crop improvement through application of novel breeding strategies or biotechnological approaches [7], although progress with long-generation crops is inherently slow. Accordingly, development of strategies enabling gene characterization is important to expedite the process of genetic improvement of cacao.

*Agrobacterium*-mediated transient and stable plant transformation techniques were developed to enable the introduction of recombinant DNA into plant cells in plants [8, 9]. Whereas transient expression is largely the result of transcription and translation of non-integrated T-DNA, stable transformation by definition implies the integration of T-DNA into the host genome [10]. Transiently transfected plants typically show a peak in expression 2–4 days after infection with *Agrobacterium* which

*Correspondence: mjg9@psu.edu
[1] The Huck Institutes of the Life Sciences, The Pennsylvania State University, 422 Life Sciences Building, University Park, PA 16802, USA
Full list of author information is available at the end of the article

subsequently declines [10], while stable transformation is typically achieved through selection and culturing of transformed tissue, and leads to persistent expression of transgenes [11]. If germ line cells are transformed, integration of T-DNA is heritable [12]. While stable transformation is essential for applications in crop improvement, transient transformation enables rapid testing of gene function, and is therefore an invaluable tool for plant genetics research. Both transformation strategies have been applied to a number of tree crops including cacao [13–20], and it has been applied to enhancement the of disease resistance, abiotic stress response, improvement of quality traits, and general study of functional genetics [21].

Traditional breeding strategies for tree crops are laborious and expensive. For cacao, generation of new varieties through breeding programs can take 15–20 years [3]. A strategy for generation of stable transgenic cacao trees was previously published [16], however even this process takes several years to produce a mature tree that could be used to assay experimentally the effect of a transgene's overexpression or knockdown. The transient transformation protocol and subsequent functional analysis described here can be performed in a week, and has been used to demonstrate effect of overexpression [13, 20] and knockdown [22] of cacao genes with roles in defense, expression of non-native phosphatidylinositol 3-phosphate binding proteins in cacao [14], and the function of a transcription factor controlling embryogenesis [19].

Here we present the protocol for *Agrobacterium*-mediated transiently transform of detached leaf tissue of *Theobroma cacao*. Growth conditions described here were extensively tested to optimize transformation efficiency. The strategy enables functional gene characterization to be performed in a matter of weeks, rather than the years that would be required to generate a stably transgenic cacao tree.

## Experimental design
The protocol described here has been used to rapidly screen vectors to measure the effect of gene overexpression or knockdown in cacao leaf tissue [13, 14, 19, 20, 22]. Prior to transformation, binary vector constructs were transferred into competent *Agrobacterium* of strain AGL1 as previously described [23]. Typically the experiment is performed using two vectors: an experimental construct and a control construct (typically pGH00.0126, GenBank: KF018690). Leaves are divided into two sections, one closer to the tip and one closer to the base, such that each leaf can be transformed with both constructs. Preliminary experiments have showed that transformation success usually does not differ significantly between the two sections of a given leaf (data not shown).

The two sections of a leaf are simultaneously infiltrated by submerging leaf discs in cultures of *Agrobacterium* and applying a vacuum. Transformation success is evaluated 48 h after infiltration by observing EGFP fluorescence. A leaf is only used for subsequent functional characterization of EGFP is uniformly present across >80 % of the surface area of the control and experimental sections of a given leaf. A workflow diagram of the transient transformation process is depicted in Fig. 1. It is important to note that efficiency of transformation varies significantly between leaves, and proper appraisal of leaf stage is critical for a successful experiment. At least 3 replicates per transgene are typically used for statistical power. In order to ensure that 3–5 leaf sections per construct are successfully transformed, we recommend infiltrating 8–10, anticipating several leaves will not pass the EGFP coverage threshold.

Cacao leaf stages were previously described [17]; however, as accurate determination of leaf stage is integral to successful transient transformation, we sought to more quantitatively describe the stages to enhance reproducibility of the protocol. In developing the protocol, we found that leaf age affected transformation efficiency, with both earlier and later developmental stages showing lower transformation success as measured by EGFP fluorescence. This resulted in our using Stage C leaves (Fig. 2a), which are expanded but still supple, for our transient transformation experiments. To demonstrate this observation, we transformed leaves of each stage, and 48 h after infiltration, photographed EGFP fluorescence (Fig. 2b–f). To measure leaf toughness, we used a force gauge and performed a punch test on leaves of stages A through E. Figure 2g shows the mean force to puncture, averaged across five leaves, for each leaf stage. Our protocols for collection and transformation and photographing of the five leaf stages, as well as the protocol for the force to puncture test, can be found in the Additional file 1. The data indicates that early in their development (through stage C), leaves do not significantly increase in rigidity. Stage D and E leaves, however, are measurably more rigid. Therefore, it is essential to take into account both leaf color (stage C leaves are bronze to light green) and rigidity to select leaves most likely to be successfully transformed.

In order to evaluate the rate at which cacao leaves infiltrated with *Agrobacterium* become transformed, we monitored expression of an EGFP transgene over a time course after infiltration. Leaves were imaged using a fluorescence stereo-microscope. Images were acquired immediately after transformation and every 3 h after bacterial infiltration (ABI) for the first 48 h, and at hours 60, 84, 108, 132, and 156. No EGFP fluorescence was detected until 18 h ABI. Fluorescence intensity increased

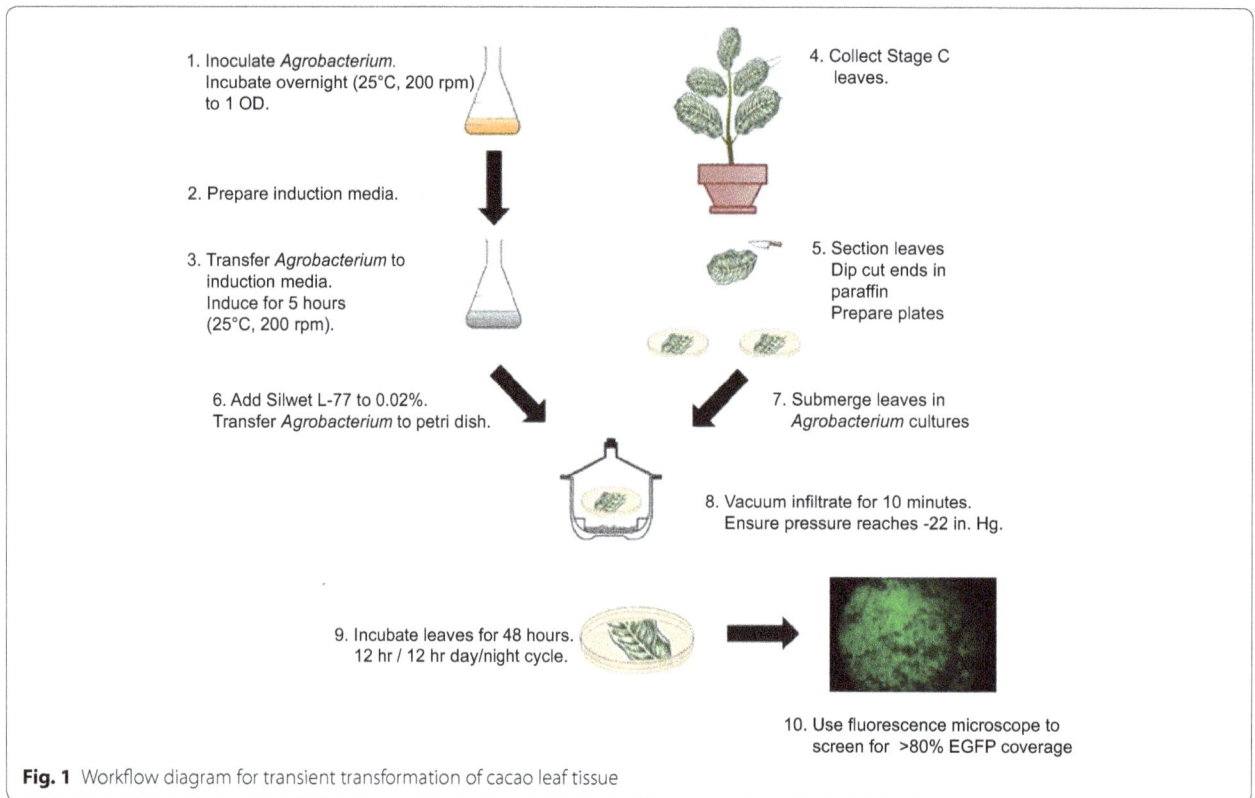

**Fig. 1** Workflow diagram for transient transformation of cacao leaf tissue

until its peak at 45 h ABI, remained high until 60 h, and then steadily declined. EGFP fluorescence was quantified using ImageJ and is graphed as a percentage of the level detected at 45 h ABI (Fig. 3). Because the intensity peaks approximately 2 days ABI, this time point was selected to evaluate transformation success before proceeding into subsequent experiments. Further, our earliest detection of transient expression at hour 18 was consistent with findings in tobacco [24], and peak expression in our time course is consistent with results from transient transformation of *Arabidopsis* [25].

While the protocol was optimized for transformation of Stage C leaves [17] from genotype Scavina 6, it can be applied to other genotypes. Figure 4 includes photographs of stage C leaves from eight genotypes (Fig. 4a), as well as representative photographs showing transformation efficiency of these genotypes (Fig. 4b–i). In Fig. 4j, the transformation efficiency of each genotype was calculated and graphed relative to that measured in the Scavina 6 genotype. Our protocol for this genotype transformation optimization test, including calculation of transformation efficiency with ImageJ [26], can be found in the Additional file 1. While Scavina 6 exhibited the highest transformation efficiency, three other genotypes (CCN51, ICS1, TSH1188) had mean transformation efficiencies greater than 80 %, suggesting that our protocol

could likely be easily applied to these varieties. Physiological differences between leaves of different genotypes may contribute to decreased efficiency, and some alterations to the protocol may be necessary to overcome low efficiencies of the transformation-recalcitrant varieties. We have also previously noted that Scavina 6 leaves appear to remain green and survive longer in petri dishes than other genotypes [13], so it may be generally more suitable to long-duration experiments.

After identifying successfully transformed leaves, subsequent experiments including RNA extractions, pathogen inoculations, and lipid extractions can be performed, as have been described [13, 14, 19, 22]. Leaves will show significant desiccation 5–7 days after being detached from plants; therefore, experiments should not require more than 3–5 days after transformation success is confirmed. Other than this limitation, the transformation strategy can be widely applied to gene characterization studies. In addition to the transformation protocol, we also provide here a detailed methodology for infection of leaves with pathogen after transformation. In addition to the transformation protocol, we also provide here a detailed methodology for infection of leaves with pathogen after transformation.

In Fig. 5, we have included additional data demonstrating the effect of transient overexpression of a previously

**Fig. 2** Leaf stages and force to puncture measurements. **a** Photograph displaying representative leaves of stages A (*leftmost*) to E (*rightmost*) collected from genotype Scavina 6. *Scale bar* represents 5 cm. **b–f** Representative photographs of EGFP fluorescence taken 48 h after infiltration of leaves (stages A–E) with *Agrobacterium*. *Scale bars* represent 1 mm. **g** Measurement of force to puncture for each leaf stage. *Bars* represent mean of five measurements, each representing one leaf from that stage. *Bars* represent standard deviation across five replicates. T test p values are shown above *bars* for Stage D and Stage E, which are comparisons of measurements of Stage C leaves with those of the older stages. Differences between Stage A and C and B and C were not significant

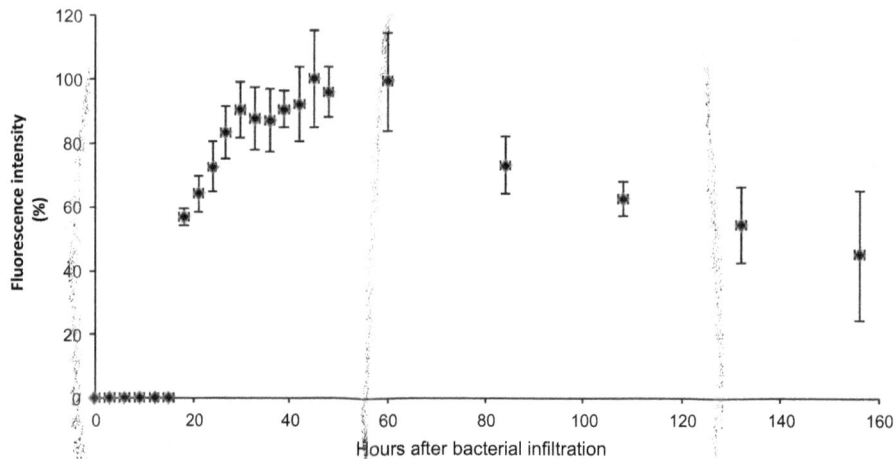

**Fig. 3** Time course of EGFP fluorescence intensity after infiltration of leaf tissue with *Agrobacterium*. Fluorescence is expressed as a percentage of the intensity measured at hour 45, the peak time point. *Error bars* represent standard deviation calculated from three biological replicates

described cacao chitinase gene [15]. Our protocol for these experiments is available in the Additional file 1. Two constructs were used for the transient transformation, pGH00.0126 (GenBank: KF018690), in which EGFP is driven by the CaMV 35S promoter, and another (pGAM00.0511, described in [15]) which has an additional cassette containing a cacao chitinase gene (Tc02_g003890) under the CaMV 35S promoter. Chitinase overexpression using this system resulted in decreased lesion size after infection with *Phytophthora tropicalis* (Fig. 5a, b), a decrease in the ratio of pathogen to cacao DNA detected in the tissue (Fig. 5c), and an approximately six-fold increase in chitinase transcript abundance as assessed by qRT-PCR (Fig. 5d).

We have previously documented differences in basal response to infection to the pathogen *Phytophthora tropicale* in leaf tissue taken from different cacao genotypes using our detached leaf infection assay [13]. While the earlier analysis focused on only one model tolerant genotype (Scavina 6) and model susceptible genotype (ICS1), here we present preliminary data expanding this analysis to 17 genotypes. Additional file 1: Fig. S1 shows box and whisker plots representing the area of infected tissue 72 h after inoculation using the detached leaf infection protocol described here. Additional file 1: Fig. S2 contains representative photographs of the infected leaf tissue from the 17 genotypes. The dramatic differences in susceptibility highlight the need for application of the transient transformation protocol to a wide range of genotypes in order to understand better the genetics underlying differential defense response.

## Reagents and equipment

For transformation:

- *Agrobacterium* is cultured in 523 media, and induced as previously described [27]. Recipes for these media can be found in Table 1.
- A Fast PES Filter unit (Thermo Scientific, Cat. No. 124-0045) is used to sterilize induction media.
- Before infiltration of leaves, Silwet L-77 (Lehle Seeds, Cat. No. VIS-01) is added to *Agrobacterium* cultures to act as a surfactant.
- Plants used for these experiments are greenhouse-grown on The Pennsylvania State University, University Park campus under previously described growth conditions [28]. They are also described in the Additional file 1.
- After leaves are infiltrated with *Agrobacterium*, they are maintained in a controlled environment at 25 °C with 50 % relative humidity and a 12 h/12 h light dark cycle. Light levels are maintained at 55 μmol m$^{-2}$ s$^{-1}$, using fluorescent bulbs 4100 K Kelvin ratings. Higher light levels did not affect transgene expression, but did lead to faster desiccation of leaves.
- Gast G582DX Vacuum Pump.
- Science-Ware vacuum desiccator (Cat# 420270000).
- Whatman grade 5 qualitative filter paper, 90 mm diameter discs (Cat# 1005-090).
- Sterile 100 mm × 20 mm petri dishes (Fisher Brand Cat# FB0875711Z).
- Paraplast Plus tissue embedding medium (McCormick Scientific Cat# 39503002).

Protocol: transient expression system for functional genomics in the tropical tree Theobroma cacao L.

29

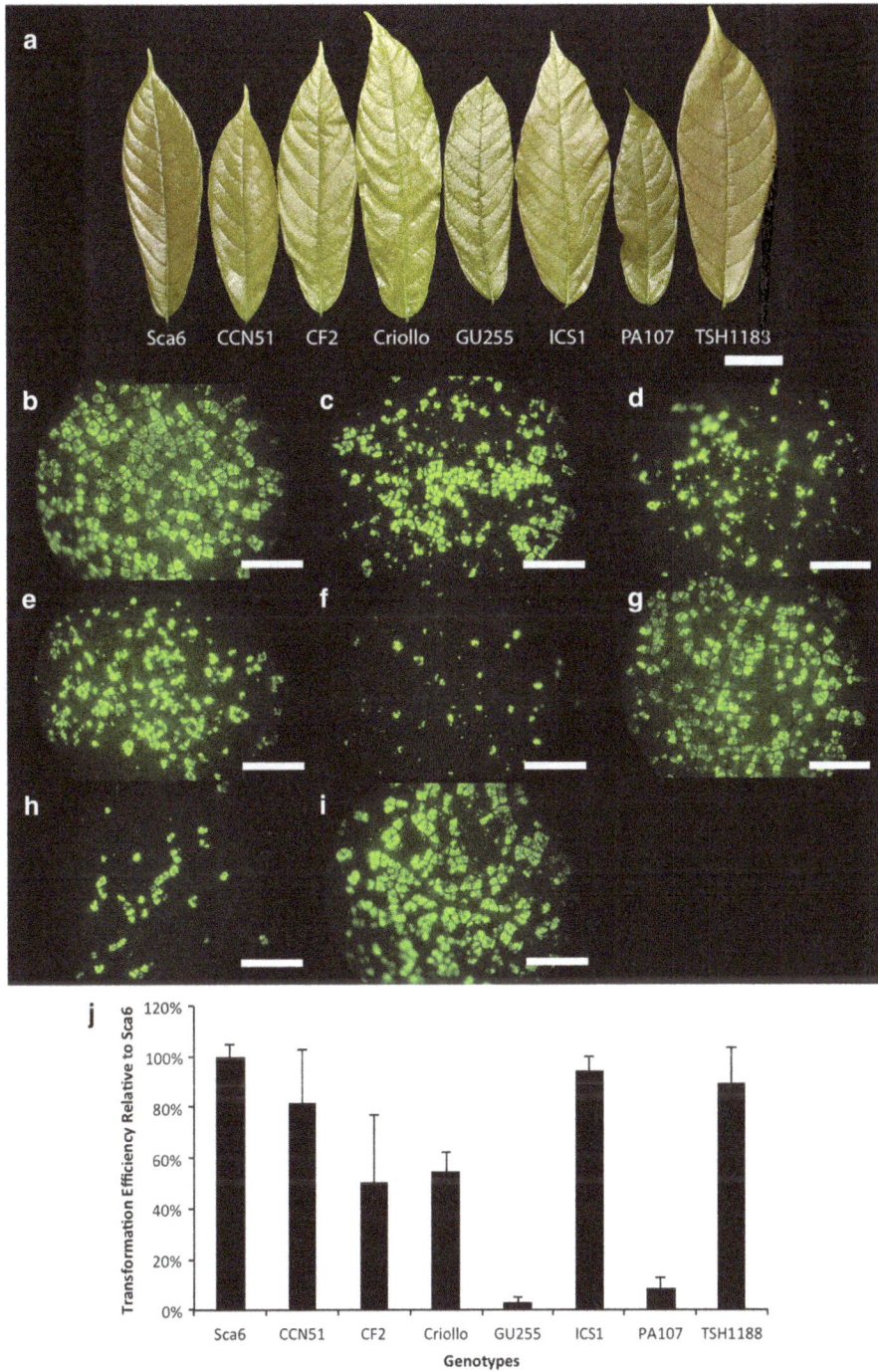

**Fig. 4** Transformation of eight cacao genotypes. **a** Photograph showing stage C leaves selected from eight cacao genotypes. Some genotype identifiers are abbreviated: *Sca6* Scavina 6, *Criollo* B97-61/B2, *ICS1* Imperial College Selection 1. *Scale bar* represents 5 mm. **b–h** Representative images of EGFP coverage 48 h after agrobacterium infiltration using the eight genotypes shown in panel **a**. *Scale bars* represent 1 mm. **b** Sca6; **c** CCN51; **d** CF2; **e** Criollo; **f** ICS1; **g** GU255; **h** PA107; **i** TSH1188. **j** *Bar graph* depicting transformation efficiency expressed as a percentage of that calculated for Scavina 6 samples. *Error bars* represent standard deviation calculated from three biological replicates

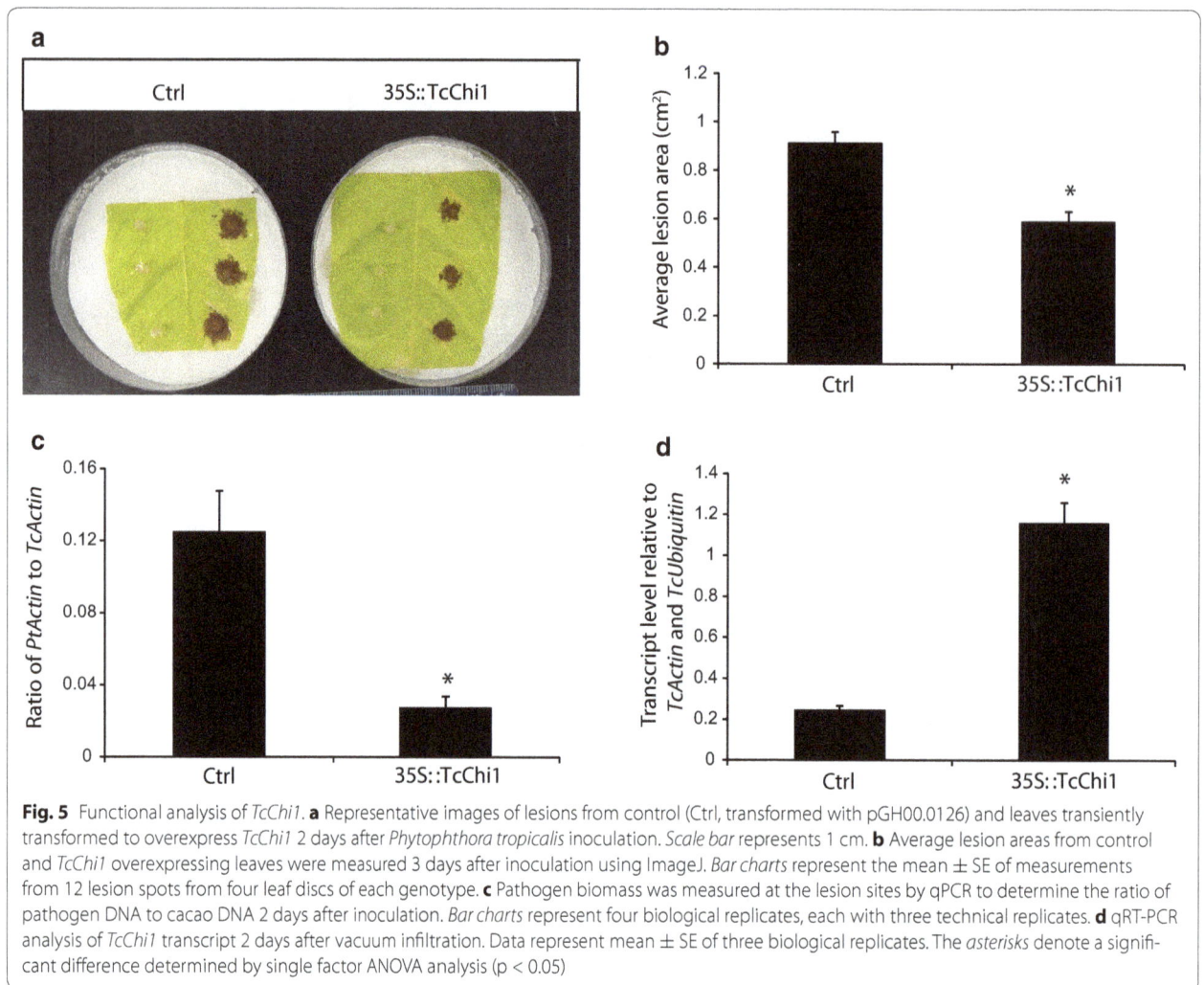

**Fig. 5** Functional analysis of *TcChi1*. **a** Representative images of lesions from control (Ctrl, transformed with pGH00.0126) and leaves transiently transformed to overexpress *TcChi1* 2 days after *Phytophthora tropicalis* inoculation. *Scale bar* represents 1 cm. **b** Average lesion areas from control and *TcChi1* overexpressing leaves were measured 3 days after inoculation using ImageJ. *Bar charts* represent the mean ± SE of measurements from 12 lesion spots from four leaf discs of each genotype. **c** Pathogen biomass was measured at the lesion sites by qPCR to determine the ratio of pathogen DNA to cacao DNA 2 days after inoculation. *Bar charts* represent four biological replicates, each with three technical replicates. **d** qRT-PCR analysis of *TcChi1* transcript 2 days after vacuum infiltration. Data represent mean ± SE of three biological replicates. The *asterisks* denote a significant difference determined by single factor ANOVA analysis ($p < 0.05$)

- Orbital shaker.
- General lab supplies: pipettors, pipette tips, Parafilm, paper towels.

For pathogen bioassay:
- Pathogen subcultures (age depends on pathogen).
- Appropriate media for pathogen growth (recipe for 20 % V8 media is listed in Table 1).
- Laminar flow hood.
- Atomizer of sterile water.
- 3 mm diameter cork borer, 6 mm diameter cork borer, 1.5 cm diameter cork borer.
- General lab supplies: forceps, probe, petri dishes.

**Protocol**
**Preparation of *Agrobacterium* working stocks for transformation**
*Timing* Approximately 1 h, plus overnight incubation

1. Prepare 523 media (see "Reagents and equipment" section).
2. *Agrobacterium* for transformations are cultured using working stocks at OD600 to ensure that cultures grow at consistent rates. To create working stocks, inoculate freezer stocks of AGL1 colony containing desired plasmid in 2 mL 523 medium with appropriate antibiotic and shake overnight at 200 rpm, 25 °C.

**Table 1 Media recipes for *Agrobacterium* growth and induction and pathogen growth**

| Reagent | Amount per liter |
| --- | --- |
| 523 Medium (1 L)[a] | |
|   Sucrose | 10 g |
|   Casein enzymatic hydrolysate | 8 g |
|   Yeast extract | 4 g |
|   $K_2HPO_4$ anhydrous | 2 g |
|   $MgSO_4$ anhydrous | 0.15 g |
| Induction medium (recipe per 30 mL volume)[b] | |
|   Liquid ED (recipe described in [30]) | 30 mL |
|   0.1 M acetosyringone (Sigma Cat. # D134406) | 30 μL |
|   L-Proline | 0.00465 g |
| 20 % V8 Media (1 L)[c] | |
|   Bacto Agar | 15 g |
|   Calcium carbonate ($CaCO_3$) | 3 g |
|   Campbell's V8 Vegetable Juice | 200 mL |

[a] Add distilled water to 1 L. Adjust pH to 7.1 and autoclave

[b] Adjust pH to 5.25–5.3 using 0.1 M KOH. Discard if pH exceeds 5.32. Do not adjust pH using HCl. Prepare induction medium on morning of leaf infiltration experiment. Use liquid ED less than 30 days old

[c] Add distilled water to 1 L. Adjust pH to 7.1 and autoclave. Shake frequently while pouring into petri dishes to maintain homogeneity of media. Pour about 20 mL of media into each plate to ensure that agar plugs and do not fall during leaf assay

3. Measure OD at 600 nm. Let the culture grow until OD600 is 1, or dilute to 1 with 523 media if above. Take 750 μL culture and transfer into a sterile 1.5 mL tube. Add 250 μL of 60 % glycerol. Mix well. Aliquot 100 μL of the mixture into cyrovial tubes. Store at −80 °C.

**Day 1: inoculation and incubation of *Agrobacterium* culture**
*Timing* Approximately 10 min, plus overnight incubation

1. Thaw a 100 μL working stock of AGL1 for each desired plasmid.
2. Inoculate 90 μl of AGL1 stock into 30 ml of 523 media with Kanamycin (50 mg/ml) in a sterile 125 ml Erlenmeyer flask covered with aluminum foil.
3. Shake in the dark overnight at 200 rpm at 25 °C (approximately 16 h).

**Day 2, part I: virulence induction of *Agrobacterium* culture**
*Timing* Approximately 1 h of active time, plus 5 h incubation

1. For every 30 mL culture of Agrobacterium, prepare 30 mL of induction media (see "Reagents and equipment" section). Vacuum sterilize the induction media using Fast PES Filter unit (Thermo Scientific, Cat. # 124-0045).

2. Measure OD of overnight cultures at 600 nm. Use 523 media as a blank. Wait for all cultures to reach OD of 1. Remove those that have passed this point from the shaker to prevent overgrowth. If OD has passed 1.3, discard cultures. If OD is between 1 and 1.3, dilute to OD 1.0 with 523 media.
3. Transfer the entire culture to a 50 mL centrifuge tube. Centrifuge the *Agrobacterium* at $1500 \times g$, 25 °C, for 17 min to pellet the bacteria.
4. Discard supernatant, gently pipette and vortex to re-suspend cultures using 30 mL of induction media and transfer to new 250 mL flasks. Ensure that pelleted bacteria are thoroughly suspended in the solution.
5. Shake at 100 rpm at 25 °C for 5 h in darkness. During this step, collect leaves and prepare plates.

**Day 2, part II: plate preparation and leaf selection**
*Timing* Approximately 1 h
*Note* Plate preparation and leaf collection will take approximately an hour, so perform these steps about 4 h after beginning *Agrobacterium* induction, typically early in the afternoon.

1. Place ten Paraplast Plus chips onto a glass petri dish and apply low heat (~56 °C) until they melt.
2. For each plate, fold a paper towel into a square, and cut off the corners to fit it into a 100 × 20 mm petri dish. Place Whatman #5 filter paper on top of the paper towel and gently press down to create a flat surface. Add 10 ml of sterile water to the plate to maintain humidity.
3. Collect Stage C leaves from greenhouse grown plants. It is essential to the success of the experiment that leaves are soft and supple, and Stage C leaves are bronze to light green in color. Cut the petiole to remove the leaf from the plant without damaging the leaf's surface area. Place the leaves in a sealable plastic bag containing wet paper towels to maintain humidity.
4. Cut leaves with a scalpel to produce leaf two sections. First, the tip and base of the leaf are removed (Fig. 6a). Next the leaf is divided into two sections of equal size. Ensure that each section is large enough to accommodate subsequent experiments (i.e. inoculation with pathogens). As leaves are cut, seal the cut edges by dipping into melted paraffin. This will limit desiccation from exposed veins. Place the leaf discs onto plates for temporary storage, abaxial side up (Fig. 6b) and close the plates. Let sit on the lab bench until induction of *Agrobacterium* is complete.

**Fig. 6** Images representing stages in leaf transformation process. **a** Stage C cacao leaf with tip and base removed. *Scale bar* represents 1 cm. **b** Two halves of a cacao leaf placed into petri dishes with wet paper towel and filter paper. *Scale bar* represents 1 cm. **c** Ideal EGFP coverage seen 48 h after vacuum infiltration of leaves with *Agrobacterium*. *Scale bar* represents 1 mm. **d** Photograph of untransformed leaf tissue using GFP filter. *Scale bar* represents 1 cm

**Day 2, part III: vacuum infiltration**

*Timing* Approximately 1–2 h, depending on replicate number

1. After 5 h in the incubator, add pure Silwet L-77 to the *Agrobacterium* culture to a final concentration of 0.02 % (for a 30 ml culture, this is 6 μL of Silwet). Silwet L-77 is necessary for successful transformation. Our preliminary results indicated that higher Silwet L-77 concentrations do not increase transformation success rates.

2. Pour induced *Agrobacterium* suspension onto 100 mm × 20 mm petri dishes labeled with the construct name on the bottom of the plate as lids are removed during infiltration.

3. With lids removed, place the petri dishes of induced *Agrobacterium* into the desiccator.

4. Select a leaf section to be placed into each dish of *Agrobacterium*, abaxial side down. The other section of the same leaf should receive the other treatment. *Agrobacterium* containing control and experimental vectors are typically infiltrated into their respective leaf sections concurrently. Place the lid on the desiccator.

5. Vacuum-infiltrate the leaves.

   (a) Turn the stopcock valve to open airflow between vacuum pump and desiccator. Start a timer as pressure begins to build.

   (b) Ensure that the pressure reaches −22 in. Hg on pressure gauge. Wait 10 min. As leaves sit in vacuum, small air bubbles should appear at the edges of the leaf.

   (c) Turn the stopcock valve to release vacuum inside desiccator.

6. Using separate tweezers and paper towels for each construct, gently remove the leaf disc from the desiccator, blot dry in paper towels, and hold up to light to look for flooding of cells to assess infiltration success.

   (a) Spots of translucence on the underside of the leaf indicate successful infiltration, which correlates with high EGFP and transgene expression on day 4.

   (b) If none of the leaves have noticeable flooding, the experiment transformation will likely be unsuccessful.

7. Place the leaf abaxial side up onto its petri dish from step 2. Ensure complete contact between the leaf and the filter paper by placing one corner down first and slowly lowering the leaf so that it adheres to the filter paper. Place lid on the petri dish, and seal it with parafilm.

8. Repeat steps 2–6 for all remaining leaf discs.

9. Incubate leaves in a growth chamber at 25 °C with 12 h:12 h light dark cycle for 2 days with a light intensity of 55 μmol m$^{-2}$ s$^{-1}$. Higher light levels were found to lead to faster deterioration of leaf tissue.

## Day 4: evaluating EGFP expression
*Timing* Approximately 1 h

1. Forty-eight hours after infiltration, gather leaf tissue.
2. Using a fluorescence microscope, scan the surface area of each leaf for EGFP as previously described [13]. In order for leaves to be useful for subsequent experiments, at least 80 % of the surface area of the leaf should fluoresce, and there should be no large patches of tissue not expressing EGFP. Representative image of EGFP fluorescence over a small area of leaf tissue is included in Fig. 6c. Coverage across the entire surface of the leaf should match this level of expression. Background fluorescence of cacao leaf tissue is minimal (6D), making transformed regions readily identifiable.
3. Leaf-to-leaf physiological variability may contribute to some variability in downstream experiments. Consequently, only pairs of leaf sections that both pass the EGFP threshold should be retained. Any pairs of leaves where either has less than 80 % of its surface showing EGFP can be discarded.

## *Phytophthora* bioassay

### Part I: subculturing pathogen
*Timing* Approximately 15 min

1. Sterilize a laminar flow hood with UV light for 2 min, and wipe the area with 70 % ethanol.
2. Sterilize the 6 mm diameter cork borer and forceps using 70 % ethanol and flame. Let cool briefly.
3. Use the cork borer to create agar plugs in a mature plate of pathogen (Fig. 7a, b).
4. Transfer agar plugs, pathogen side down, to a new plate of V8 media (Fig. 7c).
5. Incubate the leaves at 28 °C, 12:12 light/dark cycle for 48 h.

### Part II: inoculation of leaf tissue
*Timing* Approximately 1 h

1. Sterilize the laminar flow hood with UV light for 2 min, and wipe the area with 70 % ethanol.
2. Sterilize the 3 mm cork borer with ethanol and flame. Let it cool, and then use it to bore holes into a plate

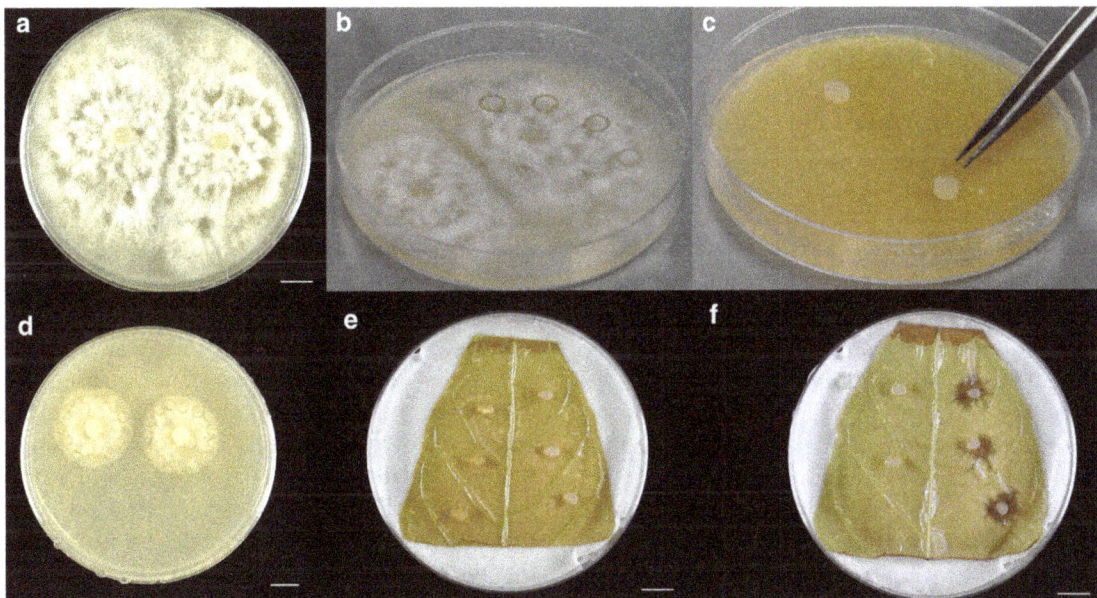

**Fig. 7** Images representing leaf infection process. **a** Mature (approximately 1 week since inoculation) plate of the cacao pathogen *Phytophthora palmivora*. **b** Plate of *P. palmivora* with four agar plugs bored into V8 media. **c** Inoculation of a new plate of 20 % V8 media by transferring agar plugs, pathogen side down, onto the media. **d** Typical size of pathogen growth 48 h after inoculation of new plate. Agar plugs are bored around the edges of the cultures to be used for leaf inoculation. **e** Inoculation of a Stage C cacao leaf with pathogen. Control (media only) plugs are placed on the *left side*, plugs containing pathogen are placed on the *right*. **f** Lesion development 48 h after inoculation. All *scale bars* represent 1 cm

of V8 media with no pathogen. These agar plugs will be used to demonstrate that placing the media on the leaves does not result in formation of a lesion.

3. Re-sterilize the cork borer and let cool briefly. Use it to bore agar plugs around the outside edges of the pathogen culture, as shown in Fig. 7d. Creating plugs from the edges of the culture ensures that the pathogen is actively growing, and that all agar plugs used will be equally virulent.

4. Place agar plugs on the leaf as shown in Fig. 7e. First, sterilize forceps and a probe, and let them cool. Use it to place three V8 agar plugs without pathogen along the left side of the leaf in a line parallel to the midvein. Place agar plugs with containing pathogen mycelia along the right side of the leaf. Ensure that the pathogen's side of the agar plug is in contact with the leaf. Avoid placing an agar plug near the primary or secondary veins as they affect the shape of lesion growth, or too close to another plug so that lesions do not coalesce. For reference, when harvesting lesions, a 1.5 cm diameter disc will be cut for each lesion.

5. Repeat the inoculation for all remaining leaf sections

6. Use the atomizer to spray each leaf with sterilized water. Ensure that the leaf was uniformly misted.

7. Seal plates with parafilm, handling carefully so as to not disturb agar plugs.

8. Incubate leaves in a growth chamber at 25 °C with 12 h:12 h light dark cycle for 2 days with a light intensity of 55 µmol m$^{-2}$ s$^{-1}$.

**Part III: leaf photography and tissue collection**
*Timing* Approximately 2 h

1. If inoculation was successful lesions will have developed after 48 h (Fig. 7f). Photograph the leaves, including a ruler as reference for measurement. Use ImageJ [26] to trace the area of the lesions, and average the three lesions on a leaf to serve as a biological replicate.

2. Remove agar plugs using forceps. Follow appropriate guidelines for disposal of the pathogen.

3. Cut lesions of each leaf using a 1.5 cm diameter cork borer with location of agar as center. Using a sharpened cork borer will prevent leaf tearing. For each leaf, place the three leaf discs into a 2 mL microfuge tube. Flash freeze the tissue with liquid nitrogen, and store at −80 °C. This tissue will be used for DNA extractions and subsequent qPCR to compare relative abundance of pathogen to host DNA within the infected tissue.

4. Use a scalpel to excise the "donut" of tissue around where the lesions developed. Again, place this tissue

in a 2 mL microfuge tube, flash freeze, and store at −80 °C. This tissue can be used for RNA extraction as previously described [19] to verify overexpression of the transgene, and to compare expression level of other genes of interest between the transgene-overexpressing samples and those treated with vector control.

## Conclusions

The transient transformation procedure described here offers a rapid means of performing functional genetic characterization studies on cacao, a long generation tree crop of significant economic importance. The strategy has already been applied to several studies [13, 14, 19, 20, 22], which were studies investigating single gene overexpression and knockdown. The cacao transient transformation protocol was first described by Shi et al. [22]. In this study, the transcription factor NPR3 was shown to be a negative regulator of the cacao defense response by using transient microRNA-mediated knockdown of the TcNPR3 transcript in cacao leaves followed by *Phytophthora* inoculation assays. The protocol was also applied to the study of cacao defense response by Mejia et al., who demonstrated that overexpression of a cacao gene induced by presence of the endophyte *Colletotrichum tropicale*, decreased susceptibility to *Phytophthora* infection [20]. This result suggested that the presence of endophytes in cacao leaves confers a mutualistic enhanced defense response to attack by pathogens [20]. The transient transformation was also applied by Fister et al. in a study demonstrating the positive role of NPR1, the master regulator of systemic acquired resistance, in cacao's response to infection by *Phytophthora* [13]. Helliwell et al. applied cacao leaf transient transformation to show that expression of phosphatidylinositol-3-phosphate binding proteins can decrease susceptibility to infection by competitively inhibiting pathogens' effector proteins' abilities to bind to host cell membranes [14]. Finally, Zhang et al. used the transient transformation strategy to characterize the role of the transcription factor TcLEC2, transiently overexpressing it in leaves to demonstrate its role in regulating genes related to embryo development [19]. While the previous work focused on transformation of only the Scavina 6 genotype, we demonstrate here that the protocol can be applied to transform other genotypes. A recent study reported that the genotypes CCN51 and TSH1188 were used as parents in a mapping population to identify resistance genes for witches' broom disease [29]. Here we demonstrate the ability to transiently transform both of these genotypes (among others), with high efficiency, which would allow screening of defense gene overexpression in the genotypes of interest. Using the infection assay described here, we have already demonstrated variable defense responses between genotypes [13], and here we provide preliminary data on a wider array of cacao genotypes. These data reflect the wide range in susceptibilities different genotypes can exhibit. Application of our transient transformation protocol will enable future work to probe the genetic mechanisms underlying these differences. Further, the development and application of this leaf transformation study enables these types of gene characterization studies to be performed rapidly and at lower cost than through the creation of stably transgenic plants. Without this strategy for rapid gene testing, similar analyses require several years and extensive resources in order to generate stably transgenic cacao trees. The transient transformation strategy is also in the process of being adopted for altering expression of multiple genes by including additional cassettes, and will also be used to develop CRISPR/CAS9-mediated genome editing in cacao leaves.

## Additional file

**Additional file 1: Fig. S1.** Multiple genotype infection graph – Box and whisker plots displaying lesions sizes 72 hours after inoculation of stage C cacao leaves of 17 genotypes with Phytophthora tropicale mycelia. **Fig. S2.** Photographs of infected leaf tissue of diverse genotypes – Representative photographs 72 hours after inoculation of leaf tissue with Phytophthora tropicale. Scale bars represent 1 cm. Supplemental Methods – Descriptions of protocols used for plant growth, transformation and photography of the five leaf stages, the force to puncture test, transformation of the eight cacao genotypes, and evaluation of effects of TcChi1 overexpression.

## Authors' contributions

ASF optimized leaf selection parameters and infiltration conditions and drafted the manuscript. ZS developed the experimental procedures and drafted a working protocol. YZ and EEH contributed to optimizing and editing the protocol. SNM and MJG oversaw experiments, contributed to optimization, and edited the manuscript. All authors read and approved the final manuscript.

## Author details

[1] The Huck Institutes of the Life Sciences, The Pennsylvania State University, 422 Life Sciences Building, University Park, PA 16802, USA. [2] Center for Applied Genetic Technologies, University of Georgia, Athens, GA 30602, USA. [3] Department of Electrical Engineering, Princeton University, Princeton, NJ 08544, USA. [4] Department of Botany and Plant Pathology, Center for Genome Research and Biocomputing, Oregon State University, Corvallis, OR 97331, USA. [5] The Department of Plant Science, The Pennsylvania State University, University Park, PA 16802, USA.

## Acknowledgements

We would like to thank Lena Sheaffer, Brian Rutkowski, and Tyler Kane for their assistance in maintaining cacao plants. We would also like to thank Siti Zulkafli, Abu Dadzie, and Dr. Désiré Pokou for their assistance in troubleshooting the transient transformation protocol. We thank Dr. Luis Mejia and Dr. Brian Bailey for developing the detached leaf inoculation methodology. This work was supported by The Pennsylvania State University's College of Agricultural Sciences, The Huck Institutes of the Life Sciences, the American Research Institute Penn State Endowed Program in Molecular Biology of Cacao and a grant from the National Science Foundation BREAD program (IOS-0965353).

## Competing interests

The authors declare that they have no competing interests.

## References

1. Motamayor JC, Lachenaud P, da Silva e Mota JW, Loor R, Kuhn DN, Brown JS, Schnell RJ. Geographic and genetic population differentiation of the Amazonian chocolate tree (*Theobroma cacao* L.). PLoS ONE. 2008;3(10):e3311.
2. Wood GAR, Lass R. Cocoa. New York: Longman Scientific & Technical; 2008.
3. Lopes UV, Monteiro WR, Pires JL, Clement D, Yamada MM, Gramacho KP. Cacao breeding in Bahia, Brazil: strategies and results. Crop Breed Appl Biotechnol. 2011;11:73–81.
4. Guiltinan M, Verica J, Zhang D, Figueira A. Genomics of *Theobroma cacao*, "The foods of the gods". In: Moore PM, Ming R, editors. Genomics of tropical crop plants. New York: Springer; 2008.
5. Argout X, Salse J, Aury J, Guiltinan M, Droc G, Gouzy J, Allegre M, Chaparro C, Legavre T, Maximova S, et al. The genome of *Theobroma cacao*. Nat Genet. 2011;43(2):101–8.
6. Motamayor J, Mockaitis K, Schmutz J, Haiminen N, Livingstone D, Cornejo O, Findley S, Zheng P, Utro F, Royaert S, et al. The genome sequence of the most widely cultivated cacao type and its use to identify candidate genes regulating pod color. Genome Biol. 2013;14(6):R53.
7. Guiltinan MJ, Maximova SN (2015) Applications of genomics to the improvement of Cacao. In: Chocolate and health: chemistry, nutrition and therapy. The Royal Society of Chemistry; 2015. pp. 67–81.
8. Janssen B-J, Gardner RC. Localized transient expression of GUS in leaf discs following cocultivation with Agrobacterium. Plant Mol Biol. 1990;14(1):61–72.
9. Schell JS. Transgenic plants as tools to study the molecular organization of plant genes. Science. 1987;237(4819):1176–83.
10. Lacroix B, Citovsky V. The roles of bacterial and host plant factors in Agrobacterium-mediated genetic transformation. Int J Dev Biol. 2013;57:467–81.
11. Křenek P, Šamajová O, Luptovčiak I, Doskočilová A, Komis G, Šamaj J. Transient plant transformation mediated by *Agrobacterium tumefaciens*: principles, methods and applications. Biotechnol Adv. 2015;33(6):1024–42.
12. Bent A. *Arabidopsis thaliana* floral dip transformation method. In: Wang K, editor. Agrobacterium protocols, vol. 343. Clifton: Humana Press; 2006. p. 87–104.
13. Fister AS, O'Neil ST, Shi Z, Zhang Y, Tyler BM, Guiltinan MJ, Maximova SN. Two *Theobroma cacao* genotypes with contrasting pathogen tolerance show aberrant transcriptional and ROS responses after salicylic acid treatment. J Exp Bot. 2015;66(20):6245–58.
14. Helliwell EE, Vega-Arreguín J, Shi Z, Bailey B, Xiao S, Maximova SN, Tyler BM, Guiltinan MJ. Enhanced resistance in *Theobroma cacao* against oomycete and fungal pathogens by secretion of phosphatidylinositol-3-phosphate-binding proteins. Plant Biotechnol J. 2016;14(3):875–86.
15. Maximova S, Marelli J-P, Young A, Pishak S, Verica J, Guiltinan M. Overexpression of a cacao class I chitinase gene in *Theobroma cacao* L. enhances resistance against the pathogen, *Colletotrichum gloeosporioides*. Planta. 2006;224(4):740–9.
16. Maximova S, Miller C, Antúnez de Mayolo G, Pishak S, Young A, Guiltinan MJ. Stable transformation of *Theobroma cacao* L. and influence of matrix attachment regions on GFP expression. Plant Cell Rep. 2003;21(9):872–83.
17. Mejia L, Guiltinan M, Shi Z, Landherr L, Maximova S. Expression of designed antimicrobial peptides in *Theobroma cacao* L. trees reduces leaf necrosis caused by *Phytophthora* spp. Small Wonders Pept Dis Control. 2012;1905:379–95.
18. Shi Z, Maximova S, Liu Y, Verica J, Guiltinan MJ. The salicylic acid receptor NPR3 Is a negative regulator of the transcriptional defense response during early flower development in Arabidopsis. Mol Plant. 2013;6(3):802–16.

19. Zhang Y, Clemens A, Maximova S, Guiltinan M. The *Theobroma cacao* B3 domain transcription factor TcLEC2 plays a duel role in control of embryo development and maturation. BMC Plant Biol. 2014;14(1):106.

20. Mejía LC, Herre EA, Sparks JP, Winter K, García MN, Van Bael SA, Stitt J, Shi Z, Zhang Y, Guiltinan MJ et al. Pervasive effects of a dominant foliar endophytic fungus on host genetic and phenotypic expression in a tropical tree. Front Microbiol. 2014;5.

21. Gambino G, Gribaudo I. Genetic transformation of fruit trees: current status and remaining challenges. Transgenic Res. 2012;21(6):1163–81.

22. Shi Z, Zhang Y, Maximova S, Guiltinan M. TcNPR3 from *Theobroma cacao* functions as a repressor of the pathogen defense response. BMC Plant Biol. 2013;13(1):204.

23. Maximova S, Miller C, De Mayolo GA, Pishak S, Young A, Guiltinan MJ. Stable transformation of *Theobroma cacao* L. and influence of matrix attachment regions on GFP expression. Plant Cell Rep. 2003;21(9):872–83.

24. Narasimhulu SB, Deng XB, Sarria R, Gelvin SB. Early transcription of Agrobacterium T-DNA genes in tobacco and maize. Plant Cell. 1996;8(5):873–86.

25. Nam J, Mysore K, Zheng C, Knue M, Matthysse A, Gelvin S. Identification of T-DNA tagged Arabidopsis mutants that are resistant to transformation by Agrobacterium. Mol General Genet MGG. 1999;261(3):429–38.

26. Schneider CA, Rasband WS, Eliceiri KW. NIH Image to ImageJ: 25 years of image analysis. Nat Methods. 2012;9(7):671–5.

27. Li Z, Traore A, Maximova S, Guiltinan M. Somatic embryogenesis and plant regeneration from floral explants of cacao (*Theobroma cacao* L.) using thidiazuron. In Vitro Cell Dev Biol Plant. 1998;34:293–9.

28. Swanson JD, Carlson JE, Guiltinan MJ. Comparative flower development in *Theobroma cacao* based on temporal morphological indicators. Int J Plant Sci. 2008;169(9):1187–99.

29. Royaert S, Jansen J, da Silva DV, de Jesus Branco SM, Livingstone DS, Mustiga G, Marelli J-P, Araújo IS, Corrêa RX, Motamayor JC. Identification of candidate genes involved in Witches' broom disease resistance in a segregating mapping population of *Theobroma cacao* L. in Brazil. BMC Genom. 2016;17(1):1–16.

30. Maximova S, Young A, Pishak S, Miller C, Traore A, Guiltinan M. Integrated system for propagation of *Theobroma cacao* L. In: Protocol for somatic embryogenesis in woody plants. 2005. pp. 209–29.

# 4

# A gene expression microarray for *Nicotiana benthamiana* based on de novo transcriptome sequence assembly

Michal Goralski[1], Paula Sobieszczanska[1], Aleksandra Obrepalska-Steplowska[2], Aleksandra Swiercz[1,3], Agnieszka Zmienko[1,3]* and Marek Figlerowicz[1,3]*

## Abstract

**Background:** *Nicotiana benthamiana* has been widely used in laboratories around the world for studying plant-pathogen interactions and posttranscriptional gene expression silencing. Yet the exploration of its transcriptome has lagged behind due to the lack of both adequate sequence information and genome-wide analysis tools, such as DNA microarrays. Despite the increasing use of high-throughput sequencing technologies, the DNA microarrays still remain a popular gene expression tool, because they are cheaper and less demanding regarding bioinformatics skills and computational effort.

**Results:** We designed a gene expression microarray with 103,747 60-mer probes, based on two recently published versions of *N. benthamiana* transcriptome (v.3 and v.5). Both versions were reconstructed from RNA-Seq data of non-strand-specific pooled-tissue libraries, so we defined the sense strand of the contigs prior to designing the probe. To accomplish this, we combined a homology search against *Arabidopsis thaliana* proteins and hybridization to a test 244k microarray containing pairs of probes, which represented individual contigs. We identified the sense strand in 106,684 transcriptome contigs and used this information to design an Nb-105k microarray on an Agilent eArray platform. Following hybridization of RNA samples from *N. benthamiana* roots and leaves we demonstrated that the new microarray had high specificity and sensitivity for detection of differentially expressed transcripts. We also showed that the data generated with the Nb-105k microarray may be used to identify incorrectly assembled contigs in the v.5 transcriptome, by detecting inconsistency in the gene expression profiles, which is indicated using multiple microarray probes that match the same v.5 primary transcripts.

**Conclusions:** We provided a complete design of an oligonucleotide microarray that may be applied to the research of *N. benthamiana* transcriptome. This, in turn, will allow the *N. benthamiana* research community to take full advantage of microarray capabilities for studying gene expression in this plant. Additionally, by defining the sense orientation of over 106,000 contigs, we substantially improved the functional information on the *N. benthamiana* transcriptome. The simple hybridization-based approach for detecting the sense orientation of computationally assembled sequences can be used for updating the transcriptomes of other non-model organisms, including cases where no significant homology to known proteins exists.

**Keywords:** *Nicotiana benthamiana*, Microarray, Coding strand, Leaf and root transcriptome

## Background

The multiple genome sequencing projects undertaken during the last decade constituted the basis for integrated, whole-genome studies of genes, gene functions and regulatory mechanisms in various organisms. Analysis of the cell's transcriptome composition and dynamics in response to specific stimuli provides important insights into the complexity of the gene regulatory network and key genetic players [1–3]. Currently, the most commonly used techniques for genome-wide expression studies are

*Correspondence: akisiel@ibch.poznan.pl; marekf@ibch.poznan.pl

[3] Institute of Computing Science, Poznan University of Technology, Piotrowo 2, 60-965 Poznan, Poland

Full list of author information is available at the end of the article

DNA microarrays and high throughput RNA sequencing (RNA-Seq) [4]. The latter technique directly reveals the sequence of transcripts and is becoming increasingly popular, as a result of continuous improvements in both the sequencing technology and the data analysis software. This increase has been marked by the development of sequencing centers and large consortia focused on specific organisms (Rice Genome Annotation Project, 1001 Arabidopsis Genomes Project, The Maize Genome Sequencing Consortium, to name just a few). These communities work on developing and standardization of protocols to facilitate aggregating and comparison of various datasets. Current RNA-Seq applications involve assembly of the transcriptome, with or without the reference genome information, gene discovery and expression analysis, identification of unknown exon junctions and alternative transcripts, measuring allele-specific expression and many more [5–8]. On the contrary, microarrays can only derive information on targets that are actually represented by the microarray probes and are sensitive to cross-hybridization, as well as display poor signal resolution and increased variation at low signal intensities [9, 10]. Despite these drawbacks, the results generated on microarray platforms are concordant with those obtained with RNA-Seq [11, 12]. Additionally, thousands of studies performed over the past decades proved that the microarrays reflect the transcriptome composition with high fidelity and that they are a rich source of biologically valuable information. Since their introduction, microarrays have been effectively used in searching for disease markers [13], alternative splicing [14], gene function prediction [15], identification of transcriptionally active regions of the nuclear, mitochondrial and chloroplast genomes [16–18] and many other applications. The microarray experiments are still much cheaper than RNA-Seq, not only regarding the price of consumables and reagents but also the computational and human resources required for data analysis and storage. The latter are often underestimated when calculating the real costs of high-throughput sequencing experiments [19]. Remarkably, extracting biological information from the RNA-Seq data requires combining computational skills with deep knowledge of the problem of interest, typically by the close cooperation of experts in each of those fields. Therefore, sequencing-based experiments may pose a substantial challenge for individual laboratories. With the small size of the resulting datasets and the relatively easy data analysis, DNA microarrays are still an attractive alternative to RNA-Seq for a variety of studies, e.g., focused on differential analysis of known genes in the conditions of study and in time-course studies, where a large number of samples are to be processed and compared in a repeatable manner. We surveyed the gene expression profiling

experiments for *Arabidopsis* and rice, deposited in Gene Expression Omnibus database in years 2012–2015. Those which utilize DNA microarrays constantly outnumber the sequencing-based studies (Additional file 1: Figure S1). Even taking into account the delays in publishing results of research projects, this comparison proves that the DNA microarrays are still used for measuring gene expression changes.

*Nicotiana benthamiana* is a plant model widely used in many laboratories around the world, especially in plant-pathogen interaction studies, due to the ease of infection by a large number of plant viruses [20–23], viroids [24], bacteria [25, 26] and fungi [27, 28]. However, the numbers of *N. benthamiana* gene expression profiling results available in public databases, such as Gene Expression Omnibus (GEO) or ArrayExpress, are surprisingly low. One possible reason for the low abundance is the lack of a microarray platform dedicated to *N. benthamiana*. The detailed design of the only *N. benthamiana*-specific oligonucleotide microarray described so far [29] based on Expressed Sequence Tags data has not been revealed (this information is available on demand from the authors [29]) and the microarray has not been widely used. Moreover, it was produced in the NimbleGen technology, which was discontinued and the NimbleGen microarrays are no longer available to purchase. Nearly 98 % of microarray experiments for *N. benthamiana* reported in GEO were carried out with microarrays specific to another *Solanaceae* species, thus employing a so-called cross-species hybridization (CSH) approach. This approach gained considerable popularity before RNA-Seq methods were introduced on a broad scale and was successfully utilized to profile gene expression in multiple organisms with limited sequence information as well as to extract valid biological information [30–33]. However, it was admitted that CSH analyses suffer from several limitations, which significantly reduce the amount of information that can be derived from the microarrays. These limitations include: higher proportion of genes with no detectable signal (due to lack of the target matching the probe), higher risk of cross-hybridization of transcripts that have similar (but not perfect) homology to the microarray probe, and the lack of probe representation for genes specific to the organism of interest. Also, several authors reported that CSH is characterized by lower mean signal intensity and disturbed spot morphology in comparison with single species hybridization, as a result of weaker binding of targets to their non-perfectly matching probes [34, 35]. All these disturbances affect the overall quality of microarray data and complicate the analysis steps. Published reports on sequencing the *N. benthamiana* genome [36, 37] and, more recently, on assembling the *N. benthamiana* transcriptome from RNA-Seq data [38,

39] now enable researchers to perform *N. benthamiana*-oriented studies in the broader, genomics context.

To facilitate these types of studies, we designed an oligonucleotide microarray (Nb-105k microarray) based on the transcriptomes assembled de novo from the short reads by Australian & New Zealand Consortium and described in detail previously [38, 39]. As those transcriptomes were derived from the non-strand-specific libraries, we used bioinformatics and experimental approaches to ensure the correct orientation of the probes on the final microarray. Our microarray design has been made public via the Agilent eArray platform. In our experience, the Nb-105k microarray showed excellent performance and high sensitivity when employed for gene expression profiling in leaves *versus* roots. We predict that it will become an appreciated and useful gene expression analysis resource for the *N. benthamiana* research community.

## Results
### Determination of the sense cDNA strand in *N. benthamiana* transcriptome v.3 unigenes

To facilitate the gene expression studies in *N. benthamiana*, we decided to design a custom microarray tool which will represent the entire *N. benthamiana* transcriptome and will be compatible with a widely employed Agilent technology platform. The microarray was primarily based on transcriptome v.3, which represented the most contemporary and comprehensive source of *N. benthamiana* transcripts available at the time our work was initiated. The transcriptome v.3 was previously created with the effort of the Australian & New Zealand Consortium, by assembling 193 million short reads generated from the RNA-Seq analysis of nine different *N. benthamiana* tissues on an Illumina sequencing platform using Abyss v1.3 and Trans-Abyss v1.1 [38]. It consists of a representative set of 119,014 unigenes with an average size of 795 bp (the longest contig size was 14,845 bp) and an extended dataset of 237,340 unique transcripts, including all spliced isoforms detected for each gene.

We expected random orientation of the unigenes because they were assembled de novo from reads that originated from non-strand-specific libraries. This posed a problem, as the oligonucleotide probe design step requires a priori knowledge of the transcript orientation. The probes targeting non-coding strands will be useless, as they will not recognize the intended transcript (represented by fluorescently labeled single-stranded cDNA/cRNA). To determine the coding strand, we performed homology searches on the amino acid level against *Arabidopsis thaliana* reference protein set. Matches were found for 61,800 *N. benthamiana* unigenes (Homology-v.3 set, see Fig. 1) and their orientation was chosen accordingly. Of the unigenes, 50.4 % needed conversion

to their reverse complement, confirming our expectations regarding strand randomness introduced by the sequencing and de novo assembly process.

Homology searches provided the strand orientation for approximately half of the unigenes. To increase this number, we turned to experimental detection of the coding strand. To accomplish this, we designed a test 244k microarray with pairs of probes representing 118,934 unigenes in both orientations (80 shortest unigenes from the v.3 transcriptome were not included, due to the microarray capacity limitations). The microarray was hybridized to a labeled RNA pool from various tissues and experimental treatments of *N. benthamiana*. The sense probe was then identified for each unigene as being the only one from the pair which produced a detectable microarray signal or had a stronger signal. We arbitrarily set the relative signal intensity of the expressed strand to be at least 4 times higher than its reverse complement to ensure high confident predictions (Fig. 2a, b). The acceptance criteria were fulfilled by 68,610 microarray probes (Hyb-high-confidency set, see Fig. 1), of which 49.1 % were in the same orientation as the transcriptome v.3 unigenes, in accordance with the expected strand randomness of the unigenes. Next, we analyzed unigenes for which the probe signal intensity ratio was not less than two but below four (Fig. 2c, d). We identified 18,134 such cases (Hyb-low-confidency set, see Fig. 1), of which 50.8 % were in the same orientation as the transcriptome v.3 contigs.

We also compared the hybridization-based and homology-based predictions. *A. thaliana* protein matches were available for 43,532 unigenes from the Hyb-high-confidency set and for 7345 unigenes from the Hyb-low-confidency set. Homology predictions identified the same coding strand as the hybridization-based approach, for 98.37 and 91.23 % of them, respectively. We concluded that the hybridization-based approach allowed us to predict the coding strand with high accuracy and that both Hyb-high-confidency and Hyb-low-confidency sets were composed of reliable data. In summary, we defined the coding strand for an additional 32,930 unigenes, with no homology-based predictions. Altogether, both approaches (homology-based and hybridization-based) defined the correct orientation of 79.6 % of the v.3 contigs. Whenever the two methods resulted in contradictory predictions, we chose the coding strand from the homology data.

### Determination of the sense cDNA strand in *N. benthamiana* transcriptome v.5 contigs

After initiating our microarray method development, a newer version of the *N. benthamiana* transcriptome (v.5) was published [39]. The updated version was generated

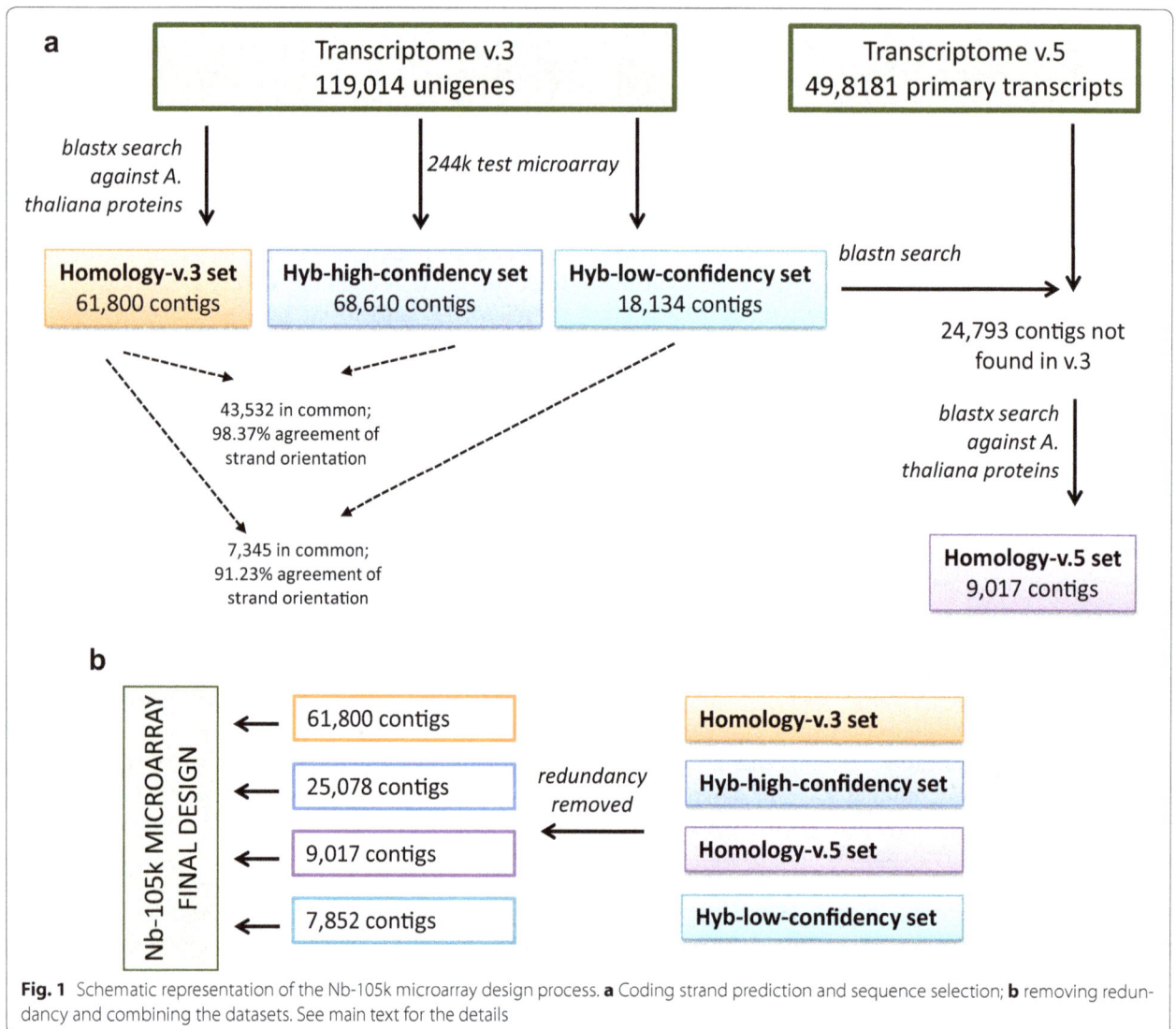

**Fig. 1** Schematic representation of the Nb-105k microarray design process. **a** Coding strand prediction and sequence selection; **b** removing redundancy and combining the datasets. See main text for the details

by increasing the amount of sequenced data and combining four de novo assemblers (TransAbyss, Trinity, SOAP-denovo-Trans, and Oases). This approach generated much longer sequence assemblies (with a mean length 1674 bp). The new transcriptome consists of 49,818 primary transcripts (representative models, usually with the longest sequence) and 184,708 alternative transcripts (cDNA isoforms other than in the primary set). We compared v.5 primary transcripts with sequences from the Homology-v.3 set, Hyb-high-confidency set and Hyb-low-confidency set and selected those without any significant matches. We then used their translated sequences in a homology search against *A. thaliana* proteins, as described above. As a result, we supplemented our dataset with 9017 additional contigs, derived from transcriptome v.5, for which the coding strand could be inferred by homology (Homology-v.5 set, see Fig. 1).

**Nb-105k microarray design and gene expression analysis**
The four sets of sequences with defined orientations (Homology-v.3, Homology-v.5, Hyb-high-confidency and Hyb-low-confidency) were used to create the final microarray (Fig. 1). The Nb-105k microarray includes 103,747 oligonucleotide probes representing *N. benthamiana* contigs and 1325 standard Agilent positive and negative controls. The performance of the new microarray was verified by comparing the gene expression in leaves vs roots of *N. benthamiana* plants. High quality data were obtained using four biological replicates (see "Methods" section). By applying restrictive filters (including a mean intensity, $A_{mean}$, of at least 6, an FDR-corrected p value < 0.0005, and a relative difference in expression of at least fourfold), we obtained 6643 probes that indicated differential gene expression with a high level of confidence. Of these, 4478 showed

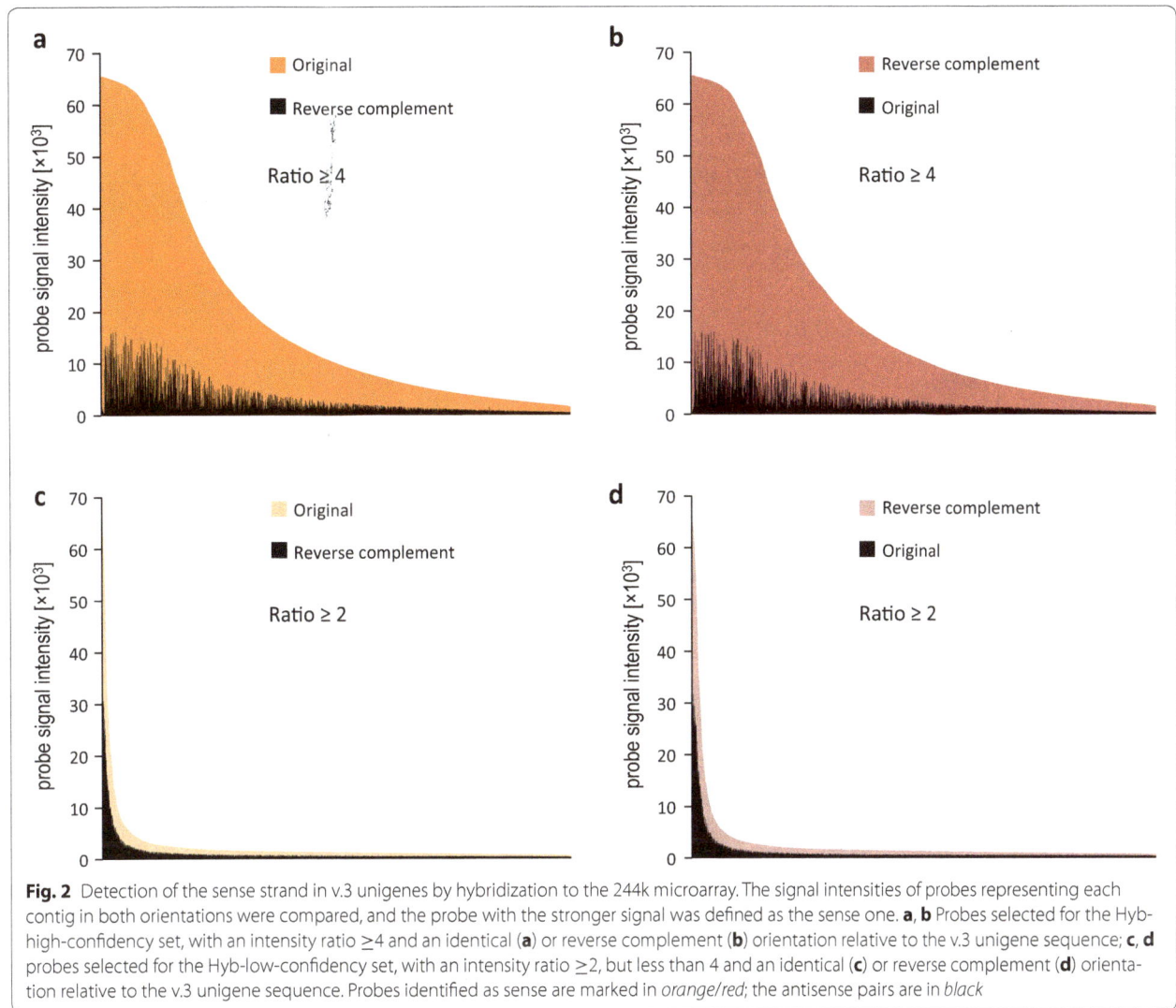

**Fig. 2** Detection of the sense strand in v.3 unigenes by hybridization to the 244k microarray. The signal intensities of probes representing each contig in both orientations were compared, and the probe with the stronger signal was defined as the sense one. **a**, **b** Probes selected for the Hyb-high-confidence set, with an intensity ratio ≥4 and an identical (**a**) or reverse complement (**b**) orientation relative to the v.3 unigene sequence; **c**, **d** probes selected for the Hyb-low-confidence set, with an intensity ratio ≥2, but less than 4 and an identical (**c**) or reverse complement (**d**) orientation relative to the v.3 unigene sequence. Probes identified as sense are marked in *orange/red*; the antisense pairs are in *black*

up-regulation and 2165 showed down-regulation in leaves.

We used available Gene Ontology (GO) information for rapid verification of the biological contents within the differentially expressed gene sets (Fig. 3). GO annotations were available for 22 % of the genes that were up-regulated in leaves. The main categories in this set encompassed genes involved in photosynthesis-related processes and functions. Among the genes that showed up-regulation in roots, 19 % possessed GO annotations. The main biological process represented in this dataset was the stress-response, and the main represented molecular function was peroxidase activity. Based on these analyses, we concluded that the Nb-105k microarray was able to correctly reflect the transcriptome specificities in the two organs under study.

### Verification of v.5 contig assembly using gene expression data generated with Nb-105k microarray

The Nb-105k microarray was designed to contain one probe per contig. However, a comparison of the two transcriptome versions found that multiple v.3 unigenes (from 2 to 173) matched the same v.5 primary transcript. One example was the v.5 primary transcript Nbv5tr6241943, matched by 63 v.3 unigenes, each represented by an individual microarray probe. Nbv5tr6241943 is annotated as Ribulose bisphosphate carboxylase (RuBisCO) small chain 8B. The respective v.3 unigenes covered the whole range of its length (Additional file 1: Figure S2A) and possessed various functional annotations (mostly annotated as RuBisCO chains: 1, 2A, 2B, 3, 3B, 8B). In our microarray experiment, all were up-regulated in leaves (20–317 times, Additional file 1: Figure S2B) and all but one fulfilled the restrictive statistical significance criteria.

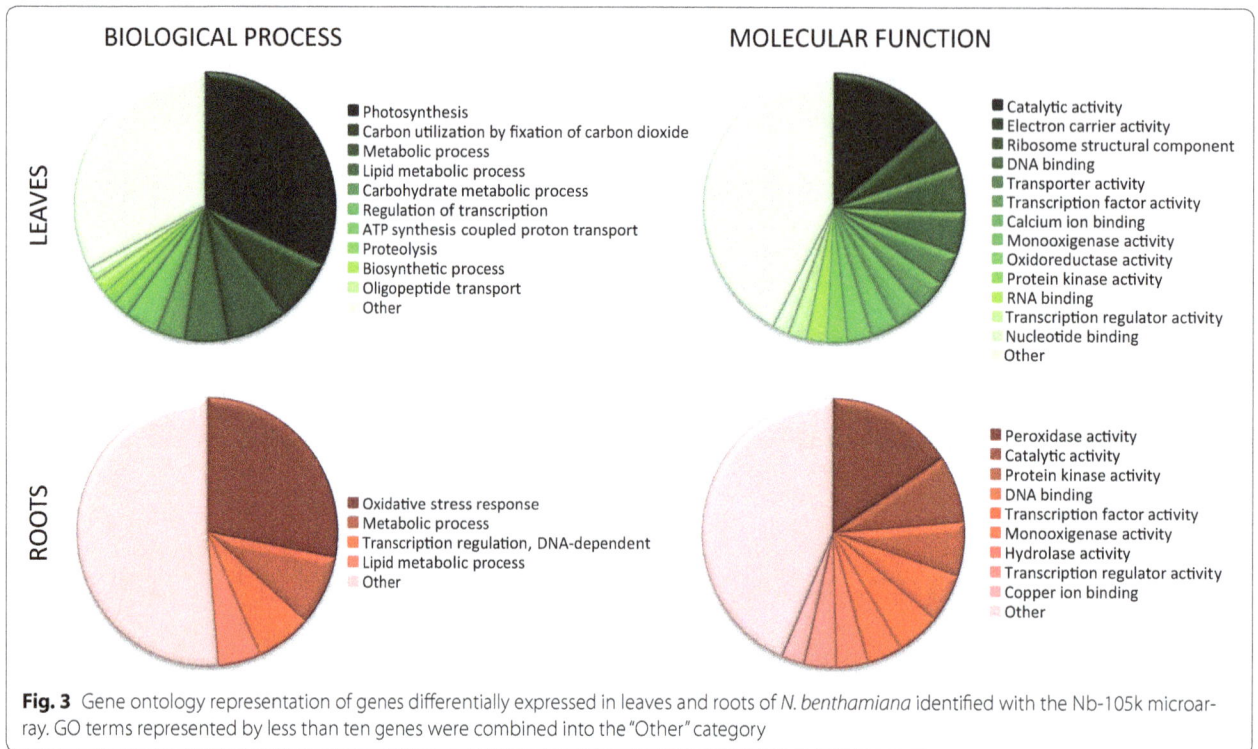

**Fig. 3** Gene ontology representation of genes differentially expressed in leaves and roots of *N. benthamiana* identified with the Nb-105k microarray. GO terms represented by less than ten genes were combined into the "Other" category

However, another cluster of 17 v.3 unigenes displayed variability in expression levels even though all of them matched the same v.5 primary transcript Nbv5tr6230285. Close inspection of Nbv5tr6230285 sequence, which is annotated as Uncharacterized RNA methyltransferase pc1998, revealed that it is probably a chimeric contig. Its 5′ end matches a conserved protein domain found in the AdoMet-dependent methyltransferases superfamily (which is encoded in open reading frame +3) and its 3′ end matches a domain of protease HtpX (which is encoded in open reading frame −3) (Additional file 1: Figure S3A). The mapping and relative orientations of v.3. unigenes (that had their coding strands defined by homology searches and hybridization to the 244k test microarray) clearly revealed that only the 3′ part of the Nbv5tr6230285 contig belongs to the gene that is up-regulated in leaves (Additional file 1: Figure S3B). When used in this manner, the gene expression data obtained with the Nb-105k microarray may also aid the improvement of *N. benthamiana* transcriptome annotation.

## Discussion

Despite the increasing availability of genomic and transcriptomic data for many species, the need for bioinformatics skills and access to large computational resources often result in the pool of "big data" being unused by biologists. Gene expression microarrays are relatively easy to explore, even by non-expert users. Dedicated databases

(e.g., Genevestigator, Plant Expression Database) provide access to integrated microarray data from multiple sources and experiments, allowing intra- and interspecies comparisons. Here, we described a *N. benthamiana* gene expression microarray (Nb-105k microarray) which reflects the current, advanced state of knowledge regarding the *N. benthamiana* genome and transcriptome. Based on our experience, commercial microarrays for plants that either have not been sequenced or do not have fully annotated genomes (e.g., Agilent Tobacco 4 × 44k or Barley 4 × 44k gene expression microarrays, which are based on 2004–2008 sequence releases) typically include multiple probes that have the incorrect orientation and, as a consequence, do not produce useful data [40, Goralski et al., unpublished]. In the present work, we used a combination of bioinformatics and experimental approaches to detect the sense orientation of the contigs prior to designing the probes. We defined the coding strand for 66.4 % of v.3 transcriptome unigenes using homology searches and for 81.3 % of v.3 transcriptome unigenes using RNA hybridization to the test microarray. The latter approach provided substantially more data and did not require sequence homology to known proteins. The efficiency of this simple, hybridization-based verification was demonstrated by the high level of agreement of outputs generated by the two methods. Ultimately, we defined the sense strand orientation of 97,667 contigs from v.3 transcriptome (82.1 %). Apart from the direct application for the Nb-105k microarray design purposes, the data we obtained will serve as a useful guide for future users of this transcriptome dataset.

Our carefully designed microarray consists of nearly 104,000 probes targeting *N. benthamiana* transcripts from de novo assemblies of RNA-Seq data. As the RNA-Seq libraries were prepared from samples representing mixed tissues, the Nb-105k microarray based on these data can be considered a versatile gene expression tool. Indeed, we observed its high sensitivity in detecting and distinguishing the regulation of genes expressed in roots and leaves of *N. benthamiana* plants. The microarray is mainly based on transcriptome v.3 [33], but we also included additional unique sequences from the more recent transcriptome v.5 [34]. The latter version consists of smaller number of contigs that are much larger, as they were assembled from the combination of multiple assemblies and include more short-read data. For this reason, some of the microarray probes derived from v.3 unigenes may map to the same primary transcript from v.5. As shown in the present study, a comparison of the expression profiles generated with such probes allows for additional experimental verification of the transcriptome assemblies. To this end, the Nb-105k microarray usage can be extended from simple gene expression analysis to increasing the accuracy and completeness of the current and future *N. benthamiana* transcriptomes. On the other hand, the future improvement of *N. benthamiana* genome and transcriptome will likely enable re-designing of the Nb-105k microarray, for example to include probes discriminating between paralogous sequences. (The current version of our microarray was not tested for discriminating between highly similar sequences).

In conclusion, we provided details on the design of an oligonucleotide microarray that reflects the current advanced state of *N. benthamiana* transcriptome research. This, in turn, will allow the *N. benthamiana* research community to take full advantage of microarray techniques for studies of gene expression in this plant.

## Methods
### Plant material
*Nicotiana benthamiana* plants were planted on Jiffy pellets and transferred to pots with soil after 2 weeks. The plants were grown in growth chambers at 22 °C/20 °C (day/night). Two sets of samples were prepared for two stages of our experiment, as follows. For detecting the coding strand using hybridization-based approach we collected roots, leaves, and stems from: 4-week old healthy plants, 10-week old healthy plants, 6-week old plants infected with peanut stunt virus, 6-week old plants infected with tomato torrado virus, 6-week old plants

wounded mechanically, 6-week old plants wounded by whitefly *Trialeurodes vaporariorum*. For the analysis of gene expression we collected leaves and roots from four independent 6-week old plants (leaves/roots from one plant constituted one biological replicate). All the material was frozen in liquid nitrogen.

### RNA extraction and labeling
Total RNA was extracted using RNeasy Plant Mini Kit QIAGEN) and DNase-digested with TURBO DNA-free kit (Ambion), according to the manufacturers' standard protocols. RNA quality was determined using Nanodrop 2000 spectrophotometer and capillary electrophoresis in 2100 Bioanalyzer (Agilent). All samples were of high quality, with $A^{260}/A^{280} \geq 2$ and no visible signs of degradation when analyzed on a capillary electrophoresis gel. For hybridization-based coding strand prediction the frozen plant material samples were mixed in equal amounts before RNA extraction ("pooled sample"). For gene expression analysis total RNA was extracted from one plant at a time, (leaves and roots separately), constituting a biological replicate: "L1", "L2", "L3", "L4" and "R1", "R2", "R3", "R4". Labeled cRNA samples were prepared from 200 ng RNA, using a Quick Amp Labeling Kit (Agilent) and quality-checked on a Nanodrop 2000 spectrophotometer.

### Homology searches of transcriptome v.3 unigenes
*Nicotiana benthamiana* unigenes (transcriptome v.3) were downloaded from a server at the School of Molecular Bioscience [40]. Each sequence was used as a query in a BLASTX homology search against the *A. thaliana* reference protein set, with e-value threshold set at 0.001. The best *A. thaliana* hit was then used to check and, if needed, correct the orientation of the query sequences. Whenever the orientation of *A. thaliana* protein matching strand was opposite to the *N. benthamiana* sequence (Plus/Minus or Minus/Plus), the latter was converted into its reverse-complement counterpart.

### Design of the 244k test microarray
Oligonucleotide probes (60-mers) representing v.3 unigenes were designed using the Agilent eArray platform. For each unigene, two probes were designed, one for each sequence orientation. During the design process, the target sequences (all v.3 unigenes, in both orientations) were combined into one reference transcriptome dataset, to ensure maximum probe specificity. Due to limited microarray capacity, pairs of microarray probes were designed for 99.93 % unigenes of the transcriptome v.3 (the longest ones). The resulting 244k microarray was purchased from Agilent.

### Hybridization and analysis of the 244k test microarray

Cy3- and Cy5-labeled cRNA obtained from the same "pooled sample" was hybridized to one 244k test microarray in A1 × 244K hybridization chamber (Tecan) on a HS 4800 Pro automatic station (Tecan), according to the manufacturer's guidelines regarding Agilent microarrays treatment. A Gene Expression Hybridization Kit (Agilent) and Gene Expression Wash Buffer Kit solutions (Agilent) were used for the hybridization and washing steps, respectively. The intensity data were collected with 4200AL GenePix scanner and processed with GenePix Pro 6.1 software using morphological opening background method. Following a quality check with background intensity plotting functions and background subtraction step ("subtract," offset = 10), implemented in R/Bioconductor limma package [41] the intensity data collected for the Cy3 and Cy5 channels were plotted against each other. They demonstrated a high linear correlation ($R^2 = 0.955$). Therefore, only the data for the Cy5 channel were used in the subsequent comparison of the signal intensities of paired probes (see the Results section). The intensity date were further compared in Microsoft Excel 2010.

### Homology searches of transcriptome v.5 primary transcripts

*Nicotiana benthamiana* primary transcripts (transcriptome v.5) were downloaded from a server at the School of Molecular Bioscience [40]. To compare v.3 and v.5 transcriptomes and to supplement our dataset with the contigs missing from the v.3 assembly, we used each v.3 contig of a previously defined strand orientation to query the v.5 dataset with BLASTN. Matches with an e-value <0.001, a gap opening penalty of 6 and a gap extension penalty of 2 were considered significant. With such parameters, the identity level of successfully aligned sequences exceeded 90 % in 93.9 % of the cases tested. The remaining v.5 sequences (those without matches in v.3 dataset) were used as queries in a BLASTX homology search against the *A. thaliana* reference protein set, with an e-value threshold set at 1,E-03 and strand orientation identified as described in the "Homology searches of transcriptome v.3 unigenes" section.

### Design of the Nb-105k microarrays

The sense probes (60-mers) were designed for each transcript from Homology-v.3, Homology-v.5, Hyb-high-confidency and Hyb-low-confidency sets on an Agilent eArray platform. The same four datasets were also combined into a reference transcriptome. This reference transcriptome was used by the Agilent software for design purposes, to ensure maximum specificity of the probes towards their targets. The microarrays in 2 × 105k format (two microarrays per slide) were then purchased from Agilent.

### Hybridization and analysis of the Nb-105k microarrays

Cy3- and Cy5-labeled cRNA from leaves and roots of *N. benthamiana* plants were hybridized to Nb-105k microarrays in the following dye-swap design: Cy3-"L1" versus Cy5-"R2", Cy3-"L2" versus Cy5-"R3", Cy3-"R4" versus Cy5-"L3" and Cy3-"R1" versus Cy5-"L4". The hybridization was performed in A2 × 105K hybridization chambers (Tecan) on a HS 4800 Pro (Tecan) automatic station, according to the manufacturer's guidelines regarding Agilent microarrays treatment. A Gene Expression Hybridization Kit (Agilent) and Gene Expression Wash Buffer Kit solutions (Agilent) were used for the hybridization and washing steps, respectively. The intensity data were collected with 4200AL GenePix scanner and processed with GenePix Pro 6.1 software using morphological opening background method. The data were then analyzed with standard analysis pipeline implemented in R/Bioconductor limma package [41], as described previously [42, 43]. Briefly, the background was subtracted from the probe intensity data. The normalization steps involved "loess" within array normalization and "Aquantile" between array normalization. Quality plots ('MA-plot' and "density plot") were generated to evaluate the data normalization performance and assess the microarray data accuracy and dynamic range. The analysis of Agilent Spike-in controls was also performed and confirmed that the generated data well reflected the theoretical Cy5/Cy3 RNA ratios across a broad range of template copy numbers. The pre-processed data were used for differential expression analysis by applying a Bayesian linear model. The significance threshold was set at a *p* value <0.0005 (after applying Benjamini and Hochberg's method to control the false discovery rate), a mean intensity ($A_{mean}$) >6 and at least a fourfold relative expression difference between roots and leaves. Gene Ontology annotations of v.3 transcripts were the same as used in [38]. Raw and normalized gene expression data from this experiment were deposited into the NCBI GEO repository [44] and are accessible via a GEO Series accession number GSE76631.

### Microarray probe mappings and availability

The Nb-105k microarray design is registered in NCBI GEO repository under Platform accession number GPL21307 and is also publically available for ordering from Agilent eArray platform under ID 066813. In this design, each microarray probe is annotated with the original ID of the appropriate contigs from transcriptome v.3 or transcriptome v.5. Whenever the reverse complement of the original sequence was used for probe design (according to the results of sense strand identification performed in the current study), the contig ID has an RC_prefix. To download the original Agilent design files, with probe sequences or to order the Nb-105k microarray, go to https://earray.chem.agilent.com/earray/ then click the "Published Designs" link button on the right and select *N. benthamiana* from the species list.

The final microarray probes were also mapped to transcriptomes v.3 and v.5 to provide the links between two transcriptomes datasets. For this, each probe sequence was used as a query in a megablast similarity search against a local database created for a transcriptome v.3 or v.5, transcriptome, respectively, with the following parameters: e-value <0.0001, word size = 7, 65 % identity, no masking for low complexity sequences. Best hit for each search was reported in Additional file 2: Table S1 in addition to the primary targets (those for which the probes were designed).

## Additional files

> **Additional File 1: Figure S1.** Proportions of array-based (*blue*) and high throughput sequencing-based (*orange*) RNA profiling experiments deposited in GEO Database in 2012-2015 for *Arabidopsis* and rice. **Figure S2.** Gene expression data for v.3 unigenes mapping v.5 primary transcript Nbv5tr6241943. **Figure S3.** Gene expression data for v.3 unigenes mapping v.5 primary transcript Nbv5tr6230285.
>
> **Additional File 2: Table S1.** Targets of Nb-105k microarray probes in Transcriptomes v.3 and v.5.

## Abbreviations
RNA-Seq: RNA sequencing; GEO: gene expression omnibus; GO: gene ontology; RuBisCO: ribulose bisphosphate carboxylase.

## Authors' contributions
AZ and MF conceived the study. AZ designed the microarrays. AOS provided plant material. MG performed RNA isolation and labeling. MG and PS performed hybridization experiments. MG, PS and AZ analyzed the data. AS and PS performed homology searches. AZ wrote the manuscript. AZ and PS prepared figures. MG, AS, AOS and MF helped to draft the manuscript. All authors read and approved the final manuscript.

## Author details
[1] Institute of Bioorganic Chemistry, Polish Academy of Sciences, Noskowskiego 12/14, 61-704 Poznan, Poland. [2] Institute of Plant Protection – National Research Institute, Wladyslawa Wegorka 20, 60-318 Poznan, Poland. [3] Institute of Computing Science, Poznan University of Technology, Piotrowo 2, 60-965 Poznan, Poland.

## Acknowledgements
We thank Kenlee Nakasugi for providing GO annotation data for v3 transcriptome dataset. The research was supported by National Science Center Grants 2012/05/B/ST6/03026 and 2011/03/B/NZ9/01577. The publication was supported by the Polish Ministry of Science and Higher Education, under the KNOW program.

## Competing interests
The authors declare that they have no competing interests.

## References
1. Dinneny JR, Long TA, Wang JY, Jung JW, Mace D, Pointer S, et al. Cell identity mediates the response of Arabidopsis roots to abiotic stress. Science. 2008;320:942–5.
2. Tohge T, Fernie AR. Co-expression and co-responses: within and beyond transcription. Front Plant Sci. 2012;3:248.
3. Zhang W, Luo J, Chen F, Yang F, Song W, Zhu A, et al. BRCA1 regulates PIG3-mediated apoptosis in a p53-dependent manner. Oncotarget. 2015;6:7608–18.
4. Mantione KJ, Kream RM, Kuzelova H, Ptacek R, Raboch J, Samuel JM, et al. Comparing bioinformatic gene expression profiling methods: microarray and RNA-Seq. Med Sci Monit Basic Res. 2014;20:138–42.
5. Garber M, Grabherr MG, Guttman M, Trapnell C. Computational methods for transcriptome annotation and quantification using RNA-seq. Nat Methods. 2011;8:469–77.
6. Gregg C, Zhang J, Weissbourd B, Luo S, Schroth GP, Haig D, et al. High-resolution analysis of parent-of-origin allelic expression in the mouse brain. Science. 2010;329:643–8.
7. McGettigan PA. Transcriptomics in the RNA-seq era. Curr Opin Chem Biol. 2013;17:4–11.
8. Oshlack A, Robinson MD, Young MD. From RNA-seq reads to differential expression results. Genome Biol. 2010;11:220.
9. Richard AC, Lyons PA, Peters JE, Biasci D, Flint SM, Lee JC, et al. Comparison of gene expression microarray data with count-based RNA measurements informs microarray interpretation. BMC Genom. 2014;15:649.
10. Ritchie ME, Silver J, Oshlack A, Holmes M, Diyagama D, Holloway A, et al. A comparison of background correction methods for two-colour microarrays. Bioinformatics. 2007;23:2700–7.
11. Wang C, Gong B, Bushel PR, Thierry-Mieg J, Thierry-Mieg D, Xu J, et al. The concordance between RNA-seq and microarray data depends on chemical treatment and transcript abundance. Nat Biotechnol. 2014;32:926–32.
12. Zhao S, Fung-Leung WP, Bittner A, Ngo K, Liu X. Comparison of RNA-Seq and microarray in transcriptome profiling of activated T cells. PLoS ONE. 2014;9:e78644.
13. Bejjani BA, Shaffer LG. Clinical utility of contemporary molecular cytogenetics. Annu Rev Genomics Hum Genet. 2008;9:71–86.
14. Zhang C, Li HR, Fan JB, Wang-Rodriguez J, Downs T, Fu XD, Zhang MQ. Profiling alternatively spliced mRNA isoforms for prostate cancer classification. BMC Bioinform. 2006;7:202.
15. Zhu M, Deng X, Joshi T, Xu D, Stacey G, Cheng J. Reconstructing differentially co-expressed gene modules and regulatory networks of soybean cells. BMC Genom. 2012;13:437.
16. Giegé P, Sweetlove LJ, Cognat V, Leaver CJ. Coordination of nuclear and mitochondrial genome expression during mitochondrial biogenesis in Arabidopsis. Plant Cell. 2005;17:1497–512.
17. Li L, Wang X, Sasidharan R, Stolc V, Deng W, He H, et al. Global identification and characterization of transcriptionally active regions in the rice genome. PLoS ONE. 2007;2:e294.
18. Żmieńko A, Guzowska-Nowowiejska M, Urbaniak R, Pląder W, Formanowicz P, Figlerowicz M. A tiling microarray for global analysis of chloroplast genome expression in cucumber and other plants. Plant Methods. 2011;7:29.
19. Muir P, Li S, Lou S, Wang D, Spakowicz DJ, Salichos L, Zhang J, Weinstock GM, Isaacs F, Rozowsky J, Gerstein M. The real cost of sequencing: scaling computation to keep pace with data generation. Genome Biol. 2016;17:53.
20. Obrępalska-Stęplowska A, Wieczorek P, Budziszewska M, Jeszke A, Renaut J. How can plant virus satellite RNAs alter the effects of plant virus infection? A study of the changes in the Nicotiana benthamiana proteome after infection by peanut stunt virus in the presence or absence of its satellite RNA. Proteomics. 2013;13:2162–75.
21. Obrępalska-Stęplowska A, Renaut J, Planchon S, Przybylska A, Wieczorek P, Barylski J, et al. Effect of temperature on the pathogenesis, accumulation of viral and satellite RNAs and on plant proteome in peanut stunt virus and satellite RNA-infected plants. Front Plant Sci. 2015;6:903.
22. Lukhovitskaya NI, Cowan GH, Vetukuri RR, Tilsner J, Torrance L, Savenkov EI. Importin-α-mediated nucleolar localization of potato mop-top virus TRIPLE GENE BLOCK1 (TGB1) protein facilitates virus systemic movement, whereas TGB1 self-interaction is required for cell-to-cell movement in Nicotiana benthamiana. Plant Physiol. 2015;167:738–52.

23. Margaria P, Miozzi L, Rosa C, Axtell MJ, Pappu HR, Turina M. Small RNA profiles of wild-type and silencing suppressor-deficient tomato spotted wilt virus infected *Nicotiana benthamiana*. Virus Res. 2015;208:30–8.

24. Dadami E, Boutla A, Vrettos N, Tzortzakaki S, Karakasilioti I, Kalantidis K. DICER-LIKE 4 but not DICER-LIKE 2 may have a positive effect on potato spindle tuber viroid accumulation in *Nicotiana benthamiana*. Mol Plant. 2013;6:232–4.

25. Lee S, Yang DS, Uppalapati SR, Sumner LW, Mysore KS. Suppression of plant defense responses by extracellular metabolites from *Pseudomonas syringae* pv. tabaci in *Nicotiana benthamiana*. BMC Plant Biol. 2013;13:65.

26. Meng F, Altier C, Martin GB. Salmonella colonization activates the plant immune system and benefits from association with plant pathogenic bacteria. Environ Microbiol. 2013;15:2418–30.

27. Rivas-San Vicente M, Larios-Zarate G, Plasencia J. Disruption of sphingolipid biosynthesis in *Nicotiana benthamiana* activates salicylic acid-dependent responses and compromises resistance to *Alternaria alternata* f. sp. lycopersici. Planta. 2013;237:121–36.

28. Ramegowda V, Senthil-Kumar M, Ishiga Y, Kaundal A, Udayakumar M, Mysore KS. Drought stress acclimation imparts tolerance to *Sclerotinia sclerotiorum* and *Pseudomonas syringae* in *Nicotiana benthamiana*. Int J Mol Sci. 2013;14:9497–513.

29. Goodin MM, Zaitlin D, Naidu RA, Lommel SA. *Nicotiana benthamiana*: its history and future as a model for plant-pathogen interactions. Mol Plant Microbe Interact. 2008;21:1015–26.

30. Rifkin SA, Kim J, White KP. Evolution of gene expression in the *Drosophila melanogaster* subgroup. Nat Genet. 2003;33:138–44.

31. Bigger CB, Brasky KM, Lanford RE. DNA microarray analysis of chimpanzee liver during acute resolving hepatitis C virus infection. J Virol. 2001;75:7059–66.

32. Schenk PM, Thomas-Hall SR, Nguyen AV, Manners JM, Kazan K, Spangenberg G. Identification of plant defence genes in canola using Arabidopsis cDNA microarrays. Plant Biol (Stuttg). 2008;10:539–47.

33. Becher M, Talke IN, Krall L, Krämer U. Cross-species microarray transcript profiling reveals high constitutive expression of metal homeostasis genes in shoots of the zinc hyperaccumulator *Arabidopsis halleri*. Plant J. 2004;37:251–68.

34. Bar-Or C, Bar-Eyal M, Gal TZ, Kapulnik Y, Czosnek H, Koltai H. Derivation of species-specific hybridization-like knowledge out of cross-species hybridization results. BMC Genom. 2006;7:110.

35. Bar-Or C, Novikov E, Reiner A, Czosnek H, Koltai H. Utilizing microarray spot characteristics to improve cross-species hybridization results. Genomics. 2007;90:636–45.

36. Bombarely A, Rosli HG, Vrebalov J, Moffett P, Mueller LA, Martin GB. A draft genome sequence of *Nicotiana benthamiana* to enhance molecular plant-microbe biology research. Mol Plant Microbe Interact. 2012;25:1523–30.

37. Naim F, Nakasugi K, Crowhurst RN, Hilario E, Zwart AB, Hellens RP, Taylor JM, Waterhouse PM, Wood CC. Advanced engineering of lipid metabolism in *Nicotiana benthamiana* using a draft genome and the V2 viral silencing-suppressor protein. PLoS ONE. 2012;7:e52717.

38. Nakasugi K, Crowhurst RN, Bally J, Wood CC, Hellens RP, Waterhouse PM. De novo transcriptome sequence assembly and analysis of RNA silencing genes of *Nicotiana benthamiana*. PLoS ONE. 2013;8:e59534.

39. Nakasugi K, Crowhurst R, Bally J, Waterhouse P. Combining transcriptome assemblies from multiple de novo assemblers in the allo-tetraploid plant *Nicotiana benthamiana*. PLoS ONE. 2014;9:e91776.

40. School of Molecular Bioscience, The University of Sydney. http://sydney.edu.au/science/molecular_bioscience/sites/benthamiana/downloads.php. Accessed 09 May 2014.

41. Limma Smyth GK. linear models for microarray data. In: Gentleman R, Carey V, Dudoit S, Irizarry R, Huber W, editors. Bioinformatics and computational biology solutions using R and bioconductor. New York: Springer; 2003. p. 397–420.

42. Zmienko A, Samelak-Czajka A, Goralski M, Sobieszczuk-Nowicka E, Kozlowski P, Figlerowicz M. Selection of reference genes for qPCR- and ddPCR-based analyses of gene expression in senescing barley leaves. PLoS ONE. 2015;10:e0118226.

43. Zmienko A, Goralski M, Samelak-Czajka A, Sobieszczuk-Nowicka E, Figlerowicz M. Time course transcriptional profiling of senescing barley leaves. Genom Data. 2015;4:78–81.

44. Barrett T, Wilhite SE, Ledoux P, Evangelista C, Kim IF, Tomashevsky M, et al. NCBI GEO: archive for functional genomics data sets—update. Nucleic Acids Res. 2013;41:D991–5.

# 5

# High-throughput phenotyping of lateral expansion and regrowth of spaced *Lolium perenne* plants using on-field image analysis

Peter Lootens[1*], Tom Ruttink[1], Antje Rohde[1,3], Didier Combes[2], Philippe Barre[2] and Isabel Roldán-Ruiz[1]

## Abstract

**Background:** Genetic studies and breeding of agricultural crops frequently involve phenotypic characterization of large collections of genotypes grown in field conditions. These evaluations are typically based on visual observations and manual (destructive) measurements. Robust image capture and analysis procedures that allow phenotyping large collections of genotypes in time series during developmental phases represent a clear advantage as they allow non-destructive monitoring of plant growth and performance. A *L. perenne* germplasm panel including wild accessions, breeding material and commercial varieties has been used to develop a low-cost, high-throughput phenotyping tool for determining plant growth based on images of individual plants during two consecutive growing seasons. Further we have determined the correlation between image analysis-based estimates of the plant's base area and the capacity to regrow after cutting, with manual counts of tiller number and measurements of leaf growth 2 weeks after cutting, respectively. When working with field-grown plants, image acquisition and image segmentation are particularly challenging as outdoor light conditions vary throughout the day and the season, and variable soil colours hamper the delineation of the object of interest in the image. Therefore we have used several segmentation methods including colour-, texture- and edge-based approaches, and factors derived after a fast Fourier transformation. The performance of the procedure developed has been analysed in terms of effectiveness across different environmental conditions and time points in the season.

**Results:** The procedure developed was able to analyse correctly 77.2 % of the 24,048 top view images processed. High correlations were found between plant's base area (image analysis-based) and tiller number (manual measurement) and between regrowth after cutting (image analysis-based) and leaf growth 2 weeks after cutting (manual measurement), with r values up to 0.792 and 0.824, respectively. Nevertheless, these relations depend on the origin of the plant material (forage breeding lines, current forage varieties, current turf varieties, and wild accessions) and the period in the season.

**Conclusions:** The image-derived parameters presented here deliver reliable, objective data, complementary to the breeders' scores, and are useful for genetic studies. Furthermore, large variation was shown among genotypes for the parameters investigated.

**Keywords:** *Lolium perenne*, Field phenotyping, Image analysis, Growth, Regrowth, Tillering

## Background

*Lolium perenne* (perennial ryegrass) is a dominant species of sown grasslands in temperate regions because of its excellent forage quality [11], and is also a primary turf species with rapid growth and establishment [22, 33]. For both applications, forage and turf, the perennial ryegrass plants are cut repeatedly throughout the season and need to resume growth from existing tillers and form new ones. Understanding these two processes, leaf growth and lateral expansion through the formation of new

*Correspondence: peter.lootens@ilvo.vlaanderen.be
[1] Plant Sciences Unit - Growth and Development, ILVO, Caritasstraat 39, 9090 Melle, Belgium
Full list of author information is available at the end of the article

tillers is therefore relevant to breed for optimal sward establishment, growth, tillering and persistence.

It is common practice during the first stages of perennial ryegrass breeding to evaluate large collections of genotypes as spaced plants in the field [11, 28]. Destructive measurements at several moments throughout the season are combined with visual categorical scores of growth, regrowth and rust infection to select elite plants. Such evaluation methods are inexpensive in terms of investments, but can be time-consuming, do not provide detailed information and, in the case of visual scorings, are prone to subjectivity. For example, regrowth is usually evaluated by visual inspection of the plants a few days or weeks after mowing, without any reference to the status of the plant just before or after cutting. It is therefore usually not known whether a good score is due to a high capacity to resume growth from tillers already formed before cutting, or by the formation of new tillers in the periphery of the plant. Because these two processes might be controlled by different genetic factors, a clearer differentiation would allow quicker genetic progress. Furthermore full exploitation of molecular tools to advance genetic improvement of perennial ryegrass depends on the availability of detailed phenotypic evaluation data [12]. For this purpose, methodologies that allow a higher level of resolution and precision in the determination of growth-related characteristics are required.

Recent advances in image analysis-based methods allow phenotyping large collections of plants in an objective, non-invasive way [30], enabling dynamic measurements of plant growth and development. While the use of automatic phenotyping platforms suitable for the evaluation of plants in growth chambers or greenhouses has become common practice [9, 30], these systems are particularly suited for the screening of young plants in experiments of short duration (weeks to months) [5, 14]. Linking results of evaluations carried out in indoor facilities and the behaviour of plants under field conditions is challenging due to differences in environmental factors, soil characteristics, soil volume, etc. [9]. Thus, field evaluation of crops has clear advantages [2]. This is of particular importance in perennial species, such as *L. perenne*, for which it can be relevant to evaluate growth-related parameters during a full growing season or even over several seasons [30]. In recent years spectacular progress has been achieved in the development of phenotyping methodologies that make use of image analysis to evaluate crop performance in the field [1]. However, the application of these methods to *L. perenne* and related species is rather limited as of today. For example, field-based image analysis has been used to determine ground cover in turf grasses (e.g., in bermudagrass overseeded with perennial ryegrass [10, 26]. More recently, Hunt

et al. [12] described a methodology for the acquisition and processing of outdoor images to estimate dry matter of spaced, 4-month-old perennial ryegrass plants. The image analysis algorithm developed was based on colour segmentation, allowing efficient discrimination between green foliage and brown soil. The performance of this algorithm with older plants, recently cut plants containing brownish sections or photographed in different seasons of the year with varying background (soil) colour was not investigated. Such a methodology is, however, required if the purpose is to estimate lateral expansion and the capacity to regrow after cutting, as this implies comparison of images of the same plant taken at different time points [12].

It is challenging to optimize and automatize an image analysis procedure, if image acquisition and image segmentation should be able to cope with plants of different ages and images acquired at different dates under varying climates. Outdoor light conditions vary throughout the day and the season. In addition, moisture level and weed or algal growth affect soil colours, and make the delineation of the object of interest in the image difficult. Furthermore, ryegrass plants can display sectors with a yellow–brown colour just after cutting or after a rainy period, which are difficult to discriminate from the soil.

Standardization of light conditions can be achieved by one of the following options: (i) avoiding external light by photographing the plants during the night using flashes; (ii) using covers to eliminate or stabilize natural light, in combination with flashes; (iii) using a NIR (Near Infrared) camera instead of a digital single lens reflex camera (DSLR); and (iv) photographing in open air using flashes to partly stabilize the white balance. Options *i* and *ii* should render relatively uniform series of images in terms of exposure and colour temperature, which are ideal for colour-based segmentation methods. While option *i* is simpler, it can be logistically difficult. Option *ii* requires the use of a mobile construction to cover the plants for photographing (see for example [3, 12]), making it rather impractical when large plants are photographed. With regards to option *iii*, commercially available NIR cameras have a relatively low resolution and are more expensive than DSLR cameras. Therefore, option *iv* is currently the most straightforward to implement. This choice has implications for image processing and segmentation because colour segmentation alone cannot be used due to unstable light conditions [17, 27].

The aim of image segmentation is to partition the image into regions that are distinct from each other, but internally uniform with respect to certain properties [17], allowing to separate the plant from the background. Segmentation methods can be divided into several types [8, 19, 23] of which colour-based, texture-based, edge detection-based and fast Fourier transformation are frequently used. Colour-based segmentation, based on colour differences between plant and background, can make use of different colour spaces but in horticulture and agriculture RGB and HSV spaces are commonly used [20]. RGB is the standard colour space used by the detector of DSLR cameras but it does not correspond to the way humans see colour [21]. The HSV space is, for human interpretation, more convenient. As the Hue channel (H) is relatively invariant to light level and shading, segmentation can be performed on the H dimension alone [31]. An alternative is to normalise the intensity levels of a colour channel using another colour channel (e.g. using the ratio of B/G in the RGB colour space) [7] or apply other transformations [20]. Although commonly used, colour-based segmentation methods might be less effective to process images taken outdoors. First, the parameters used for the colour segmentation are very sensitive to the white balance, which is affected by sun conditions (under cloudy conditions colours are perceived differently than under sunny conditions by imaging sensors). Second, although automatic light measurement by the camera generates images of the same quality (light, colour), the effective amount of light captured depends on the scene [20]. Third, light reflection on waxy leaves can result in bright or even white spots in the image, and corresponding colour values need to be included in the colour range for selection. Fourth, the presence of algal growth or weeds may hamper the correct identification of the object of interest because they are also green. Texture-based segmentation methods make use of a variance operator to identify regions with different textures (different repeated pattern of different pixel intensities), enabling the separation of objects with the same colour but different textures (for example soil covered by green algae and a green grass plant, or between brownish soil and brown plant parts) [21]. However, segmentation based on texture will be hard for thin-leaved, small plants compared to large plants displaying a more uniform texture. This problem is less pronounced in edge-based segmentation methods, which identify regions in the image where brightness changes rapidly. Ideally edge detection leads to a set of connected curves that correspond to the boundaries of the object [15]. Because this method is rather sensitive to noise, resulting in the detection of irrelevant features in the image, a Gaussian filter is usually applied before the edge detection procedure is executed [4]. Colour-, texture- and edge-based segmentation methods make use of the spatial domain of an image. Alternatively, an image can be converted by fast Fourier transformation to the so-called frequency domain showing frequency and orientation. Erasing information from a location in the frequency transformation is equivalent to removing the corresponding information in every part of the spatial domain image [18, 21]. As a result, the rough plant contours can be selected as the main feature of an image.

From the characteristics of the different segmentation approaches summarized above, it is clear that combining information generated by different methods can result in a more accurate segmentation that exploits their complementarities. It follows that a considerable improvement in segmentation can result from the combination of colour, texture, and edge information [13, 17].

Here, we present a method for high-throughput phenotyping of lateral expansion and regrowth of spaced *L. perenne* plants based on images taken in the field using a DSLR camera placed on a tripod. During image processing, information derived from different segmentation methods is compared and then combined via post-processing integration enabling robust plant object recognition. We show that the combined approach renders better results than the different segmentation methods separately and that the methodology developed is effective under different illumination conditions in outdoor environments. Based on top view images taken at specific time points, the area covered by the *L. perenne* plants was determined in a standardized, quantitative way. We show high correlations between image analysis-based estimates of plant growth and manual measurements.

## Methods

### Plant material and field trial description

A total of 501 genotypes constituting a diverse genetic and morphological collection were planted in a nursery in Melle, Belgium (N50°59′32″ E3°46′59″). Genotypes from four main sources were considered: 'forage breeding lines' (n = 117), 'current forage varieties' (n = 50), 'current turf varieties' (n = 69) and 'wild accessions' (n = 265). 'Forage breeding lines' comprises genotypes from different European breeding programs, 'current forage varieties' and 'current turf varieties' comprise genotypes from commercial varieties, and 'wild accessions' comprises genotypes from natural accessions originating from across Europe. Individual plants containing 3–5 tillers and trimmed to 5 cm were planted at 75 cm spacing within and between rows in a randomised block design with three blocks and one clonal replicate per genotype per block. The field was established on October 2009, and was maintained for three consecutive years with regular weeding and fertilisation. The plants were cut at 6-week intervals during 2010 (March 17th (Y1C1), May 4th (Y1C2), June 15th (Y1C3), July 29th (Y1C4), September 6th (Y1C5) and October 26th (Y1C6) (Additional file 1: Fig. S1). During 2011 all plants were cut on March 16th (Y2C1); in the period May–June 2011 individual plants were cut around 3 weeks after their respective heading date (Y2C2), and all plants were again cut at the same day on July 14th (Y2C3), and August 29th (Y2C4).

## Image acquisition

Top view images were taken from each individual plant directly after cutting (W0) and 1 week after (W1) for the time points Y1C2 to Y1C6 and Y2C1, Y2C3 and Y2C4 (Fig. 1; Additional file 1: Fig. S1). Images were acquired using a DSLR camera (D90, Nikon Corporation, Japan) with a 35 mm lens (AF-S NIKKOR 35 mm 1:1.8G, Nikon Corporation, Japan) or a 24 mm lens (AF-S NIKKOR 24 mm 1:2.8D, Nikon Corporation, Japan) and three wireless remote speedlights (SB-R200, Nikon Corporation, Japan) controlled with a wireless commander (SU-800,

Nikon Corporation, Japan). The images for the first three time points (Y1C2, Y1C3 and Y1C4) were saved as JPG files. For the later dates also NEF (raw) images were recorded, as this offers more possibilities for correction of the white balance, exposure or contrast without data loss. A tripod (055XPROB + 804RC2, Manfrotto, Italy) was used, with the camera placed in perpendicular orientation with respect to the soil surface. The distance between the camera objective and the soil was 90 cm. Images were taken using the Live View function of the camera for the ease and speed of working. Before taking images the white balance of the camera was adjusted to the weather conditions (cloudy or sunny), and an image was taken of a reference card (Grey card, Novoflex) to transform from pixel-scale dimensions to cm-scale dimensions, and to have a colour-neutral reference. The resolution of the images was 74.36 pixels per cm at soil surface when using the 35 mm lens and 49.57 pixels per cm when using the 24 mm lens.

## Image analysis procedure

To optimise the image analysis procedure we chose a subset of 25 genotypes (in three replicates, 75 plants) that represent the broad range of phenotypic diversity (such as base area after cutting, tiller number, plant height and increase in leaf length after cutting) present in the plant collection of 501 genotypes. This subset of 1200 images, representing clonal replicates and different time points of these 25 genotypes, covered various aspects of variation present in the complete set (501 genotypes and 24,048 images), ranging from light spectral quality and intensity, to plant size and colour, and variation in soil background conditions (dry, wet, algae coverage present, small weeds present). Once an optimised and validated procedure was established, it was applied on the total of 24,048 images of all plants and time points.

### Overall description

The images were processed using an automatic program developed in WiT (8.3 sp7, Dalsa Digital Imaging Inc., Canada) (Additional file 2: Fig. S2). Each image was segmented using eight segmentation methods available in WiT (here called C1, C2, T1, T2, E1, E2, E3 and E4) as described below. In addition, based on the resulting mask for each segmentation method, a composite mask image was constructed in which each binary mask had a weight of 1/8th of a maximum intensity of 255. For this composite mask image the pixel intensity threshold was set at 128. This means that in the composite mask, a pixel belongs to the object (i.e., the ryegrass plant) if it belongs to the masks derived from at least four (out of eight) methods [256 (possible intensities) divided by 8 (methods) multiplied by 4 (correct methods) = 128]. For the resulting

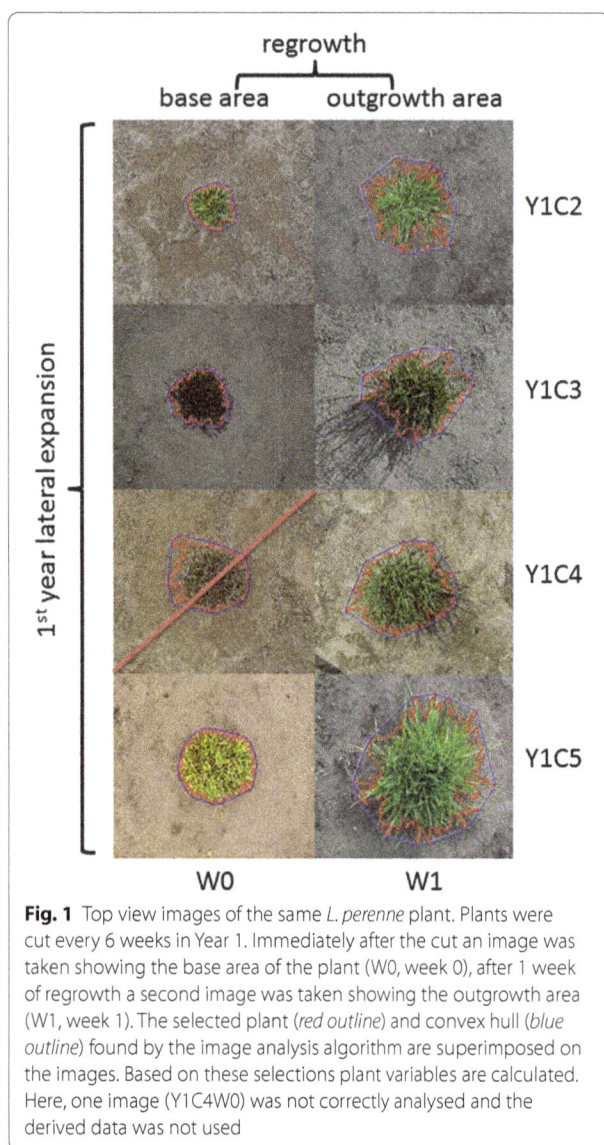

**Fig. 1** Top view images of the same *L. perenne* plant. Plants were cut every 6 weeks in Year 1. Immediately after the cut an image was taken showing the base area of the plant (W0, week 0), after 1 week of regrowth a second image was taken showing the outgrowth area (W1, week 1). The selected plant (*red outline*) and convex hull (*blue outline*) found by the image analysis algorithm are superimposed on the images. Based on these selections plant variables are calculated. Here, one image (Y1C4W0) was not correctly analysed and the derived data was not used

High-throughput phenotyping of lateral expansion and regrowth of spaced Lolium perenne plants...

51

composite mask only the largest object was kept and used to produce an object outline overlay on the original image, depicting the edge of the final selection. This image was stored. After all images had been processed automatically, we determined by visual inspection if the automatic delineation of the ryegrass plant was correct. The images where parts of the plants were not detected or mismatched were not used for the between-method comparisons described below. Examples are shown in Fig. 1.

### Segmentation methods

An overall scheme of the image analysis procedure is shown in Additional file 2: Fig. S2. Eight different segmentation methods were combined. These include colour-based methods (C1, C2), texture-based methods (T1, T2) and edge based-methods (E1–E4) (Table 1). Intensity values for the HSV colour space used within WiT range from 0 to 255 for all channels.

For all segmentation methods except C2 the RGB image was converted to an HSV colour space to enable an easier selection of the green colour: greenish leaves can easily be found between intensities of 35 and 100 of the Hue (H) colour channel. We used threshold boundaries for the H, S and V colour channels of 35–100, 50–255 and 40–226, respectively. The resulting image, which is a raw mask, was then cleaned using a dilate-erode process to remove small objects not related to the plant, and the holes within this mask were filled. This procedure of cleaning up the raw mask was the same for all methods.

For C2 the RGB image was used, and the ratio between the blue (B) and the green (G) colour channel was calculated. In this ratio image the plant appears with a lower intensity than the background. A threshold of 0.625 was used to select the plant object.

The calculations for the texture and edge based methods (T1, T2 and E1–E4) were all based on the saturation (S) channel of the HSV colour space. This colour channel was chosen because of its high contrast between background and object. In T1 a fast Fourier transform operator was used to calculate a frequency domain image. After using a cosine filter with a radius of 25, a reverse fast Fourier transform was calculated resulting in a spatial domain image. The threshold was set at 110. Intensities above this value were used to delineate the raw mask. T2 makes use of an entropy operator. This operator calculates the entropy of the input image pixel values in a specified neighbourhood ($100 \times 100$) around each pixel. In this context, the entropy value of a pixel is a measure of the disorder in the neighbouring pixels. High intensity changes within a limited area due to the transition from plant to the background results in an entropy image in which the plant gets a higher intensity than the background. The threshold was set at 4.7.

For the edge-based methods, E1–E4, a Gaussian filter with a filter size of $5 \times 5$ was used first to achieve a two-dimensional smoothing of the input image. This avoided the selection of irrelevant edges. E1 used a gradient magnitude and direction filter based on Prewitt filter weights. A threshold was set at 29. E2 used a two-dimensional convolution operator with a $3 \times 3$ kernel. The elements in the kernel were set to 1 except for the middle kernel value, which was set to 15. A scale factor by which

**Table 1 Description of the different segmentation methods [colour-based (C); texture-based (T); edge-based (E)] used during the images analysis of top view images of _L. perenne_ plants**

| Name | Segmentation method | Colour space—channel | Domain | Method (WiT) | Extra tools used | |
|------|---------------------|----------------------|--------|--------------|------------------|---|
| | | | | | Before | After |
| C1 | Colour threshold | HSV–HSV | Spatial | Threshold | | Dilate, fill holes, erode, selection of the largest object |
| C2 | Colour threshold | RGB-B/G | Spatial | Threshold | | Dilate, fill holes, erode, selection of the largest object |
| T1 | Texture | HSV-S | Frequency | Fast Fourier transform and cosine filter | | Dilate, fill holes, erode, selection of the largest object |
| T2 | Texture | HSV-S | Spatial | Entropy measurement | | Dilate, fill holes, erode, selection of the largest object |
| E1 | Edge | HSV-S | Spatial | Gradient magnitude and direction filter (Prewitt) | Gauss filter | Dilate, fill holes, erode, selection of the largest object |
| E2 | Edge | HSV-S | Spatial | Two dimensional convolution | Gauss filter | Dilate, fill holes, erode, selection of the largest object |
| E3 | Edge | HSV-S | Spatial | Prewitt edge detection | Gauss filter | Dilate, fill holes, erode, selection of the largest object |
| E4 | Edge | HSV-S | Spatial | Refine edges based on Prewitt | Gauss filter | Dilate, fill holes, erode, selection of the largest object |

all values in the convolved image are divided was set to 13. The threshold was set at an intensity of 230. For E3 an edge-detect operator was used to detect areas of high slope using a Prewitt gradient type operation. The threshold was set at 27. Finally, E4 used a refine edges operator based on a Prewitt kernel with a threshold set at 15. On the resulting image of the raw mask a threshold intensity of 1 was applied.

The parameter settings for the different methods, as described above, were determined using a selected subset of 40 images taken at different time points throughout two subsequent growing seasons, representing different light colour (depending on the time of the day that the images are taken), different background colour or texture, different plant size, etc. Histograms of sections of the images belonging to the plant or the background were extracted. Based on inspection lines showing local intensities, and based on visual interpretation of the image, values were optimised in an iterative process. Based on the visual inspection of the results of each method and the composite mask (combining information from all 8 methods), overlaid on the original image (Fig. 1), the success rate per time point was determined. Based on these results, the threshold settings of the different segmentation methods were refined. Finally, an optimum was found so that a maximum number of images was correctly analysed and with similar success rates for all time points. In the final analysis for the optimisation stage these optimal settings were used for all images (1200 in total).

### Evaluation and comparison of segmentation methods

The performance of the eight segmentation methods was evaluated according to the methodology described by Van Rijsbergen [29] and Smochina [25]. Precision ($P$), Recall ($R$), and the harmonic mean of P and R (F value, denoted as parameter $F = 2PR/(P + R)$) were determined. The parameter $P$ estimates how many pixels in the mask of the corresponding segmentation method do really belong to the object of interest. The parameter $R$ estimates how many pixels of the object (i.e. plant) are included in the mask.

Comparisons using these three parameters were carried out at two levels. First, the agreement between methods was evaluated in a pairwise fashion by using the mask generated by one of the methods as ground-truth and calculating P, R, and F for the mask generated by the other method. In total 28 comparisons ($8 \times 7/2$) were made per image, and all 1200 images are used for this analysis. Second, we selected by visual inspection a subset of images in which the overlay of the composite mask (see above) fitted precisely on the plant (Fig. 1). For these images, the image analysis procedure accurately outlined the plant object, and we used the composite mask as the ground-truth. The eight different segmentation methods were then compared to that composite mask to establish the correctness of each method. Note that because only for the correctly analysed images a ground-truth area (i.e. composite mask) was available, the correctness of the methods is slightly overestimated.

### Image analysis-based parameters

Once the image analysis procedure had been set up, the images of 501 genotypes and all time points (24,048 images in total) were processed in an automatic way. For each correctly processed image, the number of pixels contained in the composite mask corresponds to the area covered by the plant either directly after cutting (termed 'base area'; W0), or 1 week later (termed 'outgrowth area'; W1). Pixel data were converted to $cm^2$. The short term regrowth after cutting, here termed 'regrowth', was calculated as outgrowth area minus base area at a given cut (e.g. Y1C2W1–Y1C2W0, Fig. 1).

### Manual measurements

We counted the number of tillers of each individual plant prior to each cut (in what follows 'tiller number') and recorded the plant height 2 weeks after each cut by measuring with a ruler the length of the longest leaf when stretched vertically. Plant height data were used to calculate leaf growth per growing degree day for each cut (in what follows 'leaf growth'). These measurements were only carried out in Year 1. Minimum and maximum daily temperatures were measured in a weather station within a distance of 500 m from the field trial. A base temperature of 0 °C was used for the calculation of thermal time.

### Data analysis

To estimate the correspondence and complementarity between image analysis-derived data and manual measurements we compared sets of data as follows. First we determined the correlation between 'leaf growth' over 2 weeks after cutting (manual measurement) and 'regrowth' (derived from image analysis). Second, we correlated 'tiller number' (manual measurement) and 'base area' (derived from image analysis).

Finally, we plotted the changes in 'base area' over the whole growing season and calculated the total lateral expansion during the first growing season, here termed 'first year lateral expansion' (Fig. 1), as the slope of the regression for the base areas (W0) of Y1C2, Y1C3, Y1C4, and Y1C5 ($cm^2$ °C day$^{-1}$). Only plants with at least three valid observations were used to estimate 'first year lateral expansion'.

In all these calculations average values per genotype were derived from data of the three clonal replicates in the field trial. Correlations were calculated using Statistica (v12, StatSoft Inc., Tulsa, Oklahoma, US).

## Results

### Overall evaluation of the image analysis segmentation methodology

We first used colour segmentation on the earliest series of images taken. No single segmentation method resulted in a satisfactory number of images that were correctly analysed over all series (a series refers here to a set of pictures taken at a single time point, e.g. Y1C2W0), because of the large variation in light quality and intensity, background and plant characteristics among the different series of images (data not shown). Therefore, we combined the masks of eight segmentation methods into a composite mask to improve the robustness of the overall image analysis procedure. We analysed all the images of 75 plants (corresponding to 25 genotypes) across two growing seasons. We inspected visually whether the overlay composite mask correctly outlined the actual plant form in each image and found that 86.2 % of the images were correctly analysed (Fig. 2). This shows that the combination of different segmentation methods allows a correct assessment of the images, independent of the light conditions, background characteristics, and plant size throughout two growing seasons. An exception concerns the mages of Y2C1W1, of which only 22.7 % were correctly analysed. This low success rate was because at the time of image acquisition the soil had a similar texture as the plant due to suboptimal weeding. Under these

circumstances typically E4 and E1 yielded poor results. This further shows that field management is an integral part of proper image acquisition and contributes to the quality of downstream image processing.

Over all cutting periods, 84.5 % of the images of plants that were just cut (W0) were correctly analysed, while 87.5 % of the images of plants that were allowed to regrow for 1 week (W1) were correctly analysed. After elimination of Y2C1W1 data, the success rate increased from 87.5 to 96.8 %, demonstrating a higher success rate for images of plants that were allowed to regrow for 1 week and which displayed fewer brownish sectors. Indeed, after 1 week of regrowth most visible plant parts are green and are easier to delineate using image analysis (Fig. 1).

### Comparison of segmentation methods

The overlap of the masks defined by independent segmentation methods was evaluated based on all 1200 images. Per image, the resulting mask of one method was treated as ground-truth, irrespective of whether the image was correctly analysed or not, and compared to the masks defined by each of the other seven segmentation methods. Thus, per image a total of 28 pairwise comparisons were made. The average of all harmonic mean (F) values for all 1200 images was used for pairwise comparisons of methods (Table 2). The highest correspondence across all images was found between methods E1 and E3 (96.2 %), followed by E1 and E4 (92.1 %), and E3 and E4 (89.4 %). These are all edge-based. E2 displayed the lowest correspondence with all other segmentation methods (average F values between 50.6 and 59.1 %). The two colour-based methods C1 and C2 showed 77.8 % correspondence, while the texture-based methods T1 and T2 showed 66.8 % correspondence. Other combinations showed intermediate values. This clearly shows that the different methods can be complementary, even though some methods with high correspondence may be redundant.

### Performance of the different segmentation methods relative to the composite mask

Next, we compared the masks derived from the different segmentation methods to the composite mask. Since here we were primarily interested in how accurate the different methods could find the ryegrass plant in the image, we used only the 1034 images (86.2 % of the complete set) in which visual inspection confirmed that the composite mask accurately outlined the plant object.

A high Recall (R) in combination with a high Precision (P) is ideal. A high Recall in combination with a low Precision indicates a mask that overestimates the area of the ryegrass plant. The harmonic mean of P and R, also

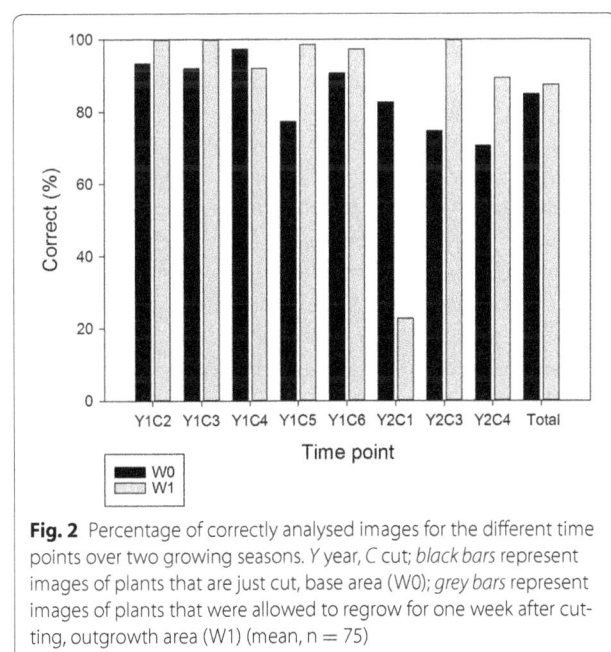

**Fig. 2** Percentage of correctly analysed images for the different time points over two growing seasons. *Y* year, *C* cut; *black bars* represent images of plants that are just cut, base area (W0); *grey bars* represent images of plants that were allowed to regrow for one week after cutting, outgrowth area (W1) (mean, n = 75)

**Table 2** Overlap of the masks derived from eight different segmentation methods [colour-based (C); texture-based (T); edge-based (E)] described in Table 1 estimated as the average of F values (=harmonic mean of precision and recall), n = 1200

| | Colour segmentation | | Texture segmentation | | Edge based segmentation | | | |
|----|------|------|------|------|------|------|------|------|
| | C1 | C2 | T1 | T2 | E1 | E2 | E3 | E4 |
| C1 | | 77.8 | 65.7 | 72.8 | 74.4 | 50.6 | 75.0 | 71.5 |
| C2 | | | 79.7 | 76.5 | 77.7 | 58.5 | 78.1 | 75.5 |
| T1 | | | | 66.8 | 65.9 | 50.9 | 66.4 | 63.3 |
| T2 | | | | | 88.6 | 56.3 | 88.1 | 86.8 |
| E1 | | | | | | 57.1 | 96.2 | 92.1 |
| E2 | | | | | | | 56.1 | 59.1 |
| E3 | | | | | | | | 89.4 |

called the F value, represents the performance of object recognition. The highest F values were found for E3 (92.8 %) and E1 (92.0 %) (Fig. 3). E3 had a slightly lower R than E1 whereas E1 had a slightly lower P. This indicates that E1 overestimates the plant area more than E3. E2, another edge-detection method, had the lowest F value (57.9 %). The method clearly finds the object (high R) but overestimates its area (low P). The colour-based methods (C1 and C2) had an intermediate F value of 78.1 and 85.0 % respectively. C2 performs better than C1 in terms of R. T1, a texture-based method, displays an intermediate performance, between E2 and the colour-based methods. In comparison to E2, the R of T1 is lower but the P is higher. T2 and E4 show slightly lower F values compared to E3 and E1. Nevertheless, both show a higher R but a lower P.

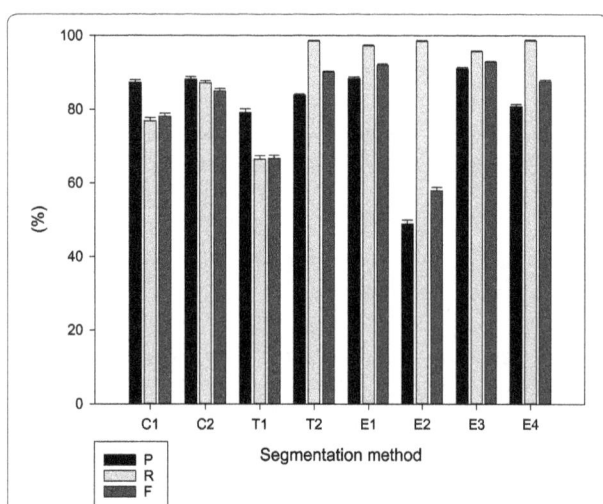

**Fig. 3** Precision (P, %), recall (R, %), and harmonic mean of precision and recall (F, %) of the different segmentation methods [colour-based (C); texture-based (T); edge-based (E)] for all time points together. The composite mask derived from eight segmentation methods was used as the ground truth in these comparisons (mean and SE, n ≤ 75)

The colour-based method C1, the texture-based method T1, and the edge-based method E2 display higher F values for W1 (Fig. 4b) than for W0 images (Fig. 4a). Segmentation methods C2, T2, E1, E3, and E4 performed in a similar way for W0 and W1 images, and were less dependent on the fine structure of the plant.

### Description of plant growth characteristics during the first year based on image analysis

A total of 24,048 images were processed using the procedure developed. For 77.2 % of these images (18,565 images) the procedure yielded correctly analysed images. Based on temporal series of images of base area (W0) and outgrowth area (W1) taken at repeated cuts during the first growing season (Y1), three different aspects of plant growth dynamics were assessed for each genotype. We present results for the whole collection of genotypes and for four subsets (forage breeding lines, current forage varieties, current turf varieties and wild accessions), expected to differ morphologically.

First, by subtracting the base area (W0) from the outgrowth area at 1 week after cutting (W1) (Fig. 1), we estimated the short-term regrowth after cutting, here termed 'regrowth'. This measurement is related to the capacity to quickly regain leaf surface and photosynthetic active biomass. 'Regrowth' varies throughout the season (Fig. 5a), consistent with general seasonal growth patterns and environmental constraints. Cut Y1C2 in spring and Y1C3 in early summer are characterised by relatively strong 'regrowth'. Y1C4 represents summer growth depression after flowering and during warmer months with lower water availability. At cut Y1C5 in autumn, plants show 'regrowth' that is on average comparable to that in spring. As expected, at all cuts, 'regrowth' of forage types (breeding forage lines and current forage varieties) was higher than that of turf types and wild accessions.

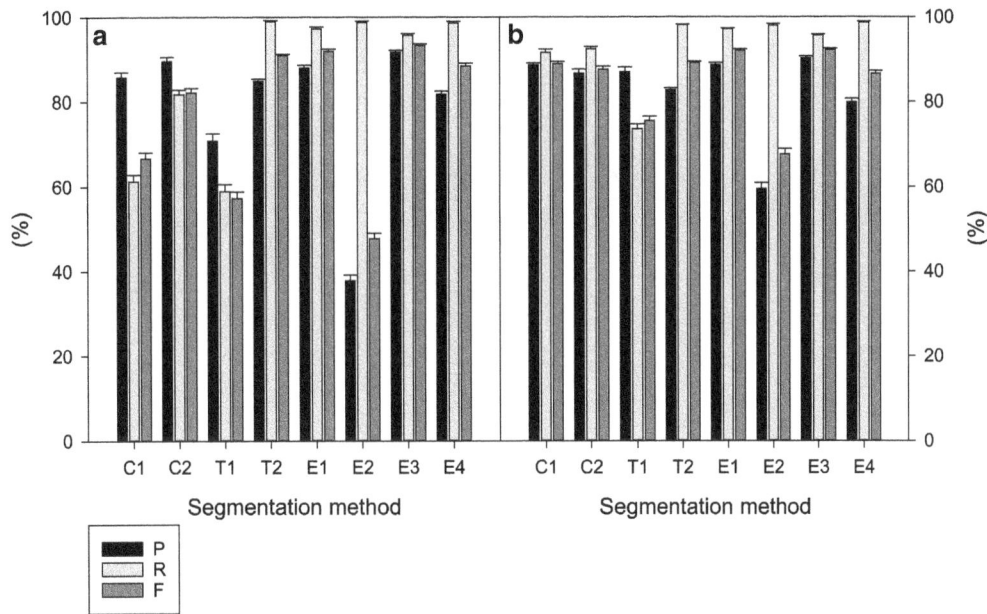

**Fig. 4** Precision (P, %), recall (R, %), and harmonic mean of precision and recall (F, %) of the different segmentation methods [colour-based (C); texture-based (T); edge-based (E)]. **a** Images of plants that were just cut (base area; W0); **b** Images of plants that have regrown for 1 week (outgrowth area; W1). The composite mask derived from eight segmentation methods was used as the ground truth in these comparisons (mean and SE, n ≤ 75)

Second, by plotting 'base area' (W0) calculated for subsequent cuts in year 1, we investigated the plant growth over a whole growing season (Fig. 6a). This increase in 'base area' is mainly due to the production of new tillers in the periphery of the plant. Strong lateral expansion growth is observed early in the season (between Y1C2 and Y1C3), followed by a period of little increase during summer after flowering (between Y1C3 and Y1C4), and further lateral expansion in autumn (between Y1C4 and Y1C5). Also in this case, lateral expansion was higher for forage types (forage breeding lines and current forage varieties), as compared to current turf varieties and wild accessions.

Third, a linear regression of the 'base area' measured directly after cutting (W0) across the entire first growing season (Fig. 1) reflects the global lateral expansion growth, here termed 'first year lateral expansion'. This is an important factor for sward closure during the first growing season. For the 501 genotypes, the first year lateral expansion was on average $0.059 \pm 0.001$ cm$^2$ °C day$^{-1}$ with a maximum of $0.141$ cm$^2$ °C day$^{-1}$. Breeding forage lines and current forage varieties show significantly higher first year lateral expansion rates than wild accessions and current turf varieties, which is to be expected as current forage varieties are selected for productivity and sward forming capacity (Fig. 6b). It should be noted here that the cutting frequency applied in this experiment was probably not sufficient to stimulate vigorous

lateral expansion in turf types, typically selected to very frequent cutting. A relatively large phenotypic variation is present in the genepool, indicating that further improvement for this trait is possible. For instance, a number of wild accessions have higher 'first year lateral expansion' rates than the average within forage breeding genotypes, current forage genotypes, or current turf genotypes.

### Correlation between on-field measurements and image analysis data

Next, we compared counts of 'tiller number' and measurements of 'leaf growth' on the one side with the image analysis derived parameters 'base area' (W0) and 'regrowth', respectively, to estimate whether image analysis could replace and/or yield complementary data for the evaluation of plant growth.

'Leaf growth' after the cut was variable throughout the season (Fig. 5b), consistent with summer depression around Y1C4 as observed in the 'regrowth' measurements based on image analysis (Fig. 5a). However, the estimated 'leaf growth' was lower at Y1C3 than at Y1C2. This was not the case for image analysis-based regrowth estimations (Fig. 5a). Probably, young plants grow initially in the vertical direction, which is not easily detected using top view images. The correlations between 'leaf growth' after cutting and 'regrowth' were 0.645, 0.608, 0.792 and 0.790, respectively, for Y1C2, Y1C3, Y1C4, and Y1C5 when all 501 genotypes were considered together

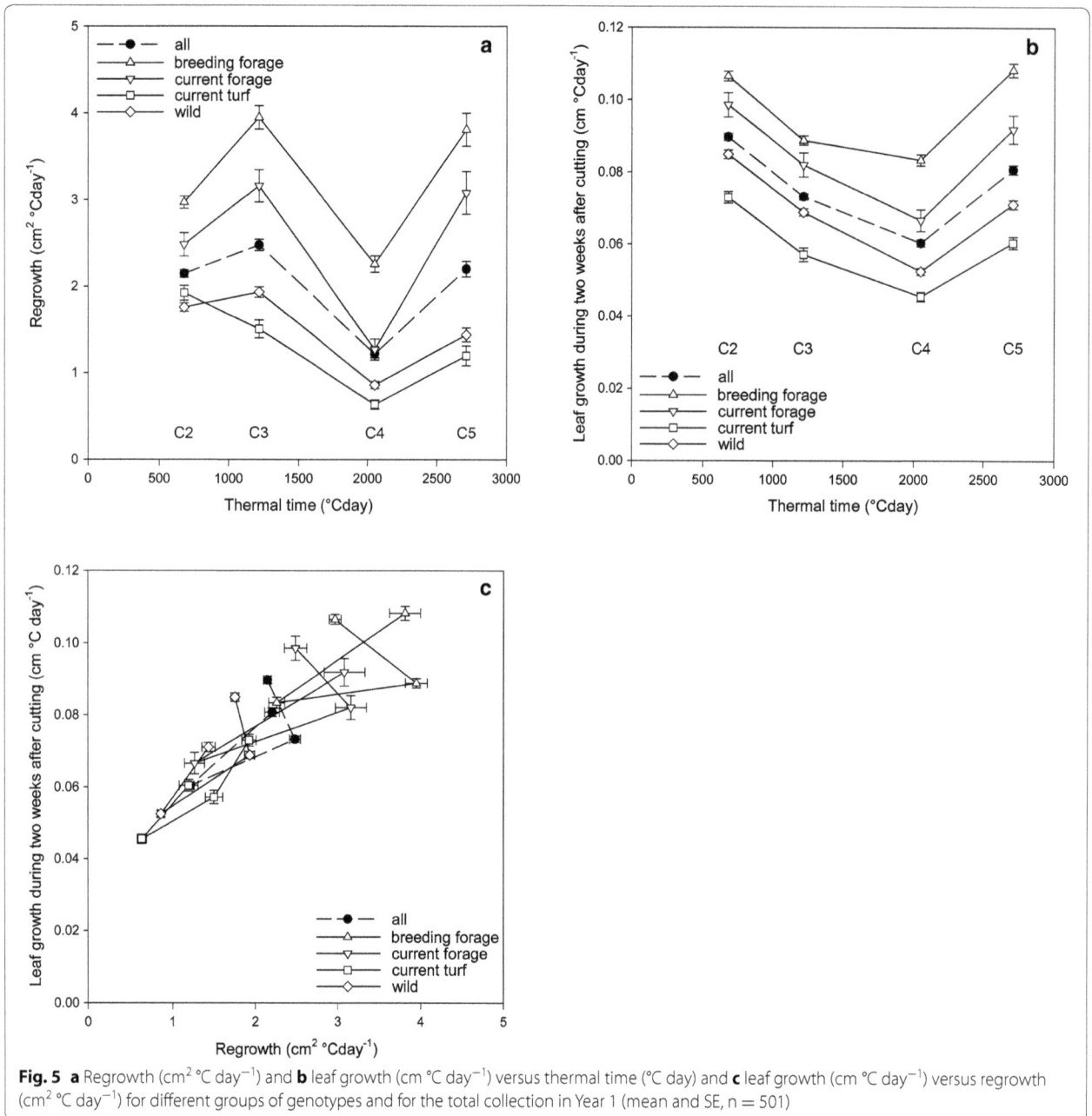

**Fig. 5 a** Regrowth (cm$^2$ °C day$^{-1}$) and **b** leaf growth (cm °C day$^{-1}$) versus thermal time (°C day) and **c** leaf growth (cm °C day$^{-1}$) versus regrowth (cm$^2$ °C day$^{-1}$) for different groups of genotypes and for the total collection in Year 1 (mean and SE, n = 501)

(Table 3; Fig. 5c). The correlations were all statistically significant. However, the correlation values are different for the different groups and depend on the cut considered (Additional file 3: Fig. S3). This illustrates that plants that are able to resume growth quickly after cutting can be identified using image analysis, but that the two ways of measurements can reflect different aspects of growth, and are therefore at least partially complementary. This is because both, variation in the rate of outgrowth of leaves as well as variation in leaf orientation (erect or prostrate),

affect the estimation of 'regrowth' estimates, while 'leaf growth' measurement concerns the longest leaf of the plant 2 weeks after cutting, irrespective of its orientation.

'Tiller number' increased rapidly between Y1C2 and Y1C3 (Fig. 6c). While 'tiller number' continued to increase throughout the first growing season in current and forage breeding lines types, it almost stabilised after Y1C3 in early summer in turf types and wild accessions. On average, wild accessions produced significantly less tillers per plant than other groups. The correlations

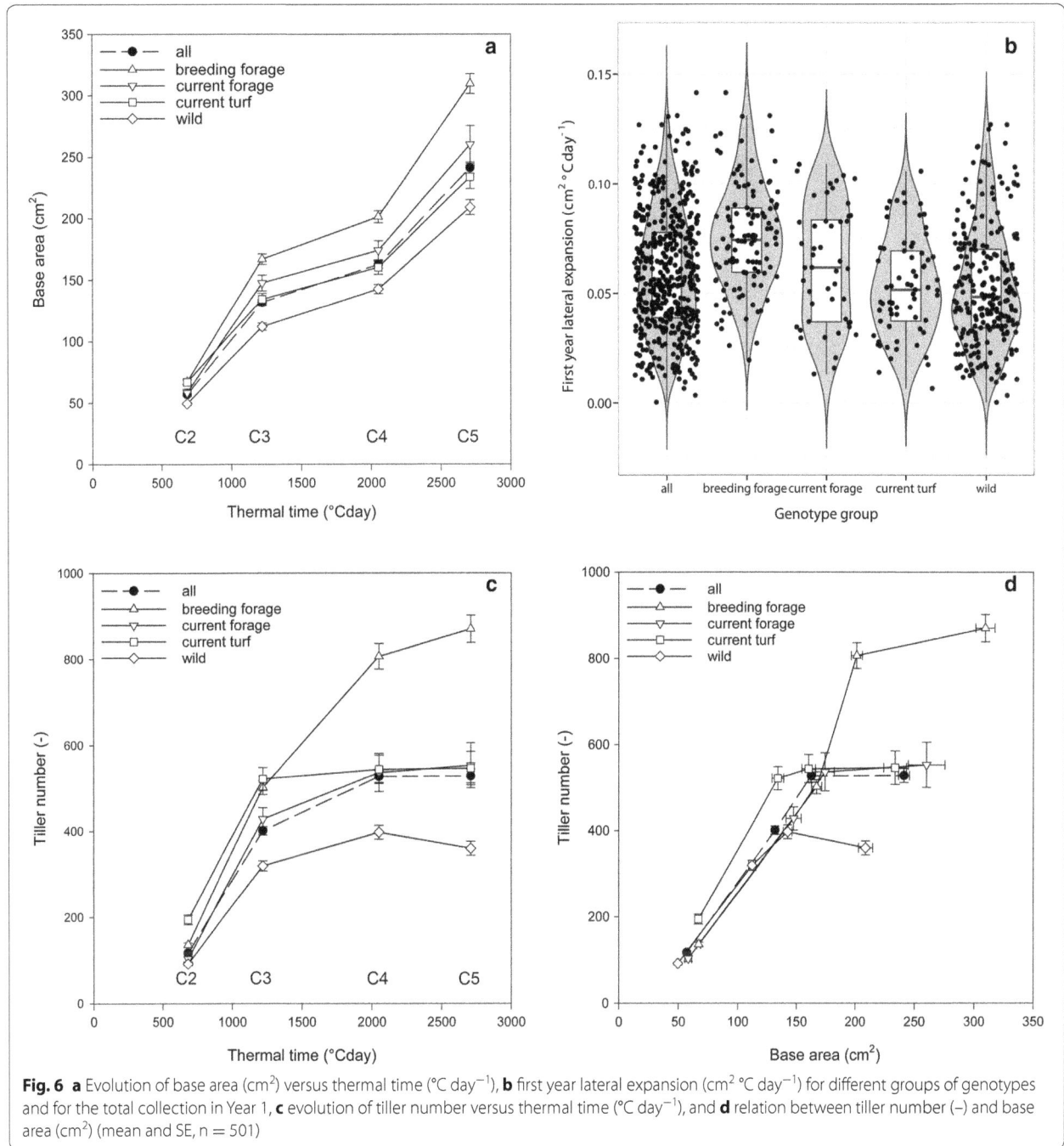

**Fig. 6** **a** Evolution of base area (cm$^2$) versus thermal time (°C day$^{-1}$), **b** first year lateral expansion (cm$^2$ °C day$^{-1}$) for different groups of genotypes and for the total collection in Year 1, **c** evolution of tiller number versus thermal time (°C day$^{-1}$), and **d** relation between tiller number (–) and base area (cm$^2$) (mean and SE, n = 501)

between 'base area' and the 'tiller number' ranged between 0.721 and 0.782, with the highest correlations at Y1C2 in spring and Y1C5 in autumn (Table 3; Additional file 4: Fig. S4). This shows that 'base area' increased due to the formation of new tillers. The relation between the 'base area' (Fig. 6a) and 'tiller number' (Fig. 6c), remained relatively constant over all groups in the first three cuts (Y1C2–Y1C4; Fig. 6d). This relation changed late in the

season (Y1C5): while the 'base area' kept on increasing, the increase in 'tiller number' became lower than earlier in the season, so that the overall tiller density decreased in autumn (Y1C5).

Taken together, our data demonstrate that image analysis provides measurements of plant growth that are relevant for the evaluation of plant performance under field conditions. These estimations are largely correlated but

**Table 3** Correlation (r) between leaf growth (cm °C day$^{-1}$) based on height at 2 weeks after cutting and regrowth (cm$^2$ °C day$^{-1}$) determined by image analysis (top panel) and between tiller number and base area (cm$^2$) determined by image analysis (bottom panel), at four consecutive cuts (between C2 and C5 of the first growing season)

| Genotype groups | Y1C2 | Y1C3 | Y1C4 | Y1C5 | Y1 all cuts |
|---|---|---|---|---|---|
| *Correlation between regrowth and leaf growth* | | | | | |
| Forage breeding lines | 0.543 | 0.425 | *0.758* | *0.769* | 0.567 |
| Current forage varieties | *0.765* | 0.508 | 0.685 | 0.523 | 0.626 |
| Current turf varieties | 0.735 | *0.682* | 0.651 | 0.753 | <u>0.770</u> |
| Wild accessions | 0.462 | 0.316 | 0.548 | 0.595 | 0.529 |
| All groups | 0.645 | 0.608 | <u>0.792</u> | 0.790 | 0.686 |
| *Correlation between base area and tiller number* | | | | | |
| Forage breeding lines | 0.758 | 0.682 | 0.636 | 0.669 | <u>0.824</u> |
| Current forage varieties | *0.847* | *0.793* | 0.686 | *0.824* | 0.799 |
| Current turf varieties | 0.769 | 0.576 | 0.461 | 0.609 | 0.638 |
| Wild accessions | 0.826 | 0.739 | *0.726* | 0.740 | 0.727 |
| All | <u>0.782</u> | 0.731 | 0.721 | 0.755 | 0.769 |

All correlations were significant (p < 0.05). The highest r values per group are depicted in italics. The highest r values for all groups or for all cuts are underlined

for some aspects are also complementary to morphological measurements. The main advantages of the image analysis procedure presented here are its objectivity and the fact that it can be applied easily to thousands of plants at repeated time points in the season.

## Discussion

### Method development

Here we present a new image analysis method to phenotype (re)growth characteristics of field-grown *L. perenne* plants. The method is robust to daily and seasonal changes in light conditions and to different background (~soil) colours and textures. When applied to a total of 24,048 images of plants that were just cut or after a short period of regrowth, and recorded at different time points throughout the season, 77.2 % of the images were correctly processed, allowing quantitative estimations of plant size ('base area' and 'outgrowth area').

Given the difficulties associated to the acquisition of images outdoor and their processing, the development of a robust methodology able to cope with differing light intensity and spectrum and with different backgrounds is not a simple task. In our experience, one operator can take a series of 500 images in about 2 h, which allows photographing a few thousand plants on a weekly basis. While images of a whole field can be captured in only a few hours, still light conditions (spectral quality and intensity) may change during this period. We suggest using speedlights and semi-automated settings on the

camera: a minimum diaphragm opening to have sufficient depth of field and a minimum shutter speed, which equals the maximum synchronisation time of the flashes. For image processing here we combine the power of eight segmentation methods in a highly automatized way. Further improvements are however possible at the level automation of the image analysis step with online extraction of parameters. Furthermore, given the high correlation detected for some pair-wise comparisons among the eight methods tested here, a subset of segmentation methods could possibly be chosen. By preference, the methods with the highest Precision (P) and Recall (R), which are fast to execute should be selected. Segmentation methods with low P and R could be eliminated if they do not have a significant contribution to the quality of the segmentation. Finally, a dynamic parameter choice using the fixed place of the plant (middle of the image) and the background (sides and corners of the image) and their local colour, texture, and edge characteristics could be used to further increase the number of correctly evaluated images. Further, for the edge based segmentation we only used Prewitt based methods as only these were available in the software, but the Canny edge detection procedure might render improved results [24]. Nevertheless, a comparison would be required as the Canny procedure involves a more complex computation and is thus processor/time demanding [24]. Moreover, our current procedure can be further optimized in terms of the number of correctly analyzed images and the processing time needed per image, which now ranges from seconds to tens of seconds.

More sophisticated field-based phenotyping platforms have been developed, such as the tractor-pulled multisensor platform BreedVision, that carries a light source allowing exclusion of environmental light [3] could increase throughput of image acquisition. Another option is using unmanned aerial vehicles (UAV), provided that the resolution is sufficient for accurate individual plant identification and characterisation. Both options, however, are expensive and not accessible to most institutes involved in forage crop breeding and research at this moment. Our system has several advantages: it is inexpensive, it has a high-throughput (it takes about 8–10 s to take an image), non-specialised staff can be involved in image acquisition, and standard DSLR cameras are sufficiently robust for common outdoor weather conditions.

### Applications and perspectives

The evaluation of *L. perenne* plants for breeding and selection purposes is typically based on the assignment of visual scores, rendering poor resolution to discriminate differential genotypic responses. Such visual observations are inexpensive but can be biased by the examiner and may not be sufficiently accurate for targeted crop improvement [16, 30]. With the recent developments in high-throughput genotyping for breeding purposes, the demand for quantitative phenotypic data is also increasing, to a point that phenotyping and not genotyping is becoming the bottleneck for genetic improvement of crops [6, 32]. We show here that image analysis-based on-field phenotyping can render quantitative evaluations of growth parameters, probably at the level of resolution required for detailed investigation of the underlying genetic mechanisms and enabling more precise selection in perennial ryegrass.

Although it was not our prime objective to define the best image analysis-derived parameters to quantify (re) growth in perennial ryegrass, the parameters 'regrowth', 'base area' and 'first year lateral expansion', as defined in this work, seem to reproduce quite well the expected overall seasonal responses and the differences between groups of genotypes from different origins. It is therefore possible to use our image-analysis based methodology to describe and follow-up the growth of individual perennial ryegrass genotypes in the field.

Further, our results demonstrate a relatively high correlation between data derived from images and manual measurements. This was the case for the whole collection of 501 genotypes, but also within the different groups considered (breeding forage lines, current forage varieties, current turf varieties and wild accessions). In general, higher correlation values were obtained across cuts and genotypes between 'base area' and 'tiller number' than between 'regrowth' and 'leaf growth'. In this latter case, top view images as considered here might not be sufficient to capture vigorous leaf elongation of plants with an erect growth pattern. Combination with side-view images or the use of UAV based technologies using digital elevations models (DEMs) could enable to estimate plant height and might render better estimates if the purpose is to obtain reliable information of 'leaf growth'. Our finding that tiller number can be estimated with a relatively high accuracy from top view images of plants that have been just cut is interesting, as counting tillers is time-consuming and not readily done in practical breeding programs. With our methodology it is therefore possible to get rather good estimates of tillering, which is an important determinant of forage yield and turf quality.

The set of plant parameters derived from image analysis could be extended in the future, using our methodology as start point. Here, we have focused on lateral expansion over a whole season and on growth after cutting over a short period of time (1 week). Growth over a period of a few weeks could be considered but the methodology presented here, in its own, is probably inefficient

for the estimation of green biomass accumulation in larger plants, as leaf density increases [12]. As mentioned above, combination of top- and side-view images or UAV derived DEMs could allow estimating the plant biovolume, helping to get accurate estimates of green mass of large plants.

Finally, although not tested here, we anticipate that the high-throughput, inexpensive image analysis procedure presented here can be easily extrapolated to other forage and turf species. In addition, the set of plant traits extracted from the images can be extended in the future, possibly in combination with side-view images able to capture information on leaf density and plant habit (erect or prostrate). This would make the estimation of dry matter accumulation in larger plants possible.

## Conclusion

We have developed a low-cost high-throughput phenotyping system and an image analysis procedure allowing a correct evaluation of 77.2 % of the top view images of field-grown perennial ryegrass plants. It was possible to quantify base area in an objective, quantitative way and to monitor lateral expansion and regrowth during a growing season under field conditions. We demonstrate that the image-derived variables are complementary to manually measured variables such as tiller number and leaf growth. This additional growth describing variables are important for genetic dissection of those traits.

## Additional files

**Additional file 1: Figure S1.** Cutting regime and measurement periods of the *L. perenne* plants in 2010 (Y1) and 2011 (Y2) related to day of the year (DOY) and growing degree days (GDD$^C$, °C day).

**Additional file 2: Figure S2.** Schematic representation of the image analysis procedure. An overlay image with the composite mask was made to allow for visual inspection. At the same time the selected object was measured and data were exported to a comma-delimited file. After visual inspection for correctness of the overlay image with the composite mask, it was decided if the corresponding measurement data were used or not. Dashed lines indicate the comparisons for the performance testing.

**Additional file 3: Figure S3.** Relation between leaf growth based and plant height measured at two weeks after cutting per cut and regrowth per group. Linear regressions are fitted and 95 % confidence intervals are shown.

**Additional file 4: Figure S4.** Relation between tiller number and base area per cut and per group. Linear regressions are fitted and 95 % confidence intervals are shown.

### Authors' contributions
PL, IRR and TR wrote this manuscript. PL, IRR, AR, TR, PB and DC were involved in the conception and design of the experiment. PL, TR and AR acquired the images. PL developed the image analysis procedure, processed the images and performed the statistical analysis of the data. PL, TR and IRR interpreted the results. PB, DC and AR critically revised the manuscript. All authors read and approved the final manuscript.

**Author details**
[1] Plant Sciences Unit - Growth and Development, ILVO, Caritasstraat 39, 9090 Melle, Belgium. [2] INRA – UR4 P3F, BP 6, 86600 Lusignan, France. [3] Present Address: Bayer CropScience, Technologiepark 38, 9052 Ghent, Belgium.

**Acknowledgements**
This research was supported by the Agency for the promotion of Innovation by Science and Technology (IWT) Flanders (Project LO-080510). We would like to thank the breeders of Barenbrug and Eurograss for providing plant material. We would also like to thank Sabine van Glabeke, Katleen Sucaet, Thomas Vanderstocken and Luc Van Gyseghem for field work and measurements, and Hilde Muylle, Gerda Cnops and Serge Carré for helpful discussions.

**Competing interests**
The authors declare that they have no competing interests.

**References**
1. Araus JL, Cairns JE. Field high-throughput phenotyping: the new crop breeding frontier. Trends Plant Sci. 2013;19(1):52–61.
2. Araus JL, Slafer GA, Royo C, Serret MD. Breeding for yield potential and stress adaptation in cereals. Crit Rev Plant Sci. 2008;27:377–412.
3. Busemeyer L, Mentrup D, Möller K, Wunder E, Alheit K, Hahn V, Maurer HP, Reif JC, Würschum T, Müller J, Rahe F, Ruckelshausen A. BreedVision—a multi-sensor platform for non-destructive field-based phenotyping in plant breeding. Sensors. 2013;13:2830–47.
4. Canny J. A computational approach to edge detection. IEEE Trans Pattern Anal Mach Intell. 1986;8:679–98.
5. Chen D, Neumann K, Friedel S, Kilian B, Chen M, Altmann T, Klukas C. Dissecting the phenotypic components of crop plant growth and drought responses based on high-throughput image analysis. Plant Cell. 2014;26:4636–55.
6. Cobb JN, De Clerck G, Greenberg A, Clark R, McCouch S. Next-generation phenotyping: requirements and strategies for enhancing our understanding of genotype-phenotype relationships and its relevance to crop improvement. TAG Theor Appl Genet (Theoretische Und Angewandte Genetik). 2013;126(4):867–87.
7. De Keyser E, Lootens P, Van Bockstaele E, De Riek J. Image analysis for QTL mapping of flower colour and leaf characteristics in pot azalea (Rhododendron simsii hybrids). Euphytica. 2012;189(3):445–60.
8. Dey V, Zhang Y, Zhong M. A review on image segmentation techniques with remote sensing perspective. In: ISPRS TC VII symposium—100 years ISPRS 2010, vol XXXVIII Part 7A, pp 31–42.
9. Fiorani F, Schurr U. Future scenarios for plant phenotyping. Annu Rev Plant Biol. 2013;64:267–91.
10. Haselbauer WD, Thoms AW, Sorochan JC, Brosnan J, Schwartz BM, Hanna WW. Evaluation of experimental Bermudagrasses under simulated athletic field traffic with perennial ryegrass overseeding. HortTechnology. 2012;22(1):94–8.
11. Humphreys M, Feuerstein U, Vandewalle M, Baert J. Ryegrasses. In: Boller B, Posselt UK, Veronesi F, editors. Fodder crops and amenity grasses. Berlin: Springer; 2010. p. 211–60.
12. Hunt CL, Jones CS, Hickey M, Hatier J-HB. Estimation in the field of individual perennial ryegrass plant position and dry matter production using a custom-made high-throughput image analysis tool. Crop Sci. 2015;55(6):2910–7.
13. Ilea DE, Whelan PF. Image segmentation based on the integration of colour-texture descriptors—a review. Pattern Recogn. 2011;44:2479–501.
14. Klukas C, Chen D, Pape J-M. Integrated analysis platform: an open-source information system for high-throughput plant phenotyping. Plant Physiol. 2014;165:506–18.
15. Marr D, Hildreth E. Theory of edge detection. Proc R Soc Lond Ser B Biol Sci. 1980;207(1167):187–217.
16. Montes JM, Technow F, Dhillon BS, Mauch F, Melchinger AE. High-throughput non-destructive biomass determination during early plant development in maize under field conditions. Field Crops Res. 2011;121:268–73.
17. Muñoz X, Freixenet J, Cufí X, Martí J. Strategies for image segmentation combining region and boundary information. Pattern Recogn Lett. 2003;24:375–92.
18. Nejati H, Azimifar Z, Zamani M. Using fast fourier fransform for weed detection in corn fields. In: International conference on systems, man and cybernetics 2008, pp 1215–19.
19. Pal NR, Pal SK. A review on image segmentation techniques. Pattern Recogn. 1993;26:1277–94.
20. Philipp I, Rath T. Improving plant discrimination in image processing by use of different colour space transformations. Comput Electron Agric. 2002;35(1):1–15.
21. Russ JC. The image processing handbook. 6th ed. Boca Raton: CRC Press; 2011.
22. Sampoux JP, Baudouin P, Bayle B, Béguier V, Bourdon P, Chosson JF, de Bruijn K, Deneufbourg F, Galbrun C, Ghesquière M, Noël D, Tharel B, Viguié A. Breeding perennial ryegrass (Lolium perenne L.) for turf usage: an assessment of genetic improvements in cultivars released in Europe, 1974–2004. Grass Forage Sci. 2012;68(1):33–48.
23. Sezgin M, Sankur B. Survey over image thresholding techniques and quantitative performance evaluation. J Electron Imaging. 2004;13:146–65.
24. Shrivakshan GT, Chandrasekar C. A comparison of various edge detection techniques used in image processing. Int J Comput Sci (IJCSI). 2012;9(5):269–76.
25. Smochina C. Image processing techniques and segmentation evaluation. Doctoral thesis. Technical University Gheorghe Asachi from Iasi; 2011. p 120.
26. Thoms AW, Sorochan JC, Brosnan JT, Samples TJ. Perennial ryegrass (Lolium perenne L.) and grooming affect bermudagrass traffic tolerance. Crop Sci. 2011;51:2204–11.
27. Uchida S. Image processing and recognition for biological images. Dev Growth Differ. 2013;55:523–49.
28. Van Bockstaele E, Baert J. Improvement of perennial ryegrass (Lolium perenne L.). Plant Sci. 2004;41:483–8.
29. Van Rijsbergen CV. Information retrieval. MA, USA: Butterworth-Heinemann Newton; 1979.
30. Walter A, Studer B, Kölliker R. Advanced phenotyping offers opportunities for improved breeding of forage and turf species. Ann Bot. 2012;110:271–1279.
31. Wang H, Suter D. Color image segmentation using global information and local homogeneity. In: Sun C, Talbot H, Ourselin S, Adriaansen T, editors. Digital Image computing: techniques and applications. Proceedings of the VIIth Biennial Australian Pattern Recognition Society Conference, DICTA 2003. Sydney; 10–12 Dec 2003. pp. 89–98.
32. White JW, Andrade-Sanchez P, Gore MA, Bronson KF, Coffelt TA, Conley MM, Feldmann KA, French AN, Heun JT, Hunsaker DJ, Jenks MA, Kimball BA, Roth RL, Strand RJ, Thorp KR, Wall GW, Wang G. Field-based phenomics for plant genetics research. Field Crops Res. 2012;133:101–12.
33. Yu X, Pijut PM, Byrne S, Asp T, Bai G, Jiang Y. Candidate gene association mapping for winter survival and spring regrowth in perennial ryegrass. Plant Sci. 2015;235:37–45.

# 6

# Simultaneous knockdown of six non-family genes using a single synthetic RNAi fragment in *Arabidopsis thaliana*

Olaf Czarnecki[1,3]*, Anthony C. Bryan[1], Sara S. Jawdy[1], Xiaohan Yang[1], Zong-Ming Cheng[2], Jin-Gui Chen[1] and Gerald A. Tuskan[1]

## Abstract

**Background:** Genetic engineering of plants that results in successful establishment of new biochemical or regulatory pathways requires stable introduction of one or more genes into the plant genome. It might also be necessary to down-regulate or turn off expression of endogenous genes in order to reduce activity of competing pathways. An established way to knockdown gene expression in plants is expressing a hairpin-RNAi construct, eventually leading to degradation of a specifically targeted mRNA. Knockdown of multiple genes that do not share homologous sequences is still challenging and involves either sophisticated cloning strategies to create vectors with different serial expression constructs or multiple transformation events that is often restricted by a lack of available transformation markers.

**Results:** Synthetic RNAi fragments were assembled in yeast carrying homologous sequences to six or seven non-family genes and introduced into pAGRIKOLA. Transformation of *Arabidopsis thaliana* and subsequent expression analysis of targeted genes proved efficient knockdown of all target genes.

**Conclusions:** We present a simple and cost-effective method to create constructs to simultaneously knockdown multiple non-family genes or genes that do not share sequence homology. The presented method can be applied in plant and animal synthetic biology as well as traditional plant and animal genetic engineering.

## Background

Targeted gene knockdown by RNA interference (RNAi) has become a powerful tool for genetic research and biotechnology in eukaryotes. Originally discovered in the nematode *Caenorhabditis elegans* [1], details of the molecular mechanisms underlying the gene silencing caused by double-stranded RNA (dsRNA) have been elucidated within the last two decades [reviewed in 2]. Briefly, dsRNA molecules are cut in pieces of 21–23 nucleotides termed small interfering RNA (siRNA), by RNAse III family endoribonucleases named DICER. ARGONAUTE proteins acting in complex with DICER as RNA-induced silencing complex (RISC) mediate unwinding of siRNA molecules where the passenger strand is released and degraded and the guide strand serves as recognition pattern to bind complementary single-stranded RNA (ssRNA) molecules (mRNA). The endonuclease activity of ARGONAUTE results in specific degradation of the homologous ssRNA molecule and RISC eventually binds to other target molecules [2–6].

The origin of dsRNA molecules can be endogenous or exogenous. Endogenous dsRNA molecules have been identified in both plants and animals and are derived from transposable elements, transcripts containing short inverted repeats or natural antisense transcripts [7–11]. Exogenous dsRNA molecules, on the other hand, are derived from viruses and trigger host defense mechanisms against viral RNA [12, 13].

Soon after its discovery, RNAi became an important tool for reverse genetics in plants, as it enables targeted gene knockdown. There are at least three advantages of RNAi: (1) the ability to knock down gene family members

*Correspondence: olaf.czarnecki@kws.com
[3] Present Address: KWS SAAT SE, Grimsehlstraße 31, 37555 Einbeck, Germany
Full list of author information is available at the end of the article

that share homologous sequences or orthologous genes in polyploid organisms relative to other methods (e.g., T-DNA insertion or EMS mutagenesis), (2) the genetically dominant mode of action of RNAi and (3) relative easiness to create transgenic plants expressing RNAi transgenes, depending on the availability of suitable RNAi plasmids and on the transformability of the organisms of interest [14–17].

In plant biotechnology, RNAi is induced by constitutive, induced or spatial/tissue-specific expression of a target gene fragment cloned as a tandem inverted repeat separated by a hairpin forming intron sequence. The transcript creates a hairpin RNA (hpRNA) that then serves as template for the RNAi machinery [18]. There are several routine plant RNAi transformation vectors available, e.g., pHANNIBAL [19], pHELLSGATE [19], pAGRIKOLA [20], pOpOff [21] and the pFGC and pGSA series [22].

There has been a tremendous amount of literature describing the modulation of transcript abundance of single genes in plants using RNAi and/or over-expression constructs. Engineering of complex metabolic pathways, however, often requires controlling more than one gene in the same or interconnected pathways [23, 24]. Current strategies to create multiple transgene plants involve sexual crossing or co-transformation and retransformation. Golden rice is an example where the entire β-carotene biosynthetic pathway was introduced into the rice endosperm by a single transformation event using three different vectors [25]. There have been several attempts to improve and create stable plant transformation vectors and cloning strategies for gene stacking, such as sophisticated use of recombinases [26–29]. Still, successful development of new or synthetic biochemical pathways in plants might not only require expression of multiple genes but also knockdown of more than one endogenous gene. In animals and cell lines, serial expression of several small hairpin RNAs (shRNA) in virus derived vectors is an established method to achieve multiple gene knockdown, but the cloning strategies to create the vectors are rather complex and time consuming [30–34]. Recently, genome editing has become an important tool to achieve a knockout of one or multiple genes in eukaryotes. For instance, TALEN or CRISPR/Cas based approaches allow targeted and highly efficient introduction of premature stop codons in the open reading frame of target genes [35–38]. However, reduction of target gene expression or gene product activity by genome editing is challenging without detailed knowledge of regulatory elements affecting gene expression or amino acid substitutions affecting protein activity.

Here we present a new consolidated method to knockdown alternant non-family genes using a single artificial

gene fragment cloned in a binary plant RNAi vector. As proof-of-concept, six- and seven-gene-RNAi vectors were introduced into *Arabidopsis thaliana* with successful down regulation of the target genes. Our technology has applications in plant and animal synthetic biology as well as traditional plant and animal genetic engineering.

## Results and discussion
### Cloning of multiple target RNAi vectors
Based on the principle that hpRNA-induced RNAi uses 21–23 bp siRNA fragments to degrade target transcripts, we developed a hypothesis that expression of a single hpRNA fragment consisting of different gene specific tags (GSTs) in Arabidopsis will eventually result in a set of siRNAs specifically degrading all transcripts targeted by the respective GSTs. To prove this hypothesis, synthetic DNA fragments consisting of six or seven alternant GSTs were assembled by means of transformation-associated recombination [TAR, 39] in *Saccharomyces cerevisiae*. This method is based on homologous recombination after simultaneous uptake of a linear double-stranded vector and double-stranded PCR products that share sequence homology or overlapping single-stranded oligonucleotides in a single transformation event [40, 41]. It has widely been used in biotechnology, including assembly of synthetic bacterial genomes [42, 43], cloning of the human mitochondrial genome [44] and joining of unrelated DNA fragments during plasmid assembly [45]. We applied this method to assemble unrelated GSTs of different target genes to synthetic hpRNA fragments (Fig. 1).

We selected seven target genes for this proof-of-concept study to be downregulated simultaneously in *A. thaliana*. Candidate genes do not share sequence homologies and are therefore considered as non-family genes (Table 1). Alignments of cDNA do not reveal any 21–23 nucleotides stretches of identical sequence (Additional 1: Fig. S1) and we assume that siRNA of one target gene do not interfere with other target genes. To our best knowledge candidate genes do not act in the same biological pathways to avoid transcript levels being affected by feedback mechanisms or pleiotropic effects. Moreover, phenotypes of knockout or loss-of-function mutants of the selected genes have been described previously [reviewed in 46]. *AtHY2* (ELONGATED HYPOCOTYL 2) is involved in biosynthesis of heme and phytochromobillin, the phytochrome chromophor and mutants develop elongated hypocotyls. *AtTRY* (TRIPTYCHON) is a negative regulator of trichome development and a knockout causes visible trichome clusters on leaves. *AtLNG1* (LONGIFOLIA1) regulates longitudinal cell elongation resulting in characteristic long leaf shapes when knocked out. *AtNPQ1* (NON-PHOTOCHEMICAL QUENCHING 1) is a violaxanthin deepoxidase

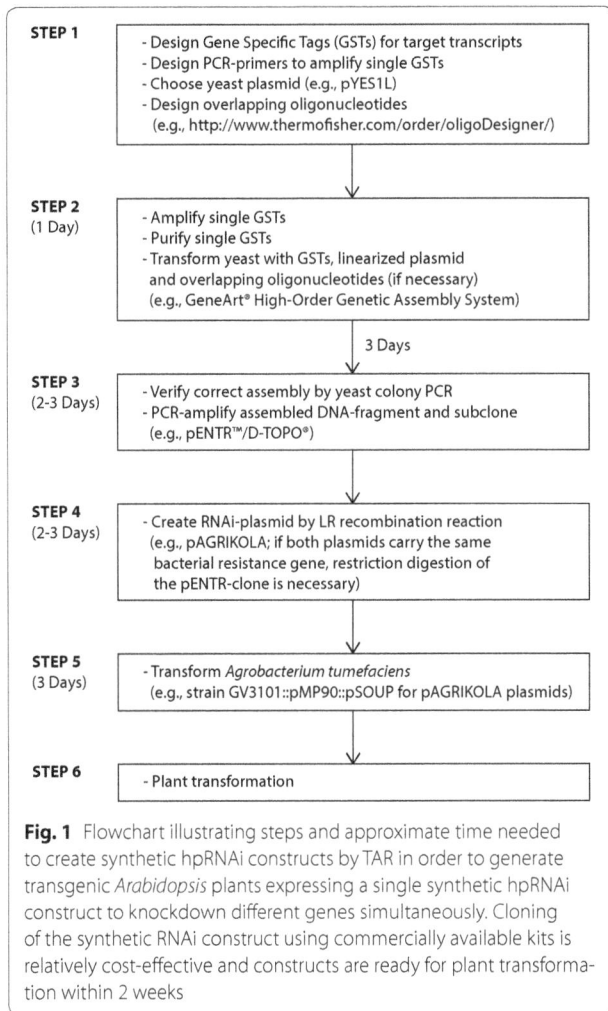

STEP 1
- Design Gene Specific Tags (GSTs) for target transcripts
- Design PCR-primers to amplify single GSTs
- Choose yeast plasmid (e.g., pYES1L)
- Design overlapping oligonucleotides
  (e.g., http://www.thermofisher.com/order/oligoDesigner/)

STEP 2
(1 Day)
- Amplify single GSTs
- Purify single GSTs
- Transform yeast with GSTs, linearized plasmid
  and overlapping oligonucleotides (if necessary)
  (e.g., GeneArt® High-Order Genetic Assembly System)

3 Days

STEP 3
(2-3 Days)
- Verify correct assembly by yeast colony PCR
- PCR-amplify assembled DNA-fragment and subclone
  (e.g., pENTR™/D-TOPO®)

STEP 4
(2-3 Days)
- Create RNAi-plasmid by LR recombination reaction
  (e.g., pAGRIKOLA; if both plasmids carry the same
  bacterial resistance gene, restriction digestion of
  the pENTR-clone is necessary)

STEP 5
(3 Days)
- Transform Agrobacterium tumefaciens
  (e.g., strain GV3101::pMP90::pSOUP for pAGRIKOLA plasmids)

STEP 6
- Plant transformation

**Fig. 1** Flowchart illustrating steps and approximate time needed to create synthetic hpRNAi constructs by TAR in order to generate transgenic Arabidopsis plants expressing a single synthetic hpRNAi construct to knockdown different genes simultaneously. Cloning of the synthetic RNAi construct using commercially available kits is relatively cost-effective and constructs are ready for plant transformation within 2 weeks

4) is a regulator of chlorophyll biosynthesis in chloroplasts and the *gun4-1* mutant showing reduced GUN4 activity suffers from reduced chlorophyll contents.

The CATMA database [47] provides GST sequences for *A. thaliana* genes that were developed using specific primer and amplicon design software [48] and undergo certain quality controls. In favor of standardization, we decided to make use of this resource. GSTs for the selected target genes *AtHY2* (H), *AtTRY* (T), *AtLNG1* (L), *AtNPQ1* (N), *AtSEX1* (S), *AtMAX3* (M) and *AtGUN4* (G), were amplified and assembled in silico to compose synthetic RNAi fragments consisting of either six or seven GSTs in alternate orders (Fig. 2). The order of the GSTs in the seven GST fragment was chosen randomly. The six GST fragments followed the same order without including the AtGUN4-GST. Six different multiple RNAi synthetic DNA fragments were designed: (a) HTLNSM, (b) NLSHMT, (c) LNTMHS, (d) GHTLNSM, (e) NLSHMTG and (f) LNGTMHS. Oligonucleotide primers to amplify the individual GSTs from an *Arabidopsis* cDNA library that contain a 5′ extension of DNA homologous to the neighboring GST in the respective synthetic DNA fragment were designed using a web based tool (http://www.thermofisher.com/order/oligoDesigner). Individual GSTs were PCR amplified and multiple RNAi synthetic DNA fragments were assembled in yeast using plasmid pYES1L as a backbone (Figs. 1, 2, 3). The assembled pYES1L plasmids served as template to PCR amplify the synthetic DNA fragments and to create respective pENTR clones (Figs. 1, 3). The cloned synthetic DNA fragments were subsequently transferred to the binary plant hpRNA vector pAGRIKOLA [20] by Gateway® Cloning [49]. The entire cloning of the multiple RNAi pAGRIKOLA vectors is cost and time-effective compared to other cloning strategies (e.g., use of restriction endonucleases, ligases and multiple subcloning steps) and can be completed within 2 weeks for any chosen GST combination (Fig. 1). Even though we used Gateway® Technology for cloning, since the assembly process itself is independent of any restriction sites, final cloning of the assembled fragment can easily be adapted to restriction enzyme based protocols, if respective target plasmids are available.

involved in xanthophyll cycle and the lack of zeaxanthin caused by a gene knockout leads to a stress phenotype under high light intensities. *AtSEX1* (STARCH EXCESS 1) is required for starch degradation in leaves and knockout mutants accumulate large amounts of starch in adult leaves. *AtMAX3* (MORE AXILLARY BRANCHING 3) plays a role in strigolactone biosynthesis and *max3* mutants display a bushy appearance and increased shoot branching. Finally, *AtGUN4* (GENOME UNCOUPLED

**Table 1 Sequence homologies of target genes. Given are the pairwise homologies of target genes in % based on cDNA sequence**

|        | AtLNG1 | AtSEX1 | AtNPQ1 | AtMAX3 | AtHY2 | AtGUN4 |
|--------|--------|--------|--------|--------|-------|--------|
| AtSEX1 | 38.06  |        |        |        |       |        |
| AtNPQ1 | 19.02  | 15.42  |        |        |       |        |
| AtMAX3 | 21.01  | 16.34  | 39.25  |        |       |        |
| AtHY2  | 14.70  | 11.88  | 26.32  | 25.55  |       |        |
| AtGUN4 | 10.20  | 8.25   | 19.58  | 18.67  | 21.36 |        |
| AtTRY  | 7.31   | 5.47   | 12.95  | 13.79  | 16.93 | 20.61  |

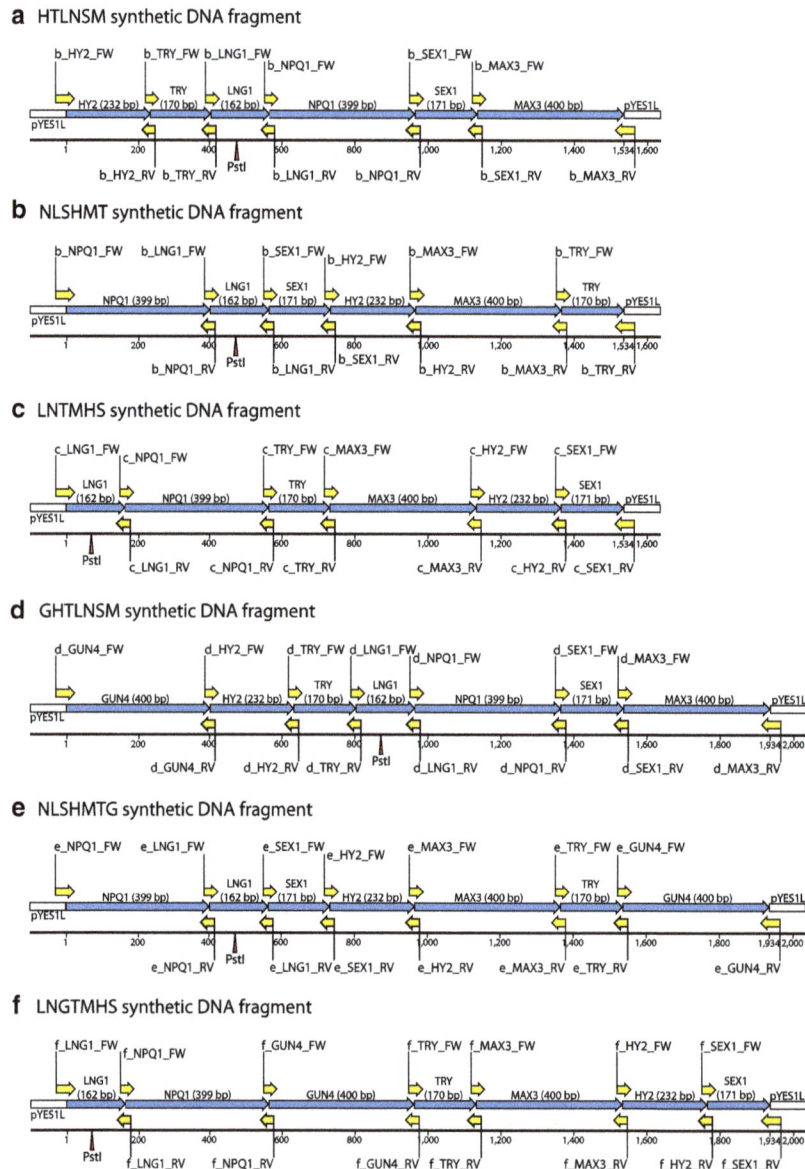

**Fig. 2** Synthetic multiple RNAi fragments assembled in pYES1L from double-stranded PCR products by a single yeast transformation event. Primers used to amplify individual gene specific tags (GSTs) are given in *yellow*. Synthetic fragments (**a–c**) contain six different GSTs in different orders, while fragments (**d–f**) carry an additional GST for the gene GUN4. Total fragment lengths are 1534 bp for the six-gene and 1934 bp for the seven-gene multiple RNAi fragment. *PstI* restriction sites are labeled by *red arrows*

After transformation of *Agrobacterium tumefaciens* with the pAGRIKOLA vectors, *A. thaliana* Col-0 plants were transformed and T1 transgenic lines were selected.

### RNAi-mediated knockdown of single target genes

In order to test whether the chosen single/individual GSTs can efficiently silence the target gene expression, *Arabidopsis* plants were transformed with respective single GST pAGRIKOLA vectors and the transcript abundance in individual RNAi lines was determined by qRT-PCR (Fig. 4). We successfully obtained *Arabidopsis* RNAi lines for the target genes *AtHY2*, *AtTRY*, *AtLNG1*, *AtNPQ1*, *AtSEX1* and *AtMAX3* and target transcript abundance declined to 20–25 % of the control (Fig. 4). RNAi-mediated down-regulation of the six target genes did not result in a visible phenotype as previously described for the respective loss of function mutants phenotype [50–55], implying that the observed 80 % reduction in transcript levels is not sufficient to cause a decrease of cellular target protein amounts as

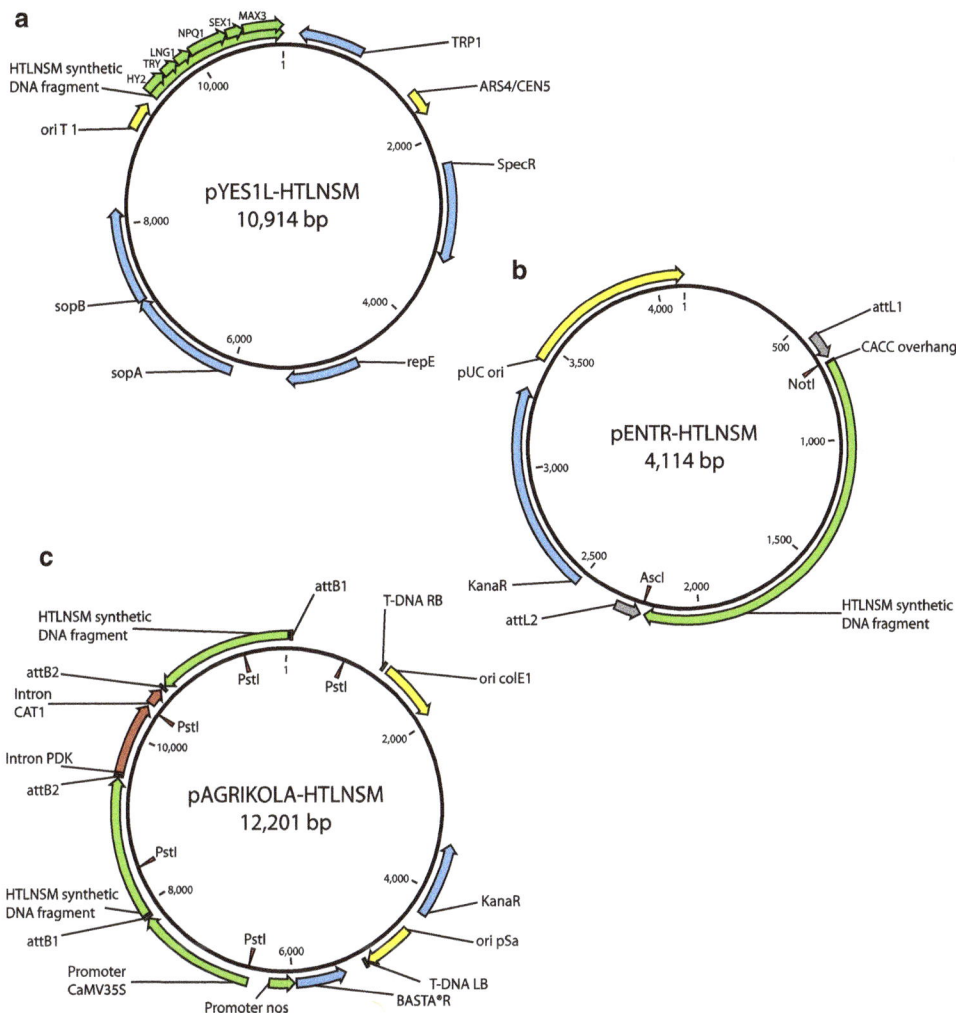

**Fig. 3** Series of vectors created to clone the HTLNSM synthetic multiple RNAi fragment. **a** Gene specific tags for *AtHY2*, *AtTRY*, *AtLNG1*, *AtNPQ1*, *AtSEX1* and *AtMAX3* were assembled to form the HTLNSM multiple RNAi fragment in yeast. **b** The HTLNSM synthetic DNA fragment was amplified by PCR using pYES1L-HTLNSM as template and the PCR product was cloned into pENTR™/D-TOPO to create pENTR-HTLNSM. The CACC-overhang was introduced to facilitate directional cloning into pENTR™/D-TOPO. **c** The HTLNSM synthetic DNA fragment was subsequently transferred to pAGRIKOLA by LR recombination reaction to create the hpRNAi vector pAGRIKOLA-HTLNSM. Refer to Thermo Fisher Scientific Inc., (www.thermofisher.com) for more information about pYES1L and pENTRY and to Hilson et al. [20] for pAGRIKOLA

seen in these mutants. Nonetheless, we showed here that with exception of AtGUN4 the selected genes could be knocked down via the RNAi approach.

Alternatively, even with several rounds of transformation we were unable to select *AtGUN4* RNAi lines. This is in contrast to the work of Du et al. [56] who successfully transformed *Arabidopsis* with a *AtGUN4* RNAi plasmid and Schwab et al. [57], who efficiently down-regulated *AtGUN4* expression using a microRNA approach. Since the two mentioned approaches used the *A. tumefaciens* mannopine synthase promoter or a set of tissue specific promoters, and we used a pAGRIKOLA that contains

the CaMV 35S promoter driving the RNAi expression, we speculated that different promoter activities may have caused the failure to select *AtGUN4* RNAi transgenic lines. We cannot rule out the possibility that we were unable to distinguish severe *AtGUN4* RNAi effects and BASTA® sensitivity of Arabidopsis seedlings during T1 selection since both of them would result in retarded growth and yellow seedlings. The originally described *Arabidopsis gun4-1* mutant's yellow leaf phenotype is caused by a missense mutation resulting in a leucine to phenylalanine exchange at position 88 of the 265 amino acid sequence [58], indicating that key changes in AtGUN4 protein result

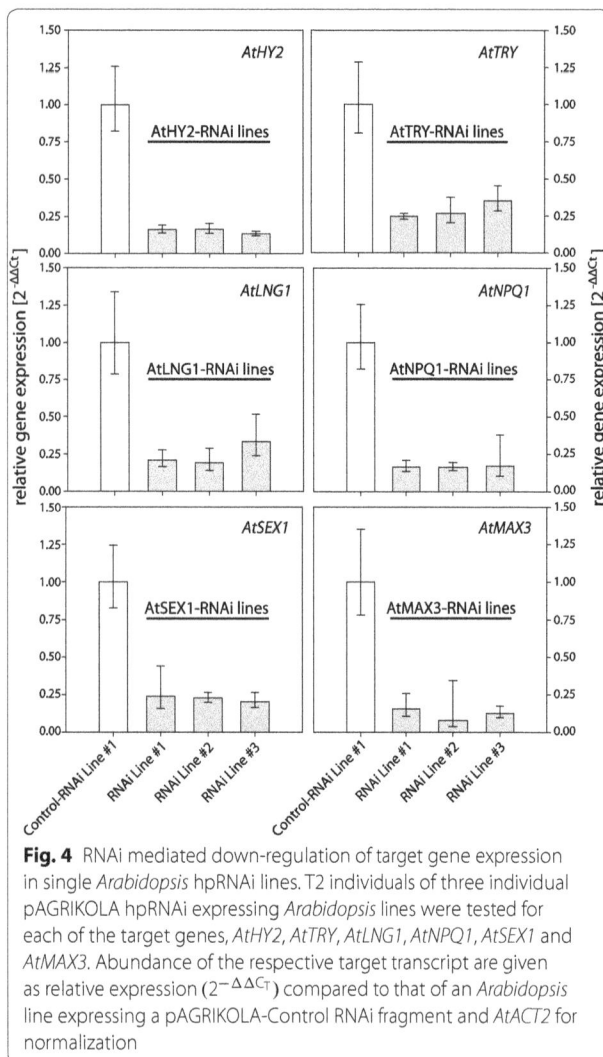

**Fig. 4** RNAi mediated down-regulation of target gene expression in single *Arabidopsis* hpRNAi lines. T2 individuals of three individual pAGRIKOLA hpRNAi expressing *Arabidopsis* lines were tested for each of the target genes, *AtHY2*, *AtTRY*, *AtLNG1*, *AtNPQ1*, *AtSEX1* and *AtMAX3*. Abundance of the respective target transcript are given as relative expression ($2^{-\Delta\Delta C_T}$) compared to that of an *Arabidopsis* line expressing a pAGRIKOLA-Control RNAi fragment and *AtACT2* for normalization

in a drastic phenotype. However, since we screened hundreds of thousands of T1 seeds for the *AtGUN4*-GST containing seven-gene RNAi fragment and eventually were able to obtain two multiple-RNAi transformants where expression of seven genes including *AtGUN4* was down regulated (described below), we did not omit *AtGUN4* as a target gene for our proof-of-concept study.

We also created *Arabidopsis* RNAi control lines that were transformed with the pAGRIKOLA-control vector and expressed an untargeted hpRNAi construct. These lines were used as standard comparators for quantitative expression analysis, as they can be treated with the herbicide BASTA® similar to the multiple RNAi lines. Moreover, it has been shown that the expression of RNAi constructs itself can result in unwanted and off-target effects on nuclear gene expression [59] and comparing gene expression of wild-type and RNAi plants can give false positive results.

## Efficient silencing of multiple transcripts by synthetic hpRNA fragments

After transformation of *Arabidopsis* with six GST constructs, we selected 10 individual lines for each of the pAGRIKOLA-LNTMHS, the pAGRIKOLA-HTLNSM, and the pAGRIKOLA-NLSHMT constructs. Figure 5 summarizes the expression of target genes as monitored in T2 individuals of four lines transformed with pAGRIKOLA-LNTMHS, two lines transformed with pAGRIKOLA-HTLNSM, and one line transformed with pAGRIKOLA-NLSHMT. Expression of target genes in leaf and flower tissue was successfully downregulated in all lines, with the exception of *AtTRY* in HTLNSM RNAi line #5 and *AtLNG1* in NLSHMT RNAi line #2 (Fig. 5). Level of downregulation ranged between 20 and 90 % compared to the *Arabidopsis* Control-RNAi line. Comparison of data obtained with the three constructs carrying the GSTs in different order reveals no obvious differences in efficiency of the RNAi effect on the respective target gene (Fig. 5).

For each of the constructs containing seven GSTs pAGRIKOLA-NLSHMTG and pAGRIKOLA-LNGTMHS, only one successful *Arabidopsis* transformant could be selected. We were unable to select a line expressing the GHTLNSM construct and we can only speculate that the low transformation efficiency is due to lethal downregulation of *AtGUN4* expression as mentioned above. Both of the seven GST lines show the similar chlorotic phenotype and had only about 10 % of the *GUN4* control transcript level remaining (Fig. 6a, b), indicating disturbed chlorophyll biosynthesis [58, 60, 61]. However, Fig. 6b shows the effects of the two seven-GST RNAi constructs on the expression of the respective target genes. Similar to the results with six individual GST constructs, with the exception of *AtMAX3*, the expression of each of the target genes is downregulated by 40 % (AtLNG1, AtSEX1) to 90 % (AtGUN4).

## Conclusions

We present a novel and cost-effective strategy to simultaneously down regulate expression of more than one target gene in plants using a single synthetic construct. Similar to gene stacking, where expression of different transgenes is triggered by a single construct, this method circumvents issues with introducing more than one transformation marker, characterizing multiple T-DNA insertion sites or laborious crossing of different mutant lines. Using the yeast assembly system reduces the cloning work and can quickly be established. We do not conceal that selecting a transformation event or transformed line that shows sufficient knock down of all targeted genes may result in screening of hundreds of individuals. Assuming that typically about one out of five *Arabidopsis*

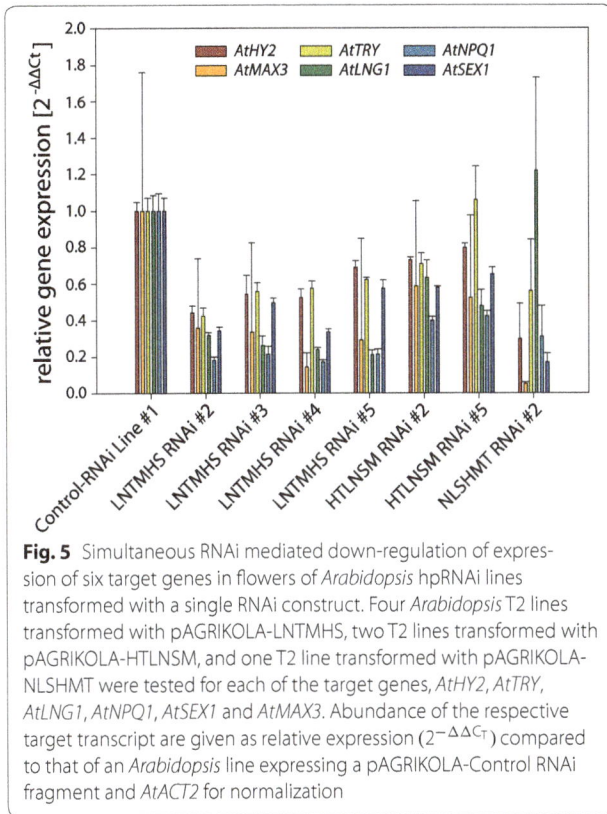

**Fig. 5** Simultaneous RNAi mediated down-regulation of expression of six target genes in flowers of *Arabidopsis* hpRNAi lines transformed with a single RNAi construct. Four *Arabidopsis* T2 lines transformed with pAGRIKOLA-LNTMHS, two T2 lines transformed with pAGRIKOLA-HTLNSM, and one T2 line transformed with pAGRIKOLA-NLSHMT were tested for each of the target genes, *AtHY2*, *AtTRY*, *AtLNG1*, *AtNPQ1*, *AtSEX1* and *AtMAX3*. Abundance of the respective target transcript are given as relative expression ($2^{-\Delta\Delta C_T}$) compared to that of an *Arabidopsis* line expressing a pAGRIKOLA-Control RNAi fragment and *AtACT2* for normalization

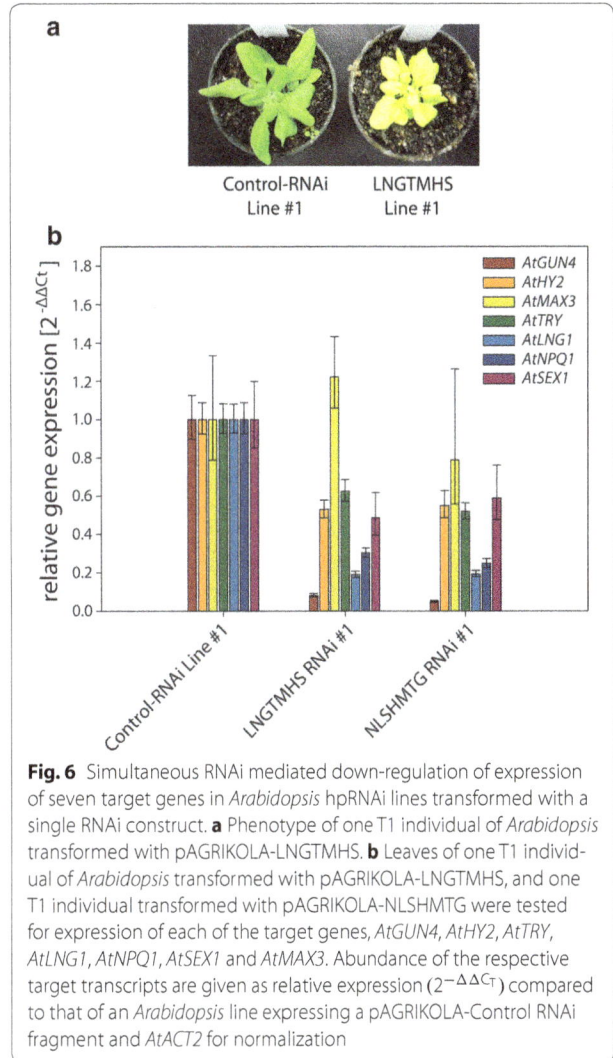

**Fig. 6** Simultaneous RNAi mediated down-regulation of expression of seven target genes in *Arabidopsis* hpRNAi lines transformed with a single RNAi construct. **a** Phenotype of one T1 individual of *Arabidopsis* transformed with pAGRIKOLA-LNGTMHS. **b** Leaves of one T1 individual of *Arabidopsis* transformed with pAGRIKOLA-LNGTMHS, and one T1 individual transformed with pAGRIKOLA-NLSHMTG were tested for expression of each of the target genes, *AtGUN4*, *AtHY2*, *AtTRY*, *AtLNG1*, *AtNPQ1*, *AtSEX1* and *AtMAX3*. Abundance of the respective target transcripts are given as relative expression ($2^{-\Delta\Delta C_T}$) compared to that of an *Arabidopsis* line expressing a pAGRIKOLA-Control RNAi fragment and *AtACT2* for normalization

RNAi lines show sufficient down regulation of a single target gene, it might be necessary to screen about $5^n$ individual transformants when n genes are targeted with our suggested approach. Comparison of RNAi efficiencies obtained with single or multiple RNAi constructs (Figs. 4, 5, 6) may lead to the conclusion that the multiple RNAi constructs act less efficiently. We cannot rule out such an interfering mechanism but also want to present a hypothesis that this might be an apparent effect based on the relatively low amount of individuals expressing the multiple RNAi constructs. In case of the single RNAi plants we were able to choose the best three out of approx. 15–20 selected transformants for each construct. We would roughly expect that screening of 15,625 ($5^6$) or even 78,125 ($5^7$) selected individuals would be necessary to identify mutants with similar levels of downregulation for all targeted genes. For our proof-of-concept study, we rather intended to provide evidence that a single synthetic RNAi fragment can affect a large panel of transcripts. In practical approaches, where e.g., up to four genes are targeted, only about 625 transformants would need to be screened for desired transgenic lines. However, the increasing amount of individual RNAi plants that need to be screened for best downregulation result is likely independent of our approach and also true for other ways to address multiple target genes (e.g.,

multiple transformations or serial cloning of constructs). Applying the presented method to artificial micro RNA (amiRNA) expression instead of hpRNAi, may solve this problem, because amiRNA are even more effective in target gene down regulation [57, 62, 63]. Multiple gene knockdown in plants using a single synthetic construct may facilitate ambitious projects dealing with establishing new biochemical pathways, e.g., transition from C3 to C4 photosynthesis [64, 65] and/or from C3 to CAM photosynthesis [66, 67], especially for rewiring the diel expression pattern of genes shared between C3 and CAM species.

## Methods

### Experimental design

Based on the work of Lloyd and Meinke [46], who provide an excellent overview about *Arabidopsis* phenotypes caused by disrupted genes, we chose six and/or seven genes that can likely be knocked down simultaneously in a single *Arabidopsis* plant without generating a lethal phenotype; they were: *AtHY2* (At3g09150) [50], *AtLNG1* (At5g15580) [51], *AtTRY* (At5g53200) [52], *AtMAX3* (At2g44990) [53], *AtNPQ1* (At1g08550) [54], *AtSEX1* (At1g10760) [55] and *AtGUN4* (At3g59400) [58]. The CATMA database [47] was used to identify specific gene sequence tags (GST) and respective cloning primer sequences for each of the seven transcripts (Additional file 2: Table S1). In order to test the functionality of the GSTs for RNAi in *Arabidopsis*, single GSTs were cloned and inserted into the binary plant hpRNAi plasmid pAGRIKOLA [20]. The expression of the resulting hpRNA-constructs contained the respective GSTs that were cloned in sense and antisense orientation flanking the *Pdk* and *Cat* introns was driven by the CaMV 35S promoter. *A. thaliana* 'Columbia-0' (Col-0) plants were transformed with these vectors and transcript levels of the respective target genes were determined in herbicide resistant T2 individuals. The same GSTs were used to assemble synthetic RNAi fragments. Three different synthetic RNAi fragments were designed for two gene sets (six or seven gene knockdown), comprising varying assembly orders to evaluate sequence effects on RNAi effectiveness. We randomly chose the order of the GSTs for the seven GST fragments. The six GST fragments followed the same order without assembling the AtGUN4-GST. This strategy enabled multiple use of assembly primers (Additional file 2: Table S1). The synthetic RNAi fragments were eventually cloned into pAGRIKOLA and after *Arabidopsis* transformation, transcript levels of the target genes were determined in T1 or T2 individuals selected for presence of the transgene by BASTA® treatment. In all experiments, *Arabidopsis* plants transformed with pAGRIKOLA carrying no GST were used as controls.

### Cloning of single RNAi plasmids and controls
Single GSTs were amplified from an *A. thaliana* Col-0 cDNA template using Phusion High-Fidelity DNA Polymerase (Thermo Fisher Scientific Inc.,) and primers listed in Additional file 2: Table S1 according to the manufacturer's protocols. 5′-CACC-3′ overhangs were introduced with the PCR primers to facilitate directional cloning into pENTR™/D-TOPO (Life Technologies). Gel purified PCR products were subcloned into pENTR™/D-TOPO and resulting pENTR clones were verified by sequencing. Since pENTR and the destination plasmid pAGRIKOLA carry the same bacterial selection marker (*nptI*), pENTR plasmids were serially digested with *PvuI* and *NruI* and the linearized plasmid backbones were gel purified before LR clonase II recombination reactions (Thermo Fisher Scientific Inc.,) and were performed to create the

pAGRIKOLA RNAi expression plasmids. The correct insertions of the GSTs into the resulting pAGRIKOLA RNAi vectors were verified by sequencing using primers listed in Additional file 2: Table S1.

An empty plasmid control was created by cloning a pENTR-control plasmid transferring the sequence 5′-CACCAAAATG-3′ into the respective destination plasmid pAGRIKOLA. Direct subcloning of the very short linker sequence into pENTR™/D-TOPO by means of annealed oligos was not successful. Therefore, one of the pENTR clones (pENTR-AtHY2) was chosen, digested with *AscI* and *NotI* and gel purified to release the inserted AtHY2 GST. Two synthetic oligos (5′-GGCCGCCC CCTT<u>CACCAAAATG</u>AAGGGTGGG-3′, 5′-CGCG CCCACCCTT<u>CATTTTGGTG</u>AAGGGGGGC-3′) were annealed to create a linker by heating 1 nmol of each oligo for 5 min at 95 °C in 100 μl buffer (10 mM Tris-HCl pH 7.5, 10 mM $MgCl_2$) and slowly cooling down at a rate of 2 °C per min. Ten pmol of the linker were subsequently ligated into 100 nmol linearized pENTR using T4 DNA Ligase (Thermo Fisher Scientific Inc.,) according to the manufacturer's protocol. Competent *E. coli* (strain NEB 5-alpha, New England Biolabs) were transformed with the ligation mixture and a correctly assembled pENTR-control plasmid was identified by restriction digestion and sequencing. An LR clonase II recombination reaction to create a pAGRIKOLA-control vector and subsequent sequencing were performed as described above.

### Assembly of multiple RNAi constructs
The GeneArt® High-Order Genetic Assembly Kit and the GeneArt® pYES1L Vector with Sapphire™ Technology (Thermo Fisher Scientific Inc.,) were used to assemble the multiple RNAi constructs in yeast. Overlapping primers to amplify GSTs were designed using the GeneArt® Primer and Construct Design Tool (http://www.thermofisher.com/order/oligoDesigner). Single GSTs to be assembled to multiple RNAi fragments were amplified from an *A. thaliana* Col-0 cDNA template using Phusion High-Fidelity DNA Polymerase and primers listed in Table S1 according to the manufacturer's protocols. PCR products were gel purified, concentrations were determined and respective PCR products and pYES1L were mixed and chemical competent yeast cells were transformed according to the manufacturer's protocol in order to create (a) pYES1L-HTLNSM, (b) pYES1L-NLSHMT, (c) pYES1L-LNTMHS, (d) pYES1L-GHTLNSM, (e) pYES1L-NLSHMTG and (f) pYES1L-LNGTMHS (Figs. 2, 3). After 3 days incubation on yeast selection medium, eight yeast colonies were picked for each assembly, plasmid DNAs were extracted according to the The GeneArt® High-Order Genetic Assembly Kit protocol and used as template for PCR proof of successful assembly using

primers given in Additional file 2: Table S1. One plasmid preparation for a given PCR product of the expected size was used as template for PCR for a subsequent cloning of the assembled synthetic fragments into pENTR™/D-TOPO. Full length sequences of the assembled fragments were obtained from the respective pENTR clones.

## Cloning of multiple RNAi plasmids

One yeast plasmid preparation for each assembled RNAi fragment creating a PCR product of the expected size was used as template for PCR for a subsequent cloning of the assembled synthetic fragments into pENTR™/D-TOPO. Therefore, synthetic RNAi fragments were amplified using Phusion High-Fidelity DNA Polymerase and primers listed in Additional file 2: Table S1. 5′-CACC-3′ overhangs were introduced with the PCR primers to facilitate directional cloning into pENTR™/D-TOPO. Gel purified PCR products were subcloned into pENTR™/D-TOPO and resulting pENTR clones were verified by sequencing. For transferring the cloned synthetic fragments into pAGRIKOLA, pENTR plasmids were serially digested with *PvuI* and *NruI* and LR clonase II recombination reactions were performed as described above. The correct insertions of the synthetic RNAi fragments into the resulting pAGRIKOLA multiple RNAi vectors were verified by restriction digestions using *PstI* and sequencing using primers listed in Additional file 2: Table S1.

## Plant growth conditions and plant transformation

Seeds of *A. thaliana* Col-0 were directly sown into soil and seedlings were transplanted to single pots after 2 weeks. Seedlings and plants were grown in growth chamber at 23 °C and approximately 200 μE light in long-day conditions (14 h of light).

Chemically competent cells [68] of *A. tumefaciens* strain GV3101(pMP90) [69] were prepared and transformed with the helper plasmid pSOUP [70]. A resulting GV3101::pMP90::pSOUP clone was again made chemically competent and transformed with the single RNAi and the multiple RNAi pAGRIKOLA vectors. *Agrobacteria* clones were used to transform *Arabidopsis* plants by floral dip [71]. Transformed T1 seedlings were selected by spraying 2-week-old soil-grown seedlings with 0.01 % (w/v) BASTA® five times in 2 days intervals. T1 seeds transformed with RNAi fragments containing the *AtGUN4* GST were surface sterilized by serial washing with 96 % (v/v) ethanol, 20 % (v/v) household bleach supplemented with 0.05 % (v/v) Tween-20, and water, and placed at 4 °C for 2 days. Seeds were subsequently plated on ½ Murashige and Skoog (MS) medium [72] supplemented with 1 % (w/v) sucrose, 0.01 % (w/v) BASTA® and 0.8 % (w/v) agar, and germinated under continuous light at approximately 90 μmol photons m$^{-2}$ s$^{-1}$. Successfully

selected lines were transplanted and grown until mature T2 seeds could be harvested. T2 seeds of pAGRIKOLA-RNAi lines, as well as pAGRIKOLA-control lines, were grown on soil and sprayed with BASTA® to remove segregating wild-type individuals.

## RNA extraction and gene expression analysis

Total RNA was extracted from leaves of 10 days and 4-week-old seedlings or flowers of 12-week-old plants, respectively using the Invisorb Spin Plant Mini Kit (Stratec Molecular). Two μg of total RNA were reversely transcribed in cDNA using an Oligo-dT$_{18}$ primer and Thermo Scientific RevertAid Reverse Transcriptase (Thermo Fisher Scientific Inc.,). Quantitative RT-PCR (qRT) was conducted to examine the transcript levels of each target gene using a StepOnePlus (Applied Biosystems), Maxima SYBR Green/ROX qPCR Master Mix (Thermo Scientific) and cDNA corresponding to 80 ng RNA in a total volume of 25 μl. Three biological replicates were used for each qRT analysis. The following cycling conditions were applied for PCR: 10 min at 95 °C, 35 cycles of 15 s at 95 °C and 60 s at 60 °C. Transcript level was normalized against *AtACTIN2* (*At3g18780.1*) and an *Arabidopsis* Col-0 pAGRIKOLA-control line and are presented as $2^{-\Delta\Delta C_T}$ values [73, 74]. Gene-specific qRT primers were designed using QuantPrime [75]. All primers used for qRT analysis are listed in Additional file 2: Table S1.

### Abbreviations
amiRNA: artificial micro RNA; dsRNA: double-stranded RNA; GST: gene specific tag; hpRNA: hairpin RNA; PTGS: post transcriptional gene silencing; qRT: quantitative RT-PCR; RNAi: RNA interference; RISC: RNA-induced silencing complex; shRNA: small hairpin RNA; siRNA: small interfering RNA; ssRNA: single-stranded RNA; TAR: transformation associated recombination.

### Authors' contributions
OC, XY, JGC and GAT conceived the project. OC, ACB, SSJ and ZMC carried out the experiments. OC, JGC and GAT analyzed the data and wrote the manuscript. All authors read and approved the final manuscript.

### Author details
[1] Biosciences Division, Oak Ridge National Laboratory, Oak Ridge, TN 37831, USA. [2] Department of Plant Sciences, University of Tennessee, Knoxville, TN 37996, USA. [3] Present Address: KWS SAAT SE, Grimsehlstraße 31, 37555 Einbeck, Germany.

### Acknowledgements
We want to thank Dr. Ian Small and Dr. Pierre Hilson for sharing the pAGRIKOLA vector and CSIRO (Canberra, Australia) for permission to use the pHELLSGATE 12 derivative. The authors wish to thank Dr. Henrique DePaoli for help comments and edits to the manuscript.

## Competing interests

The authors declare that they have no competing interests.

## Funding

This work was supported by the Department of Energy, Office of Science, Genomic Science Program (under Award Number DE-SC0008834). Oak Ridge National Laboratory is managed by UT-Battelle, LLC for the US Department of Energy (under Contract Number DE-AC05-00OR22725).

## References

1. Fire A, Xu S, Montgomery MK, Kostas SA, Driver SE, Mello CC. Potent and specific genetic interference by double-stranded RNA in *Caenorhabditis elegans*. Nature. 1998;391(6669):806–11.
2. Wilson RC, Doudna JA. Molecular mechanisms of RNA interference. Annu Rev Biophys. 2013;42:217–39.
3. Agrawal N, Dasaradhi PV, Mohmmed A, Malhotra P, Bhatnagar RK, Mukherjee SK. RNA interference: biology, mechanism, and applications. Microbiol Mol Biol Rev. 2003;67(4):657–85.
4. Hannon GJ. RNA interference. Nature. 2002;418(6894):244–51.
5. Herr AJ, Baulcombe DC. RNA silencing pathways in plants. Cold Spring Harb Symp Quant Biol. 2004;69:363–70.
6. Baulcombe D. RNA silencing in plants. Nature. 2004;431(7006):356–63.
7. Watanabe T, Totoki Y, Toyoda A, Kaneda M, Kuramochi-Miyagawa S, Obata Y, Chiba H, Kohara Y, Kono T, Nakano T, et al. Endogenous siRNAs from naturally formed dsRNAs regulate transcripts in mouse oocytes. Nature. 2008;453(7194):539–43.
8. Buchon N, Vaury C. RNAi: a defensive RNA-silencing against viruses and transposable elements. Heredity. 2006;96(2):195–202.
9. Ito H. Small RNAs and regulation of transposons in plants. Genes Genet Syst. 2013;88(1):3–7.
10. Dunoyer P, Brosnan CA, Schott G, Wang Y, Jay F, Alioua A, Himber C, Voinnet O. An endogenous, systemic RNAi pathway in plants. EMBO J. 2010;29(10):1699–712.
11. Okamura K, Lai EC. Endogenous small interfering RNAs in animals. Nat Rev Mol Cell Biol. 2008;9(9):673–8.
12. Ding SW, Voinnet O. Antiviral immunity directed by small RNAs. Cell. 2007;130(3):413–26.
13. Obbard DJ, Gordon KH, Buck AH, Jiggins FM. The evolution of RNAi as a defence against viruses and transposable elements. Philos Trans R Soc Lond B Biol Sci. 2009;364(1513):99–115.
14. Senthil-Kumar M, Mysore K. RNAi in plants: recent developments and applications in agriculture. In: Catalano A, editor. Gene silencing: theory, techniques and applications. Hauppauge: Nova Science; 2010. p. 183–99.
15. Gilchrist E, Haughn G. Reverse genetics techniques: engineering loss and gain of gene function in plants. Brief Funct Genomics. 2010;9(2):103–10.
16. Watson JM, Fusaro AF, Wang M, Waterhouse PM. RNA silencing platforms in plants. FEBS Lett. 2005;579(26):5982–7.
17. Eamens A, Wang MB, Smith NA, Waterhouse PM. RNA silencing in plants: yesterday, today, and tomorrow. Plant Physiol. 2008;147(2):456–68.
18. Fusaro AF, Matthew L, Smith NA, Curtin SJ, Dedic-Hagan J, Ellacott GA, Watson JM, Wang MB, Brosnan C, Carroll BJ, et al. RNA interference-inducing hairpin RNAs in plants act through the viral defence pathway. EMBO Rep. 2006;7(11):1168–75.
19. Wesley SV, Helliwell CA, Smith NA, Wang MB, Rouse DT, Liu Q, Gooding PS, Singh SP, Abbott D, Stoutjesdijk PA, et al. Construct design for

efficient, effective and high-throughput gene silencing in plants. Plant J. 2001;27(6):581–90.
20. Hilson P, Allemeersch J, Altmann T, Aubourg S, Avon A, Beynon J, Bhalerao RP, Bitton F, Caboche M, Cannoot B, et al. Versatile gene-specific sequence tags for Arabidopsis functional genomics: transcript profiling and reverse genetics applications. Genome Res. 2004;14(10B):2176–89.
21. Wielopolska A, Townley H, Moore I, Waterhouse P, Helliwell C. A high-throughput inducible RNAi vector for plants. Plant Biotechnol J. 2005;3(6):583–90.
22. Gendler K, Paulsen T, Napoli C. ChromDB: the chromatin database. Nucleic Acids Res. 2008;36(Database issue):D298–302.
23. Halpin C. Gene stacking in transgenic plants—the challenge for 21st century plant biotechnology. Plant Biotechnol J. 2005;3(2):141–55.
24. Naqvi S, Farre G, Sanahuja G, Capell T, Zhu C, Christou P. When more is better: multigene engineering in plants. Trends Plant Sci. 2010;15(1):48–56.
25. Ye X, Al-Babili S, Kloti A, Zhang J, Lucca P, Beyer P, Potrykus I. Engineering the provitamin A (beta-carotene) biosynthetic pathway into (carotenoid-free) rice endosperm. Science. 2000;287(5451):303–5.
26. Buntru M, Gartner S, Staib L, Kreuzaler F, Schlaich N. Delivery of multiple transgenes to plant cells by an improved version of MultiRound Gateway technology. Transgenic Res. 2013;22(1):153–67.
27. Wang Y, Yau YY, Perkins-Balding D, Thomson JG. Recombinase technology: applications and possibilities. Plant Cell Rep. 2011;30(3):267–85.
28. Vemanna RS, Chandrashekar BK, Hanumantha Rao HM, Sathyanarayana-gupta SK, Sarangi KS, Nataraja KN, Udayakumar M. A modified MultiSite gateway cloning strategy for consolidation of genes in plants. Mol Biotechnol. 2013;53(2):129–38.
29. Sun Q, Liu J, Li Y, Zhang Q, Shan S, Li X, Qi B. Creation and validation of a widely applicable multiple gene transfer vector system for stable transformation in plant. Plant Mol Biol. 2013;83(4–5):391–404.
30. Motegi Y, Katayama K, Sakurai F, Kato T, Yamaguchi T, Matsui H, Takahashi M, Kawabata K, Mizuguchi H. An effective gene-knockdown using multiple shRNA-expressing adenovirus vectors. J Control Release. 2011;153(2):149–53.
31. Xu XM, Yoo MH, Carlson BA, Gladyshev VN, Hatfield DL. Simultaneous knockdown of the expression of two genes using multiple shRNAs and subsequent knock-in of their expression. Nat Protoc. 2009;4(9):1338–48.
32. Chumakov SP, Kravchenko JE, Prassolov VS, Frolova EI, Chumakov PM. Efficient downregulation of multiple mRNA targets with a single shRNA-expressing lentiviral vector. Plasmid. 2010;63(3):143–9.
33. Zhu X, Santat LA, Chang MS, Liu J, Zavzavadjian JR, Wall EA, Kivork C, Simon MI, Fraser ID. A versatile approach to multiple gene RNA interference using microRNA-based short hairpin RNAs. BMC Mol Biol. 2007;8:98.
34. Sun D, Melegari M, Sridhar S, Rogler CE, Zhu L. Multi-miRNA hairpin method that improves gene knockdown efficiency and provides linked multi-gene knockdown. Biotechniques. 2006;41(1):59–63.
35. Sander JD, Joung JK. CRISPR–Cas systems for editing, regulating and targeting genomes. Nat Biotechnol. 2014;32(4):347–55.
36. Piatek A, Ali Z, Baazim H, Li L, Abulfaraj A, Al-Shareef S, Aouida M, Mahfouz MM. RNA-guided transcriptional regulation in planta via synthetic dCas9-based transcription factors. Plant Biotechnol J. 2015;13(4):578–89.
37. Feng Z, Zhang B, Ding W, Liu X, Yang DL, Wei P, Cao F, Zhu S, Zhang F, Mao Y, et al. Efficient genome editing in plants using a CRISPR/Cas system. Cell Res. 2013;23(10):1229–32.
38. Xie K, Yang Y. RNA-guided genome editing in plants using a CRISPR–Cas system. Mol Plant. 2013;6(6):1975–83.
39. Larionov V, Kouprina N, Graves J, Chen XN, Korenberg JR, Resnick MA. Specific cloning of human DNA as yeast artificial chromosomes by transformation-associated recombination. Proc Natl Acad Sci USA. 1996;93(1):491–6.
40. Gibson DG. Synthesis of DNA fragments in yeast by one-step assembly of overlapping oligonucleotides. Nucleic Acids Res. 2009;37(20):6984–90.
41. Orr-Weaver TL, Szostak JW, Rothstein RJ. Yeast transformation: a model system for the study of recombination. Proc Natl Acad Sci USA. 1981;78(10):6354–8.
42. Gibson DG, Benders GA, Axelrod KC, Zaveri J, Algire MA, Moodie M, Montague MG, Venter JC, Smith HO, Hutchison CA 3rd. One-step assembly of 25 overlapping DNA fragments to form a complete synthetic *Mycoplasma genitalium* genome. Proc Natl Acad Sci USA. 2008;105(51):20404–9.

43. Gibson DG, Benders GA, Andrews-Pfannkoch C, Denisova EA, Baden-Tillson H, Zaveri J, Stockwell TB, Brownley A, Thomas DW, Algire MA, et al. Complete chemical synthesis, assembly, and cloning of a *Mycoplasma genitalium* genome. Science. 2008;319(5867):1215–20.

44. Bigger BW, Liao AY, Sergijenko A, Coutelle C. Trial and error: how the unclonable human mitochondrial genome was cloned in yeast. Pharm Res. 2011;28(11):2863–70.

45. Raymond CK, Sims EH, Olson MV. Linker-mediated recombinational subcloning of large DNA fragments using yeast. Genome Res. 2002;12(1):190–7.

46. Lloyd J, Meinke D. A comprehensive dataset of genes with a loss-of-function mutant phenotype in Arabidopsis. Plant Physiol. 2012;158(3):1115–29.

47. Crowe ML, Serizet C, Thareau V, Aubourg S, Rouze P, Hilson P, Beynon J, Weisbeek P, van Hummelen P, Reymond P, et al. CATMA: a complete Arabidopsis GST database. Nucleic Acids Res. 2003;31(1):156–8.

48. Thareau V, Dehais P, Serizet C, Hilson P, Rouze P, Aubourg S. Automatic design of gene-specific sequence tags for genome-wide functional studies. Bioinformatics. 2003;19(17):2191–8.

49. Hartley JL, Temple GF, Brasch MA. DNA cloning using in vitro site-specific recombination. Genome Res. 2000;10(11):1788–95.

50. Koornneef M, Rolff E, Spruit CJP. Genetic control of light-inhibited hypocotyl elongation in *Arabidopsis thaliana* (L.) Heynh. Z Pflanzenphysiol. 1980;100(2):147–60.

51. Lee YK, Kim GT, Kim IJ, Park J, Kwak SS, Choi G, Chung WI. LONGIFOLIA1 and LONGIFOLIA2, two homologous genes, regulate longitudinal cell elongation in Arabidopsis. Development. 2006;133(21):4305–14.

52. Hulskamp M, Misra S, Jurgens G. Genetic dissection of trichome cell development in Arabidopsis. Cell. 1994;76(3):555–66.

53. Booker J, Auldridge M, Wills S, McCarty D, Klee H, Leyser O. MAX3/CCD7 is a carotenoid cleavage dioxygenase required for the synthesis of a novel plant signaling molecule. Curr Biol. 2004;14(14):1232–8.

54. Niyogi KK, Grossman AR, Bjorkman O. Arabidopsis mutants define a central role for the xanthophyll cycle in the regulation of photosynthetic energy conversion. Plant Cell. 1998;10(7):1121–34.

55. Yu TS, Kofler H, Hausler RE, Hille D, Flugge UI, Zeeman SC, Smith AM, Kossmann J, Lloyd J, Ritte G, et al. The Arabidopsis sex1 mutant is defective in the R1 protein, a general regulator of starch degradation in plants, and not in the chloroplast hexose transporter. Plant Cell. 2001;13(8):1907–18.

56. Du SY, Zhang XF, Lu Z, Xin Q, Wu Z, Jiang T, Lu Y, Wang XF, Zhang DP. Roles of the different components of magnesium chelatase in abscisic acid signal transduction. Plant Mol Biol. 2012;80(4–5):519–37.

57. Schwab R, Ossowski S, Riester M, Warthmann N, Weigel D. Highly specific gene silencing by artificial microRNAs in Arabidopsis. Plant Cell. 2006;18(5):1121–33.

58. Larkin RM, Alonso JM, Ecker JR, Chory J. GUN4, a regulator of chlorophyll synthesis and intracellular signaling. Science. 2003;299(5608):902–6.

59. Schlicke H, Hartwig AS, Firtzlaff V, Richter AS, Glasser C, Maier K, Finke-meier I, Grimm B. Induced deactivation of genes encoding chlorophyll biosynthesis enzymes disentangles tetrapyrrole-mediated retrograde signaling. Mol Plant. 2014;7(7):1211–27.

60. Wilde A, Mikolajczyk S, Alawady A, Lokstein H, Grimm B. The gun4 gene is essential for cyanobacterial porphyrin metabolism. FEBS Lett. 2004;571(1–3):119–23.

61. Peter E, Grimm B. GUN4 is required for posttranslational control of plant tetrapyrrole biosynthesis. Mol Plant. 2009;2(6):1198–210.

62. Tiwari M, Sharma D, Trivedi PK. Artificial microRNA mediated gene silencing in plants: progress and perspectives. Plant Mol Biol. 2014;86(1–2):1–18.

63. Ossowski S, Schwab R, Weigel D. Gene silencing in plants using artificial microRNAs and other small RNAs. Plant J. 2008;53(4):674–90.

64. Covshoff S, Hibberd JM. Integrating C4 photosynthesis into C3 crops to increase yield potential. Curr Opin Biotechnol. 2012;23(2):209–14.

65. von Caemmerer S, Quick WP, Furbank RT. The development of C4 rice: current progress and future challenges. Science. 2012;336(6089):1671–2.

66. Yang X, Cushman JC, Borland AM, Edwards EJ, Wullschleger SD, Tuskan GA, Owen NA, Griffiths H, Smith JA, De Paoli HC, et al. A roadmap for research on crassulacean acid metabolism (CAM) to enhance sustainable food and bioenergy production in a hotter, drier world. New Phytol. 2015;207(3):491–504.

67. DePaoli HC, Borland AM, Tuskan GA, Cushman JC, Yang X. Synthetic biology as it relates to CAM photosynthesis: challenges and opportunities. J Exp Bot. 2014;65(13):3381–93.

68. Höfgen R, Willmitzer L. Storage of competent cells for Agrobacterium transformation. Nucleic Acids Res. 1988;16(20):9877.

69. Koncz C, Schell J. The promoter of TL-DNA gene 5 controls the tissue-specific expression of chimaeric genes carried by a novel type of Agrobacterium binary vector. Mol Gen Genet (MGG). 1986;204(3):383–96.

70. Hellens RP, Edwards EA, Leyland NR, Bean S, Mullineaux PM. pGreen: a versatile and flexible binary Ti vector for Agrobacterium-mediated plant transformation. Plant Mol Biol. 2000;42(6):819–32.

71. Clough SJ, Bent AF. Floral dip: a simplified method for Agrobacterium-mediated transformation of *Arabidopsis thaliana*. Plant J. 1998;16(6):735–43.

72. Murashige T, Skoog F. A revised medium for rapid growth and bio assays with tobacco tissue cultures. Physiol Plant. 1962;15(3):473–97.

73. Schmittgen TD, Livak KJ. Analyzing real-time PCR data by the comparative C(T) method. Nat Protoc. 2008;3(6):1101–8.

74. Livak KJ, Schmittgen TD. Analysis of relative gene expression data using real-time quantitative PCR and the 2(−Delta Delta C(T)) method. Methods. 2001;25(4):402–8.

75. Arvidsson S, Kwasniewski M, Riano-Pachon DM, Mueller-Roeber B. QuantPrime—a flexible tool for reliable high-throughput primer design for quantitative PCR. BMC Bioinform. 2008;9:465.

# 7

# Detecting N-myristoylation and S-acylation of host and pathogen proteins in plants using click chemistry

Patrick C. Boyle[1,4†], Simon Schwizer[1,2†], Sarah R. Hind[1], Christine M. Kraus[1,2], Susana De la Torre Diaz[1], Bin He[3,5] and Gregory B. Martin[1,2*] (iD)

## Abstract

**Background:** The plant plasma membrane is a key battleground in the war between plants and their pathogens. Plants detect the presence of pathogens at the plasma membrane using sensor proteins, many of which are targeted to this lipophilic locale by way of fatty acid modifications. Pathogens secrete effector proteins into the plant cell to suppress the plant's defense mechanisms. These effectors are able to access and interfere with the surveillance machinery at the plant plasma membrane by hijacking the host's fatty acylation apparatus. Despite the important involvement of protein fatty acylation in both plant immunity and pathogen virulence mechanisms, relatively little is known about the role of this modification during plant-pathogen interactions. This dearth in our understanding is due largely to the lack of methods to monitor protein fatty acid modifications in the plant cell.

**Results:** We describe a rapid method to detect two major forms of fatty acylation, N-myristoylation and S-acylation, of candidate proteins using alkyne fatty acid analogs coupled with click chemistry. We applied our approach to confirm and decisively demonstrate that the archetypal pattern recognition receptor FLS2, the well-characterized pathogen effector AvrPto, and one of the best-studied intracellular resistance proteins, Pto, all undergo plant-mediated fatty acylation. In addition to providing a means to readily determine fatty acylation, particularly myristoylation, of candidate proteins, this method is amenable to a variety of expression systems. We demonstrate this using both Arabidopsis protoplasts and stable transgenic Arabidopsis plants and we leverage Agrobacterium-mediated transient expression in *Nicotiana benthamiana* leaves as a means for high-throughput evaluation of candidate proteins.

**Conclusions:** Protein fatty acylation is a targeting tactic employed by both plants and their pathogens. The metabolic labeling approach leveraging alkyne fatty acid analogs and click chemistry described here has the potential to provide mechanistic details of the molecular tactics used at the host plasma membrane in the battle between plants and pathogens.

**Keywords:** Fatty acylation, Myristoylation, Palmitoylation, Stearylation, S-acylation, Click chemistry, Plasma membrane, Pathogen effectors, Pattern recognition receptors, Resistance proteins, *Arabidopsis thaliana*, *Nicotiana benthamiana*

## Background

The covalent attachment of fatty acids to specific protein residues, a process referred to as fatty acylation, increases the hydrophobicity of the substrate protein and affects various properties, most notably subcellular localization [1, 2]. These lipid moieties often serve as hydrophobic anchors that promote protein-membrane associations [1]. There are a number of different types of protein fatty acylations, the two best characterized forms in plants being N-myristoylation and S-acylation [2].

*Correspondence: gbm7@cornell.edu
†Patrick C. Boyle and Simon Schwizer contributed equally to this work
[2] Plant Pathology and Plant–Microbe Biology Section, School of Integrative Plant Science, Cornell University, Ithaca, NY 14853, USA
Full list of author information is available at the end of the article

N-myristoylation describes the irreversible amide bond formation between myristate, a saturated 14-carbon fatty acid, and the N-terminal amine of a glycine residue exposed as a result of co-translational N-terminal methionine excision, or more rarely, post-translational proteolytic processing [1, 3, 4]. This modification is mediated by N-myristoyltransferases, cytosolic entities often associated with ribosomes since protein myristoylation is typically a co-translational modification [5–8]. In many cases, myristoylation is necessary for targeting a protein to the plasma membrane (PM), but this modification alone is not sufficient to provide permanent anchoring to the membrane and as a result myristoylation is often found in combination with other membrane interaction motifs, including polybasic domains and those involving S-acylation [1, 3, 4].

S-acylation refers to the reversible thioester bond formation between a fatty acid and a cysteine residue side chain [9]. The saturated 16-carbon palmitate is the fatty acid most frequently featured in S-acylation and therefore this modification is often termed palmitoylation [1, 10]. However, other fatty acids can be covalently attached to cysteine side chains, most notably the saturated 18-carbon stearate, and a small number of studies suggest that in plants, protein stearylation is as prevalent as palmitoylation [2, 11–13]. The enzymes responsible for S-acylation are known as S-acyltransferases, or more commonly, palmitoyl acyltransferases [14]. These enzymes are integral membrane proteins found at the PM and at the membranes of various cellular compartments, including endosomes, the Golgi apparatus, and the endoplasmic reticulum [14]. Unlike myristoylation, protein S-acylation with palmitate or stearate is sufficient for stable interaction with the membrane [2, 15]. This modification is suggested to serve roles in retaining proteins at various membranes and trafficking previously myristoylated proteins to the PM, in addition to dynamically regulating protein activity, stability, and complex assembly [1, 9, 14]. Proteins bearing both myristoylation and proximal S-acylation are said to be N-terminally dual fatty acylated and this combination of lipid modifications appears to drive stable association with the PM [1].

Fatty acylation is a form of protein modification that is conserved among eukaryotes and most of the information available about this modification is based on studies from yeast and animal systems. However, what little is known about myristoylation and S-acylation in plants suggests that this kingdom is sufficiently unique in its use of these modifications to merit independent investigation [6, 16–20]. Prediction based studies indicate that the plant proteome is proportionally more myristoylated than those of metazoans and fungi [6, 16, 17, 19]. Interestingly, many of the protein families predicted to be myristoylated exclusively in plants are implicated in stress and defense responses [6, 19].

In contrast to the absolute requirement of an N-terminal glycine for myristoylation, S-acylation does not have a clear consensus sequence beyond a requisite cysteine residue, which can occur at essentially any position in a protein [20]. The lack of an S-acylation consensus sequence has largely prevented the use of predictive bioinformatics approaches to study this modification [10, 21, 22]. However, the labile nature of thioester bonds has permitted the use of an acyl-biotin exchange (ABE) approach to identify S-acylated proteins present in the plant proteome. A recent study based on the ABE method indicated that more than 500 proteins are subject to S-acylation in Arabidopsis root suspension cells, which far exceeds the number of Arabidopsis proteins predicted to be myristoylated [20]. Similar to what has been reported with plant proteins subject to myristoylation, many of the proteins identified as being S-acylated appear to be involved in pathogen perception [20]. The prevalence for fatty acylation of proteins functioning in defense is not unexpected because these modifications are known to target proteins to the PM and this lipophilic locale constitutes the initial point of pathogen perception in plants [2]. The organization of the plant palmitoyl acyltransferases, which are present at the PM in greater proportions than in mammalian and yeast systems, suggests that the plant S-acylation apparatus is uniquely arranged for the stable recruitment of proteins to this particular membrane locale [14].

The use of host-mediated myristoylation by plant pathogen effectors supports suggestions that the plant PM is a critical interface during plant-pathogen interactions and that the fatty acylation mechanisms are distinctively organized and/or accessible in the plant cell environment [23]. The exploitation of host-mediated protein lipidation mechanisms for the spatial regulation of pathogen effectors seems to be a general virulence strategy, yet only the effectors of plant pathogenic bacteria appear to hijack the host myristoylation machinery [24–31]. To date, the reason for this observation remains unclear, but the modification is essential for the virulence activity of several bacterial effectors and is required for the recognition of many effectors in host plants armed with the appropriate intracellular sensor proteins, more commonly referred to as resistance proteins [24, 32–34].

Despite the distinct features of protein fatty acylation in plants and its importance in plant-pathogen interactions, methods to readily and decisively detect specific fatty acid modifications of host and pathogen proteins in the plant cell are currently lacking. The ability to monitor plant-mediated myristoylation has proven particularly problematic because of the irreversible nature of this modification. Traditional approaches to directly demonstrate protein fatty acylation in vivo have relied on metabolic labeling with radiolabeled fatty acids, such as $[^3H]$- or $[^{125}I]$-myristic and palmitic acids, followed by purification of the protein of interest and visualization using autoradiography [35]. This method, although effective, typically requires lengthy film exposure times to visualize fatty acylated proteins and requires the use of radioactive materials [3, 4, 34, 35]. Furthermore, radiolabeling techniques do not present any straightforward means to capture labeled proteins, preventing proteome-wide identification of fatty acylated targets [35]. Lipid modification analysis by gas chromatography coupled with mass spectrometry (GC–MS) is another approach that has advanced our understanding of candidate protein S-acylation, particularly in plants [36]. The advantages of this approach are its ability to unambiguously identify S-acylation modifications, such as palmitoylation and stearylation, and to do so without the requirement of feeding radiolabeled materials to the cells or tissues being interrogated [37]. However this technique cannot be applied to the analysis of protein myristoylation and, like radiolabeling approaches, is not amenable to whole proteome analysis. The ABE approach was developed as a relatively rapid nonradioactive alternative to study protein fatty acylation and enables proteome-wide identification of S-acylated proteins [38, 39]. ABE leverages the labile nature of thioester linkages to replace S-acylation modifications present on cysteine residues with a chosen label, most often biotin. The labeled proteins are then enriched on affinity resin and subsequently identified using mass spectrometry. Alternatively, the labeled proteins can be visualized by in-gel fluorescence or western blotting. However, the ABE method has some limitations, most notably that this approach, like the GC–MS strategy, can only be applied to study S-acylation and not myristoylation [35]. Also, due to its indirect nature the technique does not allow for the discernment of different thioester linkages, many of which are not involved in S-acylation, and therefore results in false positives [20, 35, 39–41].

Metabolic labeling approaches using fatty acid analogs containing bio-orthogonal chemical handles, which allow for the attachment of reporter or detection tags, have recently emerged as means to circumvent many of the difficulties that have impeded the study of fatty acid modifications [35, 42–45]. The strategy involves feeding cells a fatty acid analog bearing a bio-orthogonal azide or alkyne handle, resulting in metabolic incorporation of the analog into target proteins. Click chemistry is then used to react the bio-orthogonal functionality present in the fatty acid analog with a reporter tag. These

fatty acids are termed bio-orthogonal analogs because the sleek nature of the azide or alkyne handles present in terminal positions of these modified lipids interfere neither with the hydrophobic character of these molecules nor the acid moiety, preserving the ability to insert into membranes and interact with the native fatty acylation apparatus which allows their metabolic incorporation into target proteins [45]. These chemical tools enable the rapid detection of protein myristoylation and S-acylation without the need for radioactivity [35]. Click chemistry, based primarily on the Huisgen [3 + 2] Cu(I)-catalyzed azide-alkyne cycloaddition, has an ever growing number of applications and has been employed in plant systems to determine the targets of reactive small molecules, visualize cell lignification, track Golgi protein dynamics, and detect protein prenylation, but it has not yet been leveraged to study fatty acid modifications of plant proteins [46–49].

Protein fatty acylation in plants has many interesting features compared to other eukaryotes. However, many questions still remain about the role of these protein modifications in this kingdom due to the lack of techniques currently available to study fatty acylation, particularly post-translational myristoylation, in plants. Here we describe the development of a click chemistry-based method using ω-alkynyl fatty acid analogs to facilitate the study of fatty acylation of both host and pathogen effector proteins in the plant cell environment.

## Results

### General scheme for assessing fatty acylation of candidate proteins using clickable fatty acid analogs

We developed and optimized an approach to determine the fatty acylation status, especially myristoylation, of candidate proteins in plant cells using ω-alkynyl fatty acid analogs and click chemistry based largely on methods previously described [45, 50] (Fig. 1). Plant cells are transformed with a candidate gene construct, preferably encoding a commercial epitope tag, following standard protocols. The alkyne fatty acid analog for the metabolite of interest is applied to the plant cells and subsequently incorporated during protein synthesis. Total protein is extracted and the candidate protein purified using immunoprecipitation. A reporter, such as a biotin tag or a fluorescent dye, is added to the alkyne group of the fatty acid analog using click chemistry and detected by western blotting or fluorescence imaging (Fig. 1). The experimental steps outlined here can be completed within a few days and we describe below the successful application of this approach to detect fatty acid modifications in a variety of candidate proteins using different expression methods and plant systems.

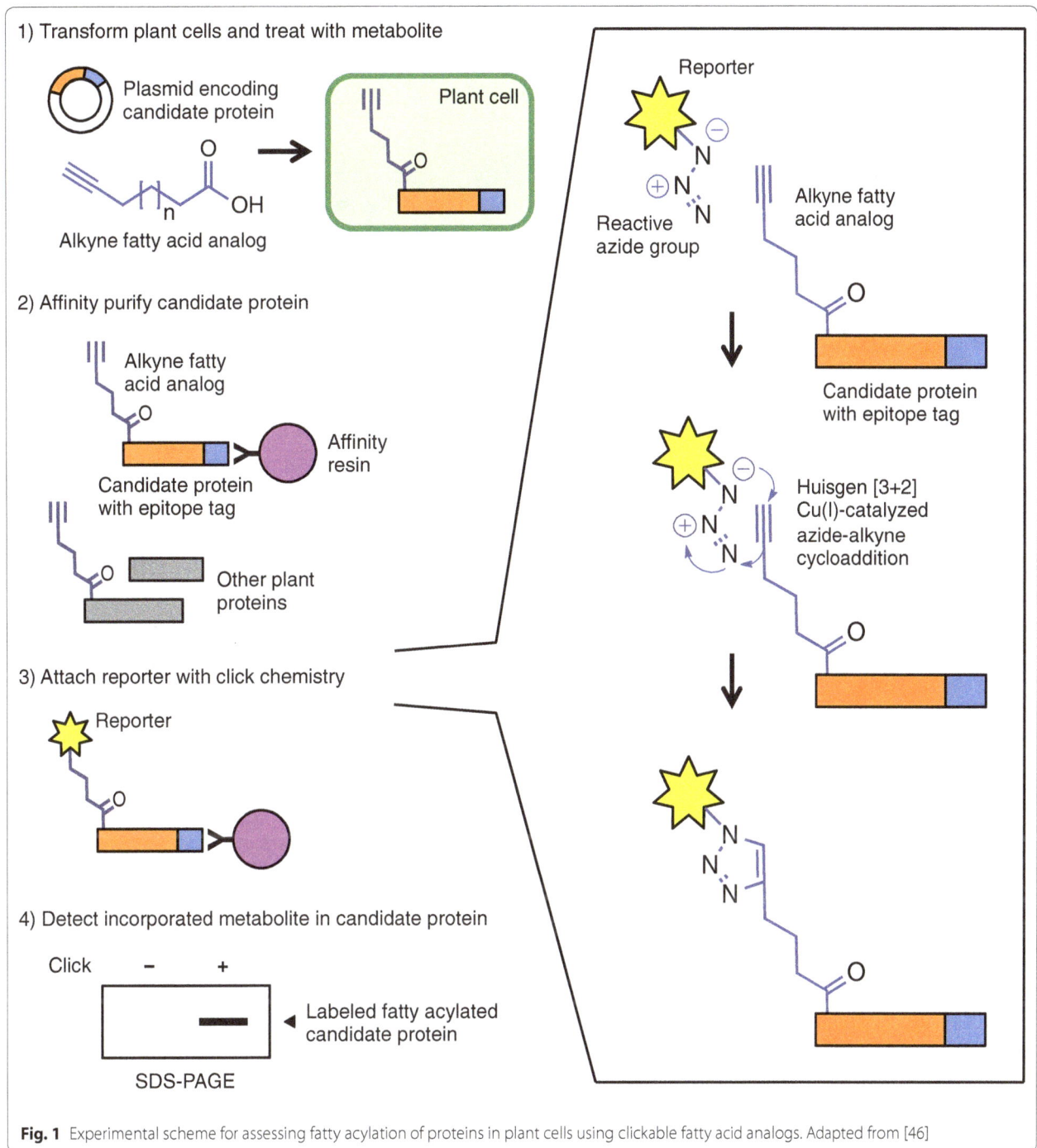

**Fig. 1** Experimental scheme for assessing fatty acylation of proteins in plant cells using clickable fatty acid analogs. Adapted from [46]

**Alkyne fatty acid analogs are better tolerated by plant cells than preparations of azide fatty acid analogs**

To assess the potential phytotoxicity of different forms of fatty acid analogs, we transformed Arabidopsis protoplasts with an expression vector encoding yellow fluorescent protein (YFP) and treated the protoplasts with either Az12 or Alk12, which are the azide- and alkyne-functionalized myristic acid analogs, respectively (Additional file 1: Figure S1). We found that even very low concentrations of Az12 strongly diminished YFP accumulation, whereas Alk12 showed inhibitory effects only at relatively high concentrations (Additional file 1: Figure S1). Therefore, we decided to perform all subsequent experiments using preparations of alkyne fatty acid analogs.

## The pattern recognition receptor FLS2 is S-acylated

The plant PM is armed with a series of sensors that function as a surveillance system to detect the presence of invading microbes and much of the machinery involved in monitoring this crucial lipophilic locale features some form of fatty acylation [23, 51]. These PM-localized pattern recognition receptors (PRRs) are able to perceive the presence of pathogens through the recognition of conserved microbe-associated molecular patterns [51]. Activation of PRRs stimulates signaling through intracellular protein kinases, resulting in the deployment of a broad-spectrum defense response referred to as pattern-triggered immunity (PTI) [52].

One of the best-characterized PRRs is Arabidopsis flagellin-sensitive 2 (FLS2), which recognizes a highly conserved 22-amino acid sequence from the N-terminal portion of bacterial flagellin [53, 54]. A recent survey of protein S-acylation in Arabidopsis using an ABE approach strongly suggested that FLS2 is S-acylated at cysteine residues 830 and 831 [55, 56]. To validate our click chemistry-based approach and to directly show incorporation of palmitic acid at these sites, we expressed wild-type *FLS2* and a mutant encoding serine substitutions at residues 830 and 831 (C830S, C831S) in Arabidopsis protoplasts in the presence of the palmitic acid analog Alk14 (Fig. 2a). Following protein extraction and immunoprecipitation, we introduced a fluorescent reporter tag using click chemistry to visualize Alk14 incorporation. We were able to detect a strong fluorescent signal only with wild-type FLS2, and not the C830S, C831S mutant. Anti-HA western blotting showed comparable accumulation of the two proteins (Fig. 2a). Importantly, given that the C830S, C831S mutant is targeted to the PM like wild-type FLS2 [20], the lack of labeling observed with the mutant indicates that fatty acid analogs are not attached to these proteins simply due to their proximity to the lipid-rich PM. Taken together, this result demonstrates that our approach is well suited to study the fatty acylation status of candidate proteins in plant cells.

## Alkyne fatty acid analogs do not appear to interfere with programmed cell death and permeate intact cells in leaf tissue

The alkyne-functionalized fatty acid analogs do not appear to interfere with protein synthesis when used at moderate concentrations and are readily incorporated into proteins in a protoplast system. However, it remained possible that in the context of whole leaf tissue these analogs could cause spurious cell death symptoms, interfere with certain immune responses, or are unable

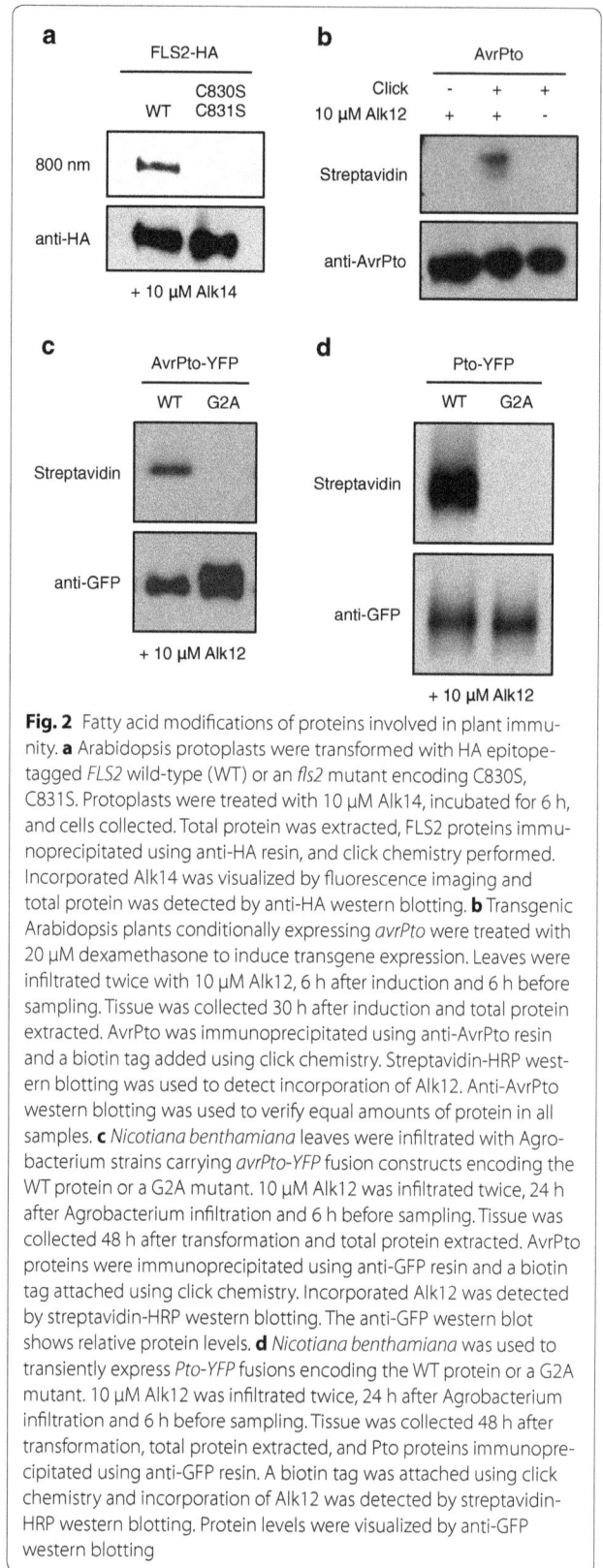

**Fig. 2** Fatty acid modifications of proteins involved in plant immunity. **a** Arabidopsis protoplasts were transformed with HA epitope-tagged *FLS2* wild-type (WT) or an *fls2* mutant encoding C830S, C831S. Protoplasts were treated with 10 μM Alk14, incubated for 6 h, and cells collected. Total protein was extracted, FLS2 proteins immunoprecipitated using anti-HA resin, and click chemistry performed. Incorporated Alk14 was visualized by fluorescence imaging and total protein was detected by anti-HA western blotting. **b** Transgenic Arabidopsis plants conditionally expressing *avrPto* were treated with 20 μM dexamethasone to induce transgene expression. Leaves were infiltrated twice with 10 μM Alk12, 6 h after induction and 6 h before sampling. Tissue was collected 30 h after induction and total protein extracted. AvrPto was immunoprecipitated using anti-AvrPto resin and a biotin tag added using click chemistry. Streptavidin-HRP western blotting was used to detect incorporation of Alk12. Anti-AvrPto western blotting was used to verify equal amounts of protein in all samples. **c** *Nicotiana benthamiana* leaves were infiltrated with Agrobacterium strains carrying *avrPto-YFP* fusion constructs encoding the WT protein or a G2A mutant. 10 μM Alk12 was infiltrated twice, 24 h after Agrobacterium infiltration and 6 h before sampling. Tissue was collected 48 h after transformation and total protein extracted. AvrPto proteins were immunoprecipitated using anti-GFP resin and a biotin tag attached using click chemistry. Incorporated Alk12 was detected by streptavidin-HRP western blotting. The anti-GFP western blot shows relative protein levels. **d** *Nicotiana benthamiana* was used to transiently express *Pto-YFP* fusions encoding the WT protein or a G2A mutant. 10 μM Alk12 was infiltrated twice, 24 h after Agrobacterium infiltration and 6 h before sampling. Tissue was collected 48 h after transformation, total protein extracted, and Pto proteins immunoprecipitated using anti-GFP resin. A biotin tag was attached using click chemistry and incorporation of Alk12 was detected by streptavidin-HRP western blotting. Protein levels were visualized by anti-GFP western blotting

to permeate intact cells. To address these concerns, we tested if the tomato resistance protein Pto, which mediates recognition of the *Pseudomonas syringae* pv. *tomato* effector AvrPto, retains the ability to trigger programmed cell death (PCD) in the presence of the different alkyne fatty acid analogs [57–59]. We transiently expressed *Pto* together with *avrPto* or an empty vector in *Nicotiana benthamiana* leaf tissue using Agrobacterium-mediated transformation, followed by infiltration of the alkyne fatty acid analogs (Additional file 2: Figure S2A). We found that neither the myristic acid analog Alk12, the palmitic acid analog Alk14, nor the stearic acid analog Alk16 produced any spurious symptoms nor did they affect PCD in response to AvrPto, even though all of the alkyne-bearing metabolites were used at high concentrations (Additional file 2: Figure S2A), suggesting that they are suitable to study the role of protein fatty acylation in plant-pathogen interactions.

To ensure that the metabolites are able to permeate intact plant cells and label fatty acylated proteins in the context of whole leaf tissue, we transiently expressed the *avrPto* effector in *N. benthamiana* leaves and syringe-infiltrated preparations of the alkyne fatty acid analogs into the transformed leaf tissue [24, 60] (Additional file 2: Figure S2B). We chose this particular protein to test the ability of the three fatty acid analogs for cell permeation and protein incorporation because AvrPto contains a predicted dual fatty acylation motif suggesting that it is subject to both plant-mediated myristoylation and S-acylation [32, 61]. We performed whole protein extraction, affinity purified the epitope-tagged AvrPto, and attached a fluorescent dye using click chemistry as described for FLS2. We visualized incorporation of the different alkyne fatty acid analogs using fluorescence imaging and detected incorporation of all three probes, although with varying signal strength (Additional file 2: Figure S2A). This demonstrates that the alkyne fatty acid analogs are able to permeate intact leaf cells and are likely incorporated by way of innate metabolic processes, making this approach suitable for the study of protein fatty acylation in leaf tissue.

### A transgenic Arabidopsis line shows incorporation of a myristic acid analog into AvrPto

Bacterial plant pathogens employ the type III secretion system to inject effector proteins directly into the plant cell to subvert PTI signaling, ultimately rendering the host susceptible to infection [52, 62]. Spatial regulation of effectors is required for their virulence function because it ensures that they engage their intended targets and enhances the local concentration of these pathogen-derived proteins, which are likely delivered into the host cell in very small amounts [30]. Several effectors have

been shown to target the PTI machinery present at the intracellular face of the plant PM and a number of these bacterial proteins appear to hijack the host fatty acylation apparatus to access this lipophilic locale [24, 32, 34, 63]. Notably, plant-mediated fatty acylation has not been decisively demonstrated for most effectors, but rather inferred from studies showing that N-terminal glycine and/or cysteine substitutions prevent PM localization and render the effectors unable to exert their virulence function or elicit an immune response, depending on the host plant [32–34]. A recent review elaborates on the exploitation of host-mediated fatty acylation by plant pathogenic effectors [23].

The AvrPto effector promotes bacterial pathogenesis by targeting the FLS2 receptor complex in order to suppress flagellin perception [64, 65]. Like the PM-associated FLS2, AvrPto was shown to localize to the cell periphery [32, 66, 67]. It is strongly suggested that targeting of this pathogen protein to the plant PM requires post-translational host-mediated myristoylation of the glycine-2 (G2) residue in AvrPto, since the G2A mutation abolishes both PM localization and virulence function of the effector [32, 66]. To test if AvrPto is indeed myristoylated in plant cells, we took advantage of a transgenic Arabidopsis line conditionally expressing *avrPto* under control of a dexamethasone-inducible system [68] (Fig. 2b). We syringe-infiltrated leaves with the myristic acid analog Alk12 following dexamethasone treatment, extracted total protein, and immunoprecipitated AvrPto using anti-AvrPto resin. The incorporated fatty acid was then biotinylated using click chemistry and visualized by western blotting. We were able to detect a specific band of the expected size in samples treated with Alk12 and subsequently subjected to click chemistry. To control for unspecific detection in the absence of an AvrPto G2A mutant, we included samples without performing click chemistry and without infiltrating the metabolite. Anti-AvrPto western blotting was used to verify equal amounts of the effector protein in all samples (Fig. 2b). This result shows that AvrPto is myristoylated in plant cells and demonstrates that our approach can be applied to assess fatty acylation of candidate proteins stably expressed in transgenic Arabidopsis leaf tissue. However, the lack of a stable Arabidopsis line expressing a G2 point mutant of AvrPto prevented us from determining whether this modification is mediated by way of canonical G2 myristoylation or not; we address this limitation in the next section.

### Transient expression in *N. benthamiana* demonstrates that AvrPto is myristoylated at its N-terminus

To establish a higher-throughput system for validation of multiple candidate proteins, we took advantage of Agrobacterium-mediated transient gene expression in *N.*

*benthamiana.* This approach also enables the use of point mutants to validate predicted fatty acylation sites and to control for nonspecific incorporation of alkyne fatty acid analogs. We transiently expressed *avrPto* variants and infiltrated the leaf tissue with the myristic acid analog Alk12 (Fig. 2c). To counter previously observed instability of the AvrPto G2A mutant, we fused the effector to YFP in an attempt to stabilize the protein. We were able to detect click-mediated biotinylation using streptavidin with wild-type AvrPto, but no band was detected with the predicted myristoylation mutant G2A despite high protein accumulation (Fig. 2c). Thus, using the *N. benthamiana* system, we were able to extend our data obtained in Arabidopsis to conclusively show typical G2-mediated myristoylation of AvrPto through the use of a specific point mutant.

As previously mentioned, the AvrPto N-terminus contains a predicted dual fatty acylation motif, MGNICVGGSR, due to the G2 and proximal cysteine-5 (C5) residues [32, 61]. While we showed labeling of this pathogen effector with Alk12, Alk14, and Alk16 (Additional file 2: Figure S2B), we were unable to map S-acylation type modifications, by Alk14 or Alk16, to the C5 position due to instability and inconsistent labeling of the AvrPto mutant forms (data not shown). However, it is likely that AvrPto is S-acylated at the C5 position because this residue is the only cysteine present in the effector protein.

### Detection of Pto myristoylation in plant cells is greatly enhanced using metabolic labeling coupled with click chemistry

Plants have evolved intracellular surveillance mechanisms to perceive the presence and activity of pathogen effectors [69]. Detection of effectors within the host cell indicates infection by an adapted pathogen and as a result the plant activates an amplified defense response referred to as effector-trigger immunity (ETI) which is often associated with PCD [69, 70]. ETI signaling is typically mediated by nucleotide-binding leucine-rich repeat (NB-LRR) proteins that are often physically partnered with either a decoy which resembles a host protein targeted by effectors, or an actual host target [71, 72]. In either case, interactions between effectors and host proteins are sensed by the associated NB-LRRs which subsequently activate ETI [72]. Regardless of the specific mode of detection, the precise localization of these surveillance mechanisms is critical to their function and because the intracellular face of the plant PM is an area intensely attacked by effectors, many of these sensors are positioned at this crucial locale by way of lipid modifications [72–77].

Some tomato accessions rely upon the Pto kinase, acting in concert with the NB-LRR protein Prf, to recognize

the fatty acylated pathogen effector AvrPto [78–82]. Pto appears to function as a decoy that mimics the structure of the kinase domains present in PRR signaling complexes, such as that of FLS2, but in contrast to the PM-spanning receptors it is proposed to mimic, Pto lacks a transmembrane domain [78]. Previous work using a radiolabeled myristic acid feeding approach showed that Pto is myristoylated in plant cells and that the G2 residue associated with myristoylation is required for full recognition of AvrPto [60, 83]. To confirm incorporation of myristic acid with our click chemistry-based approach, we transiently expressed wild-type *Pto* and a mutant encoding the G2A substitution in *N. benthamiana* (Fig. 2d). Following the strategy used for AvrPto, we fused Pto to YFP to stabilize the G2A mutant as demonstrated by the anti-GFP western blot. We were able to detect incorporation of the myristic acid analog Alk12 into wild-type Pto using streptavidin after adding a biotin tag via click chemistry. No band was detected for the G2A mutant (Fig. 2d). This experiment confirms Pto myristoylation and extends our click chemistry-based method to assess fatty acylation to include a PM receptor, a pathogen effector, and an intracellular host resistance protein.

### Myristoylome labeling using alkyne fatty acid analogs

The strict requirement of an N-terminal glycine residue coupled with the availability of plant genome sequences has enabled the prediction of myristoylated proteins across the proteome, a collection of proteins also referred to as the myristoylome [6, 16, 17, 19]. However, methods to directly validate the predicted myristoylome in plant cells are lacking. Furthermore, bioinformatic approaches are unable to predict non-canonical myristoylation, such as the post-translational protein myristic acid modification required for the virulence function and recognition of the bacterial effector AvrPphB in plant cells [34]. To begin to address these limitations, our click chemistry-based approach could be modified and applied to enrich and investigate the myristoylome and potentially enable proteome-wide analysis of other fatty acid modifications in plants [6, 16, 17]. We performed a pilot experiment using AvrPto and transiently expressed the effector in *N. benthamiana* leaf tissue and subsequently introduced the myristic acid analog Alk12 by infiltration (Additional file 3: Figure S3). Total protein was extracted, a standard methanol/chloroform precipitation performed, and click chemistry used to biotinylate the incorporated Alk12. A second methanol/chloroform precipitation was used to remove unreacted azido-biotin prior to affinity purification using streptavidin resin. We interrogated this biotinylated material for the presence of AvrPto by anti-HA western blotting and were able to detect the effector from among the multitude of biotinylated proteins (Additional file 3: Figure S3). This result demonstrates that using the described protocol it is possible to capture a myristoylated protein from among a complex plant lysate by way of its fatty acid modification. Admittedly, the AvrPto protein used in this pilot experiment was overexpressed and future work is required to determine if this method is sufficient for the labeling and enrichment of natively expressed plant proteins and if it is amenable to subsequent mass spectrometry analysis.

## Discussion

We describe the development of a click chemistry-based method using metabolic labeling with ω-alkynyl fatty acid analogs to study the fatty acylation, especially myristoylation, of both host and pathogen proteins in the plant cell. Our data directly demonstrate that the FLS2 receptor is S-acylated and the AvrPto effector that targets this sensor protein is subject to myristoylation and possibly S-acylation, supporting previous findings that strongly suggested these proteins were subject to such modifications in plants [20, 32]. Using our approach we also recapitulated an experiment demonstrating the myristoylation of Pto, the resistance protein responsible for recognition of AvrPto, that was initially performed with radiolabeled myristic acid and we were able to reduce the exposure time required to detect this fatty acid modification from a month to less than a minute [60]. Notably, these results were obtained using a combination of biotin and fluorescent reporters from an array of plant-based expression systems, including transiently transformed Arabidopsis protoplasts, stably transformed Arabidopsis plants, and transiently transformed *N. benthamiana* leaf tissue.

There are several examples demonstrating that feeding cultured cells fatty acid analogs results in their incorporation into cellular proteins through native metabolic mechanisms, without any obvious disruption to cellular processes [42, 44]. Metabolic labeling of proteins with 'clickable' fatty acid analogs has several advantages over the use of radiolabeled fatty acids, GC–MS approaches, and the ABE method for investigating protein fatty acylation. Proteins metabolically labeled with these bioorthogonal analogs can be selectively and covalently modified with a variety of secondary tags via click chemistry. These secondary tags can include various affinity purification groups, such as biotin, or fluorescent reporter dyes. The click-mediated addition of affinity purification tags enables the capture and analysis of proteins modified by a given fatty acid and unlike the ABE and GC–MS approaches this technique can be applied to the study of myristoylated proteins. A further advantage over the ABE method is the decisive nature of the

labeling offered with the fatty acid analogs. The direct tagging of proteins metabolically labeled with a given fatty acid analog, rather than the removal and replacement of all protein thioesters, can avoid much of the ambiguity and false positives associated with the ABE approach [35, 39–41].

It should be noted that the ABE method does not require feeding fatty acids to the cells or organism of interest, which is a disadvantage of metabolic labeling approaches using either radiolabeled or ω-alkynyl/-azido fatty acid analogs. However, these feeding approaches make it possible to perform pulse-chase experiments, enabling the study of S-acylation turnover dynamics which have been shown to regulate plant protein function [11, 41]. Therefore, the methods described here have the potential to enable dynamic protein S-acylation analysis in plants without the need for radioactive materials [40]. The incubation period is critical to the success of all labeling experiments and may have to be adjusted depending on the fatty acid analog, protein of interest, and metabolism of the system under investigation [40]. The major advantages of the clickable fatty acid analogs over radiolabeled fatty acids, beyond enabling proteome-wide enrichment of proteins modified by a particular form of fatty acylation, are the nonradioactive nature of these reagents and the signal strength of the click-compatible biotin and fluorescent reporters. A study comparing the detection of protein myristoylation using [3H]-myristic acid versus that produced by ω-azido myristic acid, with subsequent biotinylation, showed that the latter produced signal intensities up to one million times stronger than that of the tritiated fatty acid [84]. Similarly, using our technique we were able to detect myristoylation of Pto transiently expressed in *N. benthamiana* with exposure times of less than one minute, in contrast to the month-long exposure time required to see the signal for the fatty acylation of a similarly expressed and immunoprecipitated Pto with tritiated myristic acid [60].

Unlike myristoylation, S-acylation is a reversible modification mainly due to the thioester bond between the cysteine side chain and the fatty acid, which is less stable than the amide linkage responsible for coupling myristate to an N-terminal glycine [40, 85]. The strong amide attachment of the Alk12 myristic acid analog as a result of myristoylation-type modifications provides a stable handle for protein purification and detection, which has worked very reliably in our hands. In contrast, the labile nature of S-acylation requires more delicate handling, particularly during elution steps following enrichment via immunoprecipitation procedures [40, 85]. Whereas myristoylation labeling was resistant to 10 % 2-mercaptoethanol in the Laemmli sample buffer, in instances analyzing S-acylation the 2-mercaptoethanol concentrations were lowered to 0.1 % to preserve the thioester bonds [26, 40, 85]. We found that the use of fluorescent dyes is preferable for the study of S-acylation because it allows rapid in-gel detection and does not require blotting of the labeled proteins, a process that can lead to thioester hydrolysis and loss of the reporter molecule [40, 42]. Even so, biotin-based reporters remain an attractive option because western blotting is highly sensitive and the materials are more readily accessible compared to fluorescent dye reagents that are expensive and require somewhat specialized scanners for detection.

Metabolic labeling methods, such as the one we present here, are generally acknowledged to avoid the false positive problems inherent to ABE-type approaches; however, labeling also has the potential for false positives. The fatty acid analogs can be metabolized into the cellular lipid pools if the labeling period is too long, resulting in non-target fatty acids possessing the alkyne moiety which can yield false positives [40, 44]. For this reason it is imperative to determine the fatty acid analog incubation period for a given protein and/or plant system and note that this period might differ considerably from the times compatible with the proteins in our particular study [45, 50]. It has also been reported that Alk12 and Alk14 can participate in both myristoylation and S-acylation, which was attributed to a lack of specificity in the fatty acylation machinery rather than metabolism of the fatty acid analogs [42]. Alk16 on the other hand seems to more specifically label S-acylated proteins and might be a better choice for detecting this specific form of fatty acid modification [42, 86]. Another potential source for false positives when using labeling approaches is the addition of fatty acids to non-target amino acids. The most notable example of this phenomenon is the labeling of the G2A mutant of the mammalian membrane-associated non-tyrosine protein kinase Fyn, the native form of which is known to be subject to N-terminal dual fatty acylation, with both radioactive and alkyne bearing myristic acid analogs [42]. We also observed some instances of non-target labeling, primarily when working with the palmitic acid analog Alk14 and the stearic acid analog Alk16, which could be attributable to any of the phenomena described above (data not shown). It should be noted that the unique sensitivity of S-acyl adducts to treatments with strong reducing agents and nucleophiles such as 2-mercaptoethanol, dithiothreitol, and hydroxylamine can be leveraged to address some issues with ambiguous labeling [42]. In our experience, detection of myristoylation with the Alk12 reagent has been very specific because the G2A mutants reliably abolished labeling by the fatty acid analog. Another potential problem with labeling approaches is that the presumed overabundance of the fatty acid analogs in these feeding assays leads to unspecific incorporation at non-target residues. However, our results with FLS2 would suggest that this may not be an issue because the C830S,C831S mutant appeared to abolish incorporation of the palmitic acid analog Alk14 despite being properly localized at the PM [20]. Toxicity of these fatty acid analogs can be a concern and should be evaluated for the plant system under investigation. For example, it was found that analogs of lauric acid bearing a terminal alkyne similar to the alkyne fatty acid analogs used in our study inhibited a lauric acid ω-hydroxylase in microsome preparations from *Vicia sativa* [87]. In our experiments with Arabidopsis protoplasts we observed strong phytotoxicity with the azide fatty acid analog Az12, however, the alkyne fatty acid analog Alk12 showed no adverse effects and for this reason we decided to use the alkyne-functionalized analogs for our work. It should be noted that the reagents were prepared in accordance with their manufacturers' instructions, which called for the use of different solvents. The Az12 was prepared as a 40 mM stock in dimethyl sulfoxide (DMSO) and the Alk12 as a 50 mM stock in ethanol. The DMSO used for the Az12 stock solution could contribute to the observed toxicity, but in all treatments the stock solution was diluted at least 1000-fold, meaning that the plant cells were maximally exposed to 0.1 % DMSO. While we believe that the Az12 is most likely responsible for the observed toxicity to the protoplasts, we cannot rule out contributions from the DMSO.

Finally, our experiments were performed with overexpressed proteins and it will likely be more difficult to detect fatty acylation of natively expressed proteins. To test candidate proteins, transient overexpression with a commercial epitope tag is ideal because this enables the use of point mutants and allows for easy purification and concentration of the proteins prior to performing click chemistry. In our experience, amino acid substitutions that prevent protein fatty acylation and enable mapping of the modification to specific residues can result in protein instability and it can be helpful to employ fusions with green fluorescent protein variants to stabilize problematic substitution mutant proteins. In instances where the study of natively expressed proteins is desirable specific antibodies can be used to purify and concentrate the protein of interest, providing optimal buffer conditions for an efficient click reaction.

## Conclusions

We described the development and application of a metabolic labeling approach coupled with click chemistry to quickly and easily determine fatty acylation, especially myristoylation, of candidate proteins in plant cells. Our method can reduce the time required to assess protein fatty acid modifications from months to less than a week and relies on neither radioactivity nor mass spectrometry. We demonstrated the ability of our approach to determine the fatty acylation status of three representative proteins involved in plant-pathogen interactions using a variety of expression systems. Although presently most effective for determining protein myristoylation, this technique promises to provide mechanistic details of the molecular tactics used at the host plasma membrane in the battle between plants and pathogens. In addition, we expect that with some modifications this approach will be broadly applicable for the study of protein fatty acylation in plants and will shed light on new mechanisms not only involving plant-pathogen interactions but the wider field of plant biology.

## Methods

### Plant material
Seeds of *Arabidopsis thaliana* accession Columbia (Col-0) or the derived transgenic line conditionally expressing *avrPto* [68] were suspended in 0.1 % agarose and cold-stratified for 3 days at 4 °C. The plants were grown in a controlled environment chamber with 8 h light and 16 h dark periods at 22 °C and 20 °C, respectively, with 60 % relative humidity for 6 weeks. *Nicotiana benthamiana* accession Nb-1 [88] was grown in a controlled environment chamber with a light/dark cycle of 16 h and 8 h, respectively, with 65 % relative humidity and temperatures of 24 °C during light and 22 °C during dark periods for 4–5 weeks.

### Cloning
To generate the Gateway entry clones, complete open reading frames (ORFs) without the stop codons were amplified with Phusion DNA polymerase (cat. no. F-530S, Thermo Scientific) from existing plasmids. The ORFs were blunt-end ligated into the SmaI (cat. no. R0141S, New England Biolabs) site of pJLSmart [89] or pJM51 [90] with T4 DNA ligase (cat. no. M0202S, New England Biolabs). Point mutations were introduced using complementary custom DNA oligonucleotides (Integrated DNA Technologies) following standard protocols (e.g. Stratagene QuikChange site-directed mutagenesis kit). Entry clones were recombined into destination vectors using the LR Clonase II enzyme mix (cat. no. 11791-020, Invitrogen) following the manufacturer's protocol. Destination vectors used were HBT95 [66] for protoplast expression and the pGWB series [91] for Agrobacterium-mediated transformation. All constructs were control digested with BsrGI (cat. no. R0575S, New England Biolabs) and sequence-verified prior to use. Sequencing services were provided by the Biotechnology Resource Center at Cornell University. The pBTEX constructs have

been described previously [92, 93]. All vectors and constructs are listed in Additional file 5: Table S1.

### Protoplast isolation and transformation
Arabidopsis protoplasts were prepared and transformed as previously described [94, 95]. Briefly, the epidermis on the abaxial side of fully expanded leaves was peeled off and the leaves floated in protoplast isolation medium. The protoplasts were collected, washed, and the cell density adjusted. Plasmid DNA was added and the protoplasts were transformed by PEG-calcium transfection. Protoplasts were washed again and resuspended in the presence of the palmitic acid analog Alk14 (cat. no. 13266, Cayman Chemical) at a final concentration of 10 μM. Cells were incubated for 6 h, collected, and stored at −80 °C until further processing. For the comparison between azide and alkyne fatty acid analogs, transformed protoplasts were resuspended in the presence of the myristic acid analog Az12 (cat. no. C10268, Invitrogen) or Alk12 (cat. no. 13267, Cayman Chemical) at the indicated final concentrations. All fatty acid analogs were prepared following their manufacturers' protocols and their structures a shown in Additional file 4: Figure S4. Cells were incubated overnight, collected, and protein levels analyzed by western blotting.

### Conditional expression of *avrPto* in transgenic Arabidopsis
Transgenic Arabidopsis conditionally expressing *avrPto* under control of a dexamethasone-inducible promoter [68] were sprayed with 20 μM dexamethasone (cat. no. D1756, Sigma-Aldrich) in 0.1 % ethanol with 0.01 % Silwet L-77 (cat. no. VIS-01, Lehle Seeds) to induce gene expression. Leaves were infiltrated twice with 10 μM myristic acid analog Alk12, 6 h after induction and 6 h before sampling. Tissue was collected 30 h after induction and stored at −80 °C until further processing.

### Agrobacterium-mediated transient expression
*Agrobacterium tumefaciens* strain GV3101 with helper plasmid pMP90 [96] was transformed with the pGWB constructs; the pBTEX constructs had previously been moved into *A. tumefaciens* strain GV2260 [93]. Confirmed strains were grown on lysogeny broth (LB) plates with appropriate antibiotics at 30 °C for 36–48 h. Bacteria were then scraped from plates and resuspended in infiltration buffer containing 10 mM MgCl$_2$, 10 mM MES pH 5.7, and 200 μM acetosyringone (cat. no. D134406, Sigma-Aldrich). The OD$_{600}$ was adjusted to a final density of 0.3 for each strain and the bacteria incubated for at least 1 h at room temperature. Leaves of *N. benthamiana* were infiltrated with needleless syringes and the plants placed on a shaded growth chamber shelf. Leaves were

infiltrated twice with 10 µM Alk12, 24 h after Agrobacterium infiltration and 6 h before sampling. Tissue samples were collected 48 h after transformation. For the cell death assay, leave tissue was infiltrated once with 50 µM Alk12, Alk14, Alk16 (cat. no. 90270, Cayman Chemical), or buffer 24 h after Agrobacterium infiltration.

### Click reaction

Leaf tissue was ground with a TissueLyser II (Qiagen) and proteins extracted in 'RIPA' buffer containing $1 \times$ PBS pH 7.4, 1 % v/v Triton X-100 (cat. no. X100, Sigma-Aldrich), 0.5 % w/v sodium deoxycholate, and 0.1 % w/v SDS, with EDTA-free protease inhibitor (cat. no. 05892791001, Roche Diagnostics). Protoplasts were lysed by brief vortexing in RIPA buffer with EDTA-free protease inhibitor. Affinity resin was added to the cleared supernatant and immunoprecipitation performed. FLS2-HA was purified with anti-HA (cat. no. E6779, Sigma-Aldrich); YFP fusions were purified with anti-GFP (cat. no. gta-10, ChromoTek); and untagged AvrPto was purified with custom anti-AvrPto antibody [32] coupled to protein A resin (cat. no. P6486, Sigma-Aldrich). Agarose beads were resuspended in RIPA buffer and the following components added for the click reaction: 500 µM BTTP ligand, 250 µM $CuSO_4$, 2 mM sodium ascorbate, and 100 µM azide tag. The reaction was incubated for 1 h at room temperature or overnight at 4 °C and the beads were washed and resuspended in Laemmli sample buffer for protein detection. A detailed protocol is provided in Additional file 6: Methods S1. For the myristoylome pilot experiment, total protein was extracted in $1\times$ PBS pH 7.4 with 4 % w/v SDS and EDTA-free protease inhibitor. Standard methanol/chloroform precipitation was used to purify and concentrate the proteins [97]. Click chemistry with the crude protein extract was performed in $1\times$ PBS pH 7.4 with 4 % w/v SDS in the presence of 1 mM $CuSO_4$ to attach a biotin tag to AvrPto. We found that higher concentrations of the copper catalyst are required for efficient click reactions in crude protein extracts. A second methanol/chloroform precipitation was used to clean the sample, proteins were resuspended in $1\times$ PBS pH 7.4 with 4 % w/v SDS and EDTA-free protease inhibitor, diluted with $1\times$ PBS pH 7.4 to reduce SDS concentration to around 0.7 %, and the myristoylated proteins enriched using streptavidin resin (cat. no. 20349, Thermo Scientific). The BTTP ligand 3-[4-({bis[(1-tert-butyl-1H-1,2,3-triazol-4-yl)methyl]amino}methyl)-1H-1,2,3-triazol-1-yl] propanol was a gift from Dr. Frank C. Schroeder (Boyce Thompson Institute for Plant Research and Department of Chemistry and Chemical Biology, Cornell University) and its structure is shown in Additional file 4: Figure S4. BTTP is not commercially available at the time of writing, but it can be obtained from the Chemical Biology

Core Facility of the Albert Einstein College of Medicine (www.einstein.yu.edu/research/shared-facilities/chemical-biology/Ligands-for-CuAAC). The $N_3$-biotin reagent biotin-PEG3-azide, used as both a reporter and affinity purification handle, was purchased from Click Chemistry Tools (cat. no. AZ104-10). The infrared fluorescent reporter IRDye 800CW azide was obtained from LI-COR Biosciences (cat. no. 929-60000).

### Protein detection

All samples were brought up in Laemmli sample buffer, with Orange G (cat. no. O3756, Sigma-Aldrich) substituted for bromophenol blue for the fluorescence imaging experiments to minimize signal interference. Gel electrophoresis and western blotting was performed following standard protocols (e.g. Bio-Rad bulletin 6040 and 2895, respectively). Detection of attached infrared fluorescent dye was performed using an Odyssey infrared imager (LI-COR Biosciences) after fixing the gel by incubation in 40 % methanol and 10 % acetic acid protected from light with gentle shaking overnight at room temperature. In some instances multiple incubations in fixing buffer were required to remove background signal. Attached biotin was detected using streptavidin-HRP (cat. no. S-911, Invitrogen); untagged AvrPto was detected using custom anti-AvrPto antibody followed by anti-rabbit-HRP (cat. no. W4011, Promega); FLS2-HA and AvrPto-HA were detected using anti-HA-HRP (cat. no. 12013819001, Roche Diagnostics); YFP fusion proteins were detected using anti-GFP (cat. no. 11814460001, Roche Diagnostics) followed by anti-mouse-HRP (cat. no. sc-2005, Santa Cruz Biotechnology); and YFP-FLAG was detected with anti-FLAG-HRP (cat. no. A8592, Sigma-Aldrich).

### Additional files

**Additional file 1: Figure S1.** Azide fatty acid analogs, but not alkyne fatty acid analogs, interfere with cellular functions. **(A)** Arabidopsis protoplasts were transformed with FLAG epitope-tagged *YFP* and treated with different concentrations of the azide fatty acid analog Az12. Cells were incubated overnight and total protein extracted. Anti-FLAG western blotting was used to detect YFP accumulation. Coomassie brilliant blue (CBB) stain was used to visualize total protein and demonstrate equal loading. NT, not transformed. Black dividing lines indicate removal of irrelevant lanes from the blot and gel images. **(B)** Arabidopsis protoplasts were transformed with FLAG epitope-tagged *YFP* and treated with different concentrations of the alkyne fatty acid analog Alk12. Cells were incubated overnight and total protein extracted. Anti-FLAG western blotting was used to detect YFP accumulation. CBB stain was used to visualize total protein and demonstrate equal loading.

**Additional file 2: Figure S2.** Alkyne fatty acid analogs do not interfere with immunity mechanisms and are incorporated in fatty acylated proteins in the context of intact plant leaf tissue. **(A)** *Nicotiana benthamiana* leaves were infiltrated with Agrobacterium strains carrying *Pto* and empty vector (EV) or *avrPto*. 50 µM Alk12, Alk14, Alk16, or buffer were infiltrated 24 h after Agrobacterium infiltration. Plants were monitored for programmed cell death and pictures taken 2 days after transformation.

**(B)** *Nicotiana benthamiana* was used to transiently express HA epitope-tagged *avrPto*. 10 μM Alk12, Alk14, Alk16, or buffer was infiltrated twice, 24 h after Agrobacterium infiltration and 6 h before sampling. Tissue was collected 48 h after transformation, total protein extracted, AvrPto immunoprecipitated using anti-HA resin, and a fluorescent tag added using click chemistry. Incorporated alkyne fatty acid analogs were visualized by fluorescence imaging and total protein was detected by anti-HA western blotting. Black dividing lines indicate removal of irrelevant lanes from the blot and gel images.

**Additional file 3: Figure S3.** Protein capture by means of myristoylation provides a potential method for the enrichment and investigation of the plant myristoylome. **(A)** Modified experimental scheme to capture and enrich myristoylated proteins using AvrPto as a test protein. **(B)** *Nicotiana benthamiana* was used to transiently express HA epitope-tagged *avrPto*. 50 μM Alk12 was infiltrated twice, 24 h after Agrobacterium infiltration and 6 h before sampling. Tissue was collected 48 h after transformation, total protein extracted, and a biotin tag added using click chemistry. Streptavidin affinity purification was used to enrich biotinylated proteins and AvrPto was detected using anti-HA western blotting. Input shows AvrPto levels before affinity purification.

**Additional file 4: Figure S4.** Structures of the fatty acid analogs and ligands used in this study. **(A)** Myristic acid analog Alk12. **(B)** Palmitic acid analog Alk14. **(C)** Stearic acid analog Alk16. **(D)** BTTP ligand 3-[4-({bis[(1-tert-butyl-1H-1,2,3-triazol-4-yl)methyl]amino}methyl)-1H-1,2,3-triazol-1-yl]propanol.

**Additional file 5: Table S1.** Vectors and plasmids used in this study.

**Additional file 6: Methods S1.** Detailed click reaction protocol.

## Authors' contributions

PCB, SS, SRH, and GBM designed the research; PCB, SS, SRH, CMK, and SDD performed the research; BH contributed reagents; PCB, SS, and GBM wrote the article. All authors read and approved the final manuscript.

## Author details

[1] Boyce Thompson Institute for Plant Research, Ithaca, NY 14853, USA. [2] Plant Pathology and Plant–Microbe Biology Section, School of Integrative Plant Science, Cornell University, Ithaca, NY 14853, USA. [3] Department of Chemistry and Chemical Biology, Cornell University, Ithaca, NY 14853, USA. [4] Present Address: Monsanto Company, St. Louis, MO 63141, USA. [5] Present Address: College of Pharmacy, Guiyang Medical University, Guiyang 550004, Guizhou, China.

## Acknowledgements

We thank Dr. Frank C. Schroeder, Dr. Inish M. O'Doherty, Joshua A. Baccile, and Jason Hoki for help with the click chemistry; Dr. Johannes Mathieu for providing the *FLS2* entry clone and helpful discussions; Dr. Brendan K. Riely for contributing the *Pto-YFP* strains; and Diane Dunham, Christopher J. D'Angelo, and Paige L. Reeves for their help with sample preparation and plant care. We are also grateful to the reviewers for their helpful suggestions. This work was supported by the U.S. Department of Agriculture—National Institute of Food and Agriculture (Grant No. USDA-NIFA 2010-65108-20503 to GBM), the National Science Foundation (Grant No. IOS-1451754 to GBM), the Human Frontier Science Program (Grant No. HFSP-LT000608/2011-L to PCB), and the Triad Foundation (to PCB, SRH, and GBM).

## Competing interests

The authors declare that they have no competing interests.

## References

1. Resh MD. Fatty acylation of proteins: new insights into membrane targeting of myristoylated and palmitoylated proteins. Biochim Biophys Acta. 1999;1451:1–16.
2. Hemsley PA. The importance of lipid modified proteins in plants. New Phytol. 2015;205:476–89.
3. Wright MH, Heal WP, Mann DJ, Tate EW. Protein myristoylation in health and disease. J Chem Biol. 2010;3:19–35.
4. Martin DDO, Beauchamp E, Berthiaume LG. Post-translational myristoylation: fat matters in cellular life and death. Biochimie. 2011;93:18–31.
5. Qi Q, Rajala RV, Anderson W, Jiang C, Rozwadowski K, Selvaraj G, et al. Molecular cloning, genomic organization, and biochemical characterization of myristoyl-CoA:protein N-myristoyltransferase from *Arabidopsis thaliana*. J Biol Chem. 2000;275:9673–83.
6. Boisson B, Giglione C, Meinnel T. Unexpected protein families including cell defense components feature in the N-myristoylome of a higher eukaryote. J Biol Chem. 2003;278:43418–29.
7. Pierre M, Traverso JA, Boisson B, Domenichini S, Bouchez D, Giglione C, et al. N-myristoylation regulates the SnRK1 pathway in *Arabidopsis*. Plant Cell. 2007;19:2804–21.
8. Renna L, Stefano G, Majeran W, Micalella C, Meinnel T, Giglione C, et al. Golgi traffic and integrity depend on N-myristoyltransferase 1 in *Arabidopsis*. Plant Cell. 2013;25:1756–73.
9. Hurst CH, Hemsley PA. Current perspective on protein S-acylation in plants: more than just a fatty anchor? J Exp Bot. 2015;66:1599–606.
10. Aicart-Ramos C, Valero RA, Rodriguez-Crespo I. Protein palmitoylation and subcellular trafficking. Biochim Biophys Acta. 2011;1808:2981–94.
11. Sorek N, Poraty L, Sternberg H, Bar E, Lewinsohn E, Yalovsky S. Activation status-coupled transient S acylation determines membrane partitioning of a plant Rho-related GTPase. Mol Cell Biol. 2007;27:2144–54.
12. Batistic O, Sorek N, Schültke S, Yalovsky S, Kudla J. Dual fatty acyl modification determines the localization and plasma membrane targeting of CBL/CIPK Ca$^{2+}$ signaling complexes in Arabidopsis. Plant Cell. 2008;20:1346–62.
13. Batistic O, Rehers M, Akerman A, Schlücking K, Steinhorst L, Yalovsky S, et al. S-acylation-dependent association of the calcium sensor CBL2 with the vacuolar membrane is essential for proper abscisic acid responses. Cell Res. 2012;22:1155–68.
14. Batistic O. Genomics and localization of the Arabidopsis DHHC-cysteine-rich domain S-acyltransferase protein family. Plant Physiol. 2012;160:1597–612.
15. Shahinian S, Silvius JR. Doubly-lipid-modified protein sequence motifs exhibit long-lived anchorage to lipid bilayer membranes. Biochemistry. 1995;34:3813–22.
16. Maurer-Stroh S, Eisenhaber B, Eisenhaber F. N-terminal N-myristoylation of proteins: prediction of substrate proteins from amino acid sequence. J Mol Biol. 2002;317:541–57.
17. Podell S, Gribskov M. Predicting N-terminal myristoylation sites in plant proteins. BMC Genom. 2004;5:37.
18. Martinez A, Traverso JA, Valot B, Ferro M, Espagne C, Ephritikhine G, et al. Extent of N-terminal modifications in cytosolic proteins from eukaryotes. Proteomics. 2008;8:2809–31.
19. Traverso JA, Meinnel T, Giglione C. Expanded impact of protein N-myristoylation in plants. Plant Signal Behav. 2008;3:501–2.
20. Hemsley PA, Weimar T, Lilley K, Dupree P, Grierson C. Palmitoylation in plants: new insights through proteomics. Plant Signal Behav. 2013;8.
21. Smotrys JE, Linder ME. Palmitoylation of intracellular signaling proteins: regulation and function. Annu Rev Biochem. 2004;73:559–87.
22. Sorek N, Bloch D, Yalovsky S. Protein lipid modifications in signaling and subcellular targeting. Curr Opin Plant Biol. 2009;12:714–20.
23. Boyle PC, Martin GB. Greasy tactics in the plant-pathogen molecular arms race. J Exp Bot. 2015;66:1607–16.
24. Nimchuk Z, Marois E, Kjemtrup S, Leister RT, Katagiri F, Dangl JL. Eukaryotic fatty acylation drives plasma membrane targeting and enhances function of several type III effector proteins from *Pseudomonas syringae*. Cell. 2000;101:353–63.
25. Dean P. Functional domains and motifs of bacterial type III effector proteins and their roles in infection. FEMS Microbiol Rev. 2011;35:1100–25.
26. Hicks SW, Charron G, Hang HC, Galán JE. Subcellular targeting of *Salmonella* virulence proteins by host-mediated S-palmitoylation. Cell Host Microbe. 2011;10:9–20.
27. Méresse S. Is host lipidation of pathogen effector proteins a general virulence mechanism? Front Microbiol. 2011;2:73.
28. Feng F, Zhou J-M. Plant-bacterial pathogen interactions mediated by type III effectors. Curr Opin Plant Biol. 2012;15:469–76.

29. Geissler B. Bacterial toxin effector-membrane targeting: outside in, then back again. Front Cell Infect Microbiol. 2012;2:75.

30. Hicks SW, Galán JE. Exploitation of eukaryotic subcellular targeting mechanisms by bacterial effectors. Nat Rev Microbiol. 2013;11:316–26.

31. Ivanov SS, Roy C. Host lipidation: a mechanism for spatial regulation of Legionella effectors. Curr Top Microbiol Immunol. 2013;376:135–54.

32. Shan L, Thara VK, Martin GB, Zhou JM, Tang X. The Pseudomonas AvrPto protein is differentially recognized by tomato and tobacco and is localized to the plant plasma membrane. Plant Cell. 2000;12:2323–38.

33. Robert-Seilaniantz A, Shan L, Zhou J-M, Tang X. The Pseudomonas syringae pv. tomato DC3000 type III effector HopF2 has a putative myristoylation site required for its avirulence and virulence functions. Mol Plant Microbe Interact. 2006;19:130–8.

34. Dowen RH, Engel JL, Shao F, Ecker JR, Dixon JE. A family of bacterial cysteine protease type III effectors utilizes acylation-dependent and -independent strategies to localize to plasma membranes. J Biol Chem. 2009;284:15867–79.

35. Hannoush RN, Sun J. The chemical toolbox for monitoring protein fatty acylation and prenylation. Nat Chem Biol. 2010;6:498–506.

36. Sorek N, Yalovsky S. Analysis of protein S-acylation by gas chromatography-coupled mass spectrometry using purified proteins. Nat Protoc. 2010;5:834–40.

37. Farnsworth C, Casey P, Howald W, Glomset J, Gelb M. Structural characterization of prenyl groups attached to proteins. Methods. 1990;1:231–40.

38. Drisdel RC, Green WN. Labeling and quantifying sites of protein palmitoylation. Biotechniques. 2004;36:276–85.

39. Roth AF, Wan J, Bailey AO, Sun B, Kuchar JA, Green WN, et al. Global analysis of protein palmitoylation in yeast. Cell. 2006;125:1003–13.

40. Martin BR. Nonradioactive analysis of dynamic protein palmitoylation. Curr Protoc Protein Sci. 2013;73(Unit14):15.

41. Zhou B, An M, Freeman MR, Yang W. Technologies and challenges in poteomic analysis of protein S-acylation. J Proteomics Bioinform. 2014;7:256–63.

42. Charron G, Zhang MM, Yount JS, Wilson J, Raghavan AS, Shamir E, et al. Robust fluorescent detection of protein fatty-acylation with chemical reporters. J Am Chem Soc. 2009;131:4967–75.

43. Martin BR, Cravatt BF. Large-scale profiling of protein palmitoylation in mammalian cells. Nat Methods. 2009;6:135–8.

44. Yap MC, Kostiuk MA, Martin DDO, Perinpanayagam MA, Hak PG, Siddam A, et al. Rapid and selective detection of fatty acylated proteins using omega-alkynyl-fatty acids and click chemistry. J Lipid Res. 2010;51:1566–80.

45. Hannoush RN. Profiling cellular myristoylation and palmitoylation using ω-alkynyl fatty acids. Methods Mol Biol. 2012;800:85–94.

46. Kaschani F, Verhelst SHL, van Swieten PF, Verdoes M, Wong C-S, Wang Z, et al. Minitags for small molecules: detecting targets of reactive small molecules in living plant tissues using 'click chemistry'. Plant J. 2009;57:373–85.

47. Tobimatsu Y, Van de Wouwer D, Allen E, Kumpf R, Vanholme B, Boerjan W, et al. A click chemistry strategy for visualization of plant cell wall lignification. Chem Commun. 2014;50:12262–5.

48. Bourge M, Fort C, Soler M-N, Satiat-Jeunemaître B, Brown SC. A pulse-chase strategy combining click-EdU and photoconvertible fluorescent reporter: tracking Golgi protein dynamics during the cell cycle. New Phytol. 2015;205:938–50.

49. Dutilleul C, Ribeiro I, Blanc N, Nezames CD, Deng XW, Zglobicki P, et al. ASG2 is a farnesylated DWD protein that acts as ABA negative regulator in Arabidopsis. Plant, Cell Environ. 2016;39:185–98.

50. Yount JS, Zhang MM, Hang HC. Visualization and identification of fatty acylated proteins using chemical reporters. Curr Protoc Chem Biol. 2011;3:65–79.

51. Zipfel C. Plant pattern-recognition receptors. Trends Immunol. 2014;35:345–51.

52. Jones JDG, Dangl JL. The plant immune system. Nature. 2006;444:323–9.

53. Felix G, Duran JD, Volko S, Boller T. Plants have a sensitive perception system for the most conserved domain of bacterial flagellin. Plant J. 1999;18:1–12.

54. Gómez-Gómez L, Boller T. FLS2: an LRR receptor-like kinase involved in the perception of the bacterial elicitor flagellin in Arabidopsis. Mol Cell. 2000;5:1003–11.

55. Hemsley PA, Taylor L, Grierson CS. Assaying protein palmitoylation in plants. Plant Methods. 2008;4:2–7.

56. Hemsley PA, Weimar T, Lilley KS, Dupree P, Grierson CS. A proteomic approach identifies many novel palmitoylated proteins in Arabidopsis. New Phytol. 2012;197:805–14.

57. Scofield S, Tobias C, Rathjen J, Chang J, Lavelle D, Michelmore R, et al. Molecular basis of gene-for-gene specificity in bacterial speck disease of tomato. Science. 1996;274:2063–5.

58. Tang X, Frederick R, Zhou J, Halterman D, Jia Y, Martin GB. Initiation of plant disease resistance by physical interaction of AvrPto and Pto kinase. Science. 1996;274:2060–3.

59. Martin GB, Bogdanove AJ, Sessa G. Understanding the functions of plant disease resistance proteins. Annu Rev Plant Biol. 2003;54:23–61.

60. de Vries JS, Andriotis VME, Wu A-J, Rathjen JP. Tomato Pto encodes a functional N-myristoylation motif that is required for signal transduction in Nicotiana benthamiana. Plant J. 2006;45:31–45.

61. Maurer-Stroh S, Eisenhaber F. Myristoylation of viral and bacterial proteins. Trends Microbiol. 2004;12:178–85.

62. Dou D, Zhou J-M. Phytopathogen effectors subverting host immunity: different foes, similar battleground. Cell Host Microbe. 2012;12:484–95.

63. Zhou J, Wu S, Chen X, Liu C, Sheen J, Shan L, et al. The Pseudomonas syringae effector HopF2 suppresses Arabidopsis immunity by targeting BAK1. Plant J. 2014;77:235–45.

64. Shan L, He P, Li J, Heese A, Peck SC, Nürnberger T, et al. Bacterial effectors target the common signaling partner BAK1 to disrupt multiple MAMP receptor-signaling complexes and impede plant immunity. Cell Host Microbe. 2008;4:17–27.

65. Xiang T, Zong N, Zou Y, Wu Y, Zhang J, Xing W, et al. Pseudomonas syringae effector AvrPto blocks innate immunity by targeting receptor kinases. Curr Biol. 2008;18:74–80.

66. He P, Shan L, Lin N-C, Martin GB, Kemmerling B, Nürnberger T, et al. Specific bacterial suppressors of MAMP signaling upstream of MAPKKK in Arabidopsis innate immunity. Cell. 2006;125:563–75.

67. Göhre V, Spallek T, Häweker H, Mersmann S, Mentzel T, Boller T, et al. Plant pattern-recognition receptor FLS2 is directed for degradation by the bacterial ubiquitin ligase AvrPtoB. Curr Biol. 2008;18:1824–32.

68. Hauck P, Thilmony R, He SY. A Pseudomonas syringae type III effector suppresses cell wall-based extracellular defense in susceptible Arabidopsis plants. Proc Natl Acad Sci. 2003;100:8577–82.

69. Dodds PN, Rathjen JP. Plant immunity: towards an integrated view of plant-pathogen interactions. Nat Rev Genet. 2010;11:539–48.

70. Oh C-S, Martin GB. Effector-triggered immunity mediated by the Pto kinase. Trends Plant Sci. 2011;16:132–40.

71. van der Hoorn RAL, Kamoun S. From guard to decoy: a new model for perception of plant pathogen effectors. Plant Cell. 2008;20:2009–17.

72. Qi D, Innes RW. Recent advances in plant NLR structure, function, localization, and signaling. Front Immunol. 2013;4:348.

73. Kim MG, da Cunha L, McFall AJ, Belkhadir Y, DebRoy S, Dangl JL, et al. Two Pseudomonas syringae type III effectors inhibit RIN4-regulated basal defense in Arabidopsis. Cell. 2005;121:749–59.

74. Takemoto D, Jones DA. Membrane release and destabilization of Arabidopsis RIN4 following cleavage by Pseudomonas syringae AvrRpt2. Mol Plant Microbe Interact. 2005;18:1258–68.

75. Gao Z, Gao Z, Chung E-H, Eitas TK, Dangl JL. Plant intracellular innate immune receptor Resistance to Pseudomonas syringae pv maculicola 1 (RPM1) is activated at, and functions on, the plasma membrane. Proc Natl Acad Sci. 2011;108:7619–24.

76. Qi D, DeYoung BJ, Innes RW. Structure-function analysis of the coiled-coil and leucine-rich repeat domains of the RPS5 disease resistance protein. Plant Physiol. 2012;158:1819–32.

77. Takemoto D, Rafiqi M, Hurley U, Lawrence GJ, Bernoux M, Hardham AR, et al. N-terminal motifs in some plant disease resistance proteins function in membrane attachment and contribute to disease resistance. Mol Plant Microbe Interact. 2012;25:379–92.

78. Martin GB, Brommonschenkel SH, Chunwongse J, Frary A, Ganal MW, Spivey R, et al. Map-based cloning of a protein kinase gene conferring disease resistance in tomato. Science. 1993;262:1432–6.

79. Pedley KF, Martin GB. Molecular basis of Pto-mediated resistance to bacterial speck disease in tomato. Annu Rev Phytopathol. 2003;41:215–43.

80. Mucyn TS, Clemente A, Andriotis VME, Balmuth AL, Oldroyd GED, Staskawicz BJ, et al. The tomato NBARC-LRR protein Prf interacts with Pto kinase in vivo to regulate specific plant immunity. Plant Cell. 2006;18:2792–806.

81. Gutierrez JR, Balmuth AL, Ntoukakis V, Mucyn TS, Gimenez-Ibanez S, Jones AME, et al. Prf immune complexes of tomato are oligomeric and contain multiple Pto-like kinases that diversify effector recognition. Plant J. 2010;61:507–18.

82. Mathieu J, Schwizer S, Martin GB. Pto kinase binds two domains of AvrPtoB and its proximity to the effector E3 ligase determines if it evades degradation and activates plant immunity. PLoS Pathog. 2014;10:e1004227.

83. Balmuth A, Rathjen JP. Genetic and molecular requirements for function of the Pto/Prf effector recognition complex in tomato and *Nicotiana benthamiana*. Plant J. 2007;51:978–90.

84. Martin DDO, Vilas GL, Prescher JA, Rajaiah G, Falck JR, Bertozzi CR, et al. Rapid detection, discovery, and identification of post-translationally myristoylated proteins during apoptosis using a bio-orthogonal azido-myristate analog. FASEB J. 2008;22:797–806.

85. Bizzozero OA. Chemical analysis of acylation sites and species. Meth Enzymol. 1995;250:361–79.

86. Thinon E, Hang HC. Chemical reporters for exploring protein acylation. Biochem Soc Trans. 2015;43:253–61.

87. Helvig C, Alayrac C, Mioskowski C, Koop D, Poullain D, Durst F, et al. Suicide inactivation of cytochrome P450 by midchain and terminal acetylenes. A mechanistic study of inactivation of a plant lauric acid omega-hydroxylase. J Biol Chem. 1997;272:414–21.

88. Bombarely A, Rosli HG, Vrebalov J, Moffett P, Mueller LA, Martin GB. A draft genome sequence of *Nicotiana benthamiana* to enhance molecular plant-microbe biology research. Mol Plant Microbe Interact. 2012;25:1523–30.

89. Mathieu J, Warthmann N, Küttner F, Schmid M. Export of FT protein from phloem companion cells is sufficient for floral induction in *Arabidopsis*. Curr Biol. 2007;17:1055–60.

90. Cheng W, Munkvold KR, Gao H, Mathieu J, Schwizer S, Wang S, et al. Structural analysis of *Pseudomonas syringae* AvrPtoB bound to host BAK1 reveals two similar kinase-interacting domains in a type III effector. Cell Host Microbe. 2011;10:616–26.

91. Nakagawa T, Suzuki T, Murata S, Nakamura S, Hino T, Maeo K, et al. Improved Gateway binary vectors: high-performance vectors for creation of fusion constructs in transgenic analysis of plants. Biosci Biotechnol Biochem. 2007;71:2095–100.

92. Frederick RD, Thilmony RL, Sessa G, Martin GB. Recognition specificity for the bacterial avirulence protein AvrPto is determined by Thr-204 in the activation loop of the tomato Pto kinase. Mol Cell. 1998;2:241–5.

93. Abramovitch RB, Kim Y-J, Chen S, Dickman MB, Martin GB. *Pseudomonas type* III effector AvrPtoB induces plant disease susceptibility by inhibition of host programmed cell death. EMBO J. 2003;22:60–9.

94. Yoo S-D, Cho Y-H, Sheen J. *Arabidopsis* mesophyll protoplasts: a versatile cell system for transient gene expression analysis. Nat Protoc. 2007;2:1565–72.

95. Wu F-H, Shen S-C, Lee L-Y, Lee S-H, Chan M-T, Lin C-S. Tape-*Arabidopsis* sandwich—a simpler *Arabidopsis* protoplast isolation method. Plant Methods. 2009;5:16.

96. Hellens R, Mullineaux P, Klee H. A guide to *Agrobacterium* binary Ti vectors. Trends Plant Sci. 2000;5:446–51.

97. Wessel D, Flügge UI. A method for the quantitative recovery of protein in dilute solution in the presence of detergents and lipids. Anal Biochem. 1984;138:141–3.

# 8

# *Solanum venturii*, a suitable model system for virus-induced gene silencing studies in potato reveals St*MKK6* as an important player in plant immunity

David Dobnik[1*], Ana Lazar[1], Tjaša Stare[1], Kristina Gruden[1], Vivianne G. A. A. Vleeshouwers[2] and Jana Žel[1]

## Abstract

**Background:** Virus-induced gene silencing (VIGS) is an optimal tool for functional analysis of genes in plants, as the viral vector spreads throughout the plant and causes reduced expression of selected gene over the whole plant. Potato (*Solanum tuberosum*) is one of the most important food crops, therefore studies performing functional analysis of its genes are very important. However, the majority of potato cultivars used in laboratory experimental setups are not well amenable to available VIGS systems, thus other model plants from *Solanaceae* family are used (usually *Nicotiana benthamiana*). Wild potato relatives can be a better choice for potato model, but their potential in this field was yet not fully explored. This manuscript presents the set-up of VIGS, based on *Tobacco rattle virus* (TRV) in wild potato relatives for functional studies in potato–virus interactions.

**Results:** Five different potato cultivars, usually used in our lab, did not respond to silencing of phytoene desaturase (PDS) gene with TRV-based vector. Thus screening of a large set of wild potato relatives (different *Solanum* species and their clones) for their susceptibility to VIGS was performed by silencing PDS gene. We identified several responsive species and further tested susceptibility of these genotypes to potato virus Y (PVY) strain NTN and N. In some species we observed that the presence of empty TRV vector restricted the movement of PVY. Fluorescently tagged PVY[N]-GFP spread systemically in only five of tested wild potato relatives. Based on the results, *Solanum venturii* (VNT366-2) was selected as the most suitable system for functional analysis of genes involved in potato–PVY interaction. The system was tested by silencing two different plant immune signalling-related kinases, St*WIPK* and St*MKK6*. Silencing of St*MKK6* enabled faster spreading of the virus throughout the plant, while silencing of *WIPK* had no effect on spreading of the virus.

**Conclusions:** The system employing *S. venturii* (VNT366-2) and PVY[N]-GFP is a suitable method for fast and simple functional analysis of genes involved in potato–PVY interactions. Additionally, a set of identified VIGS responsive species of wild potato relatives could serve as a tool for general studies of potato gene function.

**Keywords:** Potato, Virus-induced gene silencing, VIGS, Potato virus Y, PVY, *Solanum venturii*, St*WIPK*, St*MKK6*, TRV

## Background

Cultivated potato (*Solanum tuberosum* L.) is, after rice, maize and wheat, the world's fourth most important food crop (http://faostat.fao.org/). Its susceptibility to

wide range of pathogens, which diminish its yield, could therefore have a great impact in the food production chain. One of the most important potato pathogens is the *potato virus Y* (PVY). The necrotic isolates of PVY are still responsible for huge agronomic and economic losses [1]. The ability of viruses to cause a disease is determined at the level of molecular interactions between the host plant counterparts and virus factors that can lead to

*Correspondence: david.dobnik@nib.si
[1] Department of Biotechnology and Systems Biology, National Institute of Biology, Večna Pot 111, 1000 Ljubljana, Slovenia
Full list of author information is available at the end of the article

compatible (sensitive) or incompatible (resistant) inter-actions [2]. The differential sensitivity of potato culti-vars lies in their different genetic background, where the resistant ones usually possess *Ry* (extreme resistance) or *Ny* (hypersensitive resistance) gene (reviewed in [3]). However, recent transcriptomic studies revealed the complexity of signalling network involved in the defence response of potato against PVY [4–8].

The use of transient transformation would be a wel-come additional tool to evaluate the role of individual genes in potato. Stable genetic transformation is a less preferred approach, since it is a lengthy process, taking at least 6 months to produce substantial number of trans-formed plants that can be used for functional experi-ments. However, to study the interaction between potato and PVY, the gene expression must be modified through-out the whole plant and not only locally, as the virus also spreads systemically. Virus-induced gene silencing (VIGS) is an optimal tool for this purpose, as the virus vector spreads throughout the plant and causes reduc-tion in activity of selected gene in the whole plant based on post-transcriptional gene silencing (PTGS) [9].

Until now, several VIGS vectors originating from RNA and DNA viruses were developed [10, 11]. The *Tobacco rattle virus* (TRV) vector is the most widely used due to its wide host range and mild symptoms [12]. Some VIGS studies with potato using TRV vector have been already described [13, 14], however they are not being used as routinely in potato as for example in model plant species *Nicotiana benthamiana*.

Tuber-bearing *Solanum* species that belong to section *Petota*, represent a large pool of potato relatives, poten-tially suitable as model species. Section *Petota* contains mostly species from North and South Americas [15, 16], e.g. *Solanum bulbocastanum*, *Solanum stoloniferum* and also the cultivated potato *Solanum tuberosum*. The resource was already used by the potato breeders to introduce the desired traits into cultivars [17]. As these are the closest relatives to cultivated potato, they would serve as a good model system with best possible data translation to cultivated potato cultivars. Large database was already constructed with information on phylogeny [18] and *Phytophthora infestans* resistance in wild potato relatives [19], but no information exist on resistance or susceptibility of these species to viruses.

In order to evaluate the potential application of TRV-based VIGS for functional analysis of potato genes we first screened cultivars for their susceptibility to TRV-based VIGS. As none of the cultivars responded to VIGS, we decided to test wild potato relatives as the closest possible model system, which could be used for potato–PVY interaction studies. We screened an assort-ment of wild potato relatives for their susceptibility to

TRV-based VIGS and selected the most responsive spe-cies for further evaluation of their response to PVY infec-tion. Finally, we were able to select good candidates for studies of gene function in potato–PVY interaction. To show that the selected system is applicable for functional evaluation of a potato genes, we selected a mitogen-acti-vated protein kinase (MAPK) gene St*WIPK* (*A. thaliana* orthologue At*MAPK3*) and a mitogen-activated protein kinase kinase (MKK), gene St*MKK6*. In several studies the members of *WIPK* family were shown to be involved in wound and pathogen responses [20–24], response to oxidative stress [25, 26], drought response and stomata development [27–30]. *WIPK* was shown to be responsive after the infection of *N. benthamiana* with potato virus X (PVX) and PVY [31]. *MKK* genes were also shown to be involved in regulation of cytokinesis [32–35] in absci-sic and salicylic acid signalling [36, 37], in salt stress response [36] and also in response to pathogen infection [37–40]. With the developed methodology, we showed that downregulation (silencing) of St*MKK6* promotes the viral spread, whereas silencing of St*WIPK* did not affect the virus.

## Results and discussion
### VIGS on different potato cultivars
TRV-based VIGS is a wide-spread system for gene silenc-ing, therefore our aim was to check its applicability on all the genotypes we use in PVY–potato interaction stud-ies in our lab (Igor, PW363, Santé, Rywal, NahG-Rywal, Désirée, Désirée Glu-III and NahG-Désirée). They are differently sensitive to PVY (Additional file 1) because of their genetic background. NahG plants were geneti-cally modified to impair salicylic acid (SA) accumulation [7, 41] and Glu-III transgenic Désirée harbours β-1,3-glucanase class III gene under control of 35S promoter [42]. No literature data existed on the susceptibility of the selected cultivars to TRV-based VIGS. Experimen-tal plants were agroinfiltrated with a mixture of pTRV1 and pTRV2:CaPDS, constructs for silencing phytoene desaturase (PDS). When silenced the carotenoid bio-synthesis is supressed, which makes plants susceptible to photobleaching [43] that can easily be visualized. *N. benthamiana* plants were used as positive control and after 12 days post agroinfiltration, plants have already started to show the photo-bleaching phenotype, which persisted throughout the experiment. To show that the original pTRV2:CaPDS construct originating from pep-per, would work also on other plant species, we have analysed the sequence identity of different PDS genes. The observed nucleotide sequence identity of *Capsicum annuum* PDS was higher than 95 % when compared to *S. tuberosum* or *S. nigrum* PDS and 92.7 % when com-pared to *N. benthamiana* PDS (Additional file 2). As *N.*

*benthamiana* plants served as positive control and were always showing photo-bleached phenotype, we concluded that pTRV2:CaPDS should also work on *Solanum* species.

Both NahG-Rywal and NahG-Désirée plants developed strong disease symptoms (leaves have dried up and fallen off) already at 10 days post agroinfiltration. Most probably the lack of SA allowed high TRV accumulation, which was devastating for the small plants. None of other cultivars showed any signs of photobleaching. Cultivar Igor responded by dropping the infiltrated leaves, what often occurs after PVY inoculation in this cultivar [44], potentially indicating the multiplication of TRV. To our knowledge, there are only two publications presenting data on VIGS in potato using TRV, showing only three cultivars of *S. tuberosum* (GT12297-4, Cara and Pentland Ivory) as susceptible [13, 14]. Noteworthy, Brigneti et al. had applied VIGS to potato plants grown from seeds [13], a system not used in many of the labs, as the propagation of potatoes in tissue culture is required for retaining the desired genotype.

## VIGS and wild potato relatives
Brigneti et al. showed that two wild potato relatives, *S. bulbocastanum* and *S. okade*, were susceptible to VIGS [13]. As *S. tuberosum* cultivars tested in our experiments were not responsive to TRV-based VIGS, we have therefore decided to find a set of VIGS susceptible wild potato relatives that could serve as model plants for studies of potato genes. We have tested 73 different clones representing 34 different species of wild potato relatives (Additional file 3), from which all were already tested for resistance/susceptibility to *P. infestans* [19]. Decision to include several clones of the same species was based on observed different resistance level against *P. infestans*. An online phylogenetic analysis to illustrate the relation between *S. tuberosum* and wild potato relatives was conducted within SolRgene [19] with clones that are available in database (Fig. 1). All tested plants were agroinfiltrated with a mixture of pTRV1 and pTRV2:CaPDS and photo-bleaching symptoms were scored 21–24 days post agroinfiltration (dpa). The extent of the recorded photo-bleaching phenotype was classified into four groups: (1) moderate to strong silencing, (2) moderate silencing, (3) low-level silencing and (4) no silencing. 15 clones (10 species) have shown moderate to strong silencing, 4 clones (4 species) have shown moderate silencing and 10 clones (10 species) have shown low-level silencing (Fig. 1). 44 clones (18 species) have shown no silencing (Additional file 3). As the results suggest, the pool of wild potato relatives is not amenable to TRV-based VIGS, with more than half of tested clones being

nonresponsive. Even with the overall highly responsive species like *Solanum venturii* or *S. stoloniferum*, there were still some non-responding clones indicating that subtle differences in genotype or epigenetic differences can affect the outcome of plant–TRV interaction. Our results also confirm previous data [13], as *S. bulbocastanum* and *S. okade* species were shown to be in the most responsive group, even though we performed our tests on the plants propagated in stem tissue cultures in contrast to the seed grown plants. When integrating this data with the phylogenetic analysis we observed that at least half of the most responsive clones (mostly *S. venturii*) were relatively closely related to *S. tuberosum* (Fig. 1), when referring to distance and position in a phylogenetic tree in comparison to *S. tuberosum and S. lycopersicon*.

## Susceptibility of selected wild potato relatives to PVY[NTN]
For the next experiment we selected all 15 clones that were classified as strongly responsive to TRV-based VIGS (Fig. 1a). In order to use the wild potato relatives in functional analysis of genes involved in potato–PVY interaction, their response to PVY must be known. Therefore, the selected clones were inoculated with highly aggressive PVY strain NTN (PVY[NTN]) and we followed its spreading to upper non-inoculated leaves. Samples were collected 14 days post inoculation (dpi), since this is the time when PVY is expected to be present throughout the plant in the case of tolerant interaction [4]. The relative viral RNA content measured in the samples is presented in Fig. 2. All of the tested clones have shown the tolerant phenotype as viral RNA was detected in upper, non-inoculated leaves in all of the clones while no symptoms of viral infection were observed (Additional file 4). *S. bulbocastanum* (BLB 331-2) had the lowest viral RNA content, however we were able to get the data only from 1 out of 3 plants, as the RNA isolation from other two samples was not successful (data not shown). *S jamesii* (JAM 355-1) and *S. lesteri* (LES 358-4) had also relatively low viral content, nevertheless the same low content was observed in individual plants of other clones (Fig. 2) indicating highly variable relative PVY[NTN] RNA content between plants and genotypes. When observing only the spreading of PVY, the viral content itself is not the true marker for the susceptibility as more susceptible plants don't necessarily have higher viral content [4]. We showed that there were no significant differences in the amount of viral RNA between clones when infected only with PVY[NTN] (Additional file 5). Most probably, the variability in the relative amount of PVY[NTN] RNA is indicating that the time after infection when the virus reaches the upper leaves differs between the individual plants of the same genotype, which was also observed in other experiments studying potato–PVY interaction [4]. These results will be added

**a** Moderate to strong silencing

BLB 331-2
HJT 349-3
JAM 355-1
LES 358-4
MCQ 186-1
OKA 970-3
PLT 378-2
PTA 767-8
SPEC 287-2
VNT 250-2
VNT 283-1
VNT 365-1
VNT 366-2
VNT 741-1
VNT 896-4

**b** Moderate silencing

HAW 634-4
PTA 765-1
SNK 293-2
STO 838-5

**c** Low-level silencing

AGF 101-1
AVL 478-2
BCP 326-3
BLB 525-1
CAP 536-1
CPH 541-2
PUR 206-1
STO 842-6
VER 393-10
VNT 367-1

**d**

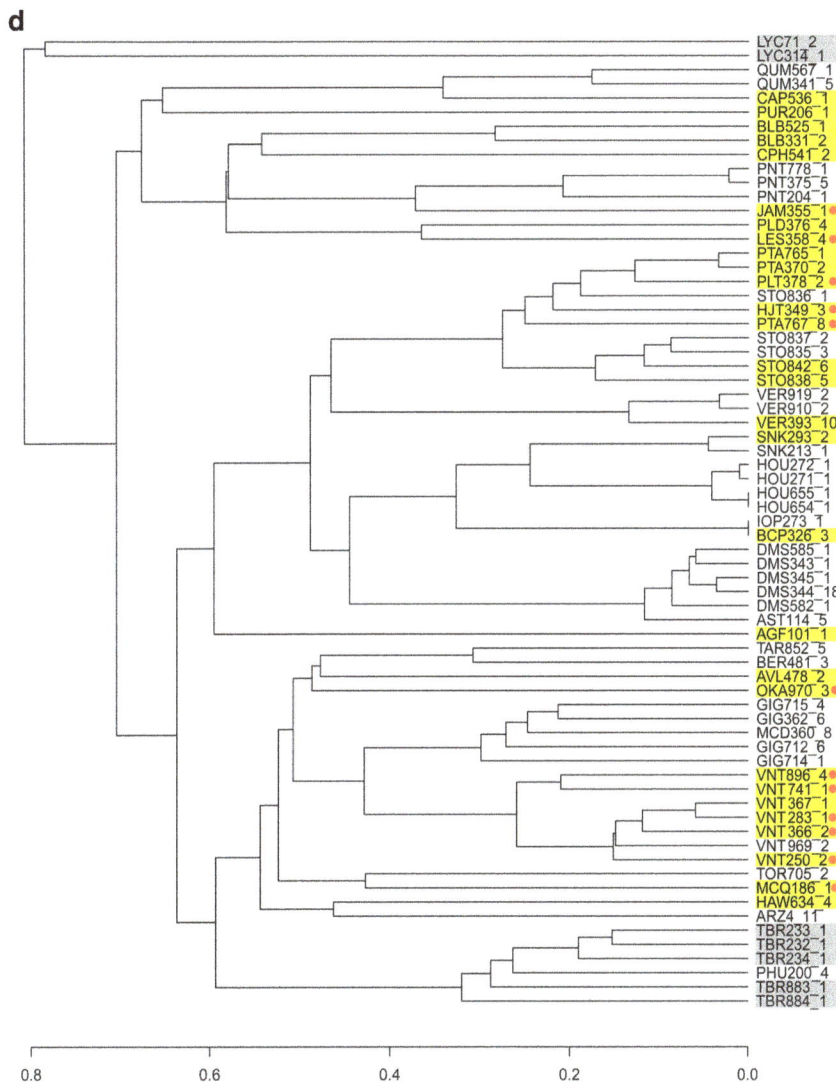

(See figure on previous page.)

**Fig. 1** Efficiency of TRV-based VIGS in different wild relatives of potato. The clones that responded to VIGS with TRV (silencing of PDS), were classified into three groups: **a** moderate to strong silencing, **b** moderate silencing and **c** low-level silencing. Photograph at the *top of the panels* **a–c** shows silencing phenotype representative for each group. Responsive clones are listed in each corresponding panel. Phylogenetic tree of 62 clones (**d**) that were available in SolRgene database was constructed with interactive online tool using UPGMA method. All VIGS responsive clones are highlighted in *yellow*. The clones from the group of moderate to strong silencing are labelled with a *red dot. Solanum lycopersicum* (LYC) and *Solanum tuberosum* (TBR) clones (highlighted in *grey*) were added to the phylogenetic tree to illustrate the relations of wild potato relatives to tomato and cultivated potato. AGF, *Solanum agrimonifolium*; AVL, *Solanum avilesii*; BCP, *Solanum brachycarpum*; BLB, *Solanum bulbocastanum*; CAP, *Solanum capsicibaccatum*; CPH, *Solanum cardiophyllum*; HAW, *Solanum hawkesianum*; HJT, *Solanum hjertingii*; JAM, *Solanum jamesii*; LES, *Solanum lesteri*; MCQ, *Solanum mochiquense*; OKA, *Solanum okadae*; PLT, *Solanum polytrichon*; PTA, *Solanum papita*; PUR, *Solanum piurana*; SNK, *Solanum schenckii*; SPEC, *Solanum species*; STO, *Solanum stoloniferum*; VER, *Solanum verrucosum*; VNT, *Solanum venturii*

to the SolRgene database and will complement the information on susceptibility to *P. infestans* in SolRgene database [19].

### The effect of TRV infection on PVY^NTN spread in selected wild potato relatives

We showed that PVY$^{NTN}$ spreads in all of the selected clones and in order to use these clones in VIGS studies, we analysed the effect of empty TRV VIGS vector on spreading of PVY$^{NTN}$. As when performing the normal VIGS experiment, we have agroinfiltrated 1 week old plants with empty TRV vector and after 3 additional weeks inoculated the plants with PVY$^{NTN}$. Again, we have tested for presence of PVY$^{NTN}$ RNA in the samples of non-inoculated leaves collected 14 days after PVY$^{NTN}$ inoculation. Although we detected significant differences in PVY$^{NTN}$ content in limited number of cases (Additional file 5), as already discussed above, rather than the relative viral RNA content, the more important indicator of susceptibility to viral infection is the number of plants with detected PVY in upper leaves [42]. In our experiments, prior agroinfiltration with empty TRV vector for VIGS reduced the number of plants successfully infected with PVY$^{NTN}$ (Fig. 2). In three (BLB 331-2, SPEC 287-2 and VNT 741-1) out of seven clones where the reduction in number of infected plants was observed (BLB 331-2, HJT 349-3, MCQ186-1, SPEC 287-2, VNT283-1, VNT 365-1 and VNT 741-1), only one out of five plants was systemically infected.

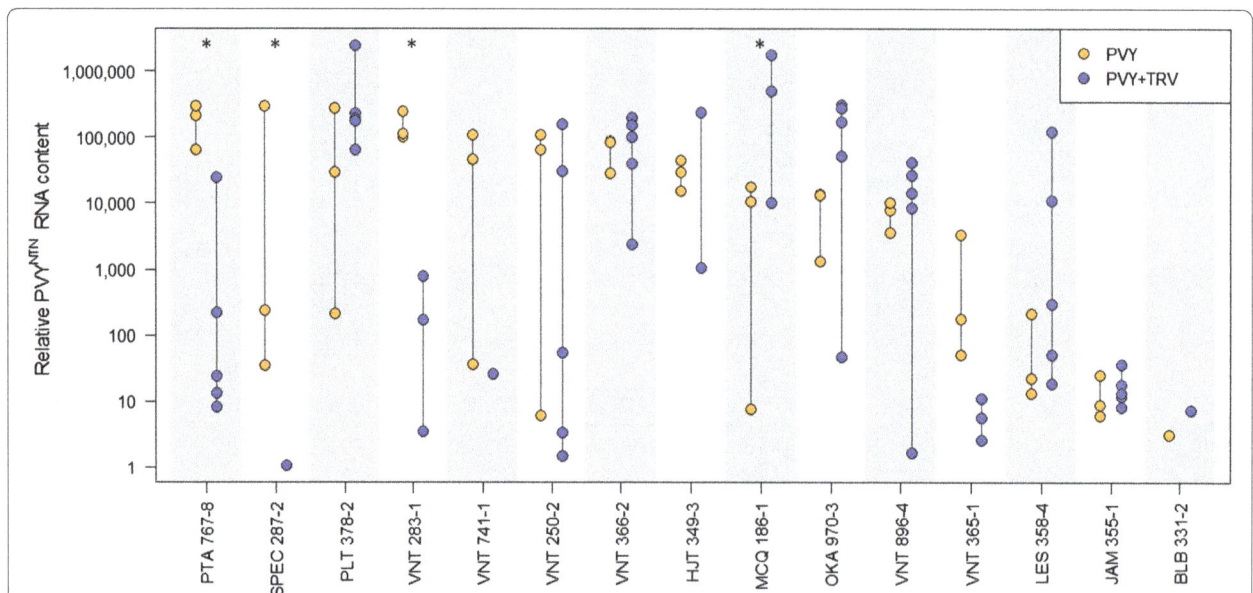

**Fig. 2** Systemic PVY$^{NTN}$ infection of selected clones. PVY$^{NTN}$ RNA concentration in samples is presented relative to the sample with lowest detected viral amount. Relative PVY$^{NTN}$ RNA content in selected clones 14 dpi after PVY infection (PVY, *orange dots*) and 14 dpi after PVY infection with prior agroinfiltration with empty TRV vector (PVY + TRV, *blue dots*). *Each dot* represents the data from individual plant. Three and five plants were originally infected in PVY and PVY + TRV group, respectively. In the case that the PVY$^{NTN}$ RNA was not detected in some of the samples, the corresponding number of *dots* is lower. *Statistically significant difference in relative PVY$^{NTN}$ RNA content due to agroinfiltration with empty TRV vector (compared between treatments of same genotype). Detailed results of statistical analysis are given in Additional file 5. BLB, *Solanum bulbocastanum*; HJT, *Solanum hjertingii*; JAM, *Solanum jamesii*; LES, *Solanum lesteri*; MCQ, *Solanum mochiquense*; OKA, *Solanum okadae*; PLT, *Solanum polytrichon*; PTA, *Solanum papita*; SPEC, *Solanum species*; VNT, *Solanum venturii*

For functional analysis studies there should be no effect of TRV on the spreading of PVY, therefore we also compared the data of PVY$^{NTN}$ content measured in both experiments (with and without prior TRV agroinfiltration) (Fig. 2, Additional file 5). We showed that the clones suitable for further studies with VIGS, in which the TRV did not affect the spreading of PVY$^{NTN}$, are *S. jamesii* (355-1), *S. lesteri* (358-4), *S. okade* (970-3), *S. polytrichon* (378-2) and *S. venturii* (250-2, 366-2 and 896-4). *S. bulbocastanum*, which was previously already used in TRV-based VIGS experiments for testing resistance genes against *P. infestans* [13], was shown not to be suitable for VIGS in the case of PVY studies, due to the influence of TRV on the course of PVY infection (Fig. 2).

### Susceptibility of selected wild potato relatives to PVY$^{N}$-GFP

Parallel to these studies, an infectious clone of PVY$^{N}$ strain fluorescently labelled with green fluorescent protein (GFP) was developed in our laboratory [45]. Detailed spatio-temporal analyses of viral multiplication are only feasible using fluorescently tagged viruses as then the virus can be monitored in a non-invasive manner using confocal microscopy. In the course of testing the developed infectious clone, also its ability to infect the wild potato relatives was analysed [45]. The green fluorescence, a trace of PVY$^{N}$-GFP multiplication, was detected in a set of TRV-based VIGS responsive wild potato relatives in inoculated leaves (at 14 dpi), indicating the virus was successfully multiplying. On the other hand, GFP fluorescence in non-inoculated leaves was detected at 14 dpi only in five clones (MCQ 186-1, PTA 767-8, VNT 283-1, VNT 365-1 and VNT366-2) [45]. When comparing this report to our results, it is obvious, that the spreading of PVY$^{N}$-GFP is not as efficient as that of PVY-$^{NTN}$, since PVY$^{N}$ is less aggressive strain of the PVY [6]. Based on these observations we have performed a more detailed study of PVY$^{N}$-GFP spread in wild potato relatives. When we evaluated the GFP fluorescence in non-inoculated leaves of wild potato relatives at 14 dpi, only VNT 366-2 and VNT283-1 showed the whole area of the non-inoculated leaves covered in green fluorescence and in other three (MCQ 186-1, PTA 767-8 and VNT 365-1) only local patches with green fluorescence were detected. The example of green fluorescence signal in the leaves of VNT 366-2 can be seen in Fig. 3. It is important to mention that some structures exhibiting green fluorescence, most probably due to accumulation of secondary metabolites, can be detected in mock inoculated leaves (Fig. 3), however the green fluorescence pattern can be clearly distinguished from true positive signal (Fig. 3). Finally, VNT 366-2 and VNT 283-1 would be the most suitable candidates for VIGS test subjects, but because in VNT 283-1 the TRV infection interfered with the speed

of PVY movement (Fig. 2), we concluded that VNT 366-2 is the most suitable clone for experiments including TRV-based VIGS and PVY$^{N}$-GFP for functional analysis of genes involved in potato–PVY interaction. However, we have also observed that the PVY$^{N}$-GFP was spreading in VNT 366-2 agroinfiltrated with empty TRV with slower rate. The green fluorescence of PVY$^{N}$-GFP in the non-inoculated leaves of an empty vector agroinfiltrated control plants was usually detected at 21 dpi or later (in individual plants green fluorescence was detected at 14 dpi, but never in all biological replicates). Nevertheless, the VIGS system with clone VNT 366-2 is applicable for functional studies of potato genes, even if using milder viral strains.

### Functional confirmation of St*MKK6* role as positive regulator of plant defence against PVY

To show the suitability of the selected wild potato relative (clone VNT366-2) for the functional analysis of genes involved in potato–PVY interaction, two different genes from the immune signalling cascade, kinases St*WIPK* and St*MKK6*, were analysed with this system. MAPKs are crucial in plant immune response, therefore disrupting MAPK signalling, by silencing its components, could lead to disturbed plant response to infection with PVY. Both selected genes have been previously reported as important components of plant response to viral infection [31, 37]. We have studied the effect of St*WIPK* and St*MKK6* silencing through observing the GFP fluorescence in the upper, non-inoculated leaves, as a marker of viral spread. Silencing of St*WIPK* had no significant effect on the viral spread in any of the observed time points. The spread of PVY$^{N}$-GFP was also observed in the silenced plants and was comparable to the non-silenced ones (Table 1). There are however several reports on *WIPK* as a positive regulator of immune response [20, 22, 24, 31, 38, 46–49], moreover, in tobacco, silencing of Nt*WIPK* led to reduced resistance to TMV [38, 50]. On the other hand, silencing of St*MKK6* showed that the kinase has important role in viral spread. Emergence of PVY$^{N}$-GFP multiplication marker in non-inoculated leaves of silenced plants preceded the emergence of PVY$^{N}$-GFP in empty vector-treated control plants. Additionally, we detected PVY in St*MKK6*-silenced plants more frequently than in control plants, suggesting that silencing of St*MKK6* renders plants more susceptible to PVY infection (Fig. 4). These results support our recent findings that St*MKK6* is involved in potato immune response against PVY and acts as positive regulator of potato defence response [37]. It has been shown that *MKK6* gene expression in potato is induced after PVY treatment in a SA-dependant manner [37]. SA is known to influence the expression of defence-related genes [4, 7], especially in response to viral

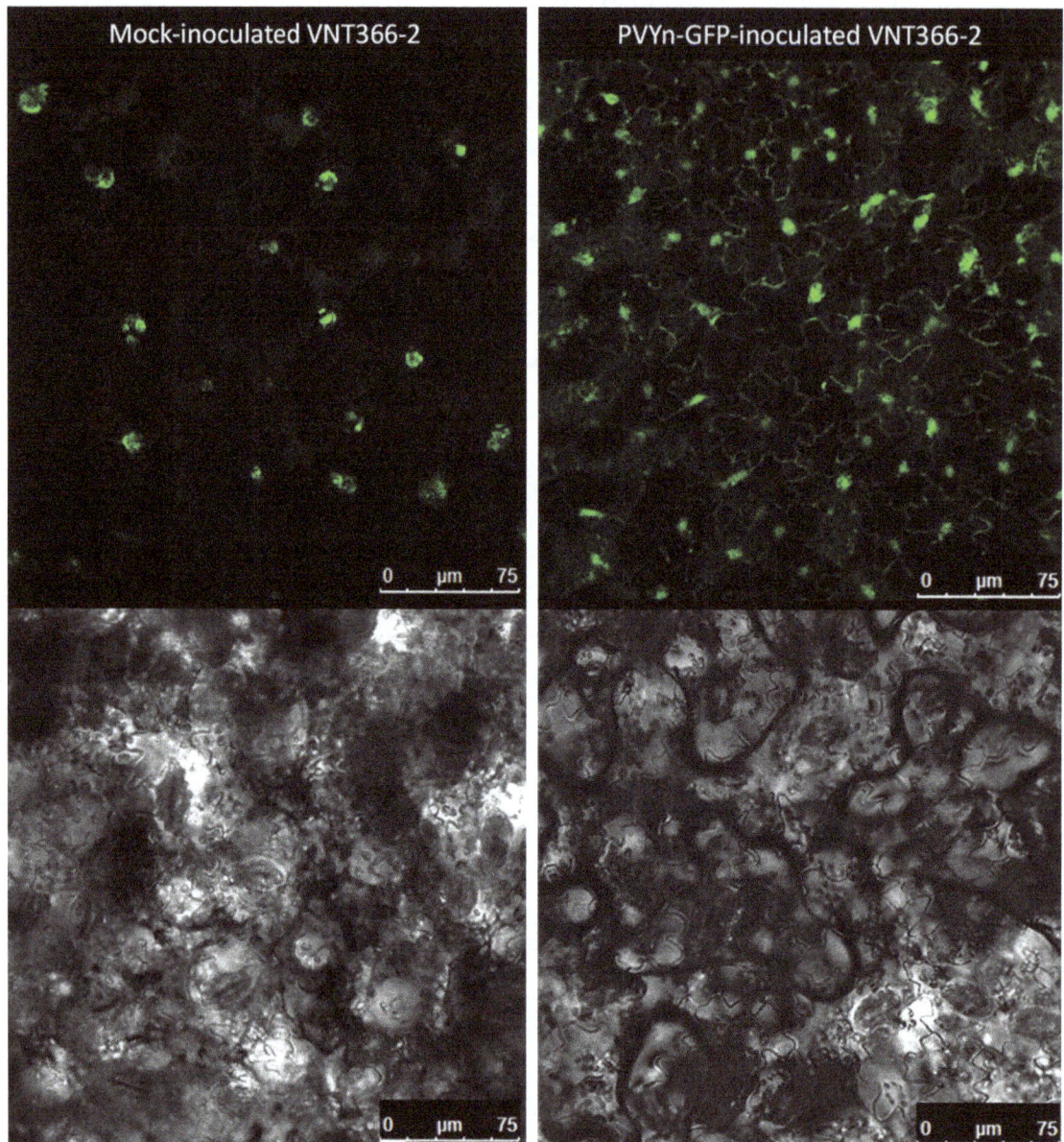

**Fig. 3** Comparison of green fluorescence signal from mock- and PVY^N-GFP-inoculated *Solanum venturii* VNT 366-2. Mock inoculated plants can show background green fluorescence signal in a form of clusters (*left side*), whereas the true positive green fluorescence signal for PVY^N-GFP can be seen as outline of the cells (*right side*), due to the presence of GFP in the cell cytoplasm. *Lower parts* of the figure (*greyscale* images) represent the corresponding transmitted light image of plant leaf section presented in the *top panel* of the figure showing fluorescence signals

pathogens, through a finely tuned signalling cascades, where many aspects of the regulation still remain undefined [51]. However, based on the results from Arabidopsis, SA signalling regulation with kinases is explained by negative feedback loop, where SA accumulation is negatively regulated by the MAP kinase MPK4 via PAD4 and EDS1 [52, 53]. It was shown that potato St*MKK6* interacts with St*MAPK4* probably through phosphorylation of St*MAPK*4 and additionally that St*MKK6* is involved

in HR of potato against PVY a process which is highly dependent on SA [37]. Our results of VIGS, together with our previous findings [37], therefore imply that St*MKK6* is involved in potato immune signalling cascade via interaction with genes involved in SA signalling. Until now, only few reports are available on the topic of *MKK6* silencing. It was shown that silencing of its orthologue has an effect on plant immune response by attenuating the resistance against TMV in *N. tabacum*, most

**Table 1  Effect of St*WIPK* silencing on spreading of PVY^N-GFP to upper non-inoculated leaves**

| dpi | Systemically infected plants (%) | | |
|---|---|---|---|
| | *WIPK* silenced | Empty vector | Positive control |
| 14 | 0 | 14.3 | 33.3 |
| 18 | 0 | 14.3 | nd |
| 22 | 85.7 | 85.7 | 100 |

dpi, days post inoculation with PVY^N -GFP; *WIPK* silenced, six plants were agroinfiltrated with TRV construct for silencing *WIPK* gene and infected with PVY^N-GFP; Empty vector, six plants were agroinfiltrated with empty TRV construct and infected with PVY^N-GFP; Positive control, three plants were infected with PVY^N-GFP at the same time as *WIPK* silenced and empty vector plants; nd, not determined at this time point

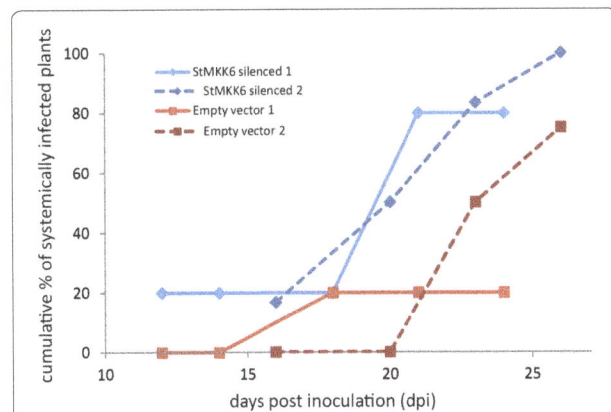

**Fig. 4** Silencing of St*MKK6* is increasing the speed of viral spread. Cumulative percentage of systemically infected plants after TRV-based VIGS in relation to days post inoculation is shown. The results of two independent time series experiments are presented. In first experiment a total of five biological replicates (plants) was used for each treatment; St*MKK6* silenced (*blue solid line*) and empty TRV vector agroinfiltrated (*red solid line*). In second experiment a total of six biological replicates were used for St*MKK6* silenced group (*blue dashed line*) and four biological replicates for empty TRV vector agroinfiltrated plants (*red dashed line*). At each time point only two or three replicates were analysed

probably through reduced activation of different transcription factors [39]. Moreover, silencing of the orthologue Nb*MEK1* prevented the hypersensitive response (HR) that is triggered via NTF6 cascade [54]. The results of silenced St*MKK6* reported here offer an additional functional confirmation of St*MKK6* as a positive regulator of plant defence against PVY and present additional piece of evidence for the above mentioned connection of SA and St*MKK6*.

## Conclusions

We established a fast screening system for evaluation of potato gene function, in particular in response to infection with PVY. As the cultivable potato is not suitable for

TRV-based VIGS, we have performed an evaluation of suitability of wild potato relatives for functional analysis of potato genes involved in potato–virus interaction using TRV-based VIGS on a wide set of different species. The most responsive species were further tested for susceptibility to two different PVY strains in order to select the best model species. All tested wild potato relatives were shown to be tolerant to PVY, suggesting they do not possess any resistance genes against this virus. As presented in the manuscript, the system employing *S. venturii* (VNT366-2) and PVY^N-GFP is the best choice for fast functional analysis of genes. Using the developed system, we have shown that St*MKK6* has a role of positive regulator in potato defence against PVY. Additionally, a set of identified TRV-based VIGS responsive species could also serve as a general tool for functional analyses.

## Methods

### Plant materials and growth conditions

Set of wild potato relatives (wild *Solanum* species, Additional file 3) used in the studies is part of collection of WUR Plant Breeding, Wageningen University and Research Centre, The Netherlands. Different *Solanum tuberosum* cultivars (Additional file 1) are part of the collection of National Institute of Biology, Slovenia. *Nicotiana benthamiana* was used in VIGS experiments as control of silencing. *N. benthamiana* plants were grown from seeds and kept in a growth chamber under controlled conditions (16 h light/8 h dark cycle at 22/20 °C respectively). In vitro potato plantlets of different *Solanum* species were propagated in sterile culture boxes containing MS medium supplemented with 3 % sucrose and 0.8 % agar and grown in a growth chamber under controlled conditions (16 h light/8 h dark cycle at 21/19 °C respectively). Two-week-old plantlets were transplanted into soil and moved to a greenhouse with 22/20 °C day/night temperature regime.

### Plasmid constructs and transformation of *Agrobacterium tumefaciens*

The basic set of TRV VIGS vectors used in the studies was previously described [55]. For PDS silencing we used the pTRV2:CaPDS that is based on pYL156 plasmid backbone with inserted 371-bp fragment (610–980 bp) of PDS gene from *Capsicum annuum* (GenBank accession X68058).

The full-length sequence of St*WIPK* was amplified from *S. tuberosum* cv. Rywal cDNA with the following primers: forward 5-ATGGTTGATGCTAATATGGGT-3 and reverse 5-GCACACAAGCTAGCACGAAC-3. The fragments were inserted into the pJET 1.2 blunt cloning vector (Thermo Scientific) and sequenced (GATC Biotech). For St*WIPK* silencing the 520-bp fragment (187–706 bp) was amplified from pJET plasmid harbouring St*WIPK* gene (GeneBank accession no. KP033231.1) with forward (5′-GAATTCTGAATGAGATGGTTG CAGTT-3′) and reverse (5′-TAAGCTCCATGAAGATG CAA-3′) primer. For St*MKK6* silencing the 453-bp fragment (157–609 bp) was amplified from plasmid harbouring St*MKK6* gene [37] (GeneBank accession no. KF837127.1) with forward (5′-GAATTCTGCCCTCAG AAACTAAGGAG-3′) and reverse (5′-TCCTTTGTGG TTCACTAGCA-3′) primer. In both cases forward primer harboured the EcoRI restriction site. The amplified fragments were inserted into the pJET 1.2 blunt cloning vector (Thermo Scientific) and sequenced (GATC Biotech). pJET plasmids harbouring St*WIPK* or St*MKK6* fragment and empty pTRV2 plasmids were restricted with EcoRI (Gibco) and XbaI (Gibco) restriction enzymes and purified from agarose gel with Wizard SV Gel and PCR Clean-Up System (Promega). Purified fragments were ligated into restricted pTRV2 with T4 DNA ligase (Fermentas). Resulting plasmids pTRV2:WIPK and pTRV2:MKK6 were sequenced (GATC Biotech) and introduced into *Agrobacterium tumefaciens* strain GV3101 by electroporation (Eppendorf Electroporator 2510) following manufacturer's procedure with voltage set to 2000 V.

## Bioinformatic analysis

Phylogenetic tree of potato and its relatives was prepared with online Interactive phylogeny tool within SolRgene database (http://www.plantbreeding.wur.nl/SolRgenes/ Phylogeny/species_select.php). In first step all the species were selected and in the second step the accessions used in our experiments were selected. 11 clones out of 73, which were used in our study for responsiveness to TRV silencing, were not available in the database and are therefore not included in the tree. Additionally two *S. lycopersicum* and six *S. tuberosum* accessions were added to the phylogenetic tree to illustrate the relations of wild potato relatives to tomato and cultivated potato. A dendrogram [average linkage (UPGMA)] was created.

To check in silico whether the CaPDS construct (part of the sequence with GenBank accession number X68058.1) would have an effect on *Solanum* species, we have aligned different PDS sequences and checked the identity in comparison to CaPDS. The parts of sequences of PDS genes that were used were from *S. tuberosum* (GenBank accession number AY484445.1), *S. nigrum* (GenBank accession number EU434622.1) and *N. benthamiana* (GenBank accession number EU165355.1). Alignment was prepared and analysed with AlignX software (a component of Vector NTI Suite 9.0.0).

## Agroinfiltration for TRV-based VIGS and inoculations with PVY

Plants, after being grown in soil for 7–10 days, were treated by co-infiltration of *Agrobacterium tumefaciens* strain GV3101 carrying pTRV1 and the various pTRV2 recombinants, in a 1:1 ratio as described by Du et al. [56]. pTRV2:CaPDS and empty pTRV2 plasmids were used in initial experiments and later on as controls. For testing VIGS susceptibility of different potato cultivars three plants were tested for each cultivar and for wild potato species two plants were tested for each species.

Five plants of the wild potato relatives were inoculated with buffered suspension of PVY$^{NTN}$ (isolate NIB-NTN, GenBank accession number AJ585342) or PVY$^N$-GFP (PVY N605-GFP [45]) infected plant sap 3 weeks after agroinfiltration with empty TRV vector or mock inoculated as described before [8, 45]. For experiments without prior agroinfiltration, three plants of each genotype were inoculated 4 weeks after they were transferred to soil. Samples (upper non-inoculated leaves) for RNA isolation (PVY$^{NTN}$ infected plants) were collected 14 days post inoculation (dpi) and immediately frozen in liquid nitrogen. Samples for confocal microscope observation (PVY$^N$-GFP infected plants) were collected 14 dpi and were stored in humid environment (petri dish with humid paper towel) until observed (maximum time from collection to observation was 4 h). For VIGS studies of St*WIPK* six biological replicates were used and four to six biological replicates (details in Additional file 6) for St*MKK6*. Upper non-inoculated leaves were sampled at different time points and observed under confocal microscope. For this purpose, leaves were cut and stored in humid environment until observed as described above.

## Real-time PCR

RNA from the samples was isolated with innuPREP Plant RNA Kit and treated with DNAse (Invitrogen, USA; 0.1 U/Dnase per μg RNA) prior to reverse transcription. 1 μg of RNA was reversely transcribed using the High Capacity cDNA Reverse Transcription Kit (Applied Biosystems, USA).

Samples were analyzed in the set-up for quantitative real-time PCR (qPCR) analysis as previously described [57], using TaqMan chemistry for determining the relative concentration of PVY$^{NTN}$ RNA [58] and cytochrome oxidase (Cox; [59]) as RNA load control. The transcript accumulation was normalized to that of Cox. The relative content of PVY$^{NTN}$ was calculated as follows: Cq value of Cox was subtracted from Cq value for PVY; the resulting value was used as exponential power on value 2; every value was finally divided by the minimum value. The resulting value was transformed with base 10 logarithm. Results were statistically evaluated using Two Way

ANOVA in SigmaPlot 13.0 software. Two independent factors were considered in analysis: Genotype (individual species or clones) and Treatment (only PVY infection or PVY infection with prior agroinfiltration with empty TRV vector). The relative PVY RNA content was the dependent variable. For multiple comparison Tukey's test was selected.

## Confocal microscopy

GFP and background chloroplast fluorescence were visualized with a Leica TCS SP5 laser-scanning microscope mounted on a Leica DMI 6000 CS inverted microscope (Leica Microsystems, Germany) with a HC PL FLUOTAR 10× objective and with Leica TCS LSI macroscope with Plan APO 1× or Plan APO 5× objective (Leica Microsystems, Germany). For excitation, the 488 nm laser line was used. Fluorescence emissions with wavelengths of 505–530 and 590–680 nm were collected simultaneously or sequentially through two channels. Images were processed by using Leica LAS AF Lite software (Leica Microsystems, Germany).

## Additional files

**Additional file 1.** List of *Solanum tuberosum* cultivars used in this study. All *S. tuberosum* cultivars used in this study are listed together with the description of its susceptibility to PVY or the genetic change for genetically modified lines.

**Additional file 2.** Sequence alignment of part of PDS gene originating from different species. Through the alignment, the similarity of the part of PDS gene used in VIGS construct to other plant species is presented.

**Additional file 3.** List of wild potato relatives used in the studies with their response to TRV-based VIGS. All wild potato relatives used in this study are listed together with their abbreviation clone number, country of origin and silencing response.

**Additional file 4.** Photographs of mock and PVY[NTN] infected set of wild potato relatives. One mock inoculated (1) and one PVY[NTN] infected (2) plant at 14 dpi are shown. Letters represent individual species/clones: a) BLB 331-2, b) HJT 349-3, c) JAM 355-1, d) MCQ 186-1, e) OKA 970-3, f) PTA 767-8, g) PLT 378-2, h) SPEC 287-2, i) VNT 250-2, j) VNT 283-1, k) VNT 365-1, l) VNT 366-2, m) VNT 741-1, n) VNT 896-4, o) LES 358-4.

**Additional file 5.** Report for Two Way ANOVA statistical analysis of relative PVY[NTN] RNA content using two independent factors. Statistical comparison (Two Way ANOVA) was conducted between groups of plants from individual clones (genotypes) and groups with different treatment (PVY or PVY + TRV). Difference in PVY[NTN] RNA content is considered statistically significant, if the *P* value is lower than 0.05 (all cases are highlighted in yellow).

**Additional file 6.** Number of infected plants at each time point in StMKK6 silenced and empty vector treated plants for two different experiments. The table shows the actual number of plants, which were used to calculate cumulative percentage of systemically infected plants after TRV-based VIGS in relation to days post inoculation, as shown in Fig. 4.

## Authors' contributions

DD performed the experimental work, acquired the results, carried out the data analysis and interpretation, and wrote the manuscript. AL and TS contributed to studies with St*WIPK* and St*MKK6* and helped in writing the manuscript.

VGAAV, KG and JŽ helped with experimental design and supervised the studies. All authors read and approved the final manuscript.

### Author details
[1] Department of Biotechnology and Systems Biology, National Institute of Biology, Večna Pot 111, 1000 Ljubljana, Slovenia. [2] Wageningen UR Plant Breeding, Wageningen University and Research Centre, P.O. Box 386, 6700 AJ Wageningen, The Netherlands.

### Acknowledgements
We thank Dr. Richard R.G.F. Visser for providing the collection of wild potato relatives. We also thank Dr. Sabine Rosahl for providing NahG-Désirée potato plants and Dr. Andrej Blejec for help with statistical analysis. We would also like to acknowledge Gerard Bijsterbosch, Juan Du and Maja Jamnik for excellent technical assistance.

### Competing interests
The authors declare that they have no competing interests.

### Funding
The study was supported by the Slovenian Research Agency (Contract Numbers 1000-07-310032, J1-4268 and P4-0165), by COST Action FA0806 and Slovenian Society of Plant Biology.

### References
1. Scholthof K-BG, Adkins S, Czosnek H, Palukaitis P, Jacquot E, Hohn T, Hohn B, Saunders K, Candresse T, Ahlquist P, Hemenway C, Foster GD. Top 10 plant viruses in molecular plant pathology. Mol Plant Pathol. 2011;12:938–54.
2. Whitham SA, Yang C, Goodin MM. Global impact: elucidating plant responses to viral infection. Mol Plant Microbe Interact. 2006;19:1207–15.
3. Kogovšek P, Ravnikar M. Physiology of the potato–potato virus Y interaction. In: Lüttge U, Beyschlag W, Francis D, Cushman J, editors. Progress in Botany SE—3, vol. 74. Berlin: Springer; 2013. p. 101–33.
4. Baebler Š, Stare K, Kovač M, Blejec A, Prezelj N, Stare T, Kogovšek P, Maruša P-N, Rosahl S, Ravnikar M, Gruden K. Dynamics of responses in compatible potato–potato virus Y interaction are modulated by salicylic acid. PLoS ONE. 2011;6:e29009.
5. Gruden K, Pompe-Novak M, Baebler Š, Krečič-Stres H, Toplak N, Hren M, Kogovšek P, Gow L, Foster GD, Boonham N, Ravnikar M. Expression microarrays in plant–virus interaction. Methods Mol Biol. 2008;451:583–613.
6. Kogovšek P, Pompe-Novak M, Baebler Š, Rotter A, Gow L, Gruden K, Foster GD, Boonham N, Ravnikar M. Aggressive and mild potato virus Y isolates trigger different specific responses in susceptible potato plants. Plant Pathol. 2010;59:1121–32.
7. Baebler Š, Witek K, Petek M, Stare K, Tušek-Žnidarič M, Pompe-Novak M, Renaut J, Szajko K, Strzelczyk-Żyta D, Marczewski W, Morgiewicz K, Gruden K, Hennig J. Salicylic acid is an indispensable component of the Ny-1 resistance-gene-mediated response against Potato virus Y infection in potato. J Exp Bot. 2014;65:1095–109.
8. Baebler S, Krecic-Stres H, Rotter A, Kogovsek P, Cankar K, Kok EJ, Gruden K, Kovac M, Zel J, Pompe-Novak M, Ravnikar M. PVY(NTN) elicits a diverse gene expression response in different potato genotypes in the first 12 h after inoculation. Mol Plant Pathol. 2009;10:263–75.
9. Baulcombe DC. Fast forward genetics based on virus-induced gene silencing. Curr Opin Plant Biol. 1999;2:109–13.
10. Unver T, Budak H. Virus-induced gene silencing, a post transcriptional gene silencing method. Int J Plant Genomics. 2009;2009:198680.
11. Faivre-Rampant O, Gilroy EM, Hrubikova K, Hein I, Millam S, Loake GJ, Birch P, Taylor M, Lacomme C. Potato virus X-induced gene silencing in leaves and tubers of potato. Plant Physiol. 2004;134:1308–16.
12. Ratcliff F, Martin-Hernandez AM, Baulcombe DC. Technical advance: Tobacco rattle virus as a vector for analysis of gene function by silencing. Plant J. 2008;25:237–45.

13. Brigneti G, Martín-Hernández AM, Jin H, Chen J, Baulcombe DC, Baker B, Jones JDG. Virus-induced gene silencing in *Solanum* species. Plant J. 2004;39:264–72.

14. Du J, Tian Z, Liu J, Vleeshouwers VGAA, Shi X, Xie C. Functional analysis of potato genes involved in quantitative resistance to *Phytophthora infestans*. Mol Biol Rep. 2013;40:957–67.

15. Hawkes J. The potato: evolution, biodiversity and genetic resources. London: Belhaven Press; 1990.

16. Spooner DM, Berg RG van den, Rodrigues A, Bamberg JB, Hijmans RJ, Lara-Cabrera S. Wild potatoes (*Solanum* section *Petota*; Solanaceae) of North and Central America. BIS, Leerstoelgroep Biosystematiek, 30: The American Society of Plant Taxonomists; 2004 (Systematic Botany Monographs: 68).

17. Berloo R, Hutten RCB, Eck HJ, Visser RGF. An online potato pedigree database resource. Potato Res. 2007;50:45–57.

18. Jacobs MM, van den Berg RG, Vleeshouwers VG, Visser M, Mank R, Sengers M, Hoekstra R, Vosman B. AFLP analysis reveals a lack of phylogenetic structure within *Solanum* section *Petota*. BMC Evol Biol. 2008;8:145.

19. Vleeshouwers VGAA, Finkers R, Budding D, Visser M, Jacobs MMJ, van Berloo R, Pel M, Champouret N, Bakker E, Krenek P, Rietman H, Huigen D, Hoekstra R, Goverse A, Vosman B, Jacobsen E, Visser RGF. SolRgene: an online database to explore disease resistance genes in tuber-bearing *Solanum* species. BMC Plant Biol. 2011;11:116.

20. Ishihama N, Yamada R, Yoshioka M, Katou S, Yoshioka H. Phosphorylation of the *Nicotiana benthamiana* WRKY8 transcription factor by MAPK functions in the defense response. Plant Cell. 2011;23:1153–70.

21. Mase K, Mizuno T, Ishihama N, Fujii T, Mori H, Kodama M, Yoshioka H. Ethylene signaling pathway and MAPK cascades are required for AAL Toxin-induced programmed cell death. Mol Plant Microbe Interact. 2012;25:1015–25.

22. Samuel MA, Hall H, Krzymowska M, Drzewiecka K, Hennig J, Ellis BE. SIPK signaling controls multiple components of harpin-induced cell death in tobacco. Plant J. 2005;42:406–16.

23. Seo S, Katou S, Seto H, Gomi K, Ohashi Y. The mitogen-activated protein kinases WIPK and SIPK regulate the levels of jasmonic and salicylic acids in wounded tobacco plants. Plant J. 2007;49:899–909.

24. Yap Y, Kodama Y, Waller F, Chung KM, Ueda H, Nakamura K, Oldsen M, Yoda H, Yamaguchi Y, Sano H. Activation of a novel transcription factor through phosphorylation by WIPK, a wound-induced mitogen activated protein kinase in tobacco plants. Plant Physiol. 2005;139:127–37.

25. Ahlfors R, Macioszek V, Rudd J, Brosché M, Schlichting R, Scheel D, Kangasjärvi J. Stress hormone-independent activation and nuclear translocation of mitogen-activated protein kinases in *Arabidopsis thaliana* during ozone exposure. Plant J. 2004;40:512–22.

26. Kovtun Y, Chiu WL, Tena G, Sheen J. Functional analysis of oxidative stress-activated mitogen-activated protein kinase cascade in plants. Proc Natl Acad Sci USA. 2000;97:2940–5.

27. Gudesblat GE, Iusem ND, Morris PC. Guard cell-specific inhibition of Arabidopsis MPK3 expression causes abnormal stomatal responses to abscisic acid and hydrogen peroxide. New Phytol. 2007;173:713–21.

28. Hamel L-PP, Nicole M-CC, Sritubtim S, Morency M-JJ, Ellis M, Ehlting J, Beaudoin N, Barbazuk B, Klessig D, Lee J, Martin G, Mundy J, Ohashi Y, Scheel D, Sheen J, Xing T, Zhang S, Seguin A, Ellis BE. Ancient signals: comparative genomics of plant MAPK and MAPKK gene families. Trends Plant Sci. 2006;11:192–8.

29. Yang Q, Hua J, Wang L, Xu B, Zhang H, Ye N, Zhang Z, Yu D, Cooke HJ, Zhang Y, Shi Q. MicroRNA and piRNA profiles in normal human testis detected by next generation sequencing. PLoS ONE. 2013;8:e66809.

30. Wang H, Ngwenyama N, Liu Y, Walker JC, Zhang S. Stomatal development and patterning are regulated by environmentally responsive mitogen-activated protein kinases in *Arabidopsis*. Plant Cell. 2007;19:63–73.

31. García-Marcos A, Pacheco R, Martiáñez J, González-Jara P, Díaz-Ruíz JR, Tenllado F. Transcriptional changes and oxidative stress associated with the synergistic interaction between potato virus X and potato virus Y and their relationship with symptom expression. Mol Plant Microbe Interact. 2009;22:1431–44.

32. Hardin SC, Wolniak SM. Molecular cloning and characterization of maize ZmMEK1, a protein kinase with a catalytic domain homologous to mitogen- and stress-activated protein kinase kinases. Planta. 1998;206:577–84.

33. Hardin SC, Wolniak SM. Expression of the mitogen-activated protein kinase kinase ZmMEK1 in the primary root of maize. Planta. 2001;213:916–26.

34. Soyano T, Nishihama R, Morikiyo K, Ishikawa M, Machida Y. NQK1/NtMEK1 is a MAPKK that acts in the NPK1 MAPKKK-mediated MAPK cascade and is required for plant cytokinesis. Genes Dev. 2003;17:1055–67.

35. Takahashi Y, Soyano T, Kosetsu K, Sasabe M, MacHida Y. HINKEL kinesin, ANP MAPKKKs and MKK6/ANQ MAPKK, which phosphorylates and activates MPK4 MAPK, constitute a pathway that is required for cytokinesis in *Arabidopsis thaliana*. Plant Cell Physiol. 2010;51:1766–76.

36. Liu YY, Zhou Y, Liu L, Sun L, Zhang M, Liu YY, Li D. Maize ZmMEK1 is a single-copy gene. Mol Biol Rep. 2012;39:2957–66.

37. Lazar A, Coll A, Dobnik D, Baebler S, Bedina-Zavec A, Zel J, Gruden K. Involvement of potato (*Solanum tuberosum* L.) MKK6 in response to potato virus Y. PLoS ONE. 2014;9:e104553.

38. Jin H, Liu Y, Yang K-YY, Kim CY, Baker B, Zhang S. Function of a mitogen-activated protein kinase pathway in N gene-mediated resistance in tobacco. Plant J. 2003;33:719–31.

39. Liu Y, Schiff M, Dinesh-Kumar SP. Involvement of MEK1 MAPKK, NTF6 MAPK, WRKY/MYB transcription factors, COI1 and CTR1 in N-mediated resistance to tobacco mosaic virus. Plant J. 2004;38:800–9.

40. Liu Y, Ren D, Pike S, Pallardy S, Gassmann W, Zhang S. Chloroplast-generated reactive oxygen species are involved in hypersensitive response-like cell death mediated by a mitogen-activated protein kinase cascade. Plant J. 2007;51:941–54.

41. Halim VA, Hunger A, Macioszek V, Landgraf P, Nürnberger T, Scheel D, Rosahl S. The oligopeptide elicitor Pep-13 induces salicylic acid-dependent and -independent defense reactions in potato. Physiol Mol Plant Pathol. 2004;64:311–8.

42. Dobnik D, Baebler S, Kogovšek P, Pompe-Novak M, Stebih D, Panter G, Janež N, Morisset D, Zel J, Gruden K. β-1,3-Glucanase class III promotes spread of PVY(NTN) and improves in planta protein production. Plant Biotechnol Rep. 2013;7:547–55.

43. Demmig-Adams B, Adams WW. Photoprotection and other responses of plants to high light stress. Annu Rev Plant Physiol Plant Mol Biol. 1992;43:599–626.

44. Pompe-Novak M, Gruden K, Baebler Š, Krečič-Stres H, Kovač M, Jongsma M, Ravnikar M. Potato virus Y induced changes in the gene expression of potato (*Solanum tuberosum* L.). Physiol Mol Plant Pathol. 2006;67:237–47.

45. Rupar M, Faurez F, Tribodet M, Gutiérrez-Aguirre I, Delaunay A, Glais L, Kriznik M, Dobnik D, Gruden K, Jacquot E, Ravnikar M. Fluorescently tagged potato virus Y: a versatile tool for functional analysis of plant–virus interactions. Mol Plant Microbe Interact. 2015;28:739–50.

46. Zhang S, Klessig DF. Resistance gene N-mediated de novo synthesis and activation of a tobacco mitogen-activated protein kinase by tobacco mosaic virus infection. Proc Natl Acad Sci USA. 1998;95:7433–8.

47. Zhang S, Klessig DF. The tobacco wounding-activated mitogen-activated protein kinase is encoded by SIPK. Proc Natl Acad Sci USA. 1998;95:7225–30.

48. Kishi-Kaboshi M, Okada K, Kurimoto L, Murakami S, Umezawa T, Shibuya N, Yamane H, Miyao A, Takatsuji H, Takahashi A, Hirochika H. A rice fungal MAMP-responsive MAPK cascade regulates metabolic flow to antimicrobial metabolite synthesis. Plant J. 2010;63:599–612.

49. Melech-Bonfil S, Sessa G. Tomato MAPKKKε is a positive regulator of cell-death signaling networks associated with plant immunity. Plant J. 2010;64:379–91.

50. Liu Y, Jin H, Yang K-Y, Kim CY, Baker B, Zhang S. Interaction between two mitogen-activated protein kinases during tobacco defense signaling. Plant J. 2003;34:149–60.

51. Alazem M, Lin N-S. Roles of plant hormones in the regulation of host–virus interactions. Mol Plant Pathol. 2015;16:529–40.

52. Brodersen P, Petersen M, Bjørn Nielsen H, Zhu S, Newman M-A, Shokat KM, Rietz S, Parker J, Mundy J. *Arabidopsis* MAP kinase 4 regulates salicylic acid- and jasmonic acid/ethylene-dependent responses via EDS1 and PAD4. Plant J. 2006;47:532–46.

53. Petersen M, Brodersen P, Naested H, Andreasson E, Lindhart U, Johansen B, Nielsen HB, Lacy M, Austin MJ, Parker JE, Sharma SB, Klessig DF, Martienssen R, Mattsson O, Jensen AB, Mundy J. *Arabidopsis* map kinase 4 negatively regulates systemic acquired resistance. Cell. 2000;103:1111–20.

54. del Pozo O, Pedley KF, Martin GB. MAPKKKalpha is a positive regulator of cell death associated with both plant immunity and disease. EMBO J. 2004;23:3072–82.

55. Liu Y, Schiff M, Dinesh-Kumar SP. Virus-induced gene silencing in tomato. Plant J. 2002;31:777–86.

56. Du J, Rietman H, Vleeshouwers VGAA. Agroinfiltration and PVX agroinfection in potato and *Nicotiana benthamiana*. J Vis Exp. 2014;83:e50971.

57. Hren M, Nikolić P, Rotter A, Blejec A, Terrier N, Ravnikar M, Dermastia M, Gruden K. "Bois noir" phytoplasma induces significant reprogramming of the leaf transcriptome in the field grown grapevine. BMC Genom. 2009;10:460.

58. Kogovšek P, Gow L, Pompe-Novak M, Gruden K, Foster GD, Boonham N, Ravnikar M. Single-step RT real-time PCR for sensitive detection and discrimination of potato virus Y isolates. J Virol Methods. 2008;149:1–11.

59. Weller SA, Elphinstone JG, Smith NC, Boonham N, Stead DE. Detection of Ralstonia solanacearum strains with a quantitative, multiplex, real-time, fluorogenic PCR (TaqMan) assay. Appl Environ Microbiol. 2000;66:2853–8.

# 9

# Development of a universal and simplified ddRAD library preparation approach for SNP discovery and genotyping in angiosperm plants

Guo-Qian Yang[1,2], Yun-Mei Chen[1,2], Jin-Peng Wang[3], Cen Guo[1,2], Lei Zhao[1,2], Xiao-Yan Wang[1], Ying Guo[1,2], Li Li[3], De-Zhu Li[1*] and Zhen-Hua Guo[1*] (iD)

## Abstract

**Background:** The double digest restriction-site associated DNA sequencing technology (ddRAD-seq) is a reduced representation sequencing technology by sampling genome-wide enzyme loci developed on the basis of next-generation sequencing. ddRAD-seq has been widely applied to SNP marker development and genotyping on animals, especially on marine animals as the original ddRAD protocol is mainly built and trained based on animal data. However, wide application of ddRAD-seq technology in plant species has not been achieved so far. Here, we aim to develop an optimized ddRAD library preparation protocol be accessible to most angiosperm plant species without much startup pre-experiment and costs.

**Results:** We first tested several combinations of enzymes by in silico analysis of 23 plant species covering 17 families of angiosperm and 1 family of bryophyta and found *AvaII + MspI* enzyme pair produced consistently higher number of fragments in a broad range of plant species. Then we removed two purifying and one quantifying steps of the original protocol, replaced expensive consumables and apparatuses by conventional experimental apparatuses. Besides, we shortened P1 adapter from 37 to 25 bp and designed a new barcode-adapter system containing 20 pairs of barcodes of varying length. This is an optimized ddRAD strategy for angiosperm plants that is economical, time-saving and requires little technical expertise or investment in laboratory equipment. We refer to this simplified protocol as *Mi*ddRAD and we demonstrated the utility and flexibility of our approach by resolving phylogenetic relationships of two genera of woody bamboos (*Dendrocalamus* and *Phyllostachys*). Overall our results provide empirical evidence for using this method on different model and non-model plants to produce consistent data.

**Conclusions:** As *Mi*ddRAD adopts an enzyme pair that works for a broad range of angiosperm plants, simplifies library constructing procedure and requires less DNA input, it will greatly facilitate designing a ddRAD project. Our optimization of this method may make ddRAD be widely used in fields of plant population genetics, phylogenetics, phylogeography and molecular breeding.

**Keywords:** RAD-seq, ddRAD, *Mi*ddRAD, Genotype-by-sequencing, Next-generation sequencing

*Correspondence: dzl@mail.kib.ac.cn; guozhenhua@mail.kib.ac.cn
[1] Germplasm Bank of Wild Species, Kunming Institute of Botany, Chinese Academy of Sciences, Kunming 650201, China
Full list of author information is available at the end of the article

## Background

Restriction-site associated DNA sequencing technology (RAD-seq) is a reduced representation sequencing technology by sampling genome-wide single enzyme loci developed on the basis of next-generation sequencing [1, 2]. The technology breaks genome into a certain size of DNA fragments by employing a restriction endonuclease (usually a low-frequency cutter) combined with the ultrasonic shearing method, then the fragmented DNA is enriched for constructing a sequencing library so that sequences beside the cleavage site can be acquired for high-throughput sequencing [3]. Because RAD-tags are DNA fragments beside a specific restriction site from the whole genome, so they can generally reflect the sequence characteristics of the entire genome. It is now possible to obtain hundreds to thousands of single nucleotide polymorphism (SNP) markers within a species or between closely related species through RAD-seq. Until now RAD-seq has been successfully applied to SNP marker development, high-density genetic map construction, QTL mapping, population genetics and phylogenetic research on eggplants, chickpeas, sesames, soybeans, cucurbit bottle gourds, bamboos, beetles, and other organisms [4–13]. But on one aspect experimental procedure of this technology is much complex and it requires a Covaris ultrasonicator and some other specialized instruments, so personnels under professional training are usually required to master the technique; on the other hand, random physical shearing methods implemented in the library construction process will result in losing lots of DNA, thereby leading to out control of the final tag number [3, 14]. So several laboratories have improved and simplified the traditional RAD-seq method, from which a variety of low cost, high throughput reduced representation sequencing methods are available. At present, reduced representation sequencing methods developed from the RAD-seq mainly includes GBS series techniques and RAD series techniques [14]. GBS and RAD-seq techniques share several basic steps while differ only in the order or details of enzyme digestion, adapter ligation, barcoding and size selection. Each alternative RAD method has both advantages and drawbacks. RAD series control number of the tags by both choosing the enzyme and size selection while GBS series techniques or close derivatives control the number of tags only by selecting different enzymes (though some GBS users may also add a size selection step in their modified GBS protocol [15], the original intention of GBS is to reduce library preparation workflow without size selection). GBS series techniques include single and double enzyme GBS [16, 17], both of which employ simple library constructing processes, but they can only enrich small fragments less than ~350 bp [18]. It's easy to sequence through the

short fragment with pair-end sequencing mode as the sequencing length is gradually becoming longer, which will result in a waste of data and the potential to discover more SNPs. Furthermore, fragments of various lengths will increase the potential for amplification bias [19, 20] and cause a decline in the data quantity and data quality. RAD series mainly includes 2b-RAD [21], ddRAD [22] and ezRAD [23]. 2b-RAD adopts a kind of type II restriction endonuclease to digest the genome, producing only ~33 bp fragments, which lack of biases due to fragment size selection but may restrict the potential for discovering more SNPs. ezRAD is the only protocol that relies on illumina authoritative kits to construct the library with customer support but the cost is still not as low as the author claimed [24]. ddRAD can tune fragments number by employing two different enzymes and size selection, and the process of constructing a library is quite simple while genomic DNA it requires is of the highest quality in all the RAD methods [24].

All RAD protocols have been proved to be powerful tools for SNP discovery and genotyping of model and non-model species. However, startup of them all usually involves pre-experiment of (1) testing candidate endonuclease that could produce a suitable RAD or GBS library [25], and (2) purchasing some relatively expensive consumables and apparatus (e.g. Agilent 2100 Bioanalyzer). This requires a significant initial investment for labs focused on traditional genotyping methods (e.g. SSR genotyping). Besides, many labs (e.g. phylogenetic bio-labs) are probably focusing on different model or non-model plant species, once a pair of enzymes selected and adapters purchased for one target species, they have to consider if these consumables could be applied to another species efficiently to be studied even some commonly used enzyme pairs could produce hundreds to thousands of markers across wide-range species. Enzyme pairs simulation of the original ddRAD protocol is mainly based on animal genomes and it is hard for us to know if performance of the enzyme pairs is good as well in plant species as Santiago et al. found that a given restriction enzyme may have strikingly variable recognition-sequence frequencies among broad eukaryotic taxonomic groups, and only phylogenetic related species could produce similar recognition-sequence frequencies [26]. In another study, Burford et al. found some enzyme pairs work more consistently than others across a wide range of taxonomic groups after optimizing ddRAD protocol and testing several restriction enzyme pairs for five genera of insects and fish [27]. Here, we sought to test the universality of several commonly used enzyme pairs across most angiosperm plants, simplify the ddRAD protocol and reduce the overall costs. Our protocol is generally according to the protocol described by Peterson et al. [22], but with some modifications as we first tested several combinations of enzymes by in silico analysis of 23 plant species covering 17 families of angiosperm (16 orders, two classes) and one family of bryophyta (one orders, one class) and found *AvaII* + *MspI* enzyme pair produced consistently higher number of fragments in a broad range of plant species. Furthermore, we removed two purifying and one quantifying steps, shortened the adapters and replaced expensive instruments by conventional experimental apparatuses which make it possible to do ddRAD sequencing with no additional investment beyond the cost of library preparation and sequencing itself.

To assess the performance of this approach, we got empirical results from the model species *Oryza sativa* L. japonica and *Zea mays* L. We also explored repeatability by testing the effectiveness of the method in non-model species *Phyllostachys edulis* and *Alloteropsis semialata* (R. Br.) Hitchc. Finally, we managed to reconstruct phylogenetic relationships of two woody bamboos genera, *Dendrocalamus* and *Phyllostachys* with data generated by the protocol. This generalized approach, using the fixed enzyme pair and standard library preparation protocol, will allow researchers to apply ddRAD-seq technology to a wide array of plants and research questions. We expect that this optimized protocol could be efficiently implemented in any small or middle-sized laboratory with few people and limited funds.

## Methods

### Plant material and DNA samples

In this project, we used *Oryza sativa* L. spp. japonica and *Z. mays* L. to estimate the robustness of our Protocol B. Besides, a total of six species of Poaceae including four temperate woody bamboo species (*Chimonocalamus pallens*, *Phyllostachys edulis*, *Phyllostachys rubicunda* T. H. Wen and *Phyllostachys vivax* McClure), one tropical woody bamboo species (*Dendrocalamus latiflorus*) and one grass species *Alloteropsis semialata* (R. Br.) Hitchc. were used in our protocols as well. Leaves of temperate woody bamboos were mostly collected from plants grown in Kunming Botanical Garden (N25°07′04.9″, E102°44′15.2″) and leaves of tropical woody bamboos, *O. sativa*, *Z. mays* and *A. semialata* were collected from plants grown in our greenhouses. All necessary permits were obtained before collecting the material. Fresh leaves of all species were obtained and then dried rapidly in silica gel. The DNA was extracted with a modified CTAB method [28].

### Choosing restriction enzymes and adapter design

At first, we selected six kinds of enzyme pairs that could recognize restriction sites of different lengths including eight bases + six bases (*SbfI* + *EcoRI*), eight bases + four bases (*SbfI* + *MluCI*), six bases + four bases (*EcoRI* + *MspI*, *PstI* + *MspI*), 4.5 bases + four bases (*AvaII* + *MspI*), four bases + four bases (*NlaIII* + *MluCI*), of which *EcoRI* + *MspI* was adopted by the original ddRAD protocol and *PstI* + *MspI* was used by the two-enzyme of GBS protocol [17, 22]. Restriction enzymes included in this study are listed in Additional file 2: Table S2. Then we in silico digested genome sequences of 23 plant species covering 17 families of angiosperm (16 orders, two classes) and one family of bryophyte (one orders, one class) of different genome size with RestrictionDigest [29]. For each enzyme pair, we recorded the total number of fragments and the number of fragments between 400–700 bp that could produce in each species. The species adopted for analysis are listed in Additional file 2: Table S1. Genome scaffolds of these species were downloaded from Plantgdb [31]. Then the distribution of DNA fragments was screened by agarose gel electrophoresis after digestion of genomic DNA of some species.

Chemosynthetic oligonucleotides of P1 and P2 adapters will account for almost half of the cost due to the need for high-performance liquid chromatography (HPLC) purification and 5′-end phosphorylation. In our protocols, original P1 adapters are shortened from 37 to 25 bp (barcode length is assumed to be 5 bp) to reduce the cost of the synthesizing DNA oligos. Besides, a different barcode-adapter system containing 20 pairs of barcodes varying in length was devised, which can be used with integer times (20 * n), rather than the original 48 kinds of barcodes with equal length (see Additional file 1). This will not only increase the flexibility of barcodes for projects with diverse samples but also improve the quality of bases near the restriction site.

### Protocols of *MiddRAD* for next-generation sequencing

We initially provided two protocols for constructing a library. Protocol A differs from protocol B only in when to select target fragments. In protocol A, selecting fragments was placed in the last step, i.e. products of all adapter-ligated restriction fragments were as templates of the PCR reaction; however in protocol B, only selected adapter-ligated restriction fragments were as templates for PCR amplification. Two non-model species *D. latiflorus* and *C. pallens* were used to construct libraries with protocol A while the model species *O. sativa* and *Z. mays* were used to construct libraries with protocol B (as data produced from protocol A contains too many adapters, we did not continue to verify this protocol in model plants). Figure 1 provides a flowchart that outlines all stages of protocol B and the original ddRAD. Protocol A and protocol B were detailed in Additional file 1. The protocol A flowchart was presented in Additional file 2: Figure S1. Sequencing of protocol A was performed on the

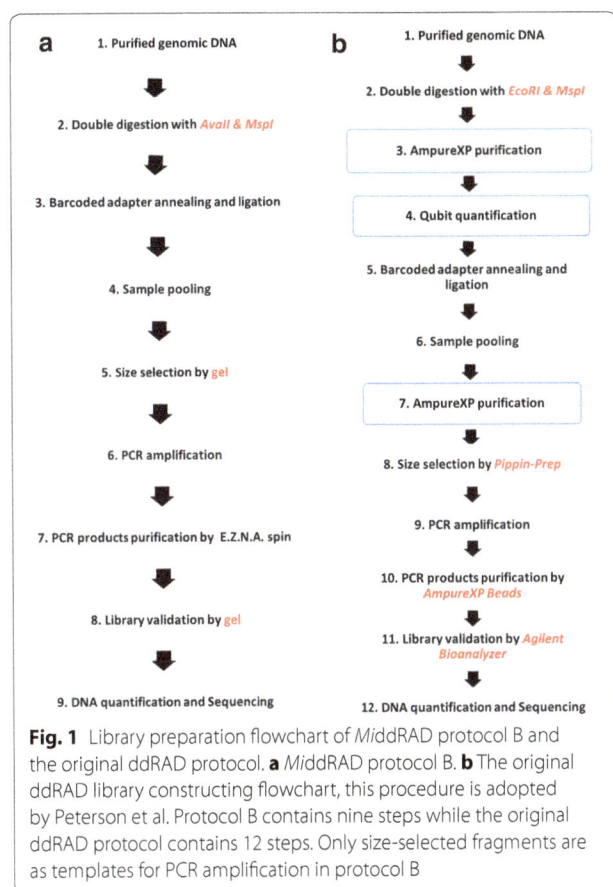

**Fig. 1** Library preparation flowchart of *Mi*ddRAD protocol B and the original ddRAD protocol. **a** *Mi*ddRAD protocol B. **b** The original ddRAD library constructing flowchart, this procedure is adopted by Peterson et al. Protocol B contains nine steps while the original ddRAD protocol contains 12 steps. Only size-selected fragments are as templates for PCR amplification in protocol B

Illumina HiSeq 2000 System (San Diego, CA, USA) using the pair read, 100 nucleotide configuration at Kunming Institute of Zoology, CAS while Sequencing of Protocol B was performed on the Illumina HiSeq X Ten System (San Diego, CA, USA) using the pair read, 150 nucleotide configuration at Cloud Health Genomics Ltd. To test the universality of *Mi*ddRAD and the restriction enzymes on more plant species, we constructed libraries for *P. edulis* and *A. semialata* with the same enzyme pairs. Libraries were constructed strictly according to Protocol B and were sent to Cloud Health Genomics Ltd. for sequencing using Illumina HiSeq X Ten (San Diego, CA, USA) with PE150 bp sequencing mode.

Then we adopted protocol B to construct libraries for three bamboo species (contains two *D. latiflorus* individuals, one *P. rubicunda* individual and one *P. vivax* individual) to explore the applicability of *Mi*ddRADseq-derived genotypes/markers in resolving phylogenetic problems. The library constructing process is according to Protocol B and fragments selected were set to 600–700 bp. Reagents and enzymes used were mainly purchased from New England Biolabs Inc. (R0153S, R0106S), Vazyme Biotech Co., Ltd. (C301-01) and SunShineBio Co., Ltd.

(SN124). Libraries were then sequenced in Cloud Health Genomics Ltd. Sequencing platform was Illumina HiSeq X Ten (San Diego, CA, USA) with sequence length PE150 bp.

To evaluate the shortened adapters and redesigned barcodes, we constructed four *Mi*ddRAD sub-libraries according to protocol B for 40 offsprings of a *D. latiflorus* F1 population and sequenced the final library with a single illumina HiSeq X ten lane (PE150 bp). The coefficient of variation (CV = standard deviation/mean) of data generated by each barcode and each sub-library were analyzed to evaluate the newly designed barcodes, indexes and shortened P1 adapters.

**Sequence quality analysis, SNP calling and genotyping**
Raw reads were demultiplexed by process_radtags program in Stacks software version 1.24 [30, 31]. Average sequence quality per read and GC–content were checked using FastQC version 0.11.3 [32]. Adapter reads were searched by Cutadapt 1.9.1 [33]. Reads containing correct restriction sites in read1 and read2 were obtained by searching restriction sites sequences in the raw reads respectively. Clean data were produced by removing the adapter reads and reads with ambiguous or low quality (below a Phred score of Q10) bases. To determine the mapping ratio of sampled reads to the genome, clean reads of *O. sativa*, *Z. mays* and *A. semialata* were mapped onto the rice, maize and sorghum genome scaffolds, CDS-DNA and repeats region respectively, while clean reads of temperate bamboo individuals onto the *P. edulis* reference genome, CDS-DNA region and repeats region and reads of tropical bamboo individuals onto the *D. latiflorus* survey genome (Zhenhua Guo et al. unpublished data) with bowtie [34]. Rice, maize and sorghum genome scaffolds, CDS-DNA and repeats region were downloaded from Plantgdb [35] while *P. edulis* reference genome, CDS region, and repeats region were downloaded from BambooGDB [36]. To obtain the number of tags, clean reads (we only used read1 for analysis) of all individuals were first trimmed 140 bp (when read length is PE150) and clustered with ustacks/pstacks program, then the reducing efficiency was determined by calculating the percentage of total tag length in total nuclear genome length.

To estimate the performance of *Mi*ddRAD protocol, tags of rice and maize produced by empirical sequencing results were compared with those predicted from in silico analysis to show how actual data meet the in silico expectations.

In order to identify SNP markers and genotypes for inferring phylogeny of three woody bamboo species, the Stacks software pipeline was implemented for the processing of Illumina sequence read data and screening

SNPs that are fixed-within a species while vary among different species. Sequence trimming was first performed using process_radtags program to remove adapter reads and reads with bases below a Phred score of Q10 within a 15 bases sliding window. Clean sequences were truncated to a final length of 140 base pairs (excluding the barcode but containing enzyme recognition site) prior to clustering. For each sample, the ustacks program was used to merge short-read sequences into tags/loci using removal algorithm and deleveraging algorithm (−m10, −M3). Then a catalog was built from all samples by the cstacks program (−n5). Tags from each sample were matched against the catalog to determine alleles with sstacks program and the populations program was used to output SNPs in Phylip format. The minimum number of taxa required for an informative unrooted phylogenetic tree is three. The major parameters m (minimum number of identical reads required to form a stack), M (maximum number of nucleotides mismatches allowed between stacks before fusing stacks into a locus) and n (number of mismatches allowed between loci in the catalog) were tuned to get the matrix with a variable number of SNPs. Furthermore, to validate the genotyping accuracy, we presented linkage map results of one *D. latiflorus* F1 mapping population (Guoqian Yang et al., unpublished data) according to *MiddRAD* protocol and 55 genotypes (eight markers, seven individuals and one genotype was missing during the SNP calling pipeline) were randomly selected and verified by independent Sanger sequencing.

### Phylogenetic tree construction of three woody bamboo species

We inferred ML phylogenies for each data matrix using RAxML version 8.0.0 [37]. ML searches were conducted in RAxML with the GTRGAMMAI model for sequence data, and a rapid bootstrapping analysis with 100 bootstrap replicates was conducted. Phylogenetic accuracy was determined by comparing inferred trees with published reference phylogenies. The reference phylogeny for woody bamboos from [38, 39], was estimated using parsimony analyses (MP) and Bayesian inference (BI).

### Results
#### Choosing universal restriction enzyme pairs across angiosperm plants

Six combinations of endonucleases identifying 4–8 nucleotide bases were tested on genomes of 23 plant species (covering 17 families of angiosperm and one family of bryophyta). The ideal combination should be able to generate a consistently higher number of sequenceable fragments across species. The four bases + eight bases enzyme pair and six bases + eight bases enzyme pair usually produced a few thousand fragments, which

were far from the requirement for large-scale genotyping (Table 1). However, the four bases + four bases enzyme pair produced up to ~23,112,695 fragments which made it difficult to control error in practice and required deep sequencing depth. Meanwhile, the four bases + six bases enzyme pair generally produced 32,319–886,527 fragments which made it easy for people to obtain a sufficient number of fragments without sequencing a large amount of data. *PstI* + *MspI* which was used in the original two-enzyme GBS protocol performed well in most plants, but only produced 3791 fragments between 400–700 bp for *Cucumis sativus* and 8258 fragments for *Carica papaya* (Table 2). *EcoRI* + *MspI* which was used in the original ddRAD protocol performed better than *PstI* + *MspI* in any simulated species by producing thousands or more tags, but only 8173 fragments fell into within 400–700 bp for *C. sativus*, which could not meet the demand for more tags in some studies. We found *AvaII* + *MspI* enzyme pair was superior to both of *EcoRI* + *MspI* and *PstI* + *MspI*. This enzyme pair could produce at least 13,958 segments between 400–700 bp in the 23 simulated species and was predicted to provide sufficient tags across diverse plant species. The largest genome (*Z. mays*, 2300 Mb) could produce 4,784,940 tags with 517,204 tags between 400–700 bp while even the smallest genome (*Prunus persica*, 226.6 Mb) could produce 237,185 tags with 34,514 tags between 400–700 bp (Fig. 2a; Table 2). Both of the enzymes are common enzymes with AvaII identifying 4.5 bases and MspI identifying four bases which are different from the combination with a common enzyme and a rare enzyme adopted in ddRAD or two-enzyme GBS. Correlation analysis showed that the total tag number is correlated positively with genome size with $R^2 = 0.9185$ and tag number between 400–700 bp is correlated positively with genome size as well with $R^2 = 0.9476$ (Fig. 2b, c). So once we get to know the genome size of one plant, the tag number produced could be estimated and the expected tag number could be tuned by selecting a proper size range. *EcoRI* + *MspI* and *PstI* + *MspI* could also be taken into consideration when designing a ddRAD project as they may produce hundreds to thousands of markers across a wide range of plant species and the total tag number or tag number between 400–700 bp is correlated positively with genome size as well (Additional file 2: Figure S2). After conducting the above simulations, we built ddRAD libraries with the *AvaII* + *MspI* enzyme pair. Fragments between 400–700 bp are highly recommended for their high sequencing efficiency on illumina system. Optimization of the ratio of sample DNA to the adapters is not required when the genome size is less than 20 Gb because we have added excess adapters in our protocol which could make each fragment be ligated with corresponding adapters (Adapter P1 contains about $3 \times 10^{12}$

**Table 1  Total number of fragments produced by in silico digestion of 23 species**

| Species | SbfI + EcoRI | SbfI + MlucI | NlaIII + MlucI | AvaII + MspI | EcoRI + MspI | PstI + MspI |
|---|---|---|---|---|---|---|
| Brassica rapa | 2352 | 2795 | 3,710,380 | 271,860 | 88,220 | 70,512 |
| Glycine max | 7498 | 9006 | 16,747,131 | 803,332 | 306,391 | 162,398 |
| Populus trichocarpa | 4876 | 5769 | 7,169,615 | 329,150 | 147,647 | 97,378 |
| Vitis vinifera | 4994 | 5903 | 8,028,433 | 380,967 | 162,506 | 91,994 |
| Brachypodium distachyon | 12,306 | 14,912 | 3,237,571 | 535,195 | 106,924 | 166,584 |
| Carica papaya | 4592 | 5255 | 4,221,765 | 245,762 | 99,163 | 59,272 |
| Physcomitrella patens | 4786 | 5413 | 7,434,539 | 329,065 | 144,069 | 102,802 |
| Cucumis sativus | 1450 | 1737 | 3,342,214 | 116,754 | 62,340 | 32,319 |
| Musa acuminata | 6599 | 7483 | 5,467,255 | 433,922 | 170,591 | 129,105 |
| Nelumbo nucifera | 9469 | 10,906 | 10,160,790 | 944,853 | 308,661 | 164,996 |
| Theobroma cacao | 3376 | 3979 | 6,027,724 | 226,779 | 114,082 | 80,088 |
| Phoenix dactylifera | 8978 | 10,334 | 7,341,396 | 716,454 | 218,488 | 177,248 |
| Amborella trichopoda | 11,702 | 13,558 | 10,008,594 | 787,796 | 234,856 | 120,412 |
| Beta vulgaris | 6458 | 7804 | 7,904,182 | 483,300 | 199,616 | 103,224 |
| Sesamum indicum | 3412 | 4135 | 4,915,092 | 247,403 | 113,511 | 78,634 |
| Eucalyptus grandis | 8364 | 9943 | 9,894,486 | 849,552 | 307,001 | 174,810 |
| Prunus persica | 3116 | 4351 | 3,505,137 | 237,185 | 90,631 | 67,280 |
| Solanum lycopersicum | 15,086 | 18,005 | 12,249,900 | 596,412 | 255,692 | 124,564 |
| Oryza sativa | 10,042 | 11,855 | 4,881,316 | 595,046 | 138,534 | 171,045 |
| Phyllostachys edulis | 51,097 | 58,916 | 23,112,695 | 3,233,281 | 734,959 | 707,912 |
| Sorghum bicolor | 21,528 | 23,978 | 8,200,062 | 1,217,504 | 301,472 | 329,922 |
| Setaria italica | 15,347 | 17,861 | 4,683,693 | 757,165 | 145,882 | 225,478 |
| Zea mays | 85,902 | 2797 | 21,096,385 | 4,784,940 | 807,008 | 886,527 |

molecules while Adapter P2 contains about $6 \times 10^{12}$ molecules). As average genome size of the angiosperm is 5.79 Gb while bryophyte is 0.66 Gb according to Plant DNA C-values Database at Kew [40], we believe that this combination of two common endonucleases may be applied to diverse plant species only by tuning the size selected.

## A comprehensive evaluation of the library quality and data quality

The performance of our protocols was evaluated from both the experimental results and data analysis results. From the experimental perspective, library concentration should meet the criteria for sequencing and fragments selected should be in the expected range. From data analysis perspective, the library should produce sufficient high-quality data for downstream analysis.

We first quantified concentration of the libraries and screened fragments distribution to evaluate the quality of protocol A and protocol B. Concentration of library A (constructed according to protocol A) was between 5–9 ng/ul, while concentration of library B (constructed according to protocol B) was between 20–30 ng/ul, both of which could meet the requirements for Illumina sequencing. Fragments distribution of library A

and library B screened by the agarose gel electrophoresis is well within the expected range (Additional file 2: Figure S3a). Fragments distribution results for library B had been further confirmed by the Agilent 2100 Bioanalyzer (Additional file 2: Figure S3b) while library A got no peaks from Agilent 2100 because its concentration is lower than 10 ng/ul. Therefore, we believe that fragments distribution can be determined by using agarose gel electrophoresis instead of the highly sensitive but expensive Agilent 2100 Bioanalyzer. Both protocols could produce libraries that can be sequenced on the Illumina sequencing platform.

Then we conducted a comprehensive data analysis including data quality distribution, GC-content, adapter reads ratio and correct restriction sites ratio of both data produced by library A and library B. As for library A, *D. latiflorus* yielded a total of 2,890,217 raw reads (i.e. 578 Mb raw data) with 58 % GC-content; read1 containing correct restriction sites accounted 95.9 % of raw reads and read2 containing the correct restriction sites accounted for 94.8 %; read1 had a ratio of 49.3 % adapter reads while read2 had a ratio of 48.0 % adapter reads. *C. pallens* yielded a total of 3,146,515 raw reads (i.e. 629 Mb raw data) with 57 % GC-content; read1 containing correct restriction sites accounted 95.5 % of raw reads and

**Table 2 Number of fragments between 400–700 bp produced by in silico digestion of 23 species**

| Species | SbfI + EcoRI | SbfI + MlucI | NlaIII + MlucI | AvaII + MspI | EcoRI + MspI | PstI + MspI |
|---|---|---|---|---|---|---|
| Brassica rapa | 132 | 287 | 42,400 | 42,261 | 14,803 | 11,512 |
| Glycine max | 521 | 524 | 51,073 | 105,780 | 42,596 | 18,519 |
| Populus trichocarpa | 441 | 226 | 20,508 | 45,611 | 23,272 | 13,128 |
| Vitis vinifera | 386 | 383 | 32,588 | 50,912 | 21,822 | 11,769 |
| Brachypodium distachyon | 796 | 1948 | 50,586 | 60,014 | 18,552 | 21,701 |
| Carica papaya | 276 | 235 | 19,405 | 32,969 | 14,169 | 8258 |
| Physcomitrella patens | 433 | 481 | 30,274 | 45,223 | 20,689 | 14,832 |
| Cucumis sativus | 103 | 83 | 10,113 | 13,958 | 8173 | 3791 |
| Musa acuminata | 389 | 913 | 51,225 | 55,068 | 26,123 | 17,209 |
| Nelumbo nucifera | 783 | 607 | 80,253 | 140,632 | 49,260 | 24,250 |
| Theobroma cacao | 242 | 209 | 16,355 | 31,259 | 17,838 | 12,145 |
| Phoenix dactylifera | 622 | 876 | 64,497 | 95,233 | 36,744 | 26,304 |
| Amborella trichopoda | 1285 | 972 | 61,985 | 121,796 | 35,348 | 16,701 |
| Beta vulgaris | 541 | 472 | 42,743 | 65,689 | 29,666 | 13,827 |
| Sesamum indicum | 277 | 152 | 15,658 | 37,530 | 17,857 | 11,229 |
| Eucalyptus grandis | 655 | 666 | 65,729 | 129,175 | 53,775 | 26,496 |
| Prunus persica | 274 | 123 | 14,466 | 34,514 | 14,194 | 9910 |
| Solanum lycopersicum | 2250 | 421 | 42,919 | 93,748 | 36,617 | 16,313 |
| Oryza sativa | 696 | 1451 | 61,098 | 75,621 | 23,537 | 24,080 |
| Phyllostachys edulis | 2309 | 7086 | 367,267 | 458,669 | 130,850 | 116,861 |
| Sorghum bicolor | 1923 | 2740 | 128,588 | 161,131 | 56,433 | 56,091 |
| Setaria italica | 896 | 2434 | 81,751 | 96,716 | 26,592 | 31,528 |
| Zea mays | 3344 | 289 | 521,797 | 517,204 | 122,210 | 131,141 |

read2 containing the correct restriction sites accounted for 94.5 %; read1 had a ratio of 40.3 % adapter reads while read2 had a ratio of 39.6 % adapter reads. Raw reads of *D. latiflorus* and *C. pallens* both had an average base Quality Score larger than 20. Furthermore, bases of restriction enzyme cutting site had an average base Quality Score larger than 30. As for library B, *O. sativa* yielded a total of 14,732,449 raw reads (i.e. 4.1 Gb raw data) with 51.5 % GC-content (Fig. 3a, c); read1 containing correct restriction sites accounted 95.80 % of raw reads and read2 containing the correct restriction sites accounted for 95.39 % (Fig. 3b); read1 had a ratio of 2.63 % adapter reads while read2 had a ratio of 3.37 % adapter reads (Fig. 3d). *Zea mays* yielded a total of 7,414,009 raw reads (i.e. 2.1 Gb raw data) with 57 % GC-content; read1 containing correct restriction sites accounted 96.18 % of raw reads and read2 containing correct restriction sites accounted for 96.37 %; read1 had 2.48 % adapter reads while read2 had 3.29 % adapter reads. Raw reads of *O. sativa* and *Z. mays* both had an average base Quality Score larger than 20 while bases of restriction enzyme cutting site had an average base Quality Score larger than 30 (Fig. 3e, f). To determine the mapping ratio of sampled reads to the reference genome, we mapped clean reads of rice and maize onto the rice and maize reference genome scaffolds,

CDS-DNA and repeats region respectively. Overall scaffolds mapping rate was 82.5–90.66 %, reads mapping to the CDS-DNA accounts 2.38–2.83 % for maize and accounts ~19.50 % for rice. Yet reads mapping on the repeats region accounted for less than 11.00 % (Table 3).

Per Bases Quality Score is a major index of the sequence quality, the higher the Quality Score, the lower probability of sequencing error occurs. Q20 and Q30 represent the sequencing error probability of 1 and 0.1 %. Illumina sequencing was found to favor the more GC-balanced regions, leading to few or no reads from the many GC-poor regions and GC bias can be introduced at several processes of Illumina sequencing, e.g. PCR amplification of the library, cluster amplification, and the sequencing step [41]. So if GC-content is around 50 %, we can conclude no bias exists in library preparation and sequencing process. Adapter reads ratio is the percentage of reads with adapters in raw reads and is an indicator of data quality. Adapter reads should be removed in the subsequent analysis. Through percentage of reads containing correct restriction sites, we can determine whether the enzyme digestion reaction works in the right way. Comprehensive analysis of data quality distribution, GC-content, adapter reads ratio and correct restriction sites ratio showed that both *Mi*ddRAD protocols could

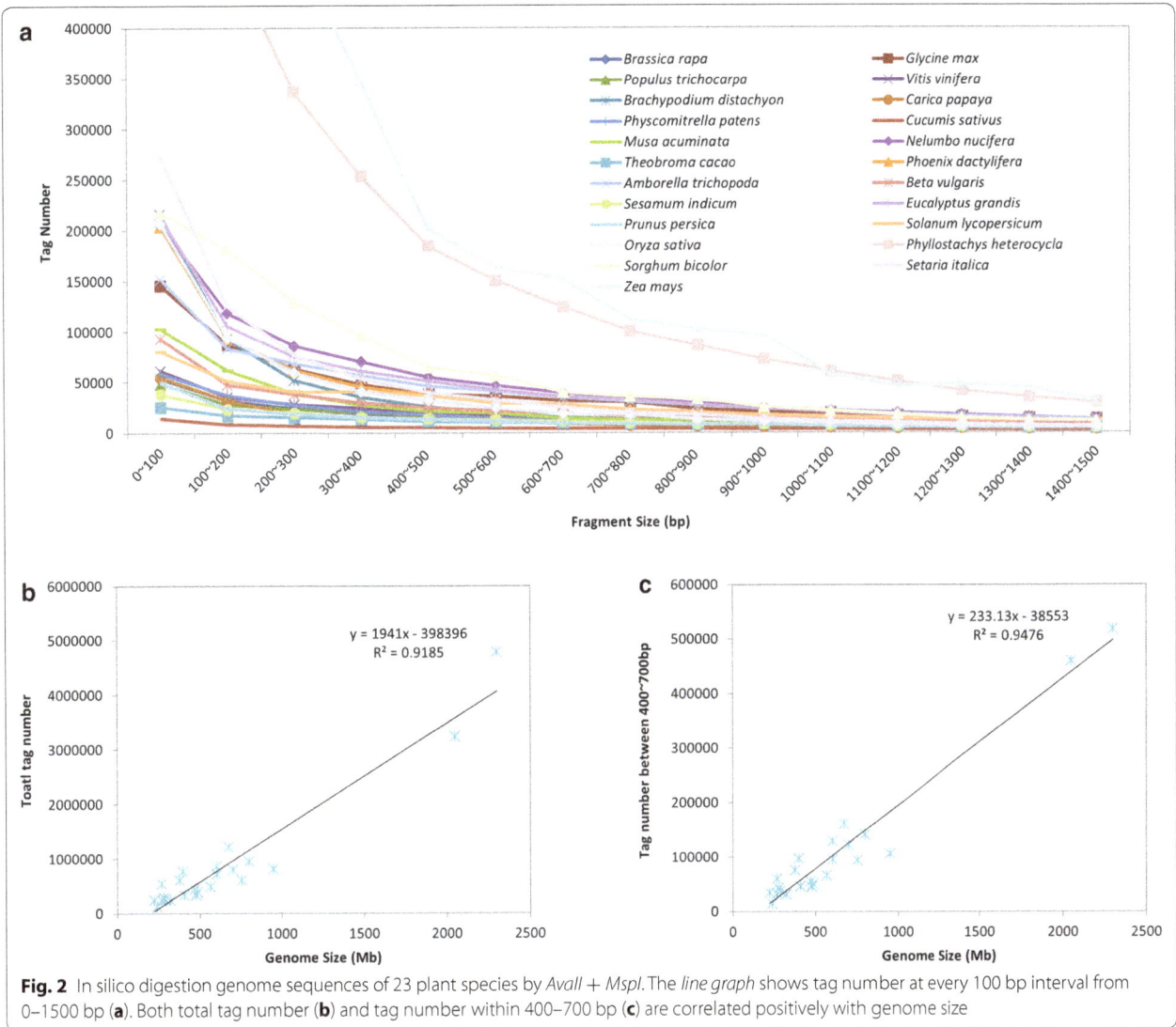

**Fig. 2** In silico digestion genome sequences of 23 plant species by *AvaII* + *MspI*. The *line graph* shows tag number at every 100 bp interval from 0–1500 bp (**a**). Both total tag number (**b**) and tag number within 400–700 bp (**c**) are correlated positively with genome size

produce high-quality data (We did not compare our data with original ddRAD data as the original ddRAD protocol did not supply their raw data, but the quality of our data is self-explaining). However, protocol A produced too many (nearly half of raw reads) reads with adapters which indicated many short fragments may exist in the selected gel, so we did not continue testing protocol A in model plants and took protocol B as the final protocol of *Mi*ddRAD.

### Comparison of empirical and simulated data and inference tags origin from the genome

A comprehensive evaluation of the protocol B was further conducted by comparison of the simulated data with the actual fragments we got. Clean data of rice and maize were clustered into tags using pstacks program.

The number of tags obtained from rice was ~66,547 with an average depth of 212.58X while maize got 290,001 tags with an average depth of 25.30X (Table 4). The expected number of tags accounted for 86.54 and 97.99 % of the actual number respectively which is similar to the results done by Sun et al.(82.86 % for rice) [42]. Then we estimated the number of fragments distributed on 12 chromosomes of rice and 10 chromosomes of maize respectively. The actual number of tags obtained was compared to the expected data to test the degree of consistency (Fig. 4). We found each of the 12 rice chromosomes was expected to produce 3521–6414 fragments while each actually generated 4121–7404 tags with the Pearson correlation coefficient r = 0.8374. The 10 maize chromosomes each was expected to produce 20,555–41,216 fragments while each was observed to

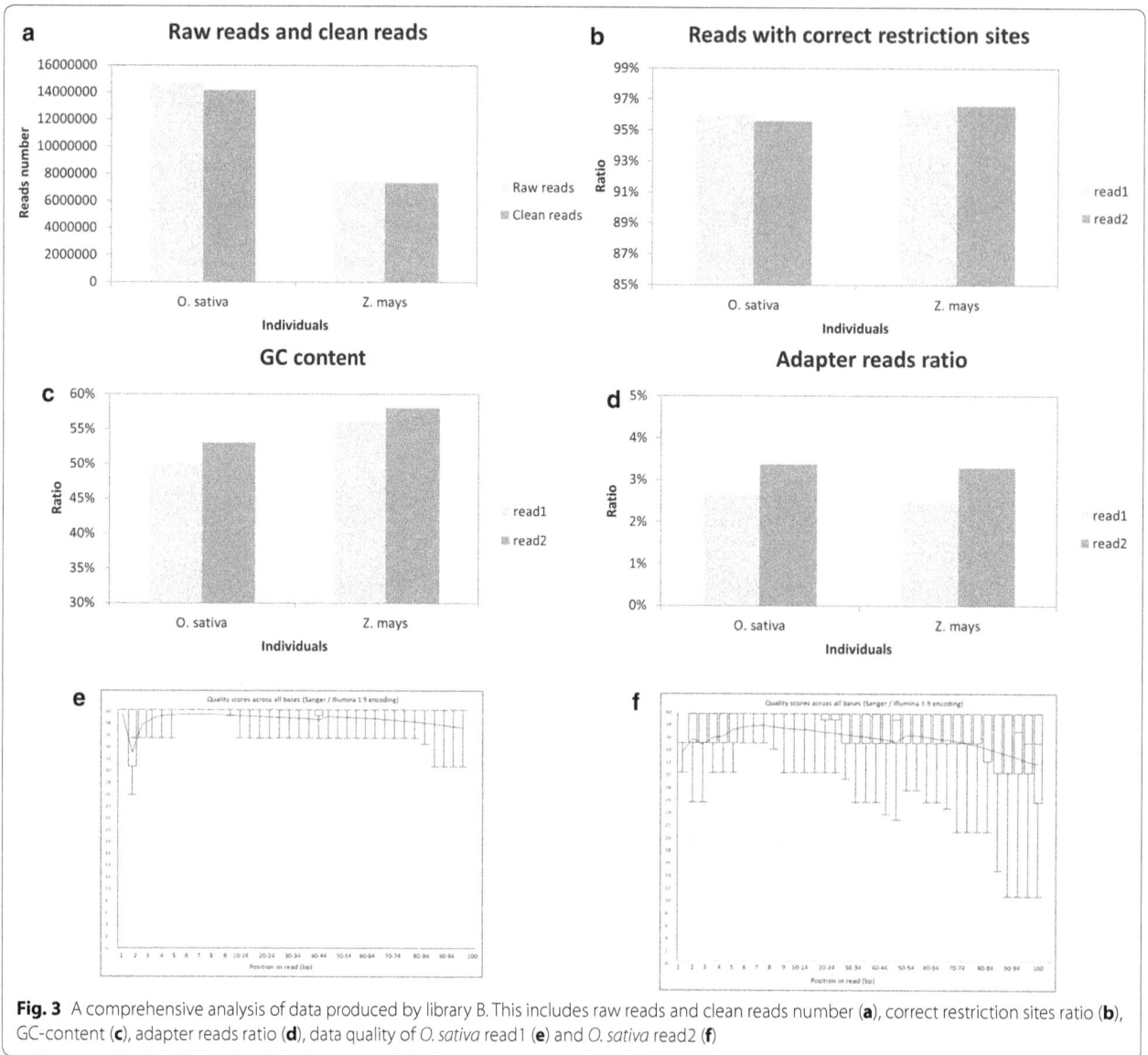

**Fig. 3** A comprehensive analysis of data produced by library B. This includes raw reads and clean reads number (**a**), correct restriction sites ratio (**b**), GC-content (**c**), adapter reads ratio (**d**), data quality of *O. sativa* read1 (**e**) and *O. sativa* read2 (**f**)

generate 21,491–42,376 tags with the Pearson correlation coefficient r = 0.9792. The actual and predicted data correlate well for maize while slightly worse for rice which is maybe due to the deviations introduced by cutting the gel. However, the observed tag number within the CDS region correlates better with expectation in rice than in maize (Table 4). It is noteworthy that while rice and maize own 39 and 85 % of repeat sequences respectively, only 15.83 and 31.44 % of tags fall into repeats region which indicates that the selected enzyme pair may be efficient in avoiding genome areas with highly repetitive DNA. It is supposed that this is because of a lack of restriction sites in some types of repetitive DNA as the two species contain more than

10 kinds of transposable elements respectively [43, 44]. As rice holds an average genome size of ~383 Mb and maize holds an average genome size of ~2300 Mb, the sampled tags only accounted for 1.77–2.43 % of the whole nuclear genome. In this sense, the efficiency of this reduced representation method on reducing genome complexity is reasonably high. From the overall mapping rate, we can infer that fragments should mainly fall into the intergenic region rather than CDS or repeats region. Our simplified approach effectively avoids repeats region in rice and maize which mainly includes transposons and retrotransposons that usually bring problems in determining orthologous fragments among different individuals.

**Table 3  Summary of alignment statistics of sequencing data**

| Individual no. | Scaffolds (%) | | CDS (%) | | Repeats (%) | |
|---|---|---|---|---|---|---|
| | Read1 | Read2 | Read1 | Read2 | Read1 | Read2 |
| *Protocol A* | | | | | | |
| *D. latiflorus-3* | 63.21 | 61.35 | – | – | – | – |
| *C. pallens* | 15.83 | 13.95 | 3.31 | 3.10 | 0.49 | 0.39 |
| *Protocol B* | | | | | | |
| *Oryza sativa* | 88.22 | 82.50 | 19.39 | 19.64 | 6.85 | 7.70 |
| *Zea mays* | 90.66 | 84.86 | 2.38 | 2.83 | 10.95 | 8.82 |
| *P. edulis* | 83.80 | 79.93 | 2.38 | 2.87 | 0.11 | 0.17 |
| *A. semialata* | 3.37 | 3.71 | 1.08 | 1.11 | 0.00 | 0.22 |
| *D. latiflorus-1* | 62.23 | 61.97 | – | – | – | – |
| *D. latiflorus-2* | 64.15 | 62.59 | – | – | – | – |
| *P. rubicunda* | 23.19 | 21.72 | 1.40 | 1.59 | 0.10 | 0.15 |
| *P. vivax* | 25.56 | 24.66 | 1.39 | 1.64 | 0.11 | 0.16 |

As CDS and repeats region were not available for *D. latiflorus* survey genome sequences, *Midd*RAD data of *D. latiflorus* individuals were only mapped to the assembled scaffolds

**Table 4  *Midd*RAD-seq data summary in rice and maize**

| Genome information | *Oryza sativa* | *Zea mays* |
|---|---|---|
| Genome size (Mb) | 383 | 2300 |
| % of repeats in genome | 39.11 | 85.00 |
| GC content (%) | 43.56 | 46.83 |
| Expected information | | |
| Enzyme pairs | *AvaII + MspI* | *AvaII + MspI* |
| Expected RAD tag size range (bp) | 460–680 | 500–680 |
| Expected no. of RAD tags | 60,925 | 284,179 |
| Tags density per 100 kb | 15.92 | 12.36 |
| % of tags in CDS | 25.10 | 3.04 |
| Observed information | | |
| Raw reads | 14,732,449 | 7,414,009 |
| Clean reads | 14,146,516 | 7,337,556 |
| Observed no. of tags | 66,547 | 290,001 |
| Tag average depth | 212.58 | 25.30 |
| Tags per 100 kb | 17.38 | 12.61 |
| Simplification ratio (%) | 2.43 | 1.77 |
| % of tags in CDS | 31.49 | 1.15 |
| % of tags in repeats | 15.83 | 31.44 |

## Evaluation of protocol B on more plant species and genotypes validation

To test the universality of our protocol and the restriction enzymes on more plant species, we used Protocol B to construct libraries for *P. edulis* and *A. semialata* which represent two subfamilies of Gramineae (Bambusoideae and Panicoideae). We first inspected library quality. The ultimate library concentration was between 20–30 ng/ul, and fragments distribution was well within the expected range when screened by the agarose gel electrophoresis. Both libraries met the qualification for sequencing.

Then we conducted the data quality analysis of both *P. edulis* and *A. semialata* as shown in Table 5. Both *P. edulis* and *A. semialata* yielded more than 8 Mb raw reads (i.e. ~2 Gb raw data) with ~52 % GC-content; read1 containing correct restriction sites accounted 96.08–97.94 % of raw reads and read2 containing the correct restriction sites accounted for 94.85–96.39 %; read1 had 1.19–2.32 % adapter reads while read2 had 2.45–3.96 % adapter reads. Raw reads of *P. edulis* and *A. semialata* both had an average base Quality Score larger than 20, and Quality Score of bases of restriction enzyme cutting site was larger than 30.

Next, we mapped clean reads of *P. edulis* onto the *P. edulis* genome scaffolds, CDS-DNA, and repeats region and clean reads of *A. semialata* onto the sorghum genome scaffolds, CDS-DNA, and repeats region respectively. For *P. edulis*, overall scaffolds mapping rate was 79.93–83.80 %, reads hits to the CDS-DNA accounted 2.38–2.87 % (Table 3). Yet reads localization on the repeats region accounted for 0.11–0.17 %. For *A. semialata*, the overall scaffolds alignment rate was 3.37–3.71 %, reads hits to the CDS-DNA accounted 1.08–1.11 %. Yet reads localization on the repeats region accounted for 0.00–0.22 %. Overall scaffolds alignment rate is relatively low for *A. semialata* is because of the sequence differences between *A. semialata* and *Sorghum bicolor*.

At last, clean data of both individuals were clustered into tags. The number of tags obtained from *P. edulis* was 128,803 with an average depth 30.18X which correlated well with the expectation. While *A. semialata* got 98,869 tags, with an average depth of 96.18X which was not within expectation as the sorghum genome was used as the reference. Since *P. edulis* has an average genome size

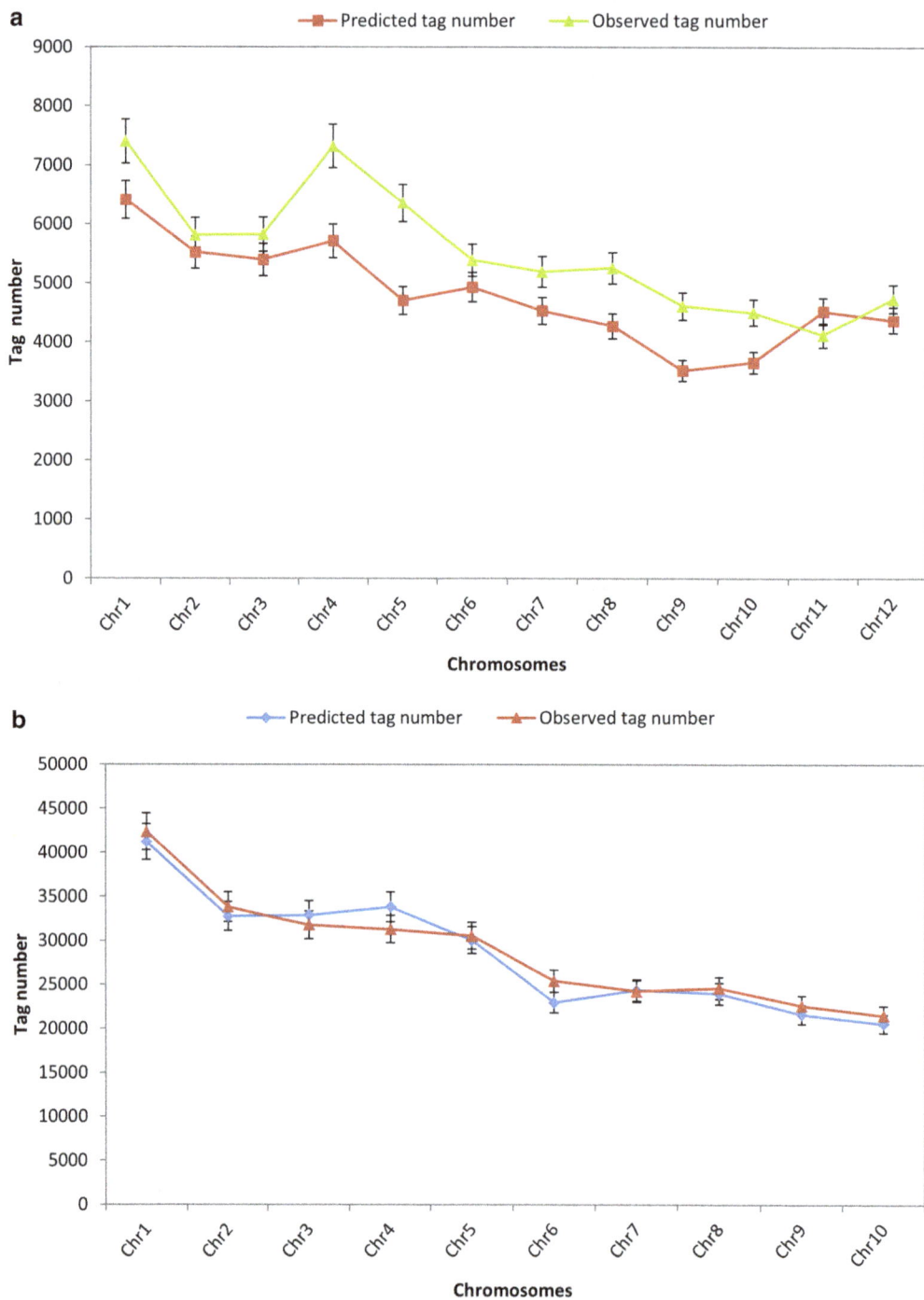

**Fig. 4** Comparison of the real sequencing data with in silico predicted results. **a** Each of the 12 rice chromosomes is expected to produce 3521–6414 fragments while each actually generates 4121–7404 tags with the Pearson correlation coefficient r = 0.8374. **b** The 10 maize chromosomes each is expected to produce 20,555–41,216 fragments while each is observed to generate 21,491–42,376 tags with the Pearson correlation coefficient r = 0.9792

**Table 5  A comprehensive data analysis of *P. edulis* and *A. semialata***

| Species | Raw reads no. | Percentage of adapter reads (%) | | Percentage of reads with correct restriction sites (%) | | GC content (%) | Clean reads no. | Tag no. | Average tag depth |
|---------|---------------|-------|-------|-------|-------|----|----------|---------|--------|
| | | Read1 | Read2 | Read1 | Read2 | | | | |
| *P. edulis* | 8,142,517 | 1.19 | 2.45 | 97.94 | 96.39 | 52 | 8,045,315 | 128,803 | 30.18 |
| *A. semialata* | 14,651,272 | 2.32 | 3.96 | 96.08 | 94.85 | 53 | 14,299,253 | 98,869 | 96.18 |

of ~2000 Mb and *A. semialata* has an average genome size of ~600 Mb (personal communications with Ms. Yang Yang), the tags we got accounted for 1.80 % of the *P. edulis* nuclear genome and 4.61 % of the *A. semialata* nuclear genome.

To verify how genotyping accuracy is maintained in the *Mi*ddRAD protocol, we have recently constructed a linkage map of *D. latiflorus* (Guoqian Yang et al., unpublished data) according to *Mi*ddRAD protocol and got a high-quality map of 2365 markers with an average map distance 1.56 cM. The 36 linkage groups generated were corresponding to the 36 haploid chromosomes of *D. latiflorus* and 52 of the 55 selected genotypes (94.55 %) were agreed with independent Sanger sequencing results (Additional file 2: Table S3). We believe that genotypes from *Mi*ddRAD-seq derived data should be of high genotyping accuracy as the fundamental of constructing a high-quality linkage map with tight map distances are the correct genotypes of most markers/loci [22, 45].

### Evaluation of shortened adapters and new barcodes

To evaluate the shortened adapters and redesigned barcodes, we constructed four *Mi*ddRAD sub-libraries containing 40 *D. latiflorus* individuals and sequenced the final library with a single Illumina lane. We used the double index strategy to distinguish each individual, which means each individual was identified by a unique barcode and index as the original ddRAD protocol implemented. We performed analysis of data generated by each barcode and found that each barcode and adapter could produce a relatively large amount of data with average 9,451,891 reads and CV value 0.0021–0.2381 (Fig. 5a). In addition, each sub-library could produce comparable amounts of data with a mean of 94,946,435 reads and CV value 0.0587 (Fig. 5b). This suggests that the newly designed barcodes and shortened P1 adapters are of high efficiency.

### Phylogenetic tree construction of three bamboo species

Maximum likelihood phylogenetic reconstruction is fully resolved with high support for all clades (2532 SNPs) (see Fig. 6). Two clades were found: the first contains the genus *Dendrocalamus* (100 % Bootstrap). *D. latiflorus*

individual 1 is sister to *D. latiflorus* individual 2. In the second clade, *P. rubicunda* is sister to *P. vivax* (100 % Bootstrap), which themselves form a monophyletic clade (100 % Bootstrap). The relationships between two genera are well resolved and the topology of the two genera in our tree agrees well with current taxonomy [38, 39]. One additional RAxML analysis using alternatives data set (1005 SNPs) from Stacks analyses displayed identical topology and minor changes in branch lengths (Additional file 2: Figure S4).

### Discussion

In this study, we tested the universality of several commonly used enzyme pairs across the angiosperm plants, simplified the ddRAD protocol and reduced the overall costs. *Mi*ddRAD library construction protocol has optimized the following areas compared with original ddRAD protocol. (1) *Mi*ddRAD protocol tests the universality of several commonly used enzyme pairs and three pairs of restriction endonucleases are maybe universal in digesting plant genomic DNA. We recommend *AvaII + MspI > EcoRI +MspI > PstI +MspI* when designing plant ddRAD projects; (2) In *Mi*ddRAD protocol several expensive consumables and apparatuses are replaced by conventional experimental apparatuses, for example, the magnetic beads purification method is replaced by a simple column purification to get rid of the dependence on the magnet, DNA fragments are selected by cutting low melting point agarose gel rather than the automatically select device Pippin-Prep, and low melting point agarose gel electrophoresis is used to screen fragments distribution instead of an expensive Agilent 2100 Bioanalyzer; (3) *Mi*ddRAD removes 3 steps in ddRAD protocol, namely purifying the enzyme-digested products, quantifying the DNA concentration before ligation and purifying ligation products after pooling samples, all of which simplify the process of constructing a library; (4) In *Mi*ddRAD protocol original P1 adapters are shortened from 37 to 25 bp (barcode is assumed to be of 5 bp length), which will reduce the cost of the synthesizing adapter oligos partially; (5) In *Mi*ddRAD a new barcode-adapter system containing 20 pairs of barcodes varying in length were devised, which can be used with integer times (20

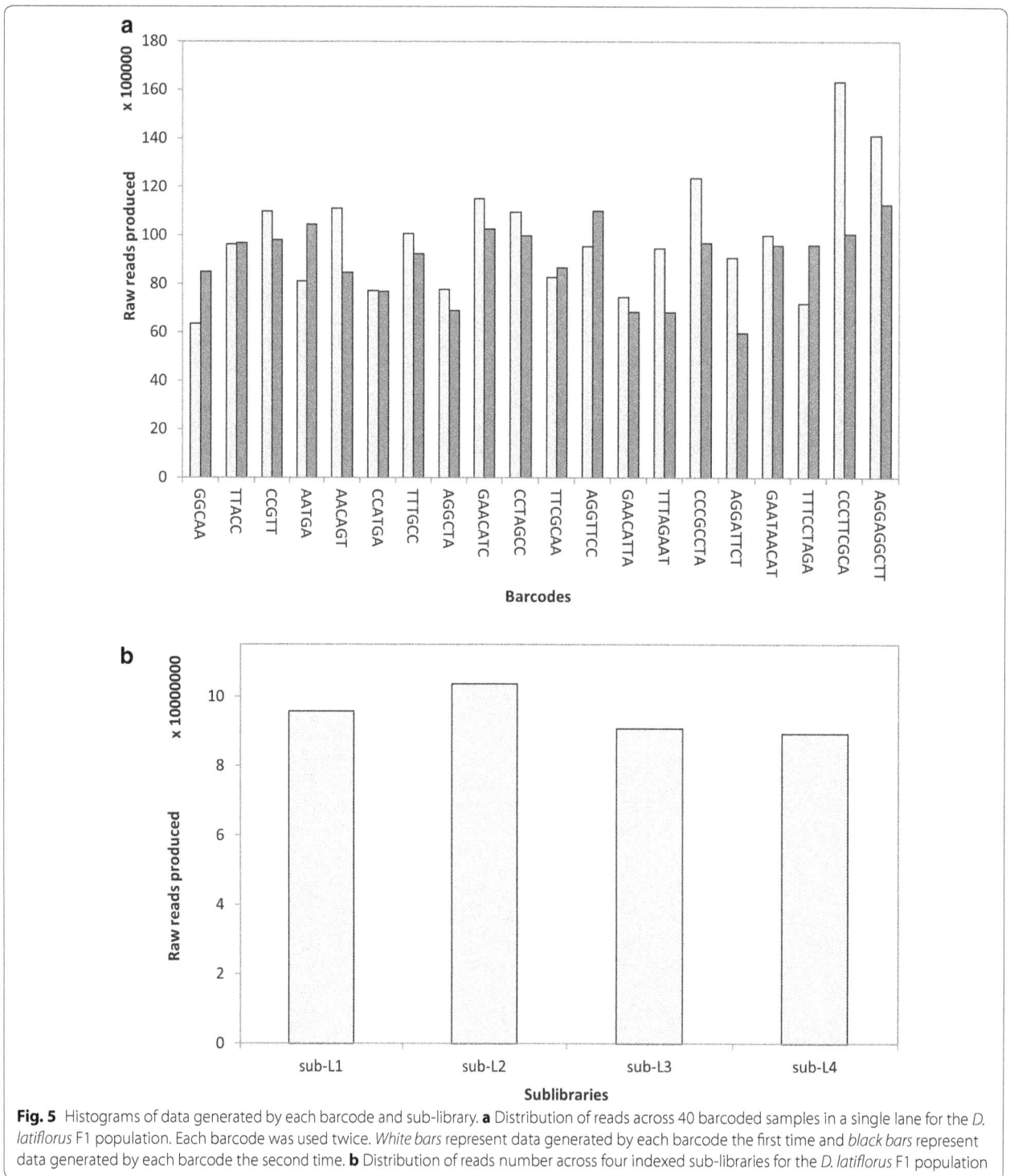

**Fig. 5** Histograms of data generated by each barcode and sub-library. **a** Distribution of reads across 40 barcoded samples in a single lane for the *D. latiflorus* F1 population. Each barcode was used twice. *White bars* represent data generated by each barcode the first time and *black bars* represent data generated by each barcode the second time. **b** Distribution of reads number across four indexed sub-libraries for the *D. latiflorus* F1 population

* n), rather than the original 48 kinds of barcode with equal length. This will not only reduce the cost of synthesizing DNA oligos but also increase the flexibility for projects with different samples and help improve the quality of bases near the restriction site. The comparison of *Mi*ddRAD with most commonly used RAD and GBS sequencing methodologies and associated costs are listed in Additional file 2: Table S4.

Since the thorough library construction process minimizes the purification times, random DNA loss is

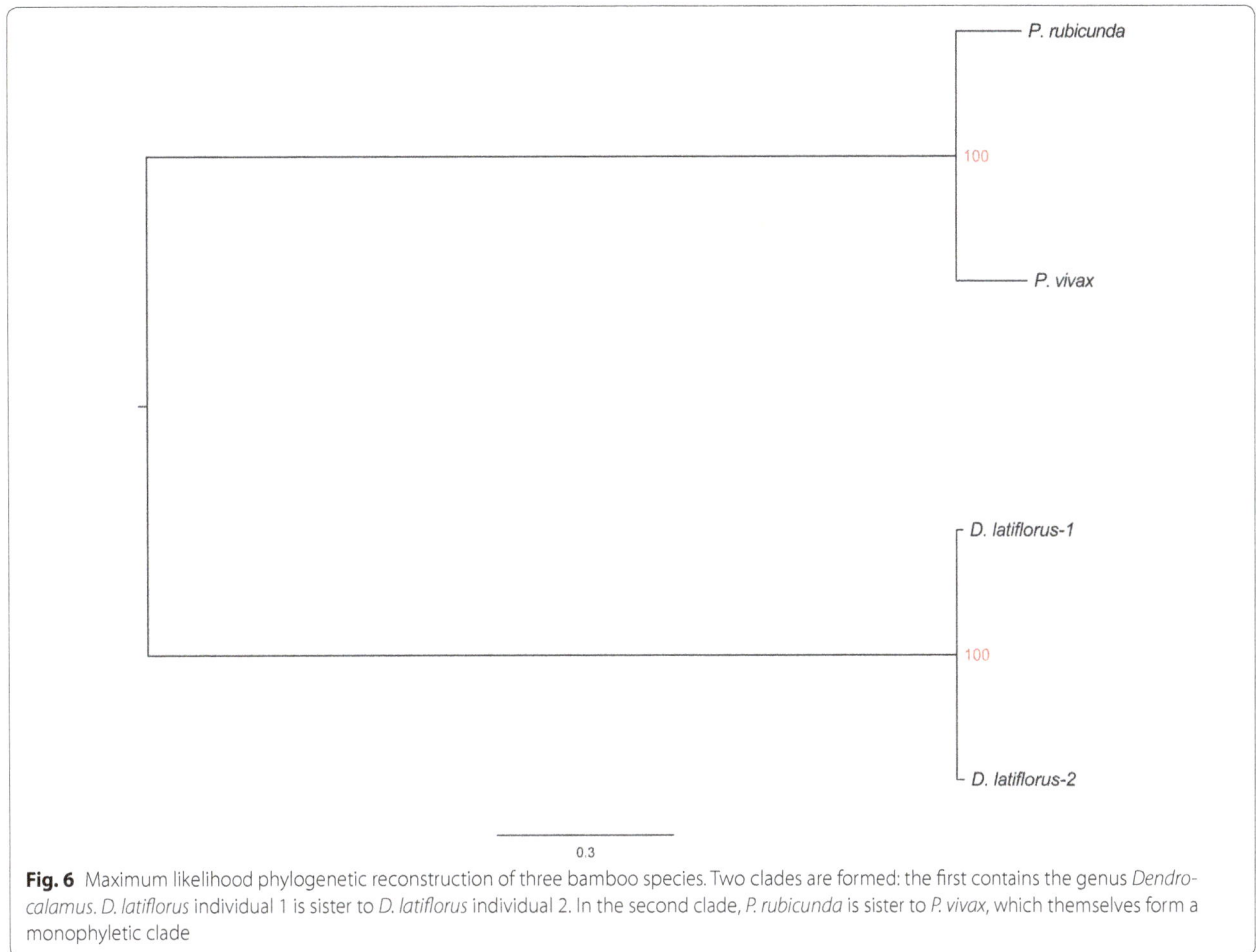

**Fig. 6** Maximum likelihood phylogenetic reconstruction of three bamboo species. Two clades are formed: the first contains the genus *Dendrocalamus. D. latiflorus* individual 1 is sister to *D. latiflorus* individual 2. In the second clade, *P. rubicunda* is sister to *P. vivax*, which themselves form a monophyletic clade

greatly reduced. The highly simplified process allows library preparation be accomplished with as low as 50 ng genomic DNA. Meanwhile, we found that reducing the two purification steps did not reduce the quality of the data by sequence quality analysis. Through data analysis of 40 individuals from a single lane, omitting the step quantifying DNA concentration of each individual before pooling samples does have influence on the amount of data among each individual (CV = 0.2146) but in our experience even if quantifying DNA concentration of each individual, pooling equal quantity DNA of each individual is still impossible as different volumes of liquid may adhere to the tips when using the pipette. As adequate data (>6 M reads) could be generated for each individual, we suggest deleting this quantifying step as the GBS protocol does (CV = 0.23) [16]. The redesigned adapters and variable length barcodes have high recognition efficiency on various individuals and could produce high-quality data which is similar to the results Burford et al. got [27]. Some people may worry that the combination of restriction endonucleases may easily cut the

repeats region with high GC-content as is shown in maize about 10 % of reads fall into repeats region. Nevertheless, we still have enough reads left for analysis. Researchers may strictly follow the protocol without re-selecting novel combination of enzymes (but have to adjust the size-selection range) if they do not want to invest too much on pilot experiments. As synthetic adapters can be used in diverse plant species and transferred across labs, our protocol will greatly reduce the overall costs.

A possible drawback of our method is that degraded DNA will not produce adequate data because once one of the enzyme sites was impaired the whole tag will be lost. Nonetheless, as long as the DNA provided shows a clear major band when detected by the agarose gel electrophoresis, it could usually generate sufficient amount of data for analysis. In addition, the final library may be a pool of tens to a hundred of samples and we only designed 20 kinds of barcodes, so it is inevitable to cut gel several times when performing the procedure. In order to maintain the consistency of the selected fragments, electrophoresis conditions must be strictly controlled, and practice cutting the gel is needed before the formal experiment begins. Besides, electrophoresis time should be long enough (1–2 h) to prevent that size selection maybe 'leaky'.

To demonstrate the applicability of *Mi*ddRADseq-derived markers in no-model species, we used *Mi*ddRAD data to resolve phylogenetic relationships of two woody bamboos genera, *Dendrocalamus* and *Phyllostachys*. *Dendrocalamus* is a tropical woody bamboo genus while *Phyllostachys* belongs to temperate woody bamboos. Our tree is congruent well with the current taxonomy. In comparison to previous studies in this clade, which used chloroplast regions [39] or nuclear DNA regions [38], the ddRAD data set is prominent for its simpleness in getting an amount of data (over 200 loci in the smallest data set). Though RAD-seq has been demonstrated to be feasible in clades as old as 40–60 million years with simulated RAD tags of Drosophila [46] and bona fide sequence of American oak [47], RAD sequences are usually considered useful for phylogenetic reconstruction in younger clades in which sufficient numbers of orthologous restriction sites are retained across species [46]. However, RAD-seq is now receiving increased attention at deeper evolutionary time scales, such as genus- or family-level phylogenetics even the problem of efficiently obtaining sequence data across many individuals exists [48]. Our study demonstrates the utility of *Mi*ddRAD data for reconstructing phylogenetic relationships in a group that spans 43–47 million-year-old divergences [49]. What we should bear in mind is that the performance of RAD or *Mi*ddRAD depends in part on the level of divergence between species. Determining orthologous RAD tags between samples should also be taken carefully in the future phylogenetic analysis with RAD sequencing [50]. The data set and analyses we provide here are a novel step forward in the use of ddRAD data to address questions in woody bamboo phylogenetic reconstruction. We show that it is possible to assemble genome-wide RAD-tags into phylogenetic matrices without the use of a reference genome.

## Conclusions

In this study, we first tested the universality of several commonly used enzyme pairs across 23 plant species and found *AvaII* + *MspI* enzyme pair produced a consistently higher number of fragments in a broad range of angiosperm plant species. Then we simplified the ddRAD protocol and designed a new barcode-adapter system that could reduce the overall costs. At last, we demonstrated the use of *Mi*ddRAD-seq data in resolving phylogenetic relationships of two woody bamboos genera. This protocol could help botanist quickly get ideal experimental data at a relatively low cost and without being specially trained. We expect that the protocol could be implemented efficiently nearly in any ordinary molecular laboratory without relying on large sequencing centers or next-generation sequencing companies.

## Additional files

**Additional file 1.** Details of *Mi*ddRAD library constructing protocol. *Mi*ddRAD contains protocol A and protocol B and we take protocol B as the final protocol. A step-by-step tutorial was presented in this file.

**Additional file 2.** A list of Additional Figures and Tables. **Figure S1.** Library preparation flowchart of *Mi*ddRAD protocol A. **Figure S2.** In silico digestion genome sequences of 23 plant species by *EcoRI* + *MspI* and *PstI* + *MspI*. **Figure S3.** Fragments distribution of library A and library B. **Figure S4.** Maximum likelihood phylogenetic reconstruction of three bamboo species. **Table S1.** Species adopted for in silico digestion and the corresponding genome size. **Table S2.** Restriction enzymes included in this study. **Table S3.** Independent Sanger sequencing for genotype validation of *Mi*ddRAD-seq genotyping. **Table S4.** Comparison of most commonly used RAD and GBS sequencing methodologies and associated costs.

### Abbreviations

RAD-seq: restriction-site associated DNA sequencing; ddRAD: double digest restriction associated DNA; *Mi*ddRAD: modified double digest restriction associated DNA; 2b-RAD: IIB restriction endonucleases restriction-site associated DNA; GBS: genotyping by sequencing; Read1: forward read of pair-end fastq data; Read2: reverse read of pair-end fastq data; CDS: coding sequences; Repeats region: repetitive sequences in a genome; SNPs: single nucleotide polymorphism; Rare-cutter enzyme: a restriction enzyme with a recognition sequence which occurs only rarely in a genome; Common-cutter enzyme: a restriction enzyme with a recognition sequence which occurs frequently in a genome; Adapter: double-stranded product by annealing two oligos; Barcodes: short DNA sequence for identifying different individuals.

### Authors' contributions

ZHG and DZL organized the project. GQY performed the experiments, analyzed the data, and wrote the paper; YMC, CG, GQY and YG harvested leaves of samples and extracted DNA. ZHG, DZL, JPW, LL, LZ and XYW reviewed and edited the manuscript. All authors read and approved the final manuscript.

### Author details

[1] Germplasm Bank of Wild Species, Kunming Institute of Botany, Chinese Academy of Sciences, Kunming 650201, China. [2] Kunming College of Life Sciences, University of Chinese Academy of Sciences, Kunming 650201, China. [3] Key Laboratory of Experimental Marine Biology, Institute of Oceanology, Chinese Academy of Sciences, Qingdao 266071, China.

### Acknowledgements

We are grateful to Ms. Yang Yang (Kunming Institute of Botany, Chinese Academy of Sciences) for kindly providing the *Oryza sativa* and *Alloteropsis semialata* material. We would like to thank Ms. Zhao-Li Ding (Kunming Institute of Zoology, Chinese Academy of Sciences) and Ms. Li Zhong (Yunnan University) for advice and supports on Illumina sequencing. We would like to thank Prof. Guo-Fan Zhang (Institute of Oceanology, Chinese Academy of Sciences) for helpful suggestions in this project. We would also like to thank Mr. Hui-Fu Zhuang, Dr. Peng-Fei Ma and Dr. Yu-Xiao Zhang at Kunming Institute of Botany for computational supports.

### Competing interests

The authors declare that they have no competing interests.

**Funding**

This project was supported by the National Natural Science Foundation of China (31470322 and 31430011).

**References**

1. Baird NA, Etter PD, Atwood TS, Currey MC, Shiver AL, Lewis ZA, Selker EU, Cresko WA, Johnson EA. Rapid SNP discovery and genetic mapping using sequenced RAD markers. PLoS One. 2008;3(10):e3376.

2. Davey JW, Hohenlohe PA, Etter PD, Boone JQ, Catchen JM, Blaxter ML. Genome-wide genetic marker discovery and genotyping using next-generation sequencing. Nat Rev Genet. 2011;12(7):499–510.

3. Etter PD, Bassham S, Hohenlohe PA, Johnson EA, Cresko WA. SNP discovery and genotyping for evolutionary genetics using RAD sequencing. In: Etter PD, Bassham S, Hohenlohe PA, Johnson EA, Cresko WA, editors. Molecular methods for evolutionary genetics. Berlin: Springer; 2011. p. 157–78.

4. Barchi L, Lanteri S, Portis E, Acquadro A, Valè G, Toppino L, Rotino GL. Identification of SNP and SSR markers in eggplant using RAD tag sequencing. BMC Genomics. 2011;12(1):304.

5. Deokar AA, Ramsay L, Sharpe AG, Diapari M, Sindhu A, Bett K, Warkentin TD, Tar'an B. Genome wide SNP identification in chickpea for use in development of a high density genetic map and improvement of chickpea reference genome assembly. BMC Genomics. 2014;15:708.

6. Wu K, Liu HY, Yang MM, Tao Y, Ma HH, Wu WX, Zuo Y, Zhao YZ. High-density genetic map construction and QTLs analysis of grain yield-related traits in Sesame (*Sesamum indicum* L.) based on RAD-Seq techonology. BMC Plant Biol. 2014;14:274.

7. Xu P, Xu SZ, Wu XH, Tao Y, Wang BG, Wang S, Qin DH, Lu ZF, Li GJ. Population genomic analyses from low-coverage RAD-Seq data: a case study on the non-model cucurbit bottle gourd. Plant J. 2014;77(3):430–42.

8. Wang XQ, Zhao L, Eaton DAR, Li DZ, Guo ZH. Identification of SNP markers for inferring phylogeny in temperate bamboos (Poaceae: Bambusoideae) using RAD sequencing. Mol Ecol Resour. 2013;13(5):938–45.

9. Cruaud A, Gautier M, Galan M, Foucaud J, Sauné L, Genson G, Dubois E, Nidelet S, Deuve T, Rasplus J-Y. Empirical assessment of RAD sequencing for interspecific phylogeny. Mol Biol Evol. 2014;31(5):1272–4.

10. DaCosta JM, Sorenson MD. ddRAD-seq phylogenetics based on nucleotide, indel, and presence-absence polymorphisms: analyses of two avian genera with contrasting histories. Mol Phylogenet Evol. 2016;94:122–35.

11. Qi ZC, Yu Y, Liu X, Pais A, Ranney T, Whetten R, Xiang QY. Phylogenomics of polyploid *Fothergilla* (Hamamelidaceae) by RAD-tag based GBS-insights into species origin and effects of software pipelines. J Syst Evol. 2015;53(5):432–47.

12. Zhang N, Zhang L, Tao Y, Guo L, Sun J, Li X, Zhao N, Peng J, Li X, Zeng L. Construction of a high density SNP linkage map of kelp (*Saccharina japonica*) by sequencing Taq I site associated DNA and mapping of a sex determining locus. BMC Genomics. 2015;16(1):189.

13. Zhou J, Wang SB, Jian J, Geng QC, Wen J, Song Q, Wu Z, Li GJ, Liu YQ, Dunwell JM. Identification of domestication-related loci associated with flowering time and seed size in soybean with the RAD-seq genotyping method. Sci Rep. 2015;5:9350.

14. Andrews KR, Good JM, Miller MR, Luikart G, Hohenlohe PA. Harnessing the power of RADseq for ecological and evolutionary genomics. Nat Rev Genet. 2016;17(2):81–92.

15. Liu H, Bayer M, Druka A, Russell JR, Hackett CA, Poland J, Ramsay L, Hedley PE, Waugh R. An evaluation of genotyping by sequencing (GBS) to map the Breviaristatum-e (ari-e) locus in cultivated barley. BMC Genomics. 2014;15:104.

16. Elshire RJ, Glaubitz JC, Sun Q, Poland JA, Kawamoto K, Buckler ES, Mitchell SE. A robust, simple genotyping-by-sequencing (GBS) approach for high diversity species. PLoS One. 2011;6(5):e19379.

17. Poland JA, Brown PJ, Sorrells ME, Jannink JL. Development of high-density genetic maps for barley and wheat using a novel two-enzyme genotyping-by-sequencing approach. PLoS One. 2012;7(2):e32253.

18. Poland JA, Rife TW. Genotyping-by-sequencing for plant breeding and genetics. Plant Genome. 2012;5(3):92–102.

19. DaCosta JM, Sorenson MD. Amplification biases and consistent recovery of loci in a double-digest RAD-seq protocol. PLoS One. 2014;9(9):e106713.

20. Quail MA, Kozarewa I, Smith F, Scally A, Stephens PJ, Durbin R, Swerdlow H, Turner DJ. A large genome center's improvements to the Illumina sequencing system. Nat Methods. 2008;5(12):1005–10.

21. Wang S, Meyer E, McKay JK, Matz MV. 2b-RAD: a simple and flexible method for genome-wide genotyping. Nat Methods. 2012;9(8):808–10.

22. Peterson BK, Weber JN, Kay EH, Fisher HS, Hoekstra HE. Double digest RADseq: an inexpensive method for de novo SNP discovery and genotyping in model and non-model species. PLoS One. 2012;7(5):e37135.

23. Toonen RJ, Puritz JB, Forsman ZH, Whitney JL, Fernandez-Silva I, Andrews KR, Bird CE. ezRAD: a simplified method for genomic genotyping in non-model organisms. PeerJ. 2013;1:e203.

24. Puritz JB, Matz MV, Toonen RJ, Weber JN, Bolnick DI, Bird CE. Demystifying the RAD fad. Mol Ecol. 2014;23(24):5937–42.

25. Glaubitz JC, Casstevens TM, Lu F, Harriman J, Elshire RJ, Sun Q, Buckler ES. TASSEL-GBS: a high capacity genotyping by sequencing analysis pipeline. PLoS One. 2014;9(2):e90346.

26. Herrera S, Reyes-Herrera PH, Shank TM. Predicting RAD-seq Marker numbers across the eukaryotic tree of life. Genome Biol Evol. 2015;7(12):3207–25.

27. Burford Reiskind MO, Coyle K, Daniels HV, Labadie P, Reiskind MH, Roberts NB, Roberts RB, Vargo EL, Schaff J. Development of a universal double-digest RAD sequencing approach for a group of non-model, ecologically and economically important insect and fish taxa. Mol Ecol Resour. 2016. doi:10.1111/1755-0998.12527.

28. Doyle JJ. A rapid DNA isolation procedure for small quantities of fresh leaf tissue. Phytochem Bull. 1987;19:11–5.

29. Wang J, Li L, Qi H, Du X, Zhang G. RestrictionDigest: a powerful Perl module for simulating genomic restriction digests. Electron J Biotechnol. 2016. doi:10.1016/j.ejbt.2016.02.003.

30. Catchen J, Hohenlohe PA, Bassham S, Amores A, Cresko WA. Stacks: an analysis tool set for population genomics. Mol Ecol. 2013;22(11):3124–40.

31. Catchen JM, Amores A, Hohenlohe P, Cresko W, Postlethwait JH. Stacks: building and genotyping loci de novo from short-read sequences. G3 (Bethesda). 2011;1(3):171–82.

32. Andrews S. Fast QC: a quality control tool for high throughput sequence data. http://www.bioinformatics.babraham.ac.uk/projects/fastqc/. Accessed 20 Dec 2015.

33. Martin M. Cutadapt removes adapter sequences from high-throughput sequencing reads. EMBnet J. 2011;17(1):10–2. doi:10.14806/ej.17.1.200.

34. Langmead B, Trapnell C, Pop M, Salzberg SL. Ultrafast and memory-efficient alignment of short DNA sequences to the human genome. Genome Biol. 2009;10(3):R25.

35. Plantgdb. http://www.plantgdb.org/. Accessed 28 Sept 2015.

36. BambooGDB. http://www.bamboogdb.org/page/download.jsp. Accessed 28 Sept 2015.

37. Stamatakis A. RAxML version 8: a tool for phylogenetic analysis and post-analysis of large phylogenies. Bioinformatics. 2014;30(9):1312–3.

38. Triplett JK, Clark LG, Fisher AE, Wen J. Independent allopolyploidization events preceded speciation in the temperate and tropical woody bamboos. New Phytol. 2014;204(1):66–73.

39. Sungkaew S, Stapleton CM, Salamin N, Hodkinson TR. Non-monophyly of the woody bamboos (Bambuseae; Poaceae): a multi-gene region phylogenetic analysis of Bambusoideae s.s. J Plant Res. 2009;122(1):95–108.

40. Plant DNA C-values database. http://www.data.kew.org/cvalues/. Accessed 25 Dec 2015.

41. Chen YC, Liu T, Yu CH, Chiang TY, Hwang CC. Effects of GC bias in next-generation-sequencing data on de novo genome assembly. PLoS One. 2013;8(4):e62856.

42. Sun XW, Liu DY, Zhang XF, Li WB, Liu H, Hong WG, Jiang CB, Guan N, Ma CX, Zeng HP, et al. SLAF-seq: an efficient method of large-scale de novo SNP discovery and genotyping using high-throughput sequencing. PLoS One. 2013;8(3):e58700.

43. International Rice Genome Sequencing Project. The map-based sequence of the rice genome. Nature. 2005;436(7052):793–800.

44. Schnable PS, Ware D, Fulton RS, Stein JC, Wei F, Pasternak S, Liang C, Zhang J, Fulton L, Graves TA, et al. The B73 maize genome: complexity, diversity, and dynamics. Science. 2009;326(5956):1112–5.

45. Cartwright DA, Troggio M, Velasco R, Gutin A. Genetic mapping in the presence of genotyping errors. Genetics. 2007;176(4):2521–7.

46. Rubin BER, Ree RH, Moreau CS. Inferring phylogenies from RAD sequence data. PLoS One. 2012;7(4):e33394.

47. Hipp AL, Eaton DA, Cavender-Bares J, Fitzek E, Nipper R, Manos PS. A framework phylogeny of the American oak clade based on sequenced RAD data. PLoS One. 2014;9(4):e93975.

48. Eaton DA. PyRAD: assembly of de novo RADseq loci for phylogenetic analyses. Bioinformatics. 2014. doi:10.1093/bioinformatics/btu121.

49. Zhang XZ, Zeng CX, Ma PF, et al. Multi-locus plastid phylogenetic biogeography supports the Asian hypothesis of the temperate woody bamboos (Poaceae: Bambusoideae). Mol Phylogenet Evol. 2016;96:118-29.

50. Takahashi T, Nagata N, Sota T. Application of RAD-based phylogenetics to complex relationships among variously related taxa in a species flock. Mol Phylogenet Evol. 2014;80:137–44.

# 10

# Protocol: optimisation of a grafting protocol for oilseed rape (*Brassica napus*) for studying long-distance signalling

Anna Ostendorp[†], Steffen Pahlow[†], Jennifer Deke, Melanie Thieß and Julia Kehr[*]🆔

## Abstract

**Background:** Grafting is a well-established technique for studying long-distance transport and signalling processes in higher plants. While oilseed rape has been the subject of comprehensive analyses of xylem and phloem sap to identify macromolecules potentially involved in long-distance information transfer, there is currently no standardised grafting method for this species published.

**Results:** We developed a straightforward collar-free grafting protocol for *Brassica napus* plants with high reproducibility and success rates. Micrografting of seedlings was done on filter paper. Grafting success on different types of regeneration media was measured short-term after grafting and as the long-term survival rate (>14 days) of grafts after the transfer to hydroponic culture or soil.

**Conclusions:** We compared different methods for grafting *B. napus* seedlings. Grafting on filter paper with removed cotyledons, a truncated hypocotyl and the addition of low levels of sucrose under long day conditions allowed the highest grafting success. A subsequent long-term hydroponic cultivation of merged grafts showed highest survival rates and best reproducibility.

**Keywords:** *Brassica napus*, Grafting, Micrografting, Rootstock, Scion, Hydroponic culture

## Background

Grafting is a well-established technique for joining vegetative tissues from two or more plants. It has been widely applied to improve the properties of different vegetables and other horticultural crop plants [1]. Key for a successful establishment of graft unions is the formation of a continuous vascular system between the grafting partners that are usually called scion (shoot part of the graft) and rootstock (root part of the graft). Therefore, grafting is most successful in dicotyledonous plants possessing a vascular cambium and more difficult or even impossible in monocotyledonous plants.

Because of the reunion of functional xylem and phloem connections, grafting has also become an excellent experimental tool to study long-distance mobility of a wide range of molecules in living plants [2]. Grafting studies have provided conclusive evidence that long-distance transport is involved in diverse, but likewise important, physiological processes. Examples are the photoperiodic regulation of flowering [3, 4], the systemic spread of viruses [5, 6], phytohormone transport and action [7], apical dominance [8], nodule formation [9], small RNA movement [10–12], systemic acquired resistance [13, 14], and systemic gene silencing [15]. Grafting methods for confirming long-distance transport of regulating molecules are established for a wide range of plant species, including *Nicotiana benthamiana* [16, 17], *Medicago truncatula* [18], and the model species *Arabidopsis thaliana* [19–24]. These techniques have been applied successfully to study signal transduction, for example by micro RNAs (miRNAs) [10, 12, 25, 26].

*Brassica napus* is a suitable plant for studying long-distance communication, because it allows obtaining phloem

*Correspondence: julia.kehr@uni-hamburg.de
†Anna Ostendorp and Steffen Pahlow contributed equally to this work
Molecular Plant Genetics, University Hamburg, Biocenter Klein Flottbek, Ohnhorststr. 18, 22609 Hamburg, Germany

and xylem exudates in sufficient amounts for analysing sap compositions. In this species that is related to the model plant *A. thaliana*, sampling is relatively easy, and sample volumes are comparably large [10–12, 27–29].

Several studies identified hundreds of proteins and small RNAs (smRNAs) in phloem sap of *B. napus* [28, 30, 31]. To verify their long-distance mobility in vivo, so far grafting studies between wild-type and mutants/overexpressor plants were performed in Arabidopsis [10, 12]. The major reasons are that Arabidopsis transformation is straightforward and unmatched genetic resources are publicly available. However, the use of Arabidopsis in grafting experiments to study phloem mobility does only allow indirect conclusions about the mobility of phloem-localised molecules, since phloem sampling and, thus, direct measurements of changing compound levels in phloem sap are hardly possible. Therefore, a system allowing the same type of analysis of phloem long-distance signalling in Brassica would be desirable. Although not as easy as in Arabidopsis, several reports describe the successful transformation of *B. napus* using *Agrobacterium tumefaciens* [32, 33] what should allow the creation of suitable transgenic plants for grafting experiments. However, up to now only a few not very detailed grafting protocols for this species have been published in international journals [34–36], and no information about efficiency and reproducibility of grafting is documented.

This study attempts to provide a robust collar-free grafting procedure for different *B. napus* cultivars that is useful to confirm long-distance phloem mobility of potential signalling compounds identified in isolated phloem sap from this species. We describe an optimised flat-surface root-to-shoot grafting protocol for *B. napus* seedlings. The established grafting method does not require a collar to support the graft union and enabled a high short-term (up to 100 %) and a reasonable long-term survival rate (70–80 %) after the transfer to hydroponic culture or soil, respectively, indicating a high ability for the formation of functional vascular connections. This was confirmed by following the movement of the phloem-specific fluorescence dye carboxyfluorescein diacetate (CFDA) through the established graft unions.

## Methods
### Reagents
- Seed sterilisation solution (see REAGENT SETUP)
- Sodium hypochloride (Applichem-Panreac, cat. no. 213322.1214)
- Tween-20 (Applichem-Panreac, cat no. A1389,0500)
- 70 and 100 % ethanol (Applichem-Panreac, cat. no. A3678,1000)

- Murashige and Skoog (MS) medium (Duchefa, cat. No. M0245.0050)
- Sucrose (Applichem-Panreac, cat. no. A3935,5000)
- Agarose (BD, cat. no. 214010)
- Sterile deionised water
- Hydroponic medium (see REAGENT SETUP)
- Carboxyfluorescein diacetate (CFDA) (Sigma-Aldrich, cat. no. 21879-100MG-F)
- Dimethylsulfoxide (Applichem-Panreac, cat. no. A1584,0500)
- Einheitserde Classic (Einheitserdewerk Uetersen).

## Equipment
- Tweezers
- Petri dishes (92 × 15 mm, Sarstedt, cat. no. 82.1473)
- Growth chamber set to 22 °C, 70 % humidity, light conditions are dependent of the step in the grafting protocol (see PROCEDURE)
- binocular (SZ40, Olympus)
- round filter paper (90 mm, GE Life Science, cat. no. 1440-090)
- razor blades (Wilkinson Sword Classic)
- 50 ml conical tubes (Sarstedt, cat. no. 62.547.254)
- fluorescence stereomicroscope (MZ FL III, Leica)
- parafilm (Bemis, cat. no. #PM999)
- laminar flow tissue culture cabinet
- 4 °C standard laboratory fridge.

## Reagent setup
**Seed sterilisation solution** 7 % (v/v) sodium hypochloride, 0.05 % (v/v) Tween-20

**Germination plate** ½ Murashige and Skoog (MS) medium, 1 % agarose

**Regeneration plates** (1) ½ MS medium, 0.5 % (w/v) sucrose, 1 % agar, (2) ½ MS, 1 % agar, (3) filter paper moistened with 0.5 % sucrose, (4) filter paper with sterile water

**Hydroponic medium** the medium is described in Buhtz et al. [12]: 0.6 mM $NH_4NO_3$, 1 mM $Ca(NO_3)_2$, 0.04 mM Fe-EDTA, 0.5 mM $K_2HPO_4$, 0.5 mM $K_2SO_4$, 0.4 mM $Mg(NO_3)_2$. Micro nutrients added: 0.8 µM $ZnSO_4$, 9 µM $MnCl_2$, 0.1 µM $Na_2MoO_4$, 23 µM $H_3BO_3$, 0.3 µM $CuSO_4$. The pH was adjusted to 4.7 with 37 % HCl

**CFDA solution** 5 mg CFDA were solved in 100 µl 100 % DMSO. For plant application a 1:100 dilution in deionized water is used.

***Brassica napus* seeds** can be obtained commercially and from various breeders and research laboratories. This protocol is optimised for the Drakkar and Licosmos cultivars.

## Procedure
### Critical step
Good sterilisation is necessary to prevent bacterial or fungal contaminations during graft recovery. All work should be done under a sterile laminar flow cabinet.

### Seed sterilisation and germination
Timing ~1 h (for sterilisation, depending on the amount of seeds), 6 days (for germination)

1. Sterilisation solution should be prepared freshly. Seeds are incubated for 2 min in 2 ml 100 % ethanol and subsequently surface sterilised with 2 ml sterilisation solution for 15 min and then washed with sterile water three times for 10 min [37]. A better washing can be achieved by shaking the reaction tube containing the seeds.
2. Transfer the sterilised seeds to the germination plate (6–8 seeds per plate) and incubate the plates at 4 °C in the dark for 3 days in a vertical orientation.
3. After 3 days incubate the plates under short day conditions (light: 8 h, dark: 16 h, light intensity 100 µmol m$^{-2}$ s$^{-1}$, 22 °C and a relative humidity of 70 %) in a growth chamber and store the plates in a vertical orientation. After approx. 6 days old 3–5 cm long seedlings were used for grafting.

### Plant grafting
Timing ~1 h (for 40 grafts), 14 days (for regeneration)

4. Cut up to four seedlings on a cutting plate (petri dish with a round Whatman filter paper moistened with sterile water) with a razor blade under a binocular microscope. Remove cotyledons as well as 1–2 cm of the middle of the hypocotyl of the seedlings (Fig. 1).
5. Join the cut plant parts on a regeneration plate with the respective regeneration conditions using forceps. Close the regeneration plates with parafilm and incubate them under short day conditions with plates in a vertical orientation (5°–10°). Check the grafts after 6, 10 and 14 days.

### Critical step
In contrast to the protocol from Marsch-Martinez [24] optimal cutting of hypocotyls and cotyledons can be achieved on a moistened filter paper, propably due to the higher stiffness of B. napus plants compared to A. thaliana. Another difference to this protocol is the removal of the central part of the hypocotyl. In Brassica, the fast longitudinal growth of the seedlings hindered a successful graft formation when this step was omitted. Attention needs to be paid when joining the cut scion and root that no water film is within the parts. Ecotypes with an increased longitudinal growth direction like Drakkar should be regenerated for 14 days since the graft junction is fairly instable. Ecotypes like Licosmos with a reduced longitudinal growth can be regenerated for only 10 days.

### Post-grafting cultivation
Timing 14 days (for regeneration), 30 min (for CFDA monitoring)

6. For hydroponic cultivation 14 day old grafts are transferred to 50 ml conical tubes (Fig. 3a) and grown in boxes using the hydroponic medium described in Buhtz et al. [12]. For this purpose, wrap black foam around the grafted plant with the graft junction located in the middle of the foam. Cover the sides of the conical tubes with aluminium foil and fill in 40 ml of hydroponic medium to reduce algal growth.
7. Place the wrapped graft in the conical tube in such a way that the foam does not get in contact with the medium, but holds the graft in place and store the tubes in a rack (Additional file 3: Figure S3).
8. Place the rack in a standard polystyrene box with at least the same height or higher than the grafted plants. A box of 32 cm × 25 cm × 17 cm (length, width, height) is sufficient for cultivation of ca. 30 plants (Additional file 3: Figure S3). Cover the box with a light permissive plastic cover. For a simple set-up use plastic wrap and puncture up to 20 small holes for ca. 12 plants to allow adequate aeration. Cultivate the covered grafts for 10–14 days.
9. For soil cultivation the plants should be well-watered. Cover the transferred grafts with a plastic cover to avoid dehydration of the plantlets. Grafts are cultivated under long day conditions in a growth chamber (light: 16 h, dark: 8 h, light intensity of 80 µmol m$^{-2}$ s$^{-1}$, 22 °C and a relative humidity of 60 %; Fig. 3). After 14 days plants can be grown in the greenhouse either hydroponically in single pots or on soil.

### Critical step
If hydroponic growth is favoured, precautions have to be taken to reduce algal growth. Therefore cover all conical tubes with aluminium foil. Also ensure that the graft junction has no contact to the hydroponic medium to minimise the formation of adventitious roots. Puncturing the cover lid or loosely covering the plants helps to achieve an appropriate aeration without letting the plantlets dry out. Chen et al. [37] describe a similar method for Arabidopsis grafts, but due to the larger size of grafted B. napus plants a floating system as described is not applicable. Furthermore we observed that taking care that the foam is not in contact with the hydroponic medium

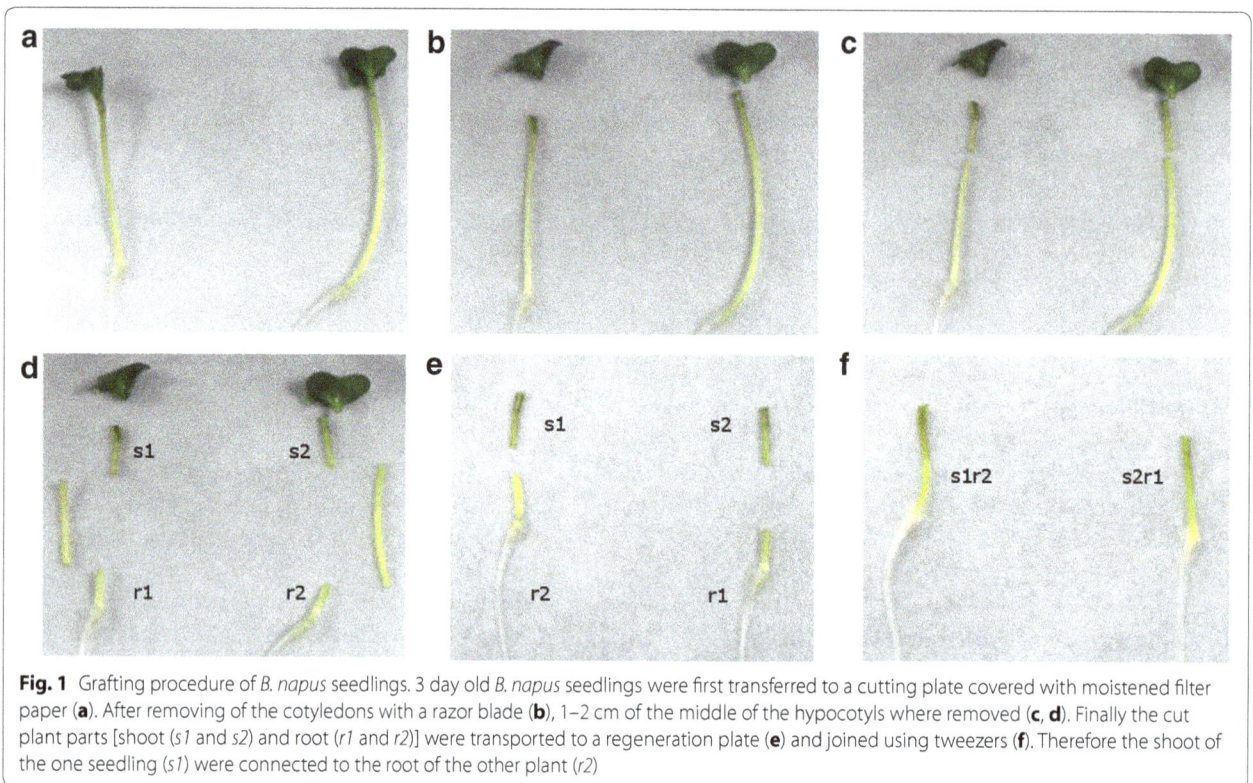

**Fig. 1** Grafting procedure of *B. napus* seedlings. 3 day old *B. napus* seedlings were first transferred to a cutting plate covered with moistened filter paper (**a**). After removing of the cotyledons with a razor blade (**b**), 1–2 cm of the middle of the hypocotyls where removed (**c**, **d**). Finally the cut plant parts [shoot (*s1* and *s2*) and root (*r1* and *r2*)] were transported to a regeneration plate (**e**) and joined using tweezers (**f**). Therefore the shoot of the one seedling (*s1*) were connected to the root of the other plant (*r2*)

avoids the formation of adventitious roots. To prevent a drying-out of freshly transferred grafts to soil or hydroponic medium it is advisable to cover those plants with a light permissive plastic cover, in addition to the high humidity set in the growth chamber.

10. The success of the formation of functional vascular connections within the grafts can be monitored using carboxyfluorescein diacetate, a phloem-specific fluorescence dye, as described in Grignon et al. 1989 [38]. Grafted plants were transferred to agar plates containing ½ MS and 1 % (w/v) agar to prevent drying-out. One leave per plant is punctured and a few microliters of a 10 μM CFDA solution are applied. After an incubation of 30 min at ambient temperature, fluorescence can be observed under a fluorescence stereomicroscope equipped with a GFP filter (Fig. 4).

## Results and discussion

Grafting conditions were optimised for *B. napus* seedlings to improve the survival rate of grafts. In contrast to other studies, we followed grafting success over a longer time period until grafted plants were successfully transferred to hydroponic culture or to soil. Obviously, success rates are lower at a later time-point than after a few days. However, since our goal was to transfer stable grafts to hydroponic culture or soil to let the plants grow until sampling of phloem and xylem sap is possible, it is the more meaningful measure in this case. Plants with a non-functional vascular system are prone to die within 2 weeks of post-grafting cultivation. Since initial experiments showed that graft formation was more successful under short day than under long day conditions, all further experiments were performed in growth chambers with 8 h light and 16 h dark.

As shown in Fig. 2b, there were significant differences in successful graft formation between conditions with and without additional sucrose. Sucrose-treated grafts showed higher survival rates (80–100 %) and a higher reproducibility when compared to sucrose-free conditions. Furthermore, the sucrose-treated grafts were much larger on agar plates and on filter paper, looked healthier, and the more stable graft connections allowed easier handling of grafts (Fig. 2a). Because of this, grafts treated with sucrose were more often successfully transferred to hydroponic culture or soil (Fig. 3a, b). Success rates ranged between 70 and 80 % (Figs. 2b, 3c). Again, regeneration without sucrose led to a high variability in survival rates after the transfer to either hydroponics or soil (Fig. 3c). The reason was that sucrose-treated grafts were less prone to breakage of the graft unions during transfer. This observation corresponds to results recently reported for *A. thaliana* where also 0.5 % sucrose

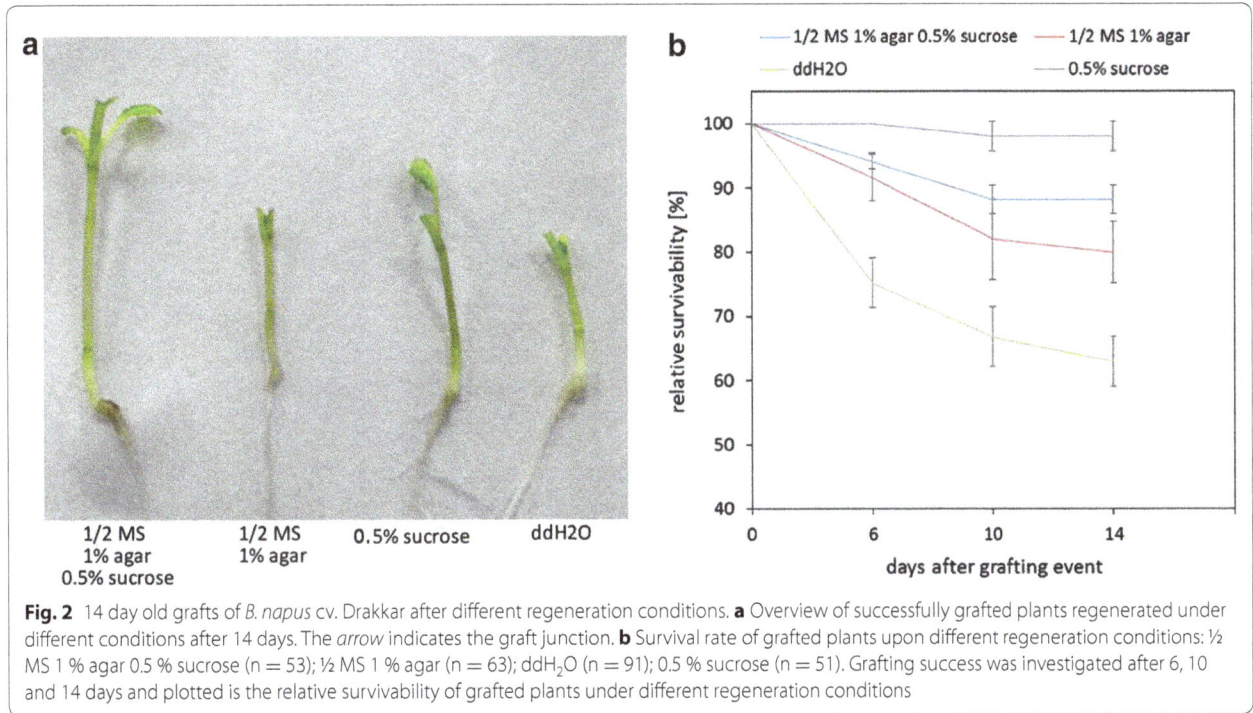

**Fig. 2** 14 day old grafts of *B. napus* cv. Drakkar after different regeneration conditions. **a** Overview of successfully grafted plants regenerated under different conditions after 14 days. The *arrow* indicates the graft junction. **b** Survival rate of grafted plants upon different regeneration conditions: ½ MS 1 % agar 0.5 % sucrose (n = 53); ½ MS 1 % agar (n = 63); ddH$_2$O (n = 91); 0.5 % sucrose (n = 51). Grafting success was investigated after 6, 10 and 14 days and plotted is the relative survivability of grafted plants under different regeneration conditions

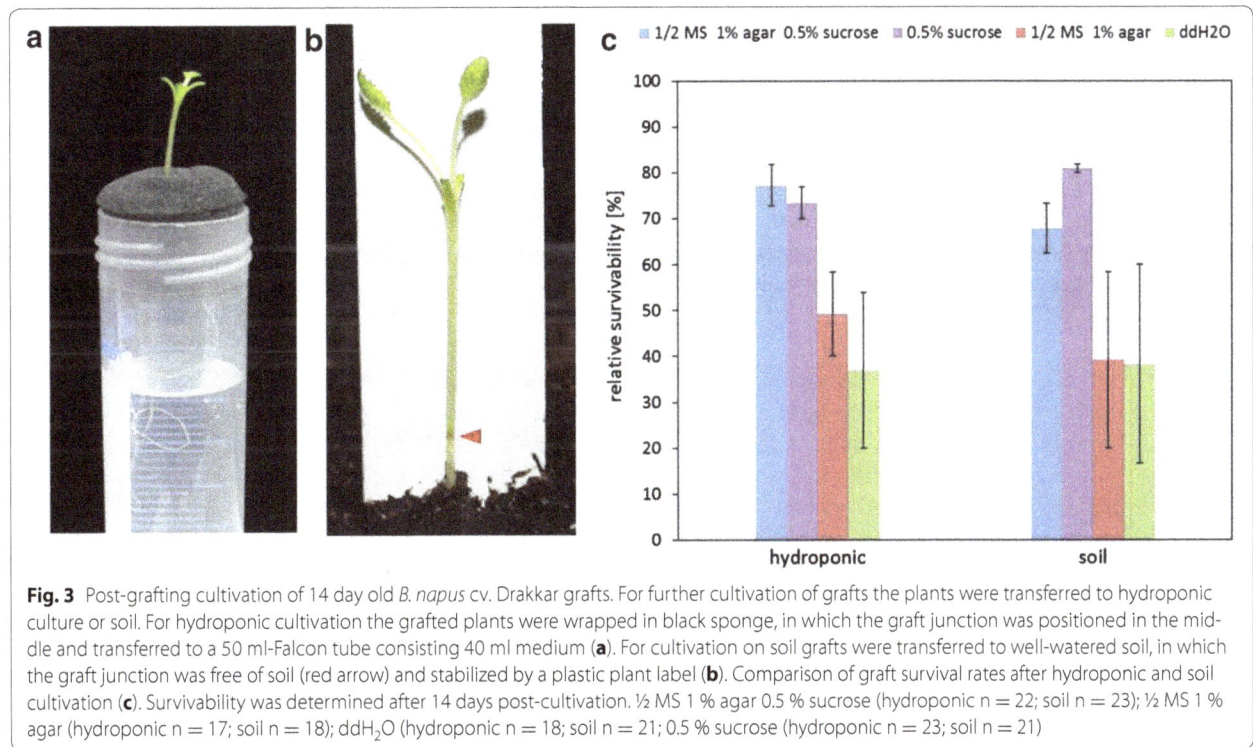

**Fig. 3** Post-grafting cultivation of 14 day old *B. napus* cv. Drakkar grafts. For further cultivation of grafts the plants were transferred to hydroponic culture or soil. For hydroponic cultivation the grafted plants were wrapped in black sponge, in which the graft junction was positioned in the middle and transferred to a 50 ml-Falcon tube consisting 40 ml medium (**a**). For cultivation on soil grafts were transferred to well-watered soil, in which the graft junction was free of soil (red arrow) and stabilized by a plastic plant label (**b**). Comparison of graft survival rates after hydroponic and soil cultivation (**c**). Survivability was determined after 14 days post-cultivation. ½ MS 1 % agar 0.5 % sucrose (hydroponic n = 22; soil n = 23); ½ MS 1 % agar (hydroponic n = 17; soil n = 18); ddH$_2$O (hydroponic n = 18; soil n = 21; 0.5 % sucrose (hydroponic n = 23; soil n = 21)

improved grafting success in cotyledon-less grafts [24]. Adventitious root formation, as described to be problematic in Arabidopsis and other species [2, 12, 39], was no problem in Brassica grafts, no matter whether they were supplied with sucrose or not. Orienting the plates slightly vertically (5°–10°) increased grafting success by reducing bending of hypocotyls and thus facilitating orientation of the grafting partners, as also observed in Arabidopsis [24,

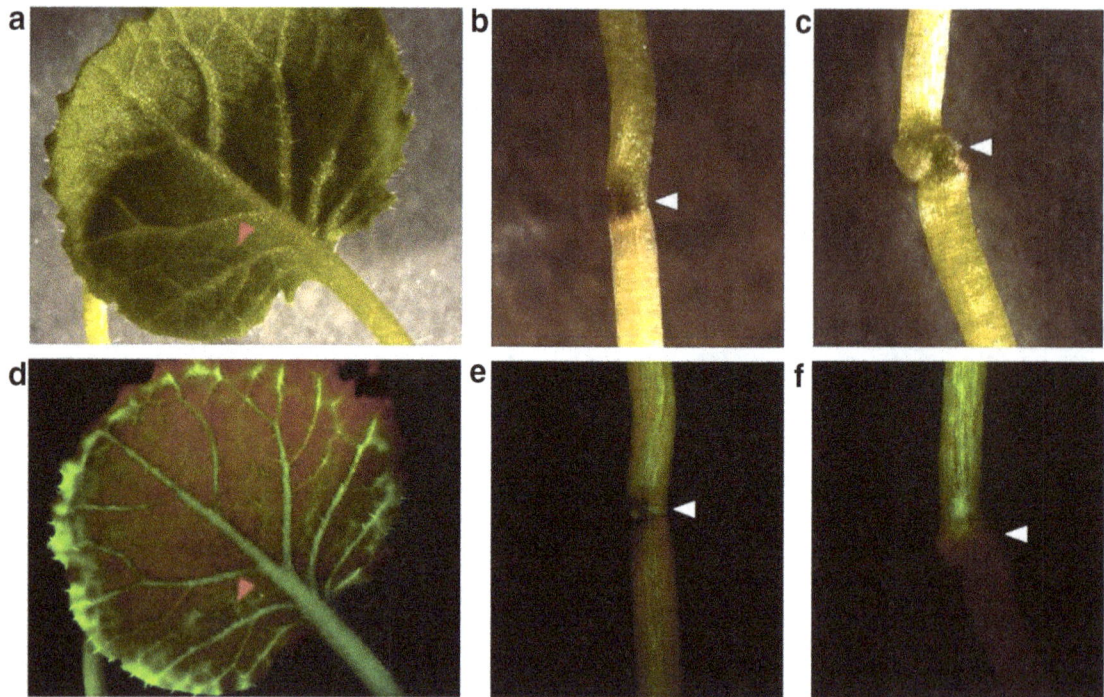

**Fig. 4** Analysing graft junction formation by CFDA staining. 14 day old grafts were transferred to a plate containing ½ MS and 1 % agar and stained with the phloem-mobile fluorochrome CFDA. Therefore a leaf of the grafts was punctured with a sterile needle (*red arrow*) and a few microliters of CFDA stain were applied at the point of lesion (**a**). After an incubation of 30 min at room temperature the plants were investigated under a fluorescence binocular using bright field (**a–c**) and fluorescence illumination with a GFP filter (**d–f**). While in successfully grafted plants the CFDA stain passed the graft junction (**e**), it accumulated in the scion part of the graft, when graft formation was not effective (**f**)

39]. Experiments performed with another *B. napus* cultivar (cv. Licosmos) confirmed the results obtained in cv. Drakkar, but Licosmos seedlings were in general smaller and even less delicate to handle (see Additional file 1: Figure S1, Additional file 2: Figure S2, Additional file 3: Figure S3).

In contrast to other published grafting procedures for *B. napus* [34–36] this work provides an easy and robust protocol for routine grafting with high success rates. Collar-free grafting is easier to handle and less laborious. When germination, grafting and graft generation are carried out under at least semi-sterile conditions, grafting success was significantly increased by minimising contamination of the graft junctions. Another advantage of our protocol is its applicability to mid- to high-throughput, as transport studies often require high numbers of grafted plants to allow statistically relevant conclusions.

In parallel experiments performed with Arabidopsis seedlings (data not shown) it could be observed that grafting on filter paper was more successful than grafting on agar plates. Also here, sucrose enhanced size and fitness of the grafts, but led to a slightly higher formation

of adventitious roots in this species. Generally, too much moisture hindered successful formation of graft unions and water films on both agar plates and filter paper in all types of grafting experiments performed. A similar observation was made in recent Arabidopsis grafting experiments [24].

The establishment of functional vascular connections was first checked using the phloem-accumulating fluorescence dye CFDA. Figure 4 shows that most viable grafts disposed of functional vascular connections that allowed movement of CFDA, while a few did not allow long-distance movement of CFDA. These grafts did not survive long-term. Since one major aim of our study was to produce grafts for xylem and phloem sampling, we tested whether 10-week-old grafted plants allow sampling of xylem and phloem sap. An example of phloem sap exuding from a grafted plant is shown in Fig. 5. Exudation can be observed after small incisions were applied with a syringe needle as first described in Giavalisco et al. [28]. We could not detect any difference in xylem or phloem sampling (concerning e.g. sap quantity or duration of exudation) when comparing grafted with

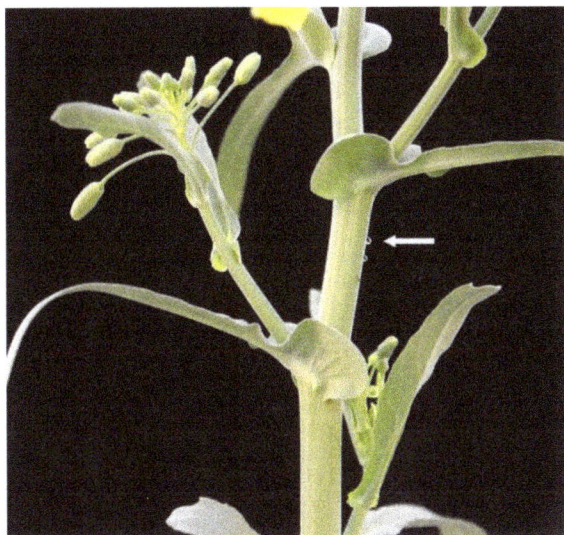

**Fig. 5** Phloem sap sampling from grafted *B. napus* plants. 10-week-old grafted plants were punctured with a hypodermic needle below the inflorescence as first described by Giavalisco et al. [28]. The *white arrow* indicates droplets of phloem sap after puncturing. For phloem sample collection the first droplet is discarded due to the danger of contamination by the content of injured cells. Several hundred microliters of phloem sap could be collected per plant independent on whether they were grafts or not

## Additional files

**Additional file 1: Figure S1.** Post-grafting cultivation of 14 day old *B. napus* cv. Licosmos grafts. Comparison of *B. napus* cv. Licosmos graft survival rates after hydroponic and soil cultivation. Survivability was determined after 14 days post-cultivation. ½ MS 1 % agar 0.5 % sucrose (hydroponic n = 40; soil n = 18); ½ MS 1 % agar (hydroponic n = 10; soil n = 10); ddH$_2$O (hydroponic n = 28; soil n = 12; 0.5 % sucrose (hydroponic n = 10; soil n = 12).

**Additional file 2: Figure S2.** Development of *B. napus* cv. Licosmos after 20 days in hydroponic culture, regenerated under different conditions.

**Additional file 3: Figure S3.** Hydroponic post-grafting cultivation of grafted *B. napus* seedlings. Plants were wrapped with foam and up to 30 could be cultivated in 50-ml conical tubes in one polystyrene boxes covered with light-permissive plastic cover to prevent desiccation.

non-grafted plants, what demonstrates the suitability of our grafting protocol for long-distance transport studies.

## Conclusions

Grafting is a versatile tool to study long-distance mobility of potential signalling compounds. The development of optimised protocols that allow reliable grafting with high success rates is essential to obtain reasonable and reproducible results. In this study we present a simple and efficient grafting procedure for *B. napus* that in routine application allows a short-term survival rate of 80–100 % and still 70–80 % after transfer to hydroponic culture or soil, respectively. This demonstrates that *B. napus* is highly suitable for performing transport studies using the easy grafting procedure presented here. That is important, because *B. napus* is already used as a suitable model plant for xylem sap and phloem sap analyses due to the relatively easy access to quite large sample volumes from both long-distance transport systems. Long-term survival rates on hydroponic culture or soil enable the growth of grafts until the time-point suitable for phloem sap sampling. Therefore, in contrast to grafting studies in model plants like Arabidopsis, Brassica grafting studies do not only allow indirect conclusions of phloem mobility of potential signalling compounds, but their direct detection in collected phloem samples.

### Authors' contributions
JK and MT contributed to conception and design of the experiments, AO, SP, JD and MT carried out the grafting experiments and analysed the results. JK has drafted the manuscript and all authors were involved in revising it critically. All authors read and approved the final manuscript.

### Acknowledgements
We would like to acknowledge the financial contribution to the research activities by a Career Integration Grant (CIG; PCIG14-GA-2013-63 0734) by the European Commission within the 7th framework programme and the grant LFF-GK06 „Deligrah" (Landesforschungsförderung Hamburg) awarded to JK.

### Competing interests
The authors declare that they have no competing interests.

### References
1. Harada T. Grafting and RNA transport via phloem tissue in horticultural plants. Sci Hortic. 2010;125:545–50.
2. Turnbull C. Grafting as a research tool. Methods Mol Biol. 2010;655:11–26.
3. Chailakhyan MK. New facts in support of the hormonal theory of plant development. Comptes Rendus (Doklady) Acad Sci URSS. 1936;13:79–83.
4. Zeevaart JAD. Flower formation as studied by grafting. Meded Landbouwhogesch Wagening. 1958;58:1–88.
5. Wisniewski LA, Powell PA, Nelson RS, Beachy RN. Local and systemic spread of Tobacco Mosaic Virus in transgenic tobacco. Plant Cell. 1990;2:559–67.
6. Bertaccini A, Bellardi MG. A rhabdovirus inducing vein yellowing in croton. Plant Pathol. 1992;41:79–82.
7. Proebsting WM, Hedden P, Lewis MJ, Croker SJ, Proebsting LN. Gibberellin concentration and transport in genetic lines of pea. Plant Physiol. 1992;100:1354–60.
8. Mapelli S, Kinet JM. Plant growth regulator and graft control of axillary bud formation and development in the TO-2 mutant tomato. Plant Growth Regul. 1992;11:385–90.
9. Lohar DP, VandenBosch KA. Grafting between model legumes demonstrates roles for roots and shoots in determining nodule type and host/rhizobia specificity. J Exp Bot. 2005;56:1643–50.
10. Pant BD, Buhtz A, Kehr J, Scheible WR. MicroRNA399 is a long-distance signal for the regulation of plant phosphate homeostasis. Plant J. 2008;53:731–8.
11. Pant BD, Musialak-Lange M, Nuc P, May P, Buhtz A, Kehr J, Walther D, Scheible W-Rd. Identification of nutrient-responsive Arabidopsis and

rapeseed microRNAs by comprehensive real-time polymerase chain reaction profiling and small RNA sequencing. Plant Physiol. 2009;150:1541–55.

12. Buhtz A, Pieritz J, Springer F, Kehr J. Phloem small RNAs, nutrient stress responses, and systemic mobility. BMC Plant Biol. 2010;10:64.

13. Jenns A, Kuc J. Graft transmission of systemic resistance of cucumber to anthracnose induced by *Colletotrichum lagenarium* and tobacco necrosis virus. Phytopathology. 1979;7:753–6.

14. Dean R, Kuc J. Induced systemic protection in cucumbers: the source of the "signal". Physiol Mol Plant Pathol. 1986;28:227–33.

15. Palauqui J-C, Elmayan T, Pollien J-M, Vaucheret H. Systemic acquired silencing: transgene-specific post-transcriptional silencing is transmitted by grafting from silenced stocks to non-silenced scions. EMBO J. 1997;16:4738–45.

16. Xu H, Iwashiro R, Li T, Harada T. Long-distance transport of Gibberellic Acid Insensitive mRNA in *Nicotiana benthamiana*. BMC Plant Biol. 2013;13:165.

17. Kasai A, Kanehira A, Harada T. miR172 can move long distances in *Nicotiana benthamiana*. Open Plant Sci J. 2010;4:1–6.

18. Kassaw T, Frugoli J. Simple and efficient methods to generate split roots and grafted plants useful for long-distance signaling studies in *Medicago truncatula* and other small plants. Plant Methods. 2012;8:38.

19. Rhee SY, Sommerville CR. Flat-surface grafting in *Arabidopsis thaliana*. Plant Mol Biol Rep. 1995;13:118–23.

20. Turnbull CGN, Booker JP, Leyser HMO. Micrografting techniques for testing long-distance signalling in Arabidopsis. Plant J. 2002;32:255–62.

21. Ayre BG, Turgeon R. Graft transmission of a floral stimulant derived from CONSTANS. Plant Physiol. 2004;135:2271–8.

22. Flaishman M, Loginovsky K, Golobowich S, Lev-Yadun S. *Arabidopsis thaliana* as a model system for graft union development in homografts and heterografts. J Plant Growth Regul. 2008;27:231–9.

23. Notaguchi M, Daimon Y, Abe M, Araki T. Adaptation of a seedling micrografting technique to the study of long-distance signaling in flowering of *Arabidopsis thaliana*. J Plant Res. 2009;122:201–14.

24. Marsch-Martinez N, Franken J, Gonzalez-Aguilera K, de Folter S, Angenent G, Alvarez-Buylla E. An efficient flat-surface collar-free grafting method for *Arabidopsis thaliana* seedlings. Plant Methods. 2013;9:14.

25. Lin SI, Chiang SF, Lin WY, Chen JW, Tseng CY, Wu PC, Chiou TJ. Regulatory network of microRNA399 and PHO$_2$ by systemic signaling. Plant Physiol. 2008;147:732–46.

26. Kuo HF, Chiou TJ. The role of miRNAs in phosphorus deficiency signaling. Plant Physiol. 2011;156:1016–24.

27. Buhtz A, Kolasa A, Arlt K, Walz C, Kehr J. Xylem sap protein composition is conserved among different plant species. Planta. 2004;219:610–8.

28. Giavalisco P, Kapitza K, Kolasa A, Buhtz A, Kehr J. Towards the proteome of *Brassica napus* phloem sap. Proteomics. 2006;6:896–909.

29. Kehr J, Buhtz A, Giavalisco P. Analysis of xylem sap proteins from *Brassica napus*. BMC Plant Biol. 2005;5:11. doi:10.1186/1471-2229-1185-1111.

30. Buhtz A, Springer F, Chappell L, Baulcombe DC, Kehr J. Identification and characterization of small RNAs from the phloem of *Brassica napus*. Plant J. 2008;53:739–49.

31. Buhtz A, Pieritz J, Springer F, Kehr J. Phloem small RNAs, nutrient stress responses, and systemic mobility. BMC Plant Biol. 2010;10:64.

32. Bhalla PL, Singh MB. Agrobacterium-mediated transformation of *Brassica napus* and *Brassica oleracea*. Nat Protoc. 2008;3:181–9.

33. Cardoza V, Steward CN. Agrobacterium-mediated transformation of canola. In: Curtis IS, editor. Transgenic crops of the world—essential protocols. Dordrecht: Kluwer Academic Publishers; 2004. p. 379–87.

34. Koeslin-Findeklee F, Becker MA, van der Graaff E, Roitsch T, Horst WJ. Differences between winter oilseed rape (*Brassica napus* L.) cultivars in nitrogen starvation-induced leaf senescence are governed by leaf-inherent rather than root-derived signals. J Exp Bot. 2015;66:3669–81.

35. Filek M, Biesaga-Kościelniak J, Marcińska I, Krekule J, Macháčková I, Dubert F. The effects of electric current on flowering of grafted scions of nonvernalized winter rape. Biol Plant. 2003;46:625–8.

36. Dubert F, Pienkowski S, Filek W. Shortening of the development cycle in winter rape (*Brassica napus* var. oleifera L.) by grafting nonvernalized scions on generative stock. Acta Agrobot. 1984;37:39–45.

37. Chen A, Komives E, Schroeder J. An improved grafting technique for mature Arabidopsis plants demonstrates long-distance shoot-to-root transport of phytochelatins in Arabidopsis. Plant Physiol. 2006;141:108–20.

38. Grignon N, Touraine B, Durand M. 6(5)Carboxyfluorescein as a tracer of phloem sap translocation. Am J Bot. 1989;76:871–7.

39. Turnbull C, Booker J, Leyser H. Micrografting techniques for testing long-distance signalling in Arabidopsis. Plant J. 2002;32:255–62.

# 11

# A method for detecting single mRNA molecules in *Arabidopsis thaliana*

Susan Duncan[2], Tjelvar S. G. Olsson[2], Matthew Hartley[2], Caroline Dean[1] and Stefanie Rosa[1*]

## Abstract

**Background:** Despite advances in other model organisms, there are currently no techniques to explore cell-to-cell variation and sub-cellular localization of RNA molecules at the single-cell level in plants.

**Results:** Here we describe a method for imaging individual mRNA molecules in *Arabidopsis thaliana* root cells using multiple singly labeled oligonucleotide probes. We demonstrate detection of both mRNA and nascent transcripts of the housekeeping gene *Protein Phosphatase 2A*. Our image analysis pipeline also enables quantification of mRNAs that reveals the frequency distribution of transcripts per cell underlying the population mean.

**Conclusion:** This method allows single molecule RNA in situ to be exploited as a powerful tool for studying gene regulation in plants.

**Keywords:** RNA, FISH, Gene expression, Fluorescence microscopy, Arabidopsis, Transcription

## Background

Quantitative real-time PCR is commonly used to analyze plant gene expression, but this method lacks potentially important information relating to sub-cellular localization of RNA and masks cell-to-cell variation [1, 2]. To effectively study these aspects of gene regulation, it is necessary to study RNA at the cellular level.

A method that has achieved this aim is in situ hybridization followed by microscopic analysis. Initially, researchers performed in situ hybridizations using radioactive probes [3]. Early improvements involved linking the probes to enzymes that catalyze chromogenic or fluorogenic reactions [4–6]. In Arabidopsis mRNA in situ hybridization has been routinely used for detailed visualization of gene expression patterns [7–9]. While this method gives good semi-quantitative spatial information, it produces images with limited cellular resolution. More recently plant researchers have used fluorescently labeled probes to directly label transcripts. This has improved cellular resolution, but relatively poor sensitivity has resigned it mainly for detection of highly repetitive RNAs

[10, 11]. Single molecule fluorescence in situ hybridization (smFISH) was developed to maximize both sensitivity and specificity by using multiple singly labelled probes to visualize RNA molecules as discrete spots of fluorescence [12]. A recent version of this method uses 48 fluorescently labeled DNA oligonucleotides (20mers) to hybridize to different portions of each transcript. This provides a balance between probe length and number that effectively reduces false positive signals (due to off-target binding) whilst maintaining single molecule sensitivity [13].

Establishment of smFISH in other model systems has led to greater understanding of transcriptional regulation for many genes [14–17]. In addition to quantifying mRNA at the single cell level, this detection method can be used to visualize sites of transcription [18] and long non-coding RNAs [19].

Optical properties of plant cells and tissues provide significant challenges for fluorescence microscopy [20]. Inherent light scattering adversely affects both the excitation and the detection efficiency; moreover plants contain many native molecules that emit high levels of background auto fluorescence compared to other organisms [20]. We chose to develop a smFISH method in

*Correspondence: stefanie.rosa@jic.ac.uk
[1] Department of Cell and Developmental Biology, John Innes Centre, Norwich Research Park, Norwich NR4 7UH, UK
Full list of author information is available at the end of the article

fixed Arabidopsis root cells as they typically allow clearer imaging than leaves or other above ground tissue.

We established our method by probing a widely expressed housekeeping gene At1G13320—the A2 scaffolding subunit of *Protein Phosphatase2A* (*PP2A*) [21]. Unlike several environmentally regulated phosphatase subunits, it exhibits mRNA levels that are relatively unperturbed by a range of abiotic and biotic stresses and is transcribed evenly across many tissue types throughout development. These robust properties led to *PP2A* being identified as a superior gene for qPCR normalization [22]. To validate our method we used smFISH to detect *PP2A* mRNAs and used an image analysis pipeline to automate transcript counting within cells. Together our smFISH protocol and image analysis algorithm provides a straightforward framework for other plant researchers to study gene expression at the single-cell and single-molecule resolution.

## Results and discussion

We designed our initial set of smFISH probes to hybridize exclusively to *PP2A* exons in order to visualize mRNA locations (Fig. 1a). We prepared our samples using a root squash method that typically yields many cells in a single-layer. This method together with the use of red and far-red dyes maximized mRNA signals whilst minimizing background fluorescence.

We observed non-specific signals in endo-reduplicated nuclei from the differentiation zone and this restricted our analysis to the meristem region (Additional file 1). Consistent with other reports, we visualized *PP2A*

mRNAs as punctate signals 250–300 nm homogeneously dispersed throughout the cytoplasm [13, 23] (Fig. 2a, b, Additional file 2). We found wide-field far superior to confocal microscopy for smFISH imaging with further improvements achieved by deconvolution (Fig. 2c). RNase treatment confirmed that our signals represent RNA locations (Additional file 3).

Next we designed 48 probes to be complimentary only to *PP2A* introns to identify sites of transcription [18] (Fig. 1a). We found that these signals were invariably restricted to the nucleus and co-localized with *PP2A* mRNA foci (Fig. 3a–f; Additional file 4). Also, consistent with RNA production being halted during cell division, we were unable to detect nascent RNA during mitosis (Fig. 3g–i).

We had equal success in imaging RNA labelled with Quasar®570 and 670 dyes, but we were unable to observe RNA labelled with FITC (data not shown). We found super-resolution structured illumination microscopy (SIM) produced high quality images of our samples; therefore it may be possible to overcome this multiplex limitation through the detection of spectrally barcoded smFISH probes [24] (Additional file 5).

We used an automated image analysis workflow to identify and quantify mRNA in our smFISH images (Fig. 4). Out-of-focus light caused background signal intensity to vary greatly through the image, making it impossible to apply a single uniform threshold level for spot counting. To overcome this we normalized image intensities for each plane of the z-stack before taking a maximum intensity projection. We applied edge

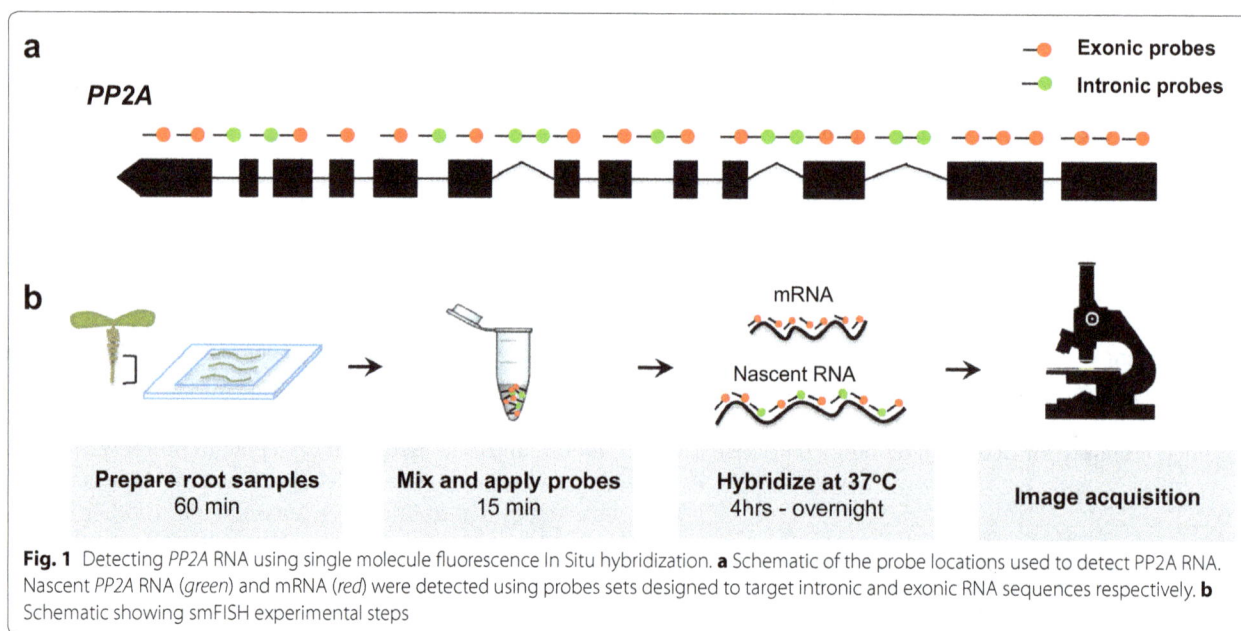

**Fig. 1** Detecting *PP2A* RNA using single molecule fluorescence In Situ hybridization. **a** Schematic of the probe locations used to detect PP2A RNA. Nascent *PP2A* RNA (*green*) and mRNA (*red*) were detected using probes sets designed to target intronic and exonic RNA sequences respectively. **b** Schematic showing smFISH experimental steps

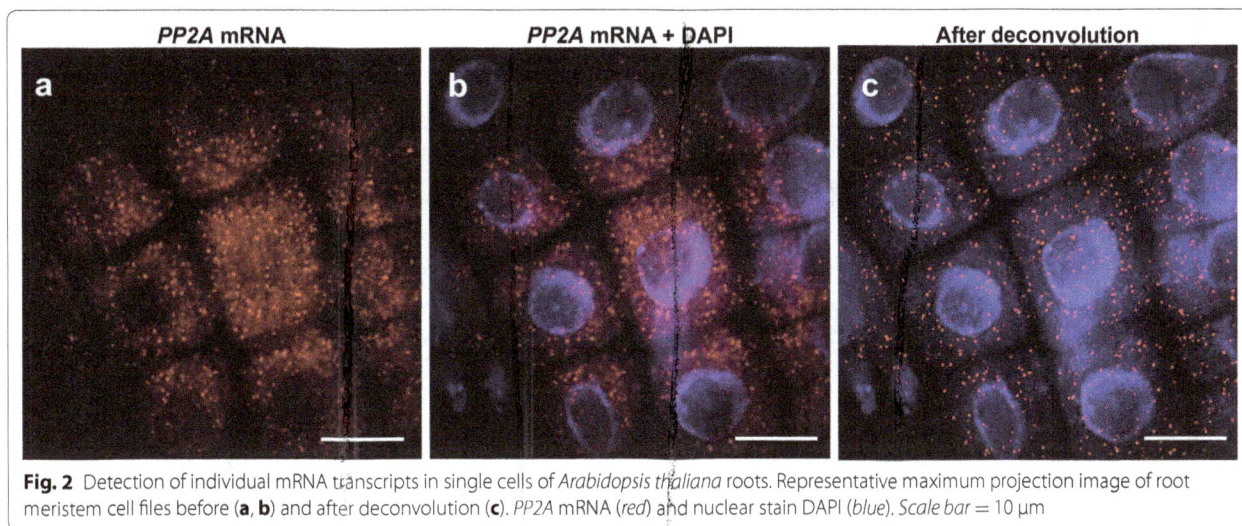

**Fig. 2** Detection of individual mRNA transcripts in single cells of *Arabidopsis thaliana* roots. Representative maximum projection image of root meristem cell files before (**a**, **b**) and after deconvolution (**c**). *PP2A* mRNA (*red*) and nuclear stain DAPI (*blue*). *Scale bar* = 10 μm

detection to this projection and then used template matching to determine the probe locations. This procedure allowed us to avoid having to determine a threshold manually for each image. To obtain cell-level transcript counts we used the watershed algorithm to segment the image into cells using seeds derived from the DAPI nuclear stain channel. We then combined this segmentation with the RNA locations within each segmented cell to generate an annotated image showing derived cell boundaries and transcript counts per cell (Fig. 5).

We chose to perform the analysis on a projection of the z-stack because it simplifies the processing considerably. For data with a higher density of mRNAs or situations where the position of the mRNAs in the z direction is of interest, the spot detection and segmentation algorithms could be implemented in three dimensions. However, for these data, spot density was not high enough to make this necessary.

Automated analysis of our images revealed that >70 % of cells contain 90 or less *PP2A* mRNA molecules whilst the remaining ~30 % contain between 90 and 220 molecules (Fig. 6a). Every cell we observed contained a minimum of 15 *PP2A* mRNA molecules (Fig. 6a) and an average of 74 mRNAs were detected in each cell (Fig. 6b). Consistent with identification of *PP2A* as a superior normalization gene [22], nascent RNA signals were observed in 84 % of cells. (Fig. 6c, d, Additional file 6).

## Conclusions

In this report we present a FISH method that allows for gene-expression profiling of transcripts in Arabidopsis roots. By characterizing cell-to-cell transcriptional variability of the housekeeping gene *PP2A* we demonstrate that smFISH can be combined with automated image analysis to quantify single RNA transcripts for the first time in plants. As smFISH has been used extensively

for RNA analysis in many model organisms [25–28] we believe that this root squash protocol can be easily adapted to suit other plant species with amendments made to the fixation and permeabilization steps as necessary. However adapting smFISH for use in green tissues is likely to represent a greater challenge due to high levels of autofluorescence. Similar issues have been overcome in other organisms through the application of tissue clearing [16] and cryosectioning [29]. We believe that similar approaches may also be employed to enable transcript imaging in other plant tissues.

In addition to quantifying mRNA and visualizing active sites of transcription at the single cell level, corroboratory qRT-PCR data has shown that smFISH can be used to calculate mRNA fold changes at the cellular level [13]. This method can also identify RNA derived from maternal and paternal gene copies [30] and, in conjunction with masking oligonucleotides, it can even distinguish RNA transcripts that differ by only a single nucleotide polymorphism [31]. Our adaptation of smFISH for use in *Arabidopsis thaliana* now opens up these exciting opportunities to the plant research community.

## Methods

### Plant material and growth conditions

Col-0 seeds were surface sterilized in 5 % v/v sodium hypochlorite for 5 min and rinsed three times in sterile distilled water before being sown on MS media minus glucose. They were stratified for 3 days at 5 °C before being transferred to a growth cabinet (Sanyo MLR-351H) 16 h light, 100 μmol m$^{-2}$ s$^{-1}$, 22 °C ± 1 °C.

### Reagents and solutions

Tables 1 and 2 list the oligonucleotide sequences used to detect *PP2A* mRNA and nascent transcripts respectively.

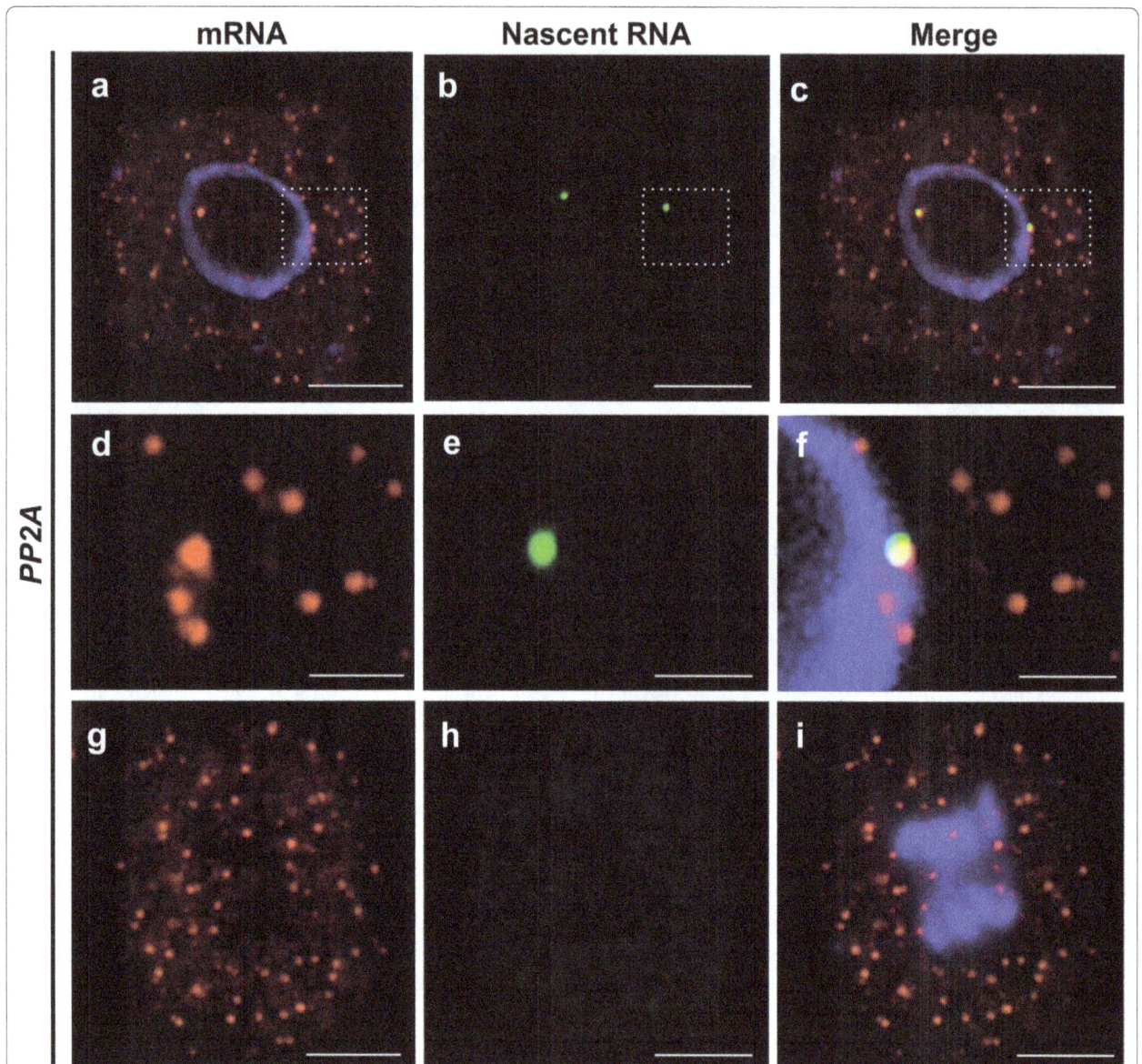

**Fig. 3** Simultaneous detection of spliced and nascent *PP2A* RNA. Representative images of cells labeled with *PP2A* mRNA (*red*) and nascent *PP2A* RNA (*green*). DNA labeled with DAPI (*blue*). **a–c** Representative image of an isolated meristem cell showing cytoplasmic mRNA (**a**); and two sites of active transcription located within the nucleus (**b**). **d–f** Magnified image from cell depicted in (**a–c**) showing co-localization of nascent *PP2A* RNA and mRNA. **g–i** Representative image of a cell during mitosis showing no transcription as judged by the absence of nascent RNA signals (**h**). *Scale Bar* = 10 μm in (**a–c**, **g–i**) and =0.5 μm in (**d–f**)

*Liquid nitrogen*

Nuclease-free water—not DEPC treated (Qiagen, Cat. No. 129117).

Paraformaldehyde (Sigma, Cat. No. P6148) freshly depolymerized, 4 % w/v in water.

Nuclease-free 10× Phosphate Buffered Saline (Thermo Scientific, Cat. No. AM9624).

70 % Ethanol (freshly made using nuclease free water).

Nuclease-free 20× saline-sodium citrate (20× SSC, Thermo Scientific, Cat. No. AM9763).

RNase A (Sigma, Cat. No. R4642) diluted to 100 μg/ml. $T_{10}E_1$ buffer (10mM Tris-HCl, 1mM EDTA, pH 8)—Sigma, Cat. No. 93283-100mL

Deionized Formamide (Sigma, Cat. No. F9037).

Dextran Sulphate (Sigma, Cat. No. Res2029D).

Nuclease free Tris HCl buffer 1 M pH8 (Thermo Scientific, Cat. No. AM9855G).

Glucose oxidase (Sigma, Cat. No. G0543).

Bovine Live Catalase (Sigma, Cat. No. C3155).

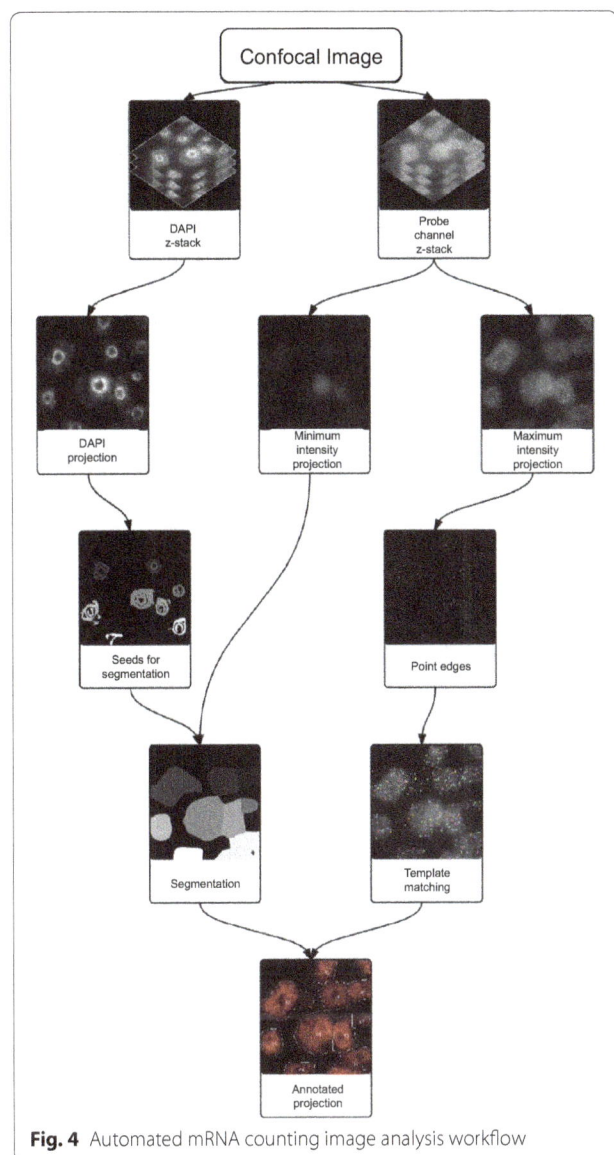

**Fig. 4** Automated mRNA counting image analysis workflow

## Wash buffer (50 ml)

5 ml nuclease free 20× SSC mixed with 5 ml nuclease free deionized formamide and nuclease free water up to 50 ml final volume. (Final composition: 10 % formamide, 2× SSC).

**DAPI** (4′, 6-Diamidino-2-phenylindole; Sigma cat. no. D9564) Diluted to 100 ng/μl in wash buffer (Final composition: 100 ng/μl, 10 % formamide, 2× SSC).

## Hybridization solution (10 ml)

Dissolve 1 g dextran sulfate in 1 ml nuclease free 20× SCC, 1 ml deionized formamide and nuclease free water up to 10 ml final volume. (Final composition: 100 mg/ml dextran sulfate and 10 % formamide in 2× SSC).

## Anti-fade GLOX buffer minus enzymes (1 ml)

40 μl 10 % glucose in nuclease-free water, 10 μl 1 M Tris–HCl, pH 8.0 and 100 μl 20× SCC was mixed with 850 μl nuclease-free water. (Final composition: 0.4 % glucose in 10 nM Tris–HCl, 2× SSC).

## Anti-fade GLOX buffer containing enzymes (100 μl)

1 μl glucose oxidase and 1 μl mildy vortexed catalase suspension added to 100 μl GLOX minus enzyme solution.

## Equipment
### Razor blades

*Forceps* Poly-L-Lysine slides (Sigma, Cat. No. PO425 or similar NOTE: these are not essential but the samples adhere better to these than untreated slides).

Low stender-form preparation dishes (VWR, Cat. No. 470144-866 or similar).

22 mm × 22 mm No.1 glass coverslips (Fisher Scientific, Cat. No. 12333128 or similar).

Coplin jar (Sigma, Cat. No. S6016 or similar).

Parafilm® M sealing film (Bemin, Cat. No. PM992).

*Orbital shaker* Hybridization chamber (or a suitable dark box with a layer of tissue moistened with water will suffice).

*37 °C incubator* Zeiss Elyra PS1 inverted microscope with cooled EM-CCD Andor iXon 897 camera.

## smFISH probe design

Since designing smFISH probes is similar to designing PCR primers most primer design software packages can be used [23] but we used the online program Stellaris® Probe Designer version 2.0 from Biosearch Technologies (http://singlemoleculefish.com). Input of *PP2A* coding sequence into the program automatically generates a set of probes complementary to the *PP2A* mRNA, optimized for binding to the target sequence. Before ordering our pre-labelled probes from Biosearch Technologies we completed a TAIR BLAST query for each sequence to ensure target specificity (https://www.arabidopsis.org/Blast/). Tables 1 and 2 list the oligonucleotide sequences used to detect *PP2A* mRNA and nascent transcripts respectively.

## Sample preparation (timing: 2 h)

Seedlings were removed from the media 10 days after germination. Root tips were dissected using a razor blade and forceps and placed into a glass dish containing 4 % paraformaldehyde to fix for 30 min at room temperature. The roots were removed from the fixative and washed twice with 1× PBS. 3–4 roots were then arranged on a slide and covered by a glass coverslip and the meristems were squashed manually by applying pressure through

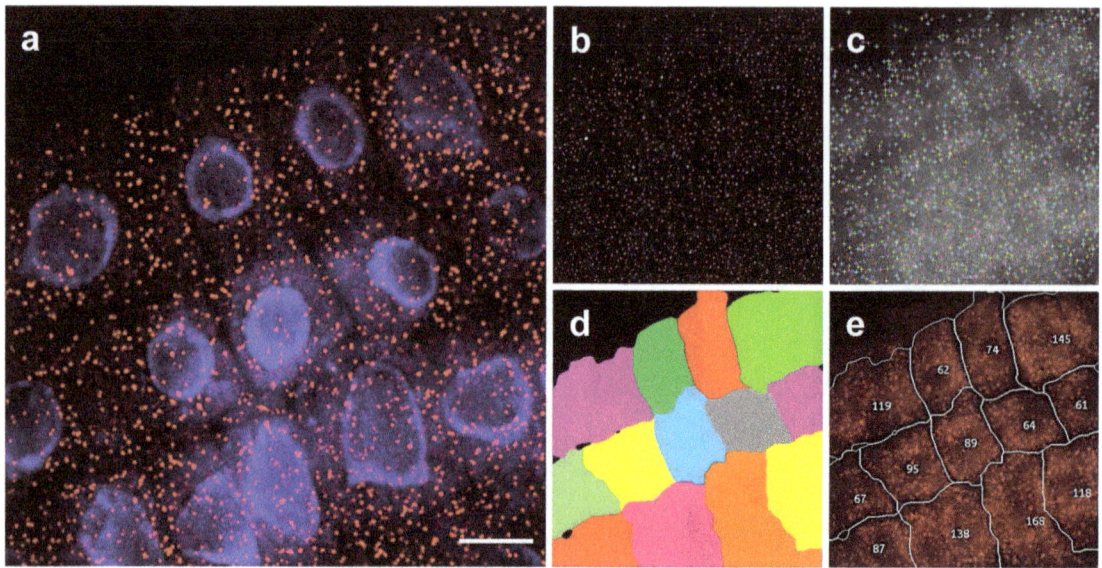

**Fig. 5** Automated image analysis of *PP2A* mRNA. **a** Representative maximum projection image of cell files labeled with PP2A mRNA probes (*red*). DNA labeled with DAPI (*blue*). **b**, **c** Screen shots showing sequential detection steps used to determine positive mRNA signals. **d** Cell segmentation, where a false-color is rendering individual cells. **e** Output image indicating the number of mRNA signals detected on each cell segmented in (**d**). *Scale bar* = 10 μm

the coverslip. The slide, together with the sample and coverslip, were then submerged briefly in liquid nitrogen (~5 s) to adhere the roots to the slide. The coverslip was then flipped off with a razor blade and the samples were left to dry at room temperature for a minimum of 30 min. Tissue permeabilization was then carried out by immersing the samples in coplin jars containing 70 % ethanol and left to shake gently for a minimum of 1 h.

Note: We ensured coplin jar lids were sealed with parafilm to prevent evaporation during the ethanol incubation period.

### Hybridization (timing: 4 h—overnight)
Residual ethanol was left to evaporate at room temperature for 5 min before 2, 2-min washes were carried out with wash buffer. 100 μl of hybridization solution with probes at a final concentration of 250 nM was then added to each slide. Coverslips were laid over the samples to prevent buffer evaporation and the probes were left to hybridize in a humid chamber at 37 °C overnight in the dark.

### Sample mounting (timing: 2 h)
Hybridization solution containing unbound probes was removed using a pipette in the morning. Each sample was then washed twice with 200 μl wash buffer and finally immersed in coplin jars containing wash buffer for 30 min at 37 °C. 100 μl of the nuclear stain DAPI was then added to each slide and left to incubate at 37 °C for 30 min. Following DAPI removal, 100 μl 2× SSC was

added samples and removed. 100 μl GLOX buffer minus enzymes was added to the samples and left to equilibrate for 2 min and then replaced with 100 μl of anti-fade GLOX buffer containing enzymes. The samples were then covered by coverslips sealed. Excess GLOX buffer was wicked away using tissue before the coverslips were sealed with nail varnish. We immediately imaged our samples as we observed a noticeable reduction in image quality around 4 h after mounting.

Note: Oxygen-scavenging GLOX buffer maximised the stability of our smFISH fluorophores and we observed rapid bleaching when it was substituted with the commercial anti-fade mounting media Vectorshield (data not shown).

### Image acquisition
A Zeiss Elyra PS1 inverted microscope was used for imaging. A 100X oil-immersion objective (1.46 NA) and cooled EM-CCD Andor iXon 897 camera (512 × 512 QE > 90 %) was used to obtain all images in the standard, rather than super-resolution mode. The following wavelengths were used for fluorescence detection: for probes labeled with Quasar®570 an excitation line of 561 nm was used and signal was detected at 570–640 nm; for probes labeled with Quasar®670 an excitation line of 642 nm and signal was detected at 655–710 nm; for DAPI an excitation line of 405 nm and signal was detected at wavelengths of 420–480 nm. For all experiments exposure times between 200–250 ms were used and a series of optical sections with z-steps of 0.2 μm were collected.

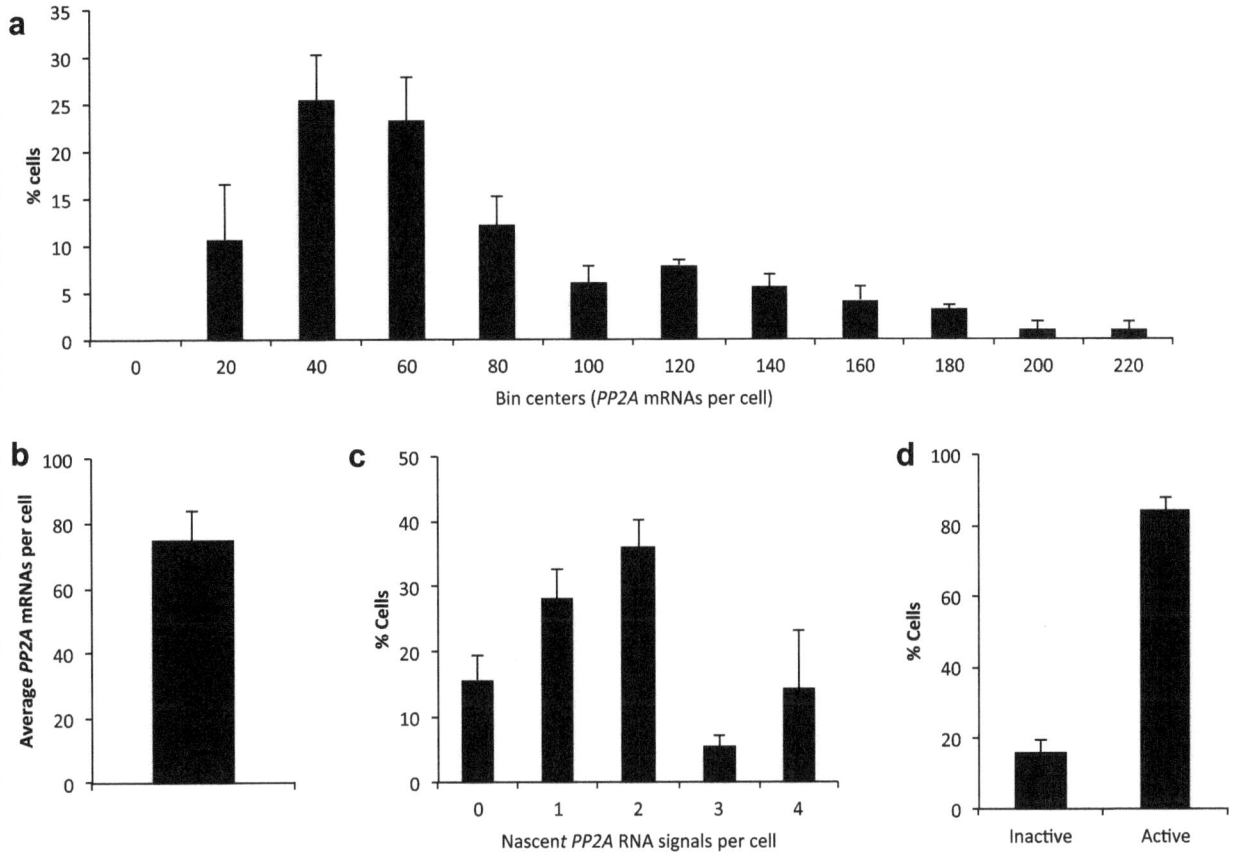

**Fig. 6** Quantification of mRNA and transcription status for *PP2A*. **a** Frequency distribution of mRNA molecules per cell. **b** Overall average mRNA number per cell. **c** Quantification of active *PP2A* transcription sites as judged by nascent RNA signals per cell. **d** Percentages of transcriptionally active versus inactive cells are shown in **d**. A total of 216 cells were analyzed. *Error bars* = +SEM

Note: When establishing this technique for the first time we recommend that the following controls be carried out: no probe (where probes are omitted from the hybridization solution) Additional file 1, and RNase A treatment (Additional file 2). To confirm RNA specificity we incubated samples with RNase for 1 h at 37 °C in a humid chamber after the ethanol permeabilization step, rinsed in 10 mM HCl for 5 min, washed twice with 2× SSC for 5 min before the protocol was continued.

Z-stacks were deconvolved using AutoQuant X2 (Media Cybernetics). Projections and analysis of 3D pictures were performed using Fiji (an implementation of ImageJ, a public domain program by W. Rasband available from http://rsb.info.nih.gov/ij/). Typically from 4 to 6 roots more than 300 cells can be obtained by this method, which were then suitable for further analysis using our automated mRNA counting programme.

### Structured illumination microscopy

A Zeiss Elyra PS1 inverted microscope was used for imaging using a 63X water objective (1.2 NA) to match samples mounted in GLOX buffer. The SIM camera used

was an EM-CCD Andor iXon 885. We collected ×5 phases at ×3 angles total 15 images per plane. Series of optical sections with z-steps of 0.2 μm were collected.

The following wavelengths were used for fluorescence detection: for probes labeled with Quasar®570 an excitation line of 561 nm was used and signal was detected at 570–640 nm; for probes labeled with Quasar®670 an excitation line of 642 nm and signal was detected at 655–710 nm; for DAPI an excitation line of 405 nm and signal was detected at wavelengths of 420–480 nm. For all experiments series of optical sections with z-steps of 0.2 μm were collected.

Images were processed using Zen Black default parameters. The images were also colour aligned using Zeiss "channel aligned" tool. Reference images of multiple coloured beads were collected in SIM mode then processed. Then an alignment matrix was generated using the SIM bead data and this was applied to the experimental SIM data.

### Image analysis

We have made our mRNA counting programme publically available at: https://github.com/JIC-CSB/

**Table 1** smFISH probe sequences used to detect PP2A mRNA

| PP2A exon probes | Sequences (5′–3′) |
| --- | --- |
| 1 | ccgagcgatctatcaatcag |
| 2 | gacatcctcaccaaaactca |
| 3 | tcgggtataaaggctcatca |
| 4 | tagctcgtcgataagcacag |
| 5 | ccaagagcacgagcaatgat |
| 6 | atcaactcttttcttgtcct |
| 7 | catcgtcattgttctcacta |
| 8 | atagccaaaagcacctcatc |
| 9 | atacagaataaaaccccca |
| 10 | caagtttcctcaacagtgga |
| 11 | tcatctgagcaccaattcta |
| 12 | tagccagaggagtgaaatgc |
| 13 | cattcaccagctgaaagtcg |
| 14 | ggaaaatcccacatgctgat |
| 15 | atattgatcttagctccgtc |
| 16 | attggcatgtcatcttgaca |
| 17 | aaattagttgctgcagctct |
| 18 | gctgattcaattgtagcagc |
| 19 | ccgaatcttgatcatcttgc |
| 20 | caaccctcaacagccaataa |
| 21 | ctccaacaatttcccaagag |
| 22 | caaccatataacgcacacgc |
| 23 | agtagacgagcatatgcagg |
| 24 | gaacttctgcctcattatca |
| 25 | cacagggaagaatgtgctgg |
| 26 | tgacgtgctgagaagagtct |
| 27 | cccattataactgatgccaa |
| 28 | tggttcacttggtcaagttt |
| 29 | tctacaatggctggcagtaa |
| 30 | cgattatagccagacgtact |
| 31 | gactggccaacaagggaata |
| 32 | catcaaagaagcctacacct |
| 33 | ttgcatgcaaagagcaccaa |
| 34 | acggattgagtgaaccttgt |
| 35 | cttcagattgtttgcagcag |
| 36 | ggaccaaactcttcagcaag |
| 37 | ggaactatatgctgcattgc |
| 38 | gtgggttgttaatcatctct |
| 39 | tgcacgaagaatcgtcatcc |
| 40 | ttactggagcgagaagcga |
| 41 | ctctgtctttagatgcagtt |
| 42 | gaacatgtgatctcggatcc |
| 43 | catcattttggccacgttaa |
| 44 | cgtatcatgttctccacaac |
| 45 | atcaacatctgggtcttcac |
| 46 | ttggagagcttgatttgcga |
| 47 | acacaattcgttgctgtctt |
| 48 | cgcccaacgaacaaatcaca |

**Table 2** smFISH probe sequences used to detect *PP2A* nascent transcripts

| PP2A intron probes | Sequences (5′–3′) |
| --- | --- |
| 1 | actattaccattcttagact |
| 2 | gaactgaaactttgtgccgt |
| 3 | tgacccattagcctctaaaa |
| 4 | ctttaaactcaattccgcct |
| 5 | tgcatacatagacaccatca |
| 6 | gtaaaccagccttatctaac |
| 7 | ttgacagagcatggaaagga |
| 8 | tcttctgtttagtggctta |
| 9 | acaattgacaaaggacccca |
| 10 | gcatatttccaaactttggg |
| 11 | acacctataaggggaacact |
| 12 | acttcaacctaccaatttcc |
| 13 | atgttctcttagatcaacca |
| 14 | aaagagcgctaaagccagag |
| 15 | tcacatacacaaccacaacc |
| 16 | acctataccgaggtatgtat |
| 17 | gcttaagtcggtttcacatt |
| 18 | acacaatgacagtgttcagt |
| 19 | cccataactaggcttgatga |
| 20 | acttgcctattacacatcag |
| 21 | tgttcaatgcagtaaccta |
| 22 | gcttaacttcagctaatggt |
| 23 | agctgagatgtagacaaccg |
| 24 | ctttcccataaagctcatca |
| 25 | agcagctcatacatatctgc |
| 26 | aacttcaaccatcactgctt |
| 27 | acctctgaagtcagtaatct |
| 28 | catggacttccaagtaccaa |
| 29 | cacactcttcttaagtgtgt |
| 30 | tggtcctttgcataatatga |
| 31 | cttagcaaacaccgacagta |
| 32 | ctacgtgtagatttataggt |
| 33 | atcggtttttaattctgctt |
| 34 | gtattcatgatatgagaggc |
| 35 | cactccaaactatagagcca |
| 36 | atctttatctctaagatgct |
| 37 | gatgacagtgactaggacga |
| 38 | ccttccaggcacagttaaaa |
| 39 | acatagtgaggttttcttat |
| 40 | atgccaagttaaaagctgca |
| 41 | gagtaacttggtcaatagca |
| 42 | acccaatgtcgtacaaagag |
| 43 | acagctcctttgaacatgtg |
| 44 | tagtcattgacttgaccaaa |
| 45 | ggacaaagaatttgctgtca |
| 46 | ctggatgattcaatgaaggt |
| 47 | ttcaagcagtagagacgaca |
| 48 | actccaataaccaatagcta |

FISHcount. Our smFISH image analysis consists of two components—cell segmentation and mRNA counting. These combine into an overall workflow that results in an image where each cell is annotated with the number of mRNA located within it (Figs. 4, 5). Bioformats [32] is used to convert the microscope image into individual z-stacks for each channel. The analysis pipeline then processes these z-stacks to produce the annotated image, and is implemented in the Python programming language [33].

## Segmentation

The Watershed algorithm is used to segment the image into regions representing cells, using the implementation provided by the scikit-image library [34]. Segmentation using the Watershed algorithm requires an input image denoting gradient magnitude, and a set of seeds for initialising the flood filling of the input image.

To identify the seeds for the Watershed algorithm each plane in the DAPI stack is normalised for intensity, then a maximum intensity projection taken. Contrast Limited Adaptive Histogram Equalization (CLAHE, [35]) is used to locally equalize the intensity of the projection. A Sobel filter is applied to the projection to find nuclear edges. Otsu's thresholding is then applied to select the nuclei. Each detected nucleus is reduced to its centroid for use as a seed for the segmentation.

The gradient magnitude input for the Watershed algorithm is generated by taking a minimum intensity projection of the probe channel, which represents the background auto-fluorescence of each cell. This projection is equalized with CLAHE and smoothed with a Gaussian filter. Taking this image as the basis for the Watershed algorithm and applying the seeds derived from the DAPI channel yields a segmented image.

## mRNA counting

To locate the spots representing RNA molecules, each z-slice in the probe channel stack is normalised, and a maximum intensity projection of the stack taken. A Sobel filter is applied to the projection to detect edges. We use scikit-image's implementation of fast normalised cross-correlation template matching to find the probe locations. This algorithm tests the correlation between a given template and the equivalently sized section of a larger image for each point in that image. It produces another image, the intensity values of which correspond to the degree to which the template correlates with the image (so that the maximum intensity value corresponds with perfect correlation, and the minimum with perfect anti-correlation). We initially apply this algorithm using a template constructed as an annular element sized to the diffraction radius of the microscope. The single closest match to this template is then taken as a second template to re-apply the correlation. We then apply a correlation threshold, correlation values above which corresponded to identified mRNA spots, yielding their locations. This threshold was chosen based on comparison to manual spot counting in test data sets, such that it gave an optimum balance between false negatives and false positives.

For validation of the results, identified spot locations and the segmentation derived from the DAPI nuclear stain and probe autofluorescence is used to produce an annotated image. This image overlays probe counts and segmentation boundaries on the projection of the probe autofluorescence channel. Each image is manually inspected to ensure that the image analysis workflow has not generated spurious results.

Graphs presented in Fig. 6 were created using Graph-Pad Prism 6 for Mac OS X software version 6.0 g (La Jolla, California).

## Additional files

**Additional File 1.** Arabidopsis root meristem cells are suitable for smFISH analysis. Representative images of nuclei from root meristem (a-c) and differentiation zone (d-f) in the absence of probe labeling. Non-specific signals were observed in endoreduplicated cells from the differentiation zone, in both red (d) and far-red channels (e). DNA labeled with DAPI (blue). Scale bar = 8 μm.

**Additional File 2.** Spot measurements. (a) Images of *PP2A* RNA spots visualized using Quasar570® and Quasar670® filter channels. (b) Line scans of fluorescent intensity corresponding to the lines shown in (a). Each line scan corresponds to the different fluorophores. The red linescan corresponds to analysis performed for a *PP2A* mRNA spot labeled with Quasar570® and the green linescan to *PP2A* unsliced RNA labeled with Quasar670® probes.

**Additional File 3.** mRNA signals are undetectable following RNase treatment. Representative images of RNase treated cells labeled with *PP2A* mRNA probes (red). DNA labeled with DAPI (blue). Scale bar = 8 μm.

**Additional File 4.** Additional examples of simultaneous detection of spliced and nascent *PP2A* RNA. Nuclei are labeled with the nuclear stain DAPI (blue), *PP2A* mRNA (red) and nascent *PP2A* RNA (green). Scale bars = 10 μm.

**Additional File 5.** *PP2A* mRNA imaged using Structured Illumination Microscopy. Nuclei are labeled with the nuclear stain DAPI (blue), *PP2A* mRNA (red) and nascent *PP2A* RNA (green). Scale bar = 10 μm.

**Additional File 6.** Raw data for PP2A mRNA and nascent counts.

**Authors' contributions**
SD, SR and CD designed the experiments. SD and SR developed the smFISH method and acquired the data. TO and MH developed the method for automated image analysis. SD, SR and CD analyzed the data. SD, TO, MH CD and SR wrote the manuscript an all authors approved the final version. All authors read and approved the final manuscript.

**Author details**
[1] Department of Cell and Developmental Biology, John Innes Centre, Norwich Research Park, Norwich NR4 7UH, UK. [2] Department of Computational and Systems Biology, John Innes Centre, Norwich Research Park, Norwich NR4 7UH, UK.

## Acknowledgements

We acknowledge Peter Shaw and Silvia Costa for comments on the manuscript, Ali Pendle for advice, as well as all members of the Dean and Howard groups for discussions. This work was supported by the UK Biotechnology and Biological Sciences Research Council (BBSRC) Grant BB/K00008X/1 and the Earth and Life Systems Alliance (a collaborative venture between John Innes Centre and University of East Anglia). C.D. acknowledges support from European Research Council Advanced Grant MEXTIM and BBSRC Institute Strategic Programme Grant BB/J004588/1.

## Competing interests

The authors declare that they have no competing interests.

## References

1. Lecuyer E, et al. Global analysis of mRNA localization reveals a prominent role in organizing cellular architecture and function. Cell. 2007;131(1):174–87.
2. Raj A, et al. Stochastic mRNA synthesis in mammalian cells. PLoS Biol. 2006;4(10):e309.
3. Gall JG. Differential synthesis of the genes for ribosomal RNA during amphibian oogenesis. Proc Natl Acad Sci U S A. 1968;60(2):553–60.
4. Akam ME. The location of Ultrabithorax transcripts in Drosophila tissue sections. EMBO J. 1983;2(11):2075–84.
5. Raap AK, et al. Ultra-sensitive FISH using peroxidase-mediated deposition of biotin- or fluorochrome tyramides. Hum Mol Genet. 1995;4(4):529–34.
6. Tautz D, Pfeifle C. A non-radioactive in situ hybridization method for the localization of specific RNAs in Drosophila embryos reveals translational control of the segmentation gene hunchback. Chromosoma. 1989;98(2):81–5.
7. Hejatko J, et al. In situ hybridization technique for mRNA detection in whole mount Arabidopsis samples. Nat Protoc. 2006;1(4):1939–46.
8. Brewer PB, et al. In situ hybridization for mRNA detection in Arabidopsis tissue sections. Nat Protoc. 2006;1(3):1462–7.
9. Javelle M, Marco CF, Timmermans M. In situ hybridization for the precise localization of transcripts in plants. J Vis Exp. 2011;57:e3328.
10. Pontes O, et al. RNA polymerase V functions in Arabidopsis interphase heterochromatin organization independently of the 24-nt siRNA-directed DNA methylation pathway. Mol Plant. 2009;2(4):700–10.
11. Pontvianne F, et al. Subnuclear partitioning of rRNA genes between the nucleolus and nucleoplasm reflects alternative epiallelic states. Genes Dev. 2013;27(14):1545–50.
12. Femino AM, et al. Visualization of single RNA transcripts in situ. Science. 1998;280(5363):585–90.
13. Raj A, et al. Imaging individual mRNA molecules using multiple singly labeled probes. Nat Methods. 2008;5(10):877–9.
14. Castelnuovo M, et al. Bimodal expression of PHO84 is modulated by early termination of antisense transcription. Nat Struct Mol Biol. 2013;20(7):851–8.
15. Neuert G, et al. Systematic identification of signal-activated stochastic gene regulation. Science. 2013;339(6119):584–7.
16. Yang B, et al. Single-cell phenotyping within transparent intact tissue through whole-body clearing. Cell. 2014;158(4):945–58.
17. Ji N, et al. Feedback control of gene expression variability in the Caenorhabditis elegans Wnt pathway. Cell. 2013;155(4):869–80.
18. Levesque MJ, Raj A. Single-chromosome transcriptional profiling reveals chromosomal gene expression regulation. Nat Methods. 2013;10(3):246–8.
19. Cabili MN, et al. Localization and abundance analysis of human lncRNAs at single-cell and single-molecule resolution. Genome Biol. 2015;16:20.
20. Frost F. Fluorescent microscopy autofluorescence: plants, fungi, bacteria. Cambridge: Cambridge University Press; 1995.
21. Lillo C, et al. Protein phosphatases PP2A, PP4 and PP6: mediators and regulators in development and responses to environmental cues. Plant Cell Environ. 2014;37(12):2631–48.
22. Czechowski T, et al. Genome-wide identification and testing of superior reference genes for transcript normalization in Arabidopsis. Plant Physiol. 2005;139(1):5–17.
23. Raj A, Tyagi S. Detection of individual endogenous RNA transcripts in situ using multiple singly labeled probes. Methods Enzymol. 2010;472:365–86.
24. Lubeck E, Cai L. Single-cell systems biology by super-resolution imaging and combinatorial labeling. Nat Methods. 2012;9(7):743–8.
25. Little SC, Tikhonov M, Gregor T. Precise developmental gene expression arises from globally stochastic transcriptional activity. Cell. 2013;154(4):789–800.
26. Raj A, van Oudenaarden A. Nature, nurture, or chance: stochastic gene expression and its consequences. Cell. 2008;135(2):216–26.
27. Raj A, van Oudenaarden A. Single-molecule approaches to stochastic gene expression. Annu Rev Biophys. 2009;38:255–70.
28. To TL, Maheshri N. Noise can induce bimodality in positive transcriptional feedback loops without bistability. Science. 2010;327(5969):1142–5.
29. Lyubimova A, et al. Single-molecule mRNA detection and counting in mammalian tissue. Nat Protoc. 2013;8(9):1743–58.
30. Hansen CH, van Oudenaarden A. Allele-specific detection of single mRNA molecules in situ. Nat Methods. 2013;10(9):869–71.
31. Levesque MJ, et al. Visualizing SNVs to quantify allele-specific expression in single cells. Nat Methods. 2013;10(9):865–7.
32. Linkert M, et al. Metadata matters: access to image data in the real world. J Cell Biol. 2010;189(5):777–82.
33. van der Walt S, Colbert SC, Varoquaux G. The NumPy array: a structure for efficient numerical computation. Comput Sci Eng. 2011;13(2):22–30.
34. van der Walt S, et al. scikit-image: image processing in Python. PeerJ. 2014;2:e453.
35. Pizer SM, et al. Adaptive histogram equalization and its variations. Comput Vis Graph Image Process. 1987;39(3):355–68.

# 12

# RADIX: rhizoslide platform allowing high throughput digital image analysis of root system expansion

Chantal Le Marié, Norbert Kirchgessner, Patrick Flütsch, Johannes Pfeifer, Achim Walter and Andreas Hund[*]

## Abstract

**Background:** Phenotyping of genotype-by-environment interactions in the root-zone is of major importance for crop improvement as the spatial distribution of a plant's root system is crucial for a plant to access water and nutrient resources of the soil. However, so far it is unclear to what extent genetic variations in root system responses to spatially varying soil resources can be utilized for breeding applications. Among others, one limiting factor is the absence of phenotyping platforms allowing the analysis of such interactions.

**Results:** We developed a system that is able to (a) monitor root and shoot growth synchronously, (b) investigate their dynamic responses and (c) analyse the effect of heterogeneous N distribution to parts of the root system in a split-nutrient setup with a throughput (200 individual maize plants at once) sufficient for mapping of quantitative trait loci or for screens of multiple environmental factors. In a test trial, 24 maize genotypes were grown under split nitrogen conditions and the response of shoot and root growth was investigated. An almost double elongation rate of crown and lateral roots was observed under high N for all genotypes. The intensity of genotype-specific responses varied strongly. For example, elongation of crown roots differed almost two times between the fastest and slowest growing genotype. A stronger selective root placement in the high-N compartment was related to an increased shoot development indicating that early vigour might be related to a more intense foraging behaviour.

**Conclusion:** To our knowledge, RADIX is the only system currently existing which allows studying the differential response of crown roots to split-nutrient application to quantify foraging behaviour in genome mapping or selection experiments. In doing so, changes in root and shoot development and the connection to plant performance can be investigated.

**Keywords:** Abiotic stress, Corn, Maize, Rhizotron, Root foraging, Split root, Root–shoot interaction, Quantitative trait loci, Imaging

## Background

The spatial distribution of a plant's root system is crucial for the success of the plant in a given environment because it determines how easily a plant can access water and nutrient resources of the soil. Root system architecture (RSA) is governed by the environment, but also by the genotype. Today it is unclear to what extent genetic variations in root system responses to spatially varying soil resources can be utilized. Closing this knowledge gap is certainly one step forward towards improving resource-use efficient crops. First studies provided evidence for an advantage of certain root architectural characteristics, such as deep rooting under drought [1, 2], shallow rooting under low phosphorus availability [3] or steep rooting angles under low nitrogen availability [4–7].

One major problem studying root growth is the fact that roots are often hidden in their growth matrix like soil unless the growth medium is transparent such as a gel. In the field, even the most advanced procedures offer only a partial glimpse of RSA. To overcome this limitation,

*Correspondence: andreas.hund@usys.ethz.ch
Institute of Agricultural Sciences, ETH Zurich, Universitätstrasse 2, 8092 Zurich, Switzerland

research was done on developing monitoring systems of root growth in greenhouse approaches that provide visibility of the roots. Various phenotyping platforms were developed to monitor root growth non-invasively in soil based systems such as rhizotrons/-boxes containing sand/soil substrates [8], via X-ray micro-tomography [9, 10] or magnetic resonance imaging [11, 12] measurements. However, to date the throughput of these techniques is still not high enough for the assessment of large sets of genotypes as needed for mapping studies of quantitative trait loci (QTL).

For fast and simple screening, germination paper has become state-of-the art in many labs focusing on roots [13–16]. The systems using filter paper as substrate are sometimes called "growth pouches" [17]. Rhizoslides are a large-scale refinement of growth pouches, which allow to study root development for a prolonged period of time [18]. The growth pouch system was developed to map QTLs in collections of about 200 genotypes in various populations of maize during the very early seedling development [19–22] whereas the rhizoslides allow studying root growth until four-leaf stage.

A limitation of the growth pouches was that only the dynamics of the embryonic roots could be studied, ignoring the most prominent root type of maize, the shoot born crown roots. Albeit embryonic roots and crown roots are under different genetic control [23, 24] and may respond distinctly different to environmental stimuli. For example, Yu et al. [25], reviewed that embryonic roots of maize seedlings respond to nitrogen-rich patches by increasing the length of lateral roots, while crown roots of adult plants increased length and density of lateral roots [26, 27].

It is widely accepted that plants can adjust to localized soil enrichment by physiological changes [28]. Yet, to date there is still a lack of understanding the utility of this response for plant improvement because of a lack of appropriate methods to quantify responses of different genotypes. For breeding research, it would be highly beneficial to select genotypes not only by their final phenotype but also by their dynamic response to temporally changing or spatially varying nutrient availability. For example, Lynch et al. [7] proposed that an "optimal" root system should ignore local resource availability in the top soil and should grow into deep soil layers to be prepared against drought and nitrogen (N) stress later in the season. Thus, a reduced response to spatial variability would be desirable. Others propose the opposite, i.e. that a better response to local variation of nutrients would be beneficial to take advantage of side banded fertilizer placements [27]. A proof-of-concept of the split-nutrient application in rhizoslides was presented by in t' Zandt et al. [29] observing a strong differential response

of crown roots to N. It is conceivable that there is an optimum response to N, which may vary depending on the species, crop management and target environment. With phenotyping systems, such as rhizoslides, it will be possible to test corresponding research hypotheses in this context. One opportunity to evaluate plant performance is to study the shoot development. The measurement of canopy development is an important component of sound root research as the shoot acts as source of carbohydrates and sink for nutrient and water. There is a close link between root and shoot growth in many crops (see [30] for a discussion). Hence, a better understanding of the relationship between root and shoot development is an important need in order to identify heritable root traits for which the additional costs of direct selection are justified as pointed out by Wissuwa et al. [31]. For example, by studying the development of rooting depth in relation to shoot development in a diverse set of maize, Grieder et al. [32] found a strong linear relationship between the two traits under well watered conditions, leaving little genetic variation to alter rooting depth without changing canopy size. The non-destructive measurement of canopy size is well established [33, 34] and the adaptation of linear models is an accepted method [35] whereas for root traits dynamic studies were not often performed and less is known about models fitting this need. Pioneer work was done by Adu et al. [13] modelling the root growth dynamics of *Brassica rapa*.

With the rhizoslides, a second generation of the growth pouch system was developed [18] and tested for its suitability for split-nutrient application (in t' Zandt et al. [29]). Here we present the RADIX platform facilitating an improved irrigation, handling and imaging of rhizoslides. The current version contains 100 slides and enables the cultivation of 200 maize plants at the same time, one on each side of the slide. The aim of the study was to demonstrate the value of the platform for phenotyping of natural variation of root traits in maize by investigating (a) root and shoot growth beyond the early seedling stage, and (b) the effect of split-nutrient application on the crown root development. This setup should provide information on genotypic variability of N uptake under spatially heterogeneous N availability.

## Methods

### Cultivation in the RADIX platform

Plants were grown in so called "rhizoslides" that consist of two PVC bars ($600 \times 60 \times 10$ mm) and a plexiglass sheet ($650 \times 530 \times 4$ mm) fixed with two screws between the bars (Fig. 1e) [18]. The dimensions of the assembled rhizoslide were $600 \times 60 \times 38$ mm. The plexiglass sheet was covered with wet germination paper ($49 \times 61$ cm) (Anchor Steel Blue Seed Germination

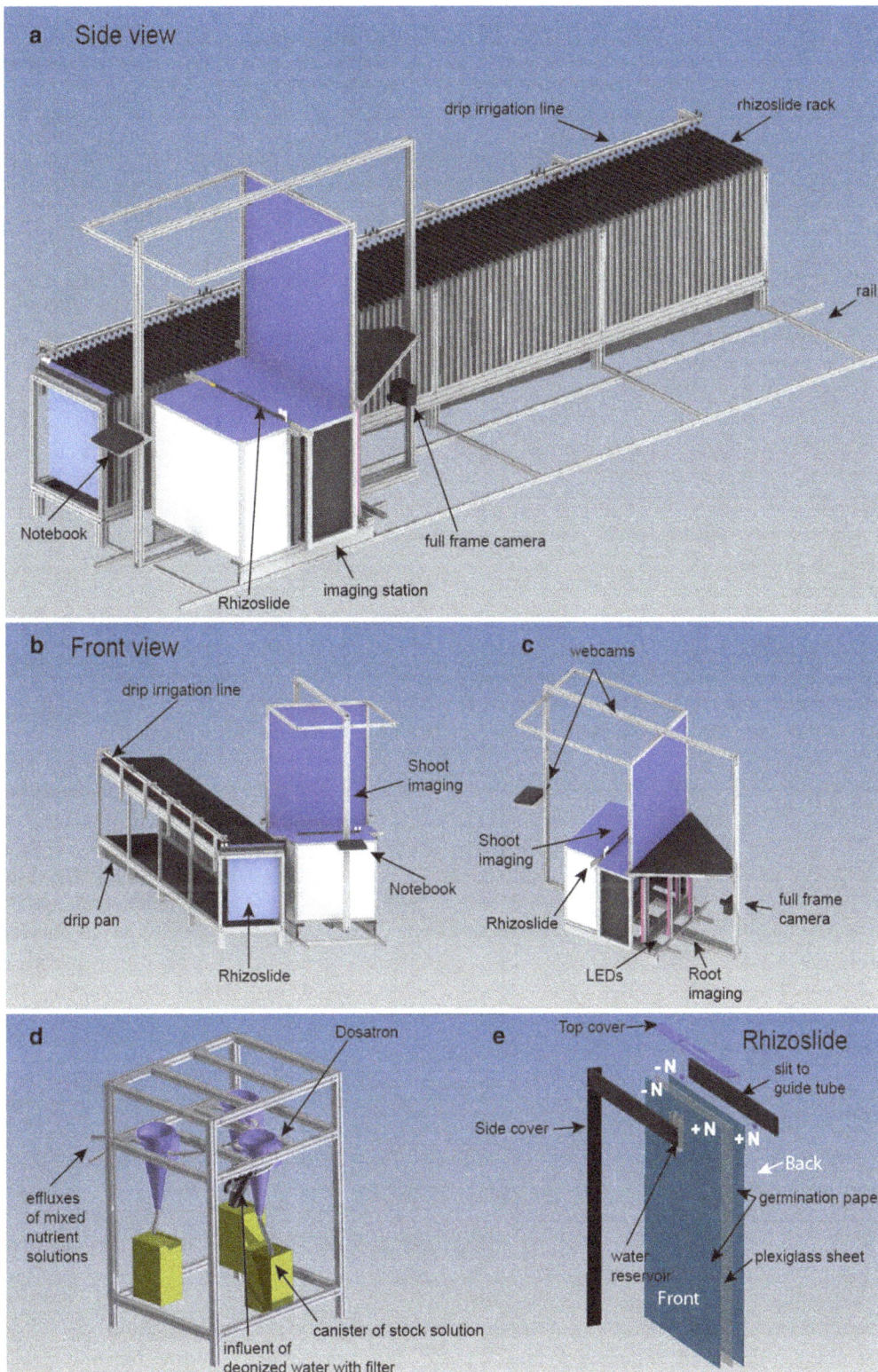

a   Side view

drip irrigation line
rhizoslide rack
rail
Notebook
full frame camera
imaging station
Rhizoslide

b   Front view

drip irrigation line
Shoot imaging
Notebook
drip pan
Rhizoslide

c   webcams

Shoot imaging
Rhizoslide
full frame camera
LEDs
Root imaging

d   Dosatron

effluxes of mixed nutrient solutions
canister of stock solution
influent of deonized water with filter

e   Top cover
Rhizoslide
-N
-N
slit to guide tube
Side cover
+N   +N
Back
germination paper
water reservoir
plexiglass sheet
Front

(See figure on previous page.)
**Fig. 1 a, b** Schematic figures of the RADIX platform with the imaging station moving on rails along a rack holding the rhizoslides. After positioning the station, the rhizoslides can be slid onto the station. Two drip irrigation lines are mounted at the backside of the rack enabling the supply of two different nutrient concentrations. **c** The imaging station holds two webcams to image the shoot from the side and from the top; a DSLR camera is mounted at 1 m distance from the slide surface to monitor the roots. Roots are illuminated by LED bars (three on each side); illumination is controlled by a microcontroller built and programmed in Arduino 1.0. Rhizoslides on the slide mount can be turned by 180° to image both sides of the rhizoslide. **d** The dosatron pump system consists of one dosatron supplying base nutrient solution that is split up to two dosatrons allowing for differential nutrient supply. **e** Scheme of a rhizoslide and the irrigation for the split root setup. High and low N concentrations were dripped into reservoirs from where tubes transported the nutrient solution to the split-nutrient germination paper separated by the wax barrier

Blotter, Anchor Papers Co, USA) on both sides, serving as substrate. These were in turn covered by a transparent, oriented polypropylene (OPP) foil with micro holes of $70 \pm 10$ μm with a distance of 105 mm (in x and y orientation) to allow for gas exchange (Maag, GmbH, Iserlohn, Germany).

For sterilization, the germination paper was heated in three cycles from room temperature to 80 °C and kept at this temperature for at least 120 min. Between the heating periods the paper was kept for 20–22 h in an oven at 37 °C and 50 % relative humidity [36]. To prevent fungal growth during the experiment, a method described by Bohn et al. [37] was used. The germination paper was moistened with deionized water containing 2.5 $gL^{-1}$ Malvin (Syngenta Agro AG, Dielsdorf, Switzerland) containing the active component Captan. Once a week, each rhizoslide was watered with 15 mL of the Malvin solution to balance for leached fungicide.

The plant material consisted of the core EURoot maize panel (www.euroot.eu), with 25 dent inbred lines B73 (1), EC169 (4), EZ11A (13), F98902 (5), FV353 (6), LH38 (14), MS153 (7), Oh33 (16), EZ37 (10), MS71 (22), Os420 RootABA− (11), Os420 RootABA+ (12), Mo17 (3), FC1890 (18), LAN496 (23), W64A (20), F7028 (8), EZ47 (9), B100 (24), N6 (25), N25 (26), A347 (32), PH207 (33), F1808 (34), B97 (36),) crossed to UH007 as common flint tester supplied by Delley seeds and plants Ltd. (DSP Ltd., Switzerland). The number in brackets indicated the EU_ID.

Seeds were surface sterilized with 0.3 % PrevicurN (Bayer CropScience GmbH, Monheim, Germany) for 30 min and became touch dry without rinsing the seed. Subsequently, seeds were incubated in Petri dishes lined on paper soaked with a soil bacteria mixture (0.0001 % RhizoPlus 42, Andermatt Biocontrol AG, Grossdietwil, Switzerland and 0.01 % FZB24, Bayer AG CropScience, Zollikofen, Switzerland) to promote the development of a healthy rhizosphere and to prevent seed-borne infections. Seeds were kept for 48 h at 26 °C in the dark for germination and were then transferred into the rhizoslides.

For this study, 100 Rhizoslides were prepared with one plant in each rhizoslide. The seed was placed on top of the germination paper enabling simultaneous study of

seminal and crown roots (for details, see [24]). A split-nutrient system was created by ironing a vertical water impermeable wax barrier into the germination paper and by a drip irrigation system (for details, see [29]). The roots could freely develop on both germination papers and on both sides of the wax barrier. Each paper was a split nutrient setup; irrigated with a high and low N solution, respectively. All 100 rhizoslides were supplied with a solution with low N (10 % of full strength ∼ high N solution) for 14 days (2–3 leaf stage). Subsequently, one side was irrigated with a thousand times lower concentration of N compared to the other side of the wax barrier (Fig. 1e). The split root treatment remained for 14 days until the end of the experiment (3–4 leaf stage). For studying root growth, images were taken every second day.

## Composition of the nutrient solutions

To prevent precipitation, two stock solutions were prepared. The solution A was identical for both treatments containing: $KH_2PO_4$: 0.5 mmol $L^{-1}$; $MgSO_4 \times 7$ $H_2O$: 2 mmol $L^{-1}$; $MnCl_2 \times 4$ $H_2O$: 9.15 μmol $L^{-1}$; $CuSO_4 \times 5$ $H_2O$: 0.2 μmol $L^{-1}$; $H_3BO_3$: 46.25 μmol $L^{-1}$; $Na_2MoO_4 \times 2$ $H_2O$: 0.58 μmol $L^{-1}$; $ZnSO_4 \times 7$ $H_2O$: 0.77 μmol $L^{-1}$; FeEDTA $(C_{10}H_{12}FeN_2NaO_8 \times H_2O)$: 0.04 mmol $L^{-1}$, (pH ∼ 6). The solution B contained: $KNO_3$, $Ca(NO_3)_2 \times 4H_2O$ and $NH_4NO_3$, (pH ∼ 6). The final concentrations in the high N treatment were: $KNO_3$: 4.5 mmol $L^{-1}$; $Ca(NO_3)_2 \times 4H_2O$: 4 mmol $L^{-1}$; $NH_4NO_3$: 1 mmol $L^{-1}$. In the low N treatment, the ammonium and nitrate concentrations were reduced. The solution contained: $KNO_3$: 5 μmol $L^{-1}$; $Ca(NO_3)_2 \times 4$ $H_2O$: 5 μmol $L^{-1}$; $NH_4NO_3$: 1 μmol $L^{-1}$ and to balance the ion activities additionally KCl: 5 μmol $L^{-1}$ and $CaCl_2 \times 2H_2O$: 5 μmol $L^{-1}$. The solutions were calculated using GEOCHEM-EZ [38]. pH was adjusted by adding KOH.

## Construction of the RADIX platform

The rhizoslides were hanging in a rack ($500 \times 65 \times 80$ cm) constructed with aluminium profiles (KANYA AG, Rüti, Schweiz) with 100 rhizoslide compartments. Each compartment was 60 cm long and 4 cm wide (Fig. 1a, Additional file 1). The distance between two neighbouring plants was 5 cm. A declining drip pan was built under the rack to collect the dripping nutrient solution (Fig. 1b) and a pump with automatic water level sensors (Gardena Comfort Tauchpumpe 9000 Aquasenor, Gardena Ulm, Germany) was used to pump down the solution. The sides of the rack were covered with a black polyethylene foil to prevent the roots against incidence of light. Each rhizoslide was covered on top with a cover plate containing holes for the shoot.

The irrigation system of every rhizoslide consisted of two PVC tubes (5/3 mm) (GVZ-Gossart AG, Otelfingen, Switzerland) and 25 mL PE tubes (Semadeni AG, Ostermundigen, Switzerland) acting as water/nutrient solution reservoirs. The reservoirs were filled via an irrigation system consisting of three Dosatron pumps (Diaphragm range 2.5 $m^3\,h^{-1}$) (Dosatron, International S.A.S, Tresses, France) to mix the stock solutions (100 times higher than final concentration) with deionized water. The stock solutions were homogenized by circulation starting ½ h before the mixing of the solutions. One Dosatron was placed in series with two parallel switched Dosatrons (Fig. 1d).

The first Dosatron in the line mixed the identical stock solution A. In the following, the line was separated and the parallel switched Dosatrons were adding solution B either with a high/or low N concentration. The two lines were supplying two different pipes, each containing 100 micro drips (Netafilm CNL-Tropfer Junior, GVZ-Rossat, Otelfingen, Schweiz); dripping into one of the two nutrient solution reservoirs with a flow of 25 ml $min^{-1}$ (Fig. 1a, b). Irrigation was controlled by an irrigation computer (Gardena, C1060 Profi, Ulm, Germany). The irrigation cycle was 4 h with duration of one minute. The amounts to 25 mL per micro drip summing up to a daily rate of 300 mL per slide.

The RADIX platform was placed in a cultivation room (720 × 330 × 250 cm). Environmental settings were a day period of 14 h light, a temperature of 28/24 °C (day/night) at seed level, approximately 50 % air humidity and a light intensity of 600 µmol $m^{-2}\,s^{-1}$ photosynthetic active radiation at plant canopy level supplied by Green-Power LED production modules (Koninklijke Philips Electronics N.V., Amsterdam, Netherlands). Sixteen LED modules were arranged in four subunits with four modules each. The subunits were hanging from the ceiling and were equipped with a motor to adjust the distance of LED lamps and plant canopy level. Each subunit was controlled separately. The dimension of the lighting system was 500 × 70 cm and each subunit was 125 × 70 cm.

## Imaging

An imaging station (~168 × 164 × 110 cm) was constructed allowing for imaging the root system and the shoot at once (Fig. 1c, Additional file 2). The imaging station was moving on a rail parallel to the slide rack and the rhizoslides were manually slid onto the imaging mount (Fig. 1a, b). The mount was rotatable allowing imaging the root system growing on both sides of the slides with only one camera (Fig. 1c, Additional file 2). The camera was fixed at a distance of 1 m from the center of the acrylic sheet (Figs. 1c, 2a, Additional file 2). For shoot imaging, two webcams (Logitech, Business) were fixed at a distance of 65 cm (increased to 80 cm in the meantime in front of the rhizoslide) and 65 cm (increased to 115 cm) on top of the imaging car (Fig. 1c). The backside of the rhizoslide mount was covered with a blue Komatex plate for an optimal contrast between shoot and background (Fig. 1c). The root zone was hanging in a box constructed with black and white Komatex plates (Röhm AG, Schweiz) at the sides to prevent light scattering in the root zone; white plates in the back of the rhizoslide and black plates in front of it. The white plates in the back area were built as a light box, reflecting the light, whereas the black plates in the front inhibited reflections into the camera focus area (Figs. 1c, 2a). Only the front view was uncovered for root imaging. Black cloth was used to cover the area between camera and box (the cloth is not shown in Figs. 1c, 2a, Additional file 2).

Images were taken with a 22.3 mega pixel full-frame digital single-lens reflex camera (EOS 5D Mark III, Canon, Tokyo, Japan) equipped with a 50 mm lens (compact macro 50 mm f/2.5, Canon, Tokyo, Japan). The resolution of the images was around 0.13 mm $pixel^{-1}$ (for details, see [18]). The flash lights were replaced by six LED bars (TRENOVA -Highpower- LED, PUR-LED GmbH & Co. KG, Selzen, Germany) allowing for a more homogenous illumination (Figs. 1a–c, 2a). Shoots were imaged with the two webcams simultaneously with the root system; one image from the front (maximal extent of the rhizoslide parallel to image plane) one image from the backside (180° turn on the horizontal axis) and images from the top view simultaneously with the side view images (in total four images). As the maximal extension of the leaf area differs between individual plants, imaging from different focal axis allows minimizing the error resulting from shoot orientation. In our setup, leaves were developing in parallel to the rhizoslide axis and therefore, imaging from the front and top rendered sufficient results and no additional imaging from the side was necessary.

## Image processing

The two root images taken with either illumination from the right or from the left side were combined to one 24 bit RGB image using Matlab (Version 7.12 The Mathworks, Natick, MA, USA) by keeping only the minimum

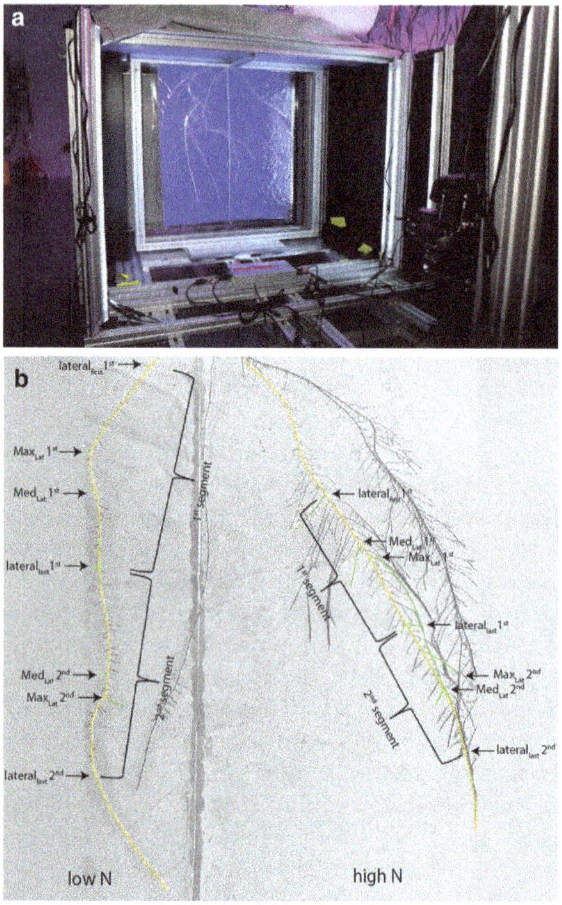

**Fig. 2 a** Imaging car subunit which facilitated analysis of the root system. The root system on a split-root rhizoslide was placed approx. in 1 m distance to the 24 megapixel DSLR-camera. LED bars were placed between camera and rhizoslide on the *right* and *left side*, respectively. **b** Root system developed on one side of the rhizoslide and treated with a split-nitrogen application 10 days after solution change. Roots highlighted in *yellow* are the crown roots which were used in this example image to investigate crown root growth. *Waved brackets* indicate the two segments used to investigate the number of lateral roots and the medium ($Med_{Lat}$) and longest ($Max_{Lat}$) lateral root length in the first ($Med_{Lat}$ 1st, $Max_{Lat}$ 1st) and in the second ($Med_{Lat}$ 2nd, $Max_{Lat}$ 2nd) segment. The first segment started at the position of the the last developed lateral root ($lateral_{first}$ 1st) at solution change and the second segment continued after the first segment ($lateral_{last}$ 1st). The lateral roots used to investigate the medium and longest lateral root length in the two segments are highlighted in *green*

two sides and top views. Shoot images were segmented from the background using background subtraction and thresholding adaption to the image material written in Matlab R2013a (The Mathworks, Natick, MA, USA). Shoot segmentation was done in a two-step method on a feature map f: $f = 2 * G\text{-}R\text{-}B$. Segmenting with a high threshold 55 (in the result pixels of f with values $\geq 55$ are set to 1 all others are set to 0) yields a coarse segmentation $s_c$ and identifies all plant regions but does not nicely determine their borders. A fine segmentation $s_f$ with Otsu's method [39] resulted in more accurate sizes of the shoot regions but also yields additional image regions which were separated from the shoot. Therefore both segmentations were combined by selecting all regions of $s_c$ and enlarging them to all therewith connected regions of $s_f$.

### Calculation of the root growth rate

Crown roots of the first whorl were traced using Smart-Root [39]. Crown roots of the first whorl were not always placed on the high and low N side. Accordingly also roots of the second whorl were used. For tracing, one crown root was selected in a high N compartment and one crown root in a low N compartment, ideally with comparable developmental stages at the start of the split root treatment. Furthermore, only crown roots were selected that did not cross the wax barrier or reach the side or bottom of the germination paper at the end of the experiment. This step was necessary to ensure that zero growth in subsequent analyses was not due to the fact that a root reached the borders of the paper and to exclude compensatory growth of lateral roots. All selected roots were healthy and undamaged until the end of the experiment. Next, root elongation of crown roots over time was calculated using R (Foundation for Statistical Computing, Vienna, Austria). Single plants did not establish traceable crown roots on both sides of the split-nutrient germination paper as sometimes crown roots on one side reached the sides or bottom of the rhizoslide before the end of the experiment or no crown roots were formed on one of the split root sides. Though, the root system (seminal and crown roots) developed over the whole germination paper and was exposed to inhomogeneous N distribution. Therefore, it was expected that plant response should be comparable to plants showing traceable crown root growth in both compartments.

### Modelling the dynamics of root and shoot elongation

For a better understanding of responses towards a heterogeneous environment, a linear model was fitted to describe changes in root and shoot growth dynamics over time.

tonal value present in each image (minimum tonal image) as described by Le Marié et al. [18].

In the current setup, the number of pixels of the front, back and top view of the maize shoot was used as proxy for canopy size. All four shoot images were segmented and used for leaf area calculation. Best results were obtained by calculating an average pixel number over the

A simple linear regression was fit to the length development of each individual crown axile root using a linear regression fit lm() in R [40]:

$$L_{Cr}(t) = a + bt \qquad (1)$$

where $L_{Cr}(t)$ is the crown root length as a function of time $t$, $a$ the estimated crown axile root length at the start of the split-nitrogen application (y-axis) and b is the slope of the regression line. The shoot growth was calculated simultaneously to the crown root length.

## Lateral root length and number

To investigate the effect of genotype and treatment on lateral root growth and number, medium and maximal lateral root length were measured and maximal lateral root number was counted at the last day of the experiment in two branching zones (Fig. 2b). The branching zone, developing after solution change, was defined by the youngest, most distal lateral root at solution change (proximal end of the branching zone) and the youngest, most distal lateral root at harvest (distal end of the branching zone) (Fig. 2b) [41]. The length of the branching zone was measured using the ruler tool in Image J (SmartRoot) and divided into two sections of equal length (first and second branching zone). A medium sized lateral root and the longest lateral root were traced with SmartRoot in both segments and the number of formed lateral roots was counted (Fig. 2b).

## Chlorophyll meter (SPAD) measurements

Chlorophyll measurements (SPAD-502, Minolta Corporations, Ramsey, NJ, USA) were done over the experimental period to control for plant health and chlorophyll content. Measurements were done on the last fully developed leaf (usually leaf 4) of every plant measuring at the shoot base, middle and tip of the leaf and averaging these values.

## Harvest

After 26 days in total, plants were harvested. The shoot was cut off at the base and leaf area and fresh weight were determined. Leaf area was determined using a leaf-area meter (LI-COR 3100, LI-COR, Lincoln, NE, USA). In order to determine shoot dry weight, shoots were dried at 60 °C for 4 days. Crown and seminal roots were separately harvested depending on the side they were growing (high or low N treatment), bagged and dried at 60 °C for 4 days. Only single lateral roots were growing into the germination paper and could not be harvested. To ensure a correct classification of crown roots, all crown and seminal roots were marked at harvest and an image was taken to compare with the last image taken in the RADIX setup.

## CN measurement

After drying, shoots were ground using a swing mill (Retsch MM200, Retsch GmbH, Haan, Germany) with a frequency of $25~s^{-1}$ for one minute. N content of the shoots was measured with a CN analyser (Flash EA, 1112 series) [42]. As reference plant material Alfalfa was used.

## Verification of the split nutrient system

To verify the separation of the two sides, a white germination paper was soaked in bromocresol green and afterwards watered with a basic ($NH_4HPO_4$) and an acidic [$Ca(NO_3)_2$] solution. A webcam was installed to image the elution. After 48 h the bromocresol was washed out and the paper was sprayed with it again to visualize the pH gradient (Additional file 3).

## Statistics

The rhizoslide experiment was a complete randomized block design with four replications. Each experimental unit consisted of one rhizoslide containing one plant. Linear mixed models were calculated in ASREML-R [43] according to the following full model for root observation in split N application:

$$y_{ijkl} = \mu + G_i + N_j + GN_{ij} + R_k + S_l + GR_{ik} + \varepsilon_{ijkl} \qquad (2)$$

where $Y_{ijk}$ is the trait measurement (slope or intercept of crown root length, length lateral root, number of children), of the root of the ith genotype (G = 1, 2, 3 ... 24) within the $jth$ N treatment (N = HN, LN) in the rth replication (k = 1, 2, 3, 4) on the lth side of the slide (l = front or back), $\mu$ is the general mean, $G_i$ is the effect of genotype, $N_j$ is the effect of the N treatment, $GN_{ij}$ is the genotype-treatment interaction, $S_l$ is the effect of the side of the slide the observed root grew on (front or back), $R_k$ is the effect of the $kth$ replicate, $GR_{ik}$ is the genotype-replication interaction identifying individual slides and $\varepsilon_{ijkl}$ is the residual error. The genotype-replication interaction was considered as random whereas all other effects were considered as fixed factors. A different variance was adjusted for each treatment level. For traits representing only one measurement per slide (SPAD, dry weight, leaf area, pixel number, slope or intercept of shoot growth), the model was reduced by only keeping the genotype and replicate as fixed factors. For root traits (Elongation rate crown roots ($ER_{Cr}$), length of crown roots at solution change (intercept; $IC_{Cr}$), length of representative lateral root ($Med_{Lat}$), maximal lateral root length ($Max_{Lat}$), number of lateral roots ($No_{Lat}$) and root dry weight ($DW_R$) a square root transformation was performed to achieve normal distribution and no outlier were removed. Root traits that were not normally distributed (length branching zone, branching density in zone 1 or 2, start branching zone) were excluded from further analysis. For shoot traits neither a transformation nor removing of outlier was necessary to achieve normal distribution.

Best linear unbiased estimators (BLUPs) were used to investigate the phenotypic diversity in the set [44]. Tukey's honestly significant differences (Tukey HSD) were used as post hoc test and calculated as:

$$TukeyHSD = q * \sqrt{\frac{MSE}{n}} \qquad (3)$$

where $q$ is the critical value according to the chosen significance level and degrees of freedom, $MSE$ is the mean square error calculated from the average standard error of the difference (avsed) supplied by the predict function of ASReml-R and $n$ is the number of treatment levels.

### Heritability

For heritability estimations across genotypes the model described by Eq. (2) was used, but the genotype and the genotype-treatment interaction were set as random factors. To estimate the heritability depending on the nitrogen treatment applied to the root system, a heterogeneous variance model with regard to the nitrogen-effect and its interactions was fit in ASReml-R following the recommendations of Butler et al. [43].

Mean-based heritability was calculated for each root trait within each nitrogen level and for each shoot trait as follows:

$$H^2 = \frac{\sigma_g^2}{\sigma_g^2 + \frac{\sigma_e^2}{r}} \qquad (4)$$

where $\sigma_g^2$ and $\sigma_e^2$ are the genotypic and residual error variance respectively obtained from the linear mixed model fitting (Eq. 2) and $r$ is the number of replications.

## Results and discussion

### RADIX: high throughput analysis of root elongation and dynamics of root topology

The established RADIX platform, which is based on rhizoslides as individual monitoring units, enables a high-throughput screening of RSA and root topology dynamics. One major advantage compared to existing phenotyping platforms such as GROWSCREEN-Rhizo [8] is the visibility of all targeted roots and a sufficient contrast for image processing. No targeted roots are hidden in deeper soil layers like it is the case in rhizotrons and limitations due to contrast between soil and root are usually a minor problem. The advantage of soil-based systems like rhizotrons and CT is that they are closer to natural conditions [8, 10]. Another advantage of rhizoslides is the high achievable throughput. The definition of high throughput usually depends on the analysed traits.

Root accessibility and the measurement of growth and topology are certainly the most rate-limiting aspects studying root elongation and dynamics. Here, we define high throughput as the ability to evaluate QTL or association mapping populations of at least 200 genotypes within a minimum of three independent replications within half a year. The imaging process in rhizoslides takes approximately 1 min per slide (front and back side). Image processing in SmartRoot is still the time-limiting step (approximately 80 h to mark axile roots and the segments of lateral roots for a time series of 1400 images) but delivers detailed information of the development of root topology. Adu et al. [13] developed a screening method with medium throughput based on germination paper (n = 24). A scanner unit is fixed in front of each paper with roots growing on it. With this setup, higher temporal resolutions can be realized. However, a mobile imaging unit as realized in the RADIX balances between temporal resolution and throughput. As the response of the root system to stimuli, such as N, happens within days rather than minutes [29], daily or half daily scans are considered sufficient. With the current setup of the RADIX, the number of genotypes that can be evaluated remains the bottleneck. The highest throughput realized to date to study root dynamics in existing platforms is achieved in hydro- or aeroponic setups, yet without the possibility of split-nutrient application. It remains to be elucidated, under which conditions and for which root traits one or the other phenotyping systems is most suitable.

To our knowledge, RADIX is the only system currently existing which allows studying root growth responses under heterogeneous nutrient availabilities with a throughput high enough for QTL and association studies. The importance of root system architecture for N-uptake is supported by several positive QTL collocations between the two traits [45, 46]. However, to date it is still unclear, if increased uptake efficiency, increased N utilization efficiency [47, 48] or an optimal housekeeping with internal N resources [49], is most promising for optimal plant performance [46]. For example, under high N, the genetic variation in nutrient uptake seems to be related to uptake efficiency, whereas at low N, it seems to be related to utilization efficiency [48]. The RADIX setup could shed light onto this topic by studying individual plants under control as well as under stress conditions. In doing so, changes in root and shoot development and the connection to plant performance can be investigated.

### Dynamics of shoot development: an imperative for sound root research

The platform was developed to allow assessing both root and shooting growth. In the current setup, the number of pixels of the front and back view of the maize shoot was used as proxy for canopy size. Shoots were successfully segmented from the background using a two-step method of coarse and fine segmentation (Fig. 3c). The number of pixels correlated well with leaf area $r^2 = 0.91$***; Fig. 3a) as well as with shoot biomass ($r^2 = 0.87$**; Fig. 3a). Best correlation could be achieved by calculating the mean value of the two images taken from the sides (rhizoslide parallel to camera focal axis in 0° (front) and 180° (back) and from the top (in total two front views and two top images). In a previous experiment, different imaging methods for the shoot were investigated. Best results were obtained by using the here presented combination of side and top view. This is in contradiction with investigations made in the GROWSCREEN-Rhizo facility where best results were obtained with images from the front side (0°) and a 90° horizontal rotation [8]. An explanation for this phenomenon could be the combination of the alternate leaf orientation in monocot species and a parallel orientation to the rhizoslide influenced by neighbouring plants (distance between two neighbouring plants of 5 cm) [45]. As leaves exceeded the imaging area 10 days after start of the treatment, the shoots dynamics could be assessed only up to this point. For future experiments, this was improved by constructing a larger vertical screen for the imaging station (already displayed in Fig. 1).

### Evaluation of shoot traits

Genotypes differed with respect to leaf area expansion (measured as pixel numbers), leaf area at harvest and shoot dry weight at harvest (Table 1a, b). For example, genotype A347 grew almost two times faster than genotype Oh33 (Additional file 4, Additional file 5B). As the traits, measured at final harvest, were strongly correlated with leaf area expansion, it can be concluded that final leaf area at this stage was driven by differences in vigour rather than differences in germination (Fig. 3b). Furthermore, the shoot pixel count at solution change was not correlated with the shoot pixel count development (Fig. 4, Additional file 6). No genotype-specific differences and no heritability were observed for chlorophyll content investigated by means of SPAD measurements (Table 1a, b). However, the genotypes differed in their N content (Table 1a). CN measurements revealed significant differences in N content between genotypes with a maximal N proportion of 5.07 % and a minimal proportion of 3.61 % (Table 1a, b). A high formation of biomass resulted in a reduction of leaf N content (Fig. 4) although this correlation was not significant (Additional file 6).

### Crown root development

The elongation of crown axile roots was described best by linear models based on the observation of original data and residual plots (Additional files 7, 8, 9). In

a previous study, best results were obtained by fitting a logistic model for the elongation rates of the crown roots on the high-N side and a regression model with segmented relationships for roots grown under no N [29]. The need for two different modelling approaches in the previous study did not enable the examination of interaction effects in a common linear mixed model. As already observed by in 't Zandt et al. [29], crown root elongation tended to decrease in the low-N compartment (Additional file 8) while the elongation rate tended to increase in the high-N compartment (Additional file 7). However, the trend was far less pronounced in the present experiment enabling the fit of a common linear model (Additional file 9). The lower effects were possibly due to the replacement of the no-N treatment [29] by a low-N treatment. Unfortunately, the standard genotype UH007 × B73, used in the study of in 't Zandt et al. [29], could not be investigated here, as it did not germinate.

First crown roots established on the germination paper after 2 weeks and their response to split-nutrient application could be measured for another 2 weeks. By then, they had reached the bottom or sides of the rhizoslides and the plants had reached four fully developed leaves. The realization of longer cultivation phases allowing the study of crown root dynamics is a major advantage compared to the growth pouch system used in past studies [17]. Genotypes differed in their developmental speed. While some genotypes started to form crown roots already seven days after start of the experiment, no crown roots were visible on the germination paper for others at the time point of solution change. The evaluation of the dynamic development of individual roots is certainly a way to account for such developmental differences.

For the genotypes F1808, EZ37, FC1890 and W64A only one or two replicates could be evaluated under low or high N as only those did develop crown roots until the time point of solution change (Additional files 7, 8). Twelve individual plants did not develop a traceable crown root on the high N side until the time point of solution change and seventeen on the low N side (Additional files 7, 8). We used a linear mixed model approach which handles missing values better than classical linear models [50]. One opportunity to increase the number of traceable crown roots would be to postpone the treatment to later developmental stages and another option would be to increase the number of replications.

Differences in the amount of roots growing at the low or high N side might influence the morphological responses of the roots to local high nitrate. However, in 't Zandt et al. [29] found no direct dependence of the dynamic of roots exposed to differential N supply

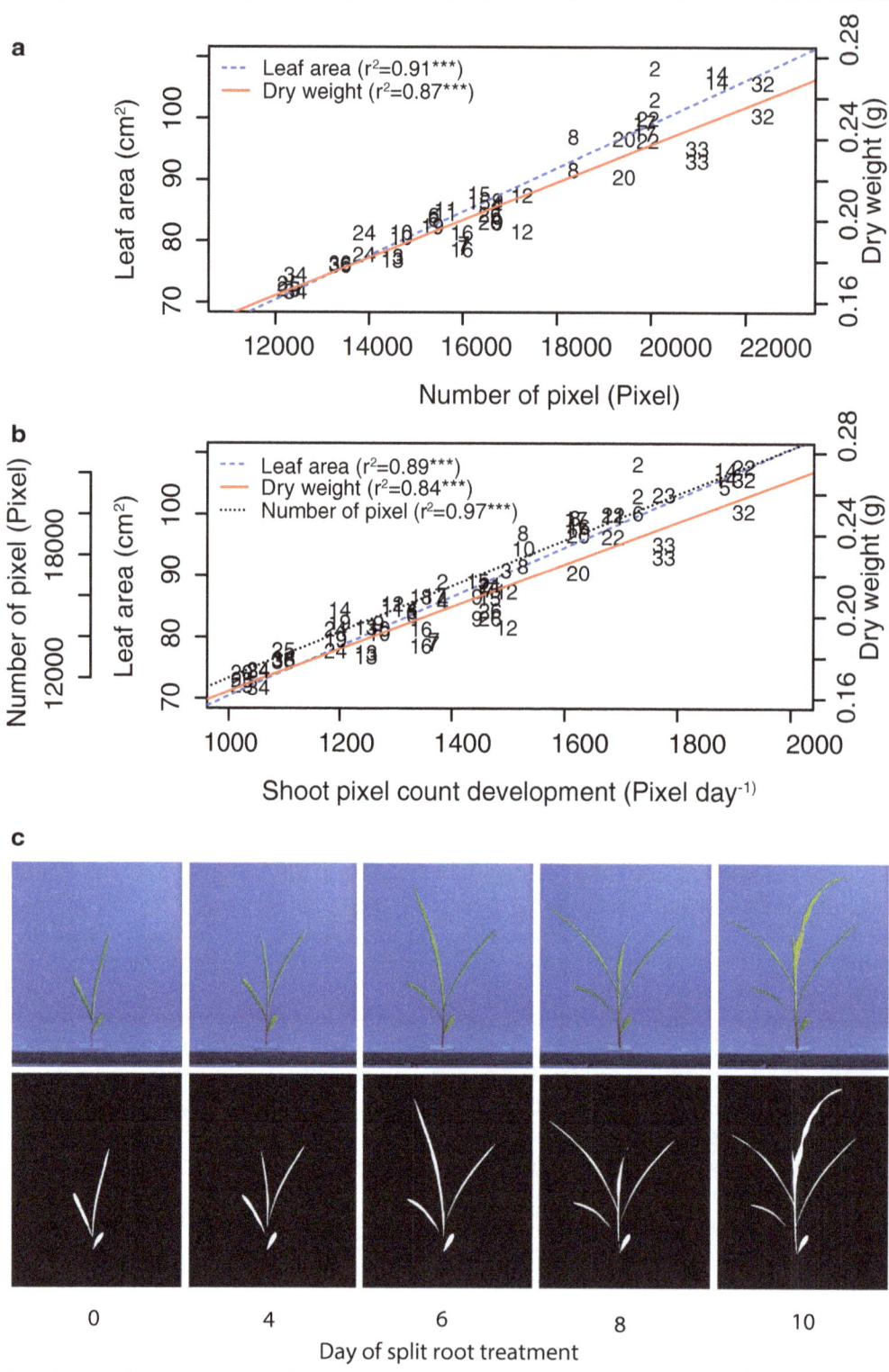

**Fig. 3** **a** Correlation between the average number of pixel (Pixel) of the *top* and *side* images and the leaf area (cm$^{-2}$) or shoot dry weight (g) measured destructively at the end of experiment. **b** Correlation of shoot pixel count development (Pixel day$^{-1}$) expressed as the slope best fitting the linear model and the number of pixels (Pixel), leaf area (cm$^{-2}$) and shoot dry weight (g). **a**, **b** *Number*s represent the average of all replicates of a single genotype. For the corresponding genotypes see "Cultivation in the RADIX platform" in the "Methods" section. The standard error of the difference was 3392 pixels for the average number of pixel, 18.01 cm$^{-2}$ for the leaf area, 0.05 g for the shoot dry weight and 315 pixels for the shoot pixel count development. **c** Growth series of an exemplary shoot and images of the segmented shoot

**Table 1 (a) Results of the analysis of variance, (b) maximal, minimal and median values and heritability of traits**

| a | | Root traits | | | | | | | | | | | Shoot traits | | | | | | |
|---|---|---|---|---|---|---|---|---|---|---|---|---|---|---|---|---|---|---|---|
| | | ER_Cr | IC_Cr | Med_Lat | | Max_Lat | | No_Lat | | DW_R | | ER_Lf | IC_S | LA_Pix | LA_m | DW_S | SPAD | N |
| | | | | 1st | 2nd | 1st | 2nd | 1st | 2nd | Embryonic | Crown | | | | | | | |
| | | cm/d | cm | cm | cm | cm | cm | Counts | Counts | mg | mg | Pixel/d | Pixel | Pixel | cm² | g | Relative | % |
| Genotype | | n.s. | n.s. | ** | * | n.s. | *** | *** | *** | n.s. | ** | *** | ** | *** | * | * | n.s. | ** |
| Treatment | | *** | n.s. | *** | *** | *** | *** | *** | *** | *** | *** | NA | NA | NA | NA | NA | NA | NA |
| G:T | | n.s. | n.s. | ** | *** | * | ** | n.s. | n.s. | n.s. | *** | NA | NA | NA | NA | NA | NA | NA |

| b | Treatment | Root traits | | | | | | | | | | Shoot traits | | | | | | |
|---|---|---|---|---|---|---|---|---|---|---|---|---|---|---|---|---|---|---|
| | | ER_Cr | IC_Cr | Med_Lat | | Max_Lat | | No_Lat | | DW_R [mg] | | ER_Lf | IC_S | LA_Pix | LA_m | DW_S | SPAD | N |
| | | | | 1st | 2nd | 1st | 2nd | 1st | 2nd | Embryonic | Crown | | | | | | | |
| | | cm/d | cm | cm | cm | cm | cm | Counts | Counts | mg | mg | Pixel/d | Pixel | Pixel | cm² | g | Relative | % |
| Max | High | 1.77 | 12.71 | 1.21 | 0.77 | 1.87 | 2.58 | 61.93 | 47.40 | 29.42 | 41.89 | 1914 | 1998 | 22,331 | 108.18 | 0.343 | 29.59 | 5.07 |
| | Low | 1.31 | 13.42 | 0.49 | 0.84 | 1.85 | 2.75 | 38.14 | 50.57 | 22.53 | 35.50 | | | | | | | |
| Min | High | 1.44 | 8.46 | 1.04 | 0.15 | 1.08 | 0.60 | 41.33 | 12.92 | 28.32 | 29.92 | 1023 | 527 | 12,261 | 71.75 | 0.117 | 29.59 | 3.61 |
| | Low | 1.06 | 9.17 | 0.32 | 0.24 | 1.06 | 0.71 | 17.54 | 16.37 | 20.79 | 15.79 | | | | | | | |
| Median | High | 1.65 | 11.12 | 1.09 | 0.47 | 1.45 | 1.72 | 50.56 | 28.90 | 28.73 | 34.63 | 1434 | 1301 | 16,556 | 86.4 | 0.192 | 29.59 | 4.30 |
| | Low | 1.16 | 11.83 | 0.37 | 0.46 | 1.48 | 1.77 | 26.77 | 33.30 | 21.42 | 22.10 | | | | | | | |
| h²mean | High | 0.41 | 0.03 | 0.72 | 0.61 | 0.41 | 0.69 | 0.53 | 0.57 | 0.07 | 0.15 | 0.58 | 0.53 | 0.59 | 0.45 | 0.46 | 0 | 0.52 |
| | Low | 0 | 0.03 | 0.26 | 0.42 | 0 | 0.59 | 0.34 | 0.30 | 0.19 | 0.57 | | | | | | | |

*ER_Cr*, Elongation rate crown roots; *intercept, IC_Cs* length of crown roots at solution change; *Med_Lat* length of representative lateral root; *Max_Lat* maximal lateral root length; *No_Lat* number of lateral roots; *DW_R* root dry weight; *ER_S* shoot pixel count development; *intercept, IC_Cr* shoot pixel count at solution change; *LA_Pix* digital leaf area; *LA_m* measured leaf area; *DW_S* shoot dry weight; *SPAD* chlorophyll measurements, *N* content in the leaf in % of total dry weight. Significance level: *** ≤0.001; ** ≤0.01; * ≤0.05

in the rhizoslides concluding that the responses on either side occurred largely independent from each other.

### Genotype-by-nitrogen interaction for crown root development

The crown root elongation under low N was on average 26 % reduced compared to the high N side (Table 1b). The minimum observed elongation in the high N compartment was 1.44 cm d$^{-1}$ and the maximal elongation was 1.77 cm d$^{-1}$ (Table 1b). There were differences for the intercept, i.e. the crown root length at the time of solution change.

There was a significant genotype-by-treatment interaction for the response of crown root growth to split-nutrient application. More specifically, the crown axile roots at different N-concentrations showed genotype-dependent differential response. For example, crown roots of the genotype PH207 grew 39 % slower under low N conditions than under high N conditions whereas the genotype LAN 496 showed a much lower reduction in

growth (15 %) under low N conditions (Table 1b, Additional file 7, 8). Interestingly, the heritability was high ($h^2 = 0.41$) for roots grown under high N but zero for roots grown under low N conditions (Table 1b). This means that it is not possible to select genotypes based on their rooting behaviour on the low-N side of a split-nutrient setup. By contrast, the fast exploration of N rich patches and selective root placement is highly heritable and could be an advantageous trait with respect to inhomogeneity in the field or patch wise application of fertilizers e.g. by row fertilization. According to a model by Dunbabin et al. [51], selective root placement could lead to a more than two fold higher N uptake efficiency throughout the whole growing season, compared to a root system that is only poorly capable to respond to N. Already the study of in 't Zandt et al. [29] demonstrated a oppositional response in the low N and high N compartment. However the study was limited to one genotype. To our knowledge, this is the first time that such a selective response could be observed for a larger set of maize genotypes.

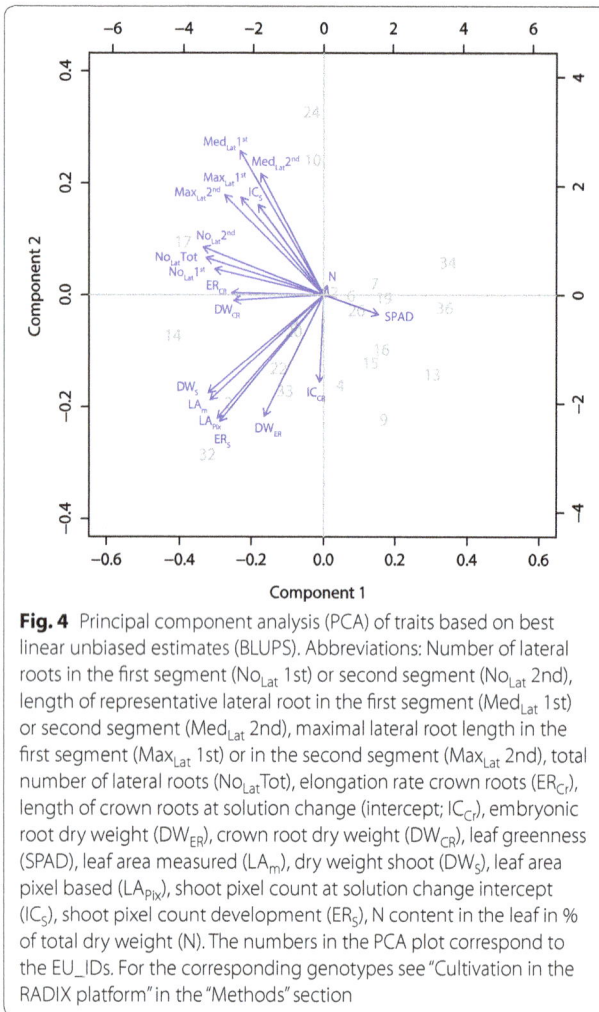

**Fig. 4** Principal component analysis (PCA) of traits based on best linear unbiased estimates (BLUPS). Abbreviations: Number of lateral roots in the first segment (No$_{Lat}$1st) or second segment (No$_{Lat}$2nd), length of representative lateral root in the first segment (Med$_{Lat}$1st) or second segment (Med$_{Lat}$2nd), maximal lateral root length in the first segment (Max$_{Lat}$1st) or in the second segment (Max$_{Lat}$2nd), total number of lateral roots (No$_{Lat}$Tot), elongation rate crown roots (ER$_{Cr}$), length of crown roots at solution change (intercept; IC$_{Cr}$), embryonic root dry weight (DW$_{ER}$), crown root dry weight (DW$_{CR}$), leaf greenness (SPAD), leaf area measured (LA$_m$), dry weight shoot (DW$_S$), leaf area pixel based (LA$_{Pix}$), shoot pixel count at solution change intercept (IC$_S$), shoot pixel count development (ER$_S$), N content in the leaf in % of total dry weight (N). The numbers in the PCA plot correspond to the EU_IDs. For the corresponding genotypes see "Cultivation in the RADIX platform" in the "Methods" section

## Length and density of lateral roots

Investigations of lateral root number and length were only done at the end of the experiment as in the study of in 't Zandt [29] the maximal number of lateral roots in the segment was already reached two days after solution change in the high N compartment and after four days in the low N compartment. Thus, it was assumed that the final number of lateral roots is representative for the branching intensity in these segments. Furthermore, the number of laterals in the zone present before solution change did not significantly increase during the two weeks of treatment, hypothesizing that only newly developed root tissues are able to respond to N availability (Additional file 10).

In general, lateral roots became longer under high N compared to low N conditions (Table 1b, Additional file 5A-C). This finding is in close agreement with previous studies reporting a selective root placement within the high N compartment by stronger lateral root formation and growth [52, 53]. However, some genotypes

were characterized by an opposed trend highlighting the opportunity to identify contrasting genotypes. Past studies already observed different responses for different species and even different genotypes within a species to the same environmental stimulus [54], but to date no study tried to use contrasting genotypic responses for mapping of quantitative traits.

The length of the branching zone (Tab 1. LBrZ) did not differ among genotypes but the number of lateral roots (Table 1, No$_{Lat}$1st and No$_{Lat}$2nd) in this zone was heritable and correlated with the axile root growth [r = 0.47* (No$_{Lat}$1st) and r = 0.49* (No$_{Lat}$2nd)]. Interestingly, the genotypes did not differ for the linear density of lateral roots per unit axile root length. Hence, the formation of lateral roots seemed to be mainly driven by a stronger axile root elongation rather than a higher branching density as no significant treatment effects could be observed for the branching density or the length of the branching zone. Although the branching intensity was the only trait that was correlated with leaf N content (Additional file 6), the missing heritability exclude the integration of this trait into breeding schemes.

Genotypic differences were observed for all root traits except the maximal lateral root length in the first segment and the dry weight of embryonic roots. Accordingly, the heritability of the number and length of lateral roots was consistently moderate to high in both segments except for the maximal length in the first segment [$h^2 = 0$ (low N); $h^2 = 0.41$ (high N)] whereas the heritability was higher in the high N ($0.41 \leq h^2 \geq 0.72$) than in the low N compartment ($0.03 \leq h^2 \geq 0.59$) in general (Table 1b).

## Genotype-by-nitrogen interaction for lateral root characteristics

A significant genotype-by-N placement interaction was observed for the medium and maximum length of lateral roots. The ability of genotypes to form long first-order lateral roots and its genetic control is a very interesting research topic. Eventually such roots are able to replace some of the functions of their parental axile root, thus allowing a more flexible response of a root system.

## Relationship between root and shoot development

We found a strong connection between lateral root formation and shoot performance. We used correlations based on genotypic mean values to evaluate the dependencies between roots formed on the high and low N side and between roots and shoots. Root traits were strongly correlated among themselves (Fig. 4) and two major clusters of root traits could be identified by principal component analysis: the lateral root length and the number of lateral roots (Fig. 4). Interestingly, the number of laterals was closely correlated with the elongation of crown roots (Fig. 4) supporting the hypothesis that the number of laterals was mainly driven by axial root elongation rather than the branching density.

The leaf area was positively correlated to the medium length and number of lateral roots (Figs. 4, 5, Additional file 6) whereas the medium length was only under high N conditions correlated to shoot growth (Fig. 5). Lengths and densities of lateral roots measured in the second segment were stronger positively correlated with shoot traits than the first segment (Additional file 6). This inconsistency might be related to a feedback loop. N starvation during the establishment phase may have led to a limited link between shoot and root growth whereas in the second segment, the roots already profit from a higher N uptake resulting in a positive feedback from the shoot. This observation supports the hypothesis of de Kroon et al. [55] suggesting a coupling of nutrient sensing and coordinated growth of the root components and is in line with observations of in 't Zandt et al. [29] demonstrating a dynamic differential response of crown and lateral root growth under either zero N or high N conditions. Furthermore, it indicates a signalling between roots and shoot transmitting the information about N availability that itself is transformed into differential growth e.g. by the provision of carbohydrates. Indeed, Tabata et al. [56] observed a signalling from the root to the shoot triggering the activity of nitrate transporters within root regions with high nitrate availability in *Arabidopsis thaliana*. An alternative explanation of this phenomenon could be a better N supply by the roots after the formation of laterals in the first segment resulting in an increasing root surface area and a simultaneously increased uptake of N after the change of the nutrient solution. Furthermore, a very high N stress level can lead to an inhibition of lateral root formation whereas a moderate stress can induce it [57].

## Considerations concerning a maize ideotype for improved foraging behaviour

There are many open questions with respect to the optimal root ideotype of maize. Foraging of N in nutrient-rich patches is mainly achieved by changing root system architecture. While, in general, the root length within the nutrient-rich patch increases, root growth outside the patch decreases [52, 53, 58, 59]. Thus, there is a trade-off between intensive foraging in patches and the overall direction of the root system to an available resource which might become important at critical stages of development, e.g. water and N at depth during grain filling. Accordingly, the proposed steep, cheap and deep ideotype to optimize water and N acquisition at depth to capture leached nitrate [7] includes "unresponsiveness of lateral branching to localized resource availability". Paper-based systems offer a huge potential to establish rapid screens for N responsiveness.

This screening already provides an insight into the diversity of responses of genotypes towards inhomogeneous nutrient distributions and gives a strong indication that the responsiveness is closely linked to shoot development. However, further investigations with contrasting genotypes under field conditions are necessary, before drawing a conclusion on the utility of the presented screening method for selection purposes.

## Conclusion

The RADIX platform allowed studying dynamic changes in root-system architecture to split-root application of nitrogen. This is the first study evaluating such a differential response using a larger set of maize genotypes. A stronger selective root placement in the high N-compartment was related to an increased shoot development. This indicates that high early vigour might be related to a more intense foraging behaviour. In ongoing experiments, we aim to (1) verify these results under field conditions and (2) map the genomic regions controlling

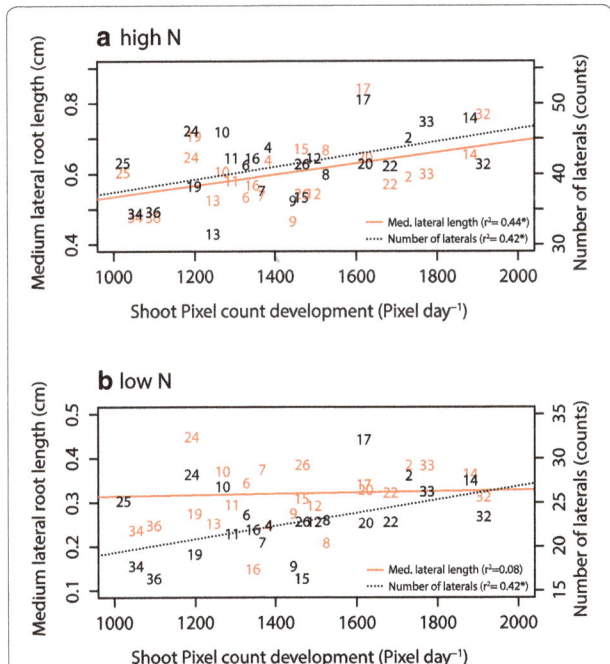

**Fig. 5** Correlation of shoot pixel count development (Pixel day$^{-1}$) expressed as the slope best fitting the linear model and medium lateral root length (cm) or the number of lateral roots (counts) either under high N (**a**) or low N conditions (**b**). Significance levels: $\leq$0.001***, $\leq$0.01**, $\leq$0.05*. Correlation was done based on best linear unbiased predictors (BLUPS). The standard error of the difference was 323 pixels for the shoot and 0.1 cm for the lateral root length under high as well as under low N conditions

these responses using an association panel. Apart from studying nutrient-use efficiency, the system may be also used to evaluate responses to stresses like extreme pH, agrochemicals or microbes.

## Additional files

**Additional file 1.** Constructional drawing of the RADIX platform (units:mm).

**Additional file 2.** Constructional drawing of the imaging station (units:mm).

**Additional file 3.** The video shows the eluation of the germination paper soaked in bromocresol green and afterwards watered with a basic ($NH_4HPO_4$) and an acidic ($Ca(NO_3)_2$) solution. A webcam was installed to image the elution. After 48 hours the bromocresol was washed out and the paper was sprayed with it again to visualize the pH gradient.

**Additional file 4.** Shoot growth of 24 different genotypes quantified by the increase in number of pixels after the start of the split root treatment.

**Additional file 5.** A) Best linear unbiased prediction (BLUPS) of parameters with significant treatment effect, but no genotype effect or genotype: treatment interaction. B) Prediction of mean values of parameters with significant genotype effect, but no genotype: treatment interaction. C) Prediction of mean values of parameters with significant genotype: treatment interaction. Abbreviations: Number of lateral roots in the first segment ($No_{Lat} 1^{st}$) or second segment ($No_{Lat} 2^{nd}$), length of representative lateral root in the first segment ($Med_{Lat} 1^{st}$) or second segment ($Med_{Lat} 2^{nd}$), maximal lateral root length in the first segment ($Max_{Lat} 1^{st}$) or in the second segment ($Max_{Lat} 2^{nd}$), elongation rate crown roots ($ER_{Cr}$), length of crown roots at solution change (intercept; $IC_{Cr}$), embryonic root dry weight ($DW_{ER}$), crown root dry weight ($DW_{CR}$), chlorophyll measurements (SPAD, measured leaf area ($LA_m$), shoot dry weight ($DW_S$), number of pixels specifying the leaf area ($LA_{Pix}$), shoot pixel count at solution change (intercept; $IC_S$), shoot pixel count development ($ER_S$), N content in the leaf in % of total dry weight (N in %).

**Additional file 6.** Correlation coefficients (Pearson's r) of traits. Correlation was done based on best linear unbiased estimates (BLUPS). Abbreviations: Number of lateral roots in the first segment ($No_{Lat} 1^{st}$) or second segment ($No_{Lat} 2^{nd}$), length of representative lateral root in the first segment ($Med_{Lat} 1^{st}$) or second segment ($Med_{Lat} 2^{nd}$), maximal lateral root length in the first segment ($Max_{Lat} 1^{st}$) or in the second segment ($Max_{Lat} 2^{nd}$), branching density in the first segment ($Br_{Lat} 1^{st}$) or in the second segment ($Br_{Lat} 2^{nd}$), branching density across both segments ($Br_{Lat} Tot$), length of the branching zone (LBrZone), total number of lateral roots ($No_{Lat} Tot$), elongation rate crown roots ($ER_{Cr}$), length of crown roots at solution change (intercept; $IC_{Cr}$), embryonic root dry weight ($DW_{ER}$), crown root dry weight ($DW_{CR}$), leaf greenness (SPAD), leaf area measured ($LA_m$), dry weight shoot (DWS), leaf area pixel based ($LA_{Pix}$), shoot pixel count at solution change intercept ($IC_S$), shoot pixel count development ($ER_S$), N content in the leaf in % of total dry weight (N). Significance level: $\leq 0.001$ ***, $\leq 0.01$ **, $\leq 0.05$ *.

**Additional file 7.** Increase in crown root length of 24 genotypes under high nitrogen after the start of the split root treatment.

**Additional file 8.** Increase in crown root length of 24 genotypes under low nitrogen after the start of the split root treatment.

**Additional file 9.** Residual vs. fitted crown root length of multiple linear models to determine intercept and slope of crown root development after solution change.

**Additional file 10.** Increase of lateral roots after solution change above the last formed lateral.

## Abbreviations
RSA: root system architecture; N: nitrogen; QTL: quantitative trait loci.

## Authors' contributions
CLM and AH, PF JP, NK and AW designed the RADIX system, PF constructed the system, NK designed the illumination unit and programmed the imaging software. CLM and JP performed the experiment. CLM and NK did the image analysis. AH and CLM did the statistical analysis and drafted the manuscript supported by AW. All authors read and approved the final manuscript.

## Acknowledgements
The authors thank Brigitta Herzog, Hansueli Zellweger and Frank Liebisch from ETH Zürich for their support handling the experiment. Thanks to Delley Seeds and Plants Ltd. for hybrid production. We kindly thank the donors of the genetic material: Department of Agroenvironmental Science and Technologies (DiSTA), University of Bologna, Italy (RootABA lines); Estación Experimental de Aula Dei (CSIC), Spain (EZ47, EZ11A, EZ37); Centro Investigaciones Agrarias de Mabegondo (CIAM), Spain (EC169); University of Hohenheim, Versuchsstation für Pflanzenzüchtung, Germany (UH007, UH250); and INRA CNRS UPS AgroParisTech, France (supply of the remaining, public lines).

## Competing interests
The authors declare that they have no competing interests.

## Funding
This research received funding from the European Community Seventh Framework Programme FP7-KBBE-2011-5 under Grant Agreement No. 289300.

## References
1. Hund A, Ruta N, Liedgens M. Rooting depth and water use efficiency of tropical maize inbred lines, differing in drought tolerance. Plant Soil. 2009;318:311–25.
2. Watt M, Magee LJ, McCully ME. Types, structure and potential for axial water flow in the deepest roots of field-grown cereals. New Phytol. 2008;178:135–46.
3. Zhu J, Kaeppler SM, Lynch JP. Topsoil foraging and phosphorus acquisition efficiency in maize (Zea mays). Funct Plant Biol. 2005;32:749–62.
4. Smith S, De Smet I. Root system architecture: insights from Arabidopsis and cereal crops. Philos Trans R Soc Lond B Biol Sci. 2012;367:1441–52.
5. Burton AL, Brown KM, Lynch JP. Phenotypic diversity of root anatomical and architectural traits in species. Crop Sci. 2013;53:1042–55.
6. Worku M, Bänziger M, Schulte auf'm Erley G, Friesen D, Diallo AO, Horst WJ. Nitrogen efficiency as related to dry matter partitioning and root system size in tropical mid-altitude maize hybrids under different levels of nitrogen stress. Field Crop Res. 2012;130:57–67.
7. Lynch JP. Steep, cheap and deep: an ideotype to optimize water and N acquisition by maize root systems. Ann Bot Lond. 2013;112:347–57.
8. Nagel KA, Putz A, Gilmer F, Heinz K, Fischbach A, Pfeifer J, Faget M, Blossfeld S, Ernst M, Dimaki C, et al. GROWSCREEN-Rhizo is a novel phenotyping robot enabling simultaneous measurements of root and shoot growth for plants grown in soil-filled rhizotrons. Funct Plant Biol. 2012;39:891–904.
9. Heeraman D, Hopmans J, Clausnitzer V. Three dimensional imaging of plant roots in situ with X-ray computed tomography. Plant Soil. 1997;189:167–79.
10. Gregory PJ, Hutchison D, Read D, Jenneson P, Gilboy W, Morton E. Non-invasive imaging of roots with high resolution X-ray micro-tomography. Plant Soil. 2003;255:351–9.

11. Blossfeld S, Le Marié C, Van Dusschoten D, Suessmilch S, Kuhn A. Non-invasive investigation of root growth via NMR imaging. Commun Agric Appl Biol Sci. 2011;76:11.

12. Rascher U, Blossfeld S, Fiorani F, Jahnke S, Jansen M, Kuhn AJ, Matsubara S, Märtin LLA, Merchant A, Metzner R. Non-invasive approaches for phenotyping of enhanced performance traits in bean. Funct Plant Biol. 2011;38:968–83.

13. Adu MO, Chatot A, Wiesel L, Bennett MJ, Broadley MR, White PJ, Dupuy LX. A scanner system for high-resolution quantification of variation in root growth dynamics of Brassica rapa genotypes. J Exp Bot. 2014;65:2039–48.

14. Planchamp C, Balmer D, Hund A, Mauch-Mani B. A soil-free root observation system for the study of root-microorganism interactions in maize. Plant Soil. 2013;367:605–14.

15. Zhu J, Kaeppler SM, Lynch JP. Mapping of QTLs for lateral root branching and length in maize (Zea mays L.) under differential phosphorus supply. Theor Appl Genet. 2005;111:688–95.

16. Liao H, Rubio G, Yan X, Cao A, Brown KM, Lynch JP. Effect of phosphorus availability on basal root shallowness in common bean. Plant Soil. 2001;232:69–79.

17. Hund A, Trachsel S, Stamp P. Growth of axile and lateral roots of maize: I development of a phenotying platform. Plant Soil. 2009;325:335–49.

18. Le Marié C, Kirchgessner N, Marschall D, Walter A, Hund A. Rhizoslides: paper-based growth system for non-destructive, high throughput phenotyping of root development by means of image analysis. Plant Methods. 2014;10:13.

19. Ruta N, Liedgens M, Fracheboud Y, Stamp P, Hund A. QTLs for the elongation of axile and lateral roots of maize in response to low water potential. Theor Appl Genet. 2009;120:621–31.

20. Ruta N, Stamp P, Liedgens M, Fracheboud Y, Hund A. Collocation of QTLs for seedling traits and yield components of tropical maize under limited water availability. Crop Sci. 2010;50:1385–92.

21. Trachsel S, Messmer R, Stamp P, Hund A. Mapping of QTLs for lateral and axile root growth of tropical maize. Theor Appl Genet. 2009;119:1413–24.

22. Reimer R. Responses of maize (Zea mays L.) seedlings to low and high temperature: association mapping of root growth and photosynthesis-related traits. PhD Thesis. Diss. ETH No. 18807, AGRL; 2010.

23. Hochholdinger F, Park WJ, Sauer M, Woll K. From weeds to crops: genetic analysis of root development in cereals. Trends Plant Sci. 2004;9:42–8.

24. Hund A, Reimer R, Messmer R. A consensus map of QTLs controlling the root length of maize. Plant Soil. 2011;344:143–58.

25. Yu P, White PJ, Hochholdinger F, Li C. Phenotypic plasticity of the maize root system in response to heterogeneous nitrogen availability. Planta. 2014;240:667–78.

26. Yu P, Li X, Yuan L, Li C. A novel morphological response of maize (Zea mays) adult roots to heterogeneous nitrate supply revealed by a split-root experiment. Physiol Plant. 2014;150:133–44.

27. Peng Y, Li X, Li C. Temporal and spatial profiling of root growth revealed novel response of maize roots under various nitrogen supplies in the field. PLoS ONE. 2012;7:e37726.

28. Jackson R, Manwaring J, Caldwell M. Rapid physiological adjustment of roots to localized soil enrichment. Nature. 1990;344:58–60.

29. in 't Zandt D, Le Marié C, Kirchgessner N, Visser E, Hund A. High-resolution quantification of root dynamics in split-nutrient rhizoslides reveals rapid and strong proliferation of maize roots in response to local high nitrogen. J Exp Bot. 2015;66:5507–17.

30. Vadez V. Root hydraulics: the forgotten side of roots in drought adaptation. Field Crop Res. 2014;165:15–24.

31. Wissuwa M, Mazzola M, Picard C. Novel approaches in plant breeding for rhizosphere-related traits. Plant Soil. 2009;321:409–30.

32. Grieder C, Trachsel S, Hund A. Early vertical distribution of roots and its association with drought tolerance in tropical maize. Plant Soil. 2014;377:295–308.

33. Hartmann A, Czauderna T, Hoffmann R, Stein N, Schreiber F. HTPheno: an image analysis pipeline for high-throughput plant phenotyping. BMC Bioinform. 2011;12:148.

34. Hairmansis A, Berger B, Tester M, Roy S. Image-based phenotyping for non-destructive screening of different salinity tolerance traits in rice. Rice. 2014;7:16.

35. Reymond M, Muller B, Leonardi A, Charcosset A, Tardieu F. Combining quantitative trait loci analysis and an ecophysiological model to analyze the genetic variability of the responses of maize leaf growth to temperature and water deficit. Plant Physiol. 2003;131:664–75.

36. Freund H. Sterilisieren der fertigen Ampullen. In: Freund H, editor. Die Ampullenfabrikation. Berlin: Springer; 1916. p 61–72.

37. Bohn M, Novais J, Fonseca R, Tuberosa R, Grift T. Genetic evaluation of root complexity in maize. Acta Agron Hung. 2006;54:291–303.

38. Shaff JE, Schultz BA, Craft EJ, Clark RT, Kochian LV. GEOCHEM-EZ: a chemical speciation program with greater power and flexibility. Plant Soil. 2010;330:207–14.

39. Otsu N. A threshold selection method from gray-level histograms. IEEE Trans Syst Man Cybern. 1979;9:62–6.

40. R Core Team: R: a language and environment for statistical computing. R Foundation for Statistical Computing, Vienna, Austria 2015. http://www.R-project.org/.

41. NPTEL. http://nptel.ac.in/courses/102103016/module1/lec1/3.html. Accessed 17 June 2016.

42. Abbeddou S, Diekmann J, Rischkowsky B, Kreuzer M, Oberson A. Unconventional feeds for small ruminants in dry areas have a minor effect on manure nitrogen flow in the soil–plant system. Nutr Cycl Agroecosyst. 2013;95:87–101.

43. Butler D: asreml: asreml() fits the linear mixed model. R package version 3.0. www.vsni.co.uk. 2009.

44. Piepho H-P, Möhring J. Computing heritability and selection response from unbalanced plant breeding trials. Genetics. 2007;177:1881–8.

45. Coque M, Martin A, Veyrieras J, Hirel B, Gallais A. Genetic variation for N-remobilization and postsilking N-uptake in a set of maize recombinant inbred lines. 3. QTL detection and coincidences. Theor Appl Genet. 2008;117:729–47.

46. Garnett T, Conn V, Kaiser BN. Root based approaches to improving nitrogen use efficiency in plants. Plant Cell Environ. 2009;32:1272–83.

47. Clark RB. Physiology of cereals for mineral nutrient uptake, use, and efficiency. In: Baligar VC, Duncan RR, editors. Crops as enhancers of nutrient use. San Diego: Academic Press, Inc; 1990. p. 131–209.

48. Moll R, Kamprath E, Jackson W. Analysis and interpretation of factors which contribute to efficiency of nitrogen utilization. Agron J. 1982;74:562–4.

49. Lynch J. Root architecture and plant productivity. Plant Physiol. 1995;109:7.

50. Piepho H, Möhring J, Melchinger A, Büchse A. BLUP for phenotypic selection in plant breeding and variety testing. Euphytica. 2008;161:209–28.

51. Dunbabin V, Rengel Z, Diggle A. Simulating form and function of root systems: efficiency of nitrate uptake is dependent on root system architecture and the spatial and temporal variability of nitrate supply. Funct Ecol. 2004;18:204–11.

52. Drew M. Comparison of the effects of a localised supply of phosphate, nitrate, ammonium and potassium on the growth of the seminal root system, and the shoot, in barley. New Phytol. 1975;75:479–90.

53. Drew MC, Saker LR, Ashley TW. Nutrient supply and the growth of the seminal root system in barley: I. The effect of nitrate concentration on the growth of axes and laterals. J Exp Bot. 1973;24:1189–202.

54. Forde B, Lorenzo H. The nutritional control of root development. Plant Soil. 2001;232:51–68.

55. De Kroon H, Visser EJ, Huber H, Mommer L, Hutchings MJ. A modular concept of plant foraging behaviour: the interplay between local responses and systemic control. Plant, Cell Environ. 2009;32:704–12.

56. Tabata R, Sumida K, Yoshii T, Ohyama K, Shinohara H, Matsubayashi Y. Perception of root-derived peptides by shoot LRR-RKs mediates systemic N-demand signaling. Science. 2014;346:343–6.

57. Forde BG. Nitrogen signalling pathways shaping root system architecture: an update. Curr Opin Plant Biol. 2014;21:30–6.

58. Granato T, Raper C. Proliferation of maize (Zea mays L.) roots in response to localized supply of nitrate. J Exp Bot. 1989;40:263–75.

59. Hackett C. A method of applying nutrients locally to roots under controlled conditions, and some morphological effects of locally applied nitrate on the branching of wheat roots. Aust J Biol Sci. 1972;25:1169–80.

# 13

# The utility of flow sorting to identify chromosomes carrying a single copy transgene in wheat

Petr Cápal[1], Takashi R. Endo[1,2], Jan Vrána[1], Marie Kubaláková[1], Miroslava Karafiátová[1], Eva Komínková[1], Isabel Mora-Ramírez[3], Winfriede Weschke[3] and Jaroslav Doležel[1*]

## Abstract

**Background:** Identification of transgene insertion sites in plant genomes has practical implications for crop breeding and is a stepping stone to analyze transgene function. However, single copy sequences are not always easy to localize in large plant genomes by standard approaches.

**Results:** We employed flow cytometric chromosome sorting to determine chromosomal location of barley sucrose transporter construct in three transgenic lines of common wheat. Flow-sorted chromosomes were used as template for PCR and fluorescence in situ hybridization to identify chromosomes with transgenes. The chromosomes carrying the transgenes were then confirmed by PCR using DNA amplified from single flow-sorted chromosomes as template.

**Conclusions:** Insertion sites of the transgene were unambiguously localized to chromosomes 4A, 7A and 5D in three wheat transgenic lines. The procedure presented in this study is applicable for localization of any single-copy sequence not only in wheat, but in any plant species where suspension of intact mitotic chromosomes suitable for flow cytometric sorting can be prepared.

**Keywords:** Transgene localization, Flow cytometric sorting, Single chromosome amplification, *Triticum aestivum*, *Hordeum vulgare*, HvSUT1

## Background

During the past 30 years, many cultivars of agricultural crops beneficial to humankind have been developed by means of genetic engineering, including plants resistant to herbicides, pests or viruses, bearing fruits with prolonged shelf life and products more suited for industrial processing [for review see 1]. Wheat ranks 5th in the commodities produced worldwide and is the second most-produced food crop occupying more than 50 % of the world crop area (http://faostat3.fao.org/). In the light of climate change and world population growth, future challenges for the increase of crop production have constantly been discussed. However, FAO statistics show that the wheat production is reaching a plateau and is severely affected by climate change. This is a consequence of a slowdown in wheat yield increase, accounting for only 0.5 % per year in the last decade [2].

Breeding improved cultivars with increased tolerance to adverse climatic conditions and with increased yield and quality could be facilitated by genetic engineering and introduction of beneficial genes from other organisms. The insertion site of a transgene is of great importance for the transgene function [3, 4] which is also influenced by its position on the chromosome, including the flanking DNA sequences [5]. However, transgene localization is not easy by routine approaches, like fluorescence in situ hybridization (FISH), or Southern blotting. A prevalent method for detection of transgenes in animals and plants is FISH, which has its pros and cons [6]. In barley and common wheat, FISH enables cytological localization of cDNAs, as short as 1.5 kb, on a

*Correspondence: dolezel@ueb.cas.cz
[1] Institute of Experimental Botany, Centre of the Region Haná for Biotechnological and Agricultural Research, Šlechtitelů 31, 78371 Olomouc, Czech Republic
Full list of author information is available at the end of the article

chromosome or chromosomes that had already been known to carry the cDNAs [7, 8]. Although some authors succeeded in localizing transgenes on plant chromosomes using FISH [9–11], this approach has not become a routine application.

Weichert et al. [12] obtained transgenic lines (HOSUT) of hexaploid wheat carrying barley (*Hordeum vulgare*) sucrose transporter HvSUT1 (SUT) gene that is overexpressed under the control of the endosperm-specific Hordein B1 promoter (HO). The HOSUT lines were found to increase grain yield significantly as compared to control non-transformed plants [13]. However, the genomic location of the transgene in these lines was not known. In the present work we employed a novel approach for unambiguous identification of chromosomes carrying the transgene in three HOSUT lines. The protocol takes the advantage of the availability of a procedure for flow cytometric chromosome sorting in wheat and the fact that flow-sorted chromosomes are suitable as templates for PCR and FISH [14]. Moreover, a protocol has been developed recently for representative DNA amplification from single copies of chromosomes [15]. By combining these approaches we could assign the transgene to particular chromosomes in three HOSUT lines of wheat.

## Results and discussion

The experimental workflow is shown on Fig. 1. As the first step, we prepared liquid suspensions of intact mitotic chromosomes from all five lines of wheat (see "Methods" section) and analyzed them by flow cytometry. Monovariate flow karyotypes (histograms of relative fluorescence intensity) were obtained after the analysis of DAPI-stained chromosomes, and bivariate flow karyotypes obtained after the analysis of DAPI-stained chromosomes with FITC-labelled GAA microsatellites. We observed differences between flow karyotypes of the HOSUT lines and the model hexaploid wheat cultivar Chinese Spring. The alterations concerned the profiles of major composite peaks on monovariate flow karyotypes (Additional file 1: Figure S1) and the distribution of chromosome populations on bivariate flow karyotypes (Fig. 2). This observation reflected the differences in karyotypes (chromosome polymorphism) between the cultivar Certo, used to produce the HOSUT lines (data not shown), and Chinese Spring. On the other hand, flow karyotypes of the three HOSUT lines were indistinguishable from each other.

In order to identify chromosomes carrying the transgenes, we first used the approach described by Vrána et al. [16]. Fractions of 200 chromosomes were sorted from different regions of monovariate flow karyotypes as shown on Additional file 1: Figure S1, and

DNA of the sorted chromosomes was used as template for PCR. This analysis identified one region (sort gate) in each line as representing chromosomes bearing a transgene. As each region (sort gate) on a monovariate flow karyotype may represent more than one chromosome type, in the next step we sorted chromosomes from regions delineated on bivariate flow karyotypes (Fig. 2). The sort gates were designed to include chromosome populations corresponding to the positive sort gates on monovariate flow karyotypes. From these regions, and also from nearby regions, chromosomes were sorted into PCR tubes (100 chromosomes per tube) and immediately afterwards also onto microscopic slides (ca. 1000 chromosomes per slide). The results obtained by PCR with primers amplifying HvSUT-RT sequence (Fig. 3) and identification of chromosomes from sort gates for each transgenic line by FISH with probes targeting Afa-repeat family and GAA-microsatellites (Fig. 4) are summarized in Table 1.

FISH analysis showed that more than 90 % of chromosomes flow-sorted from the region defined by the green rectangle consisted of one type of chromosome in each of the HOSUT lines. This fact together with the results of PCR suggested that the transgene was located on chromosome 7A in HOSUT 12/44, on chromosome 5D in HOSUT 20/6 and on chromosome 4A in HOSUT 24/31. In the former two lines, the critical type of chromosome was not found among the chromosomes flow-sorted from the region defined by red rectangles. However, chromosome 4A was found to represent 12.39 % of chromosomes flow-sorted from the red region in HOSUT 24/31. This was probably due to the similarities in size and the amount of GAA-FITC fluorescence of chromosomes 4A and 7A. Due to this similarity, mixture of the two chromosomes 4A and 7A was also observed in the chromosome fraction sorted from the green region in HOSUT 12/44.

To confirm chromosomal locations of the transgene and avoid ambiguous results due to possible contamination of flow-sorted fraction by other chromosomes, PCR was done on DNA amplified from single flow-sorted chromosomes. As each time only one copy of chromosome is sorted, the DNA cannot be contaminated by other chromosomes. Five single chromosomes were sorted from the green sort regions of the HOSUT lines and their DNA was separately amplified using multiple displacement amplification (MDA). Out of the five sorted chromosomes, whole genome amplification was successful with three chromosomes in HOSUT 12/44, two chromosomes in HOSUT 20/6 and four chromosomes in HOSUT 24/31. The successful amplification was defined by the production of measurable amount of DNA after MDA and by the presence of at least one marker for the

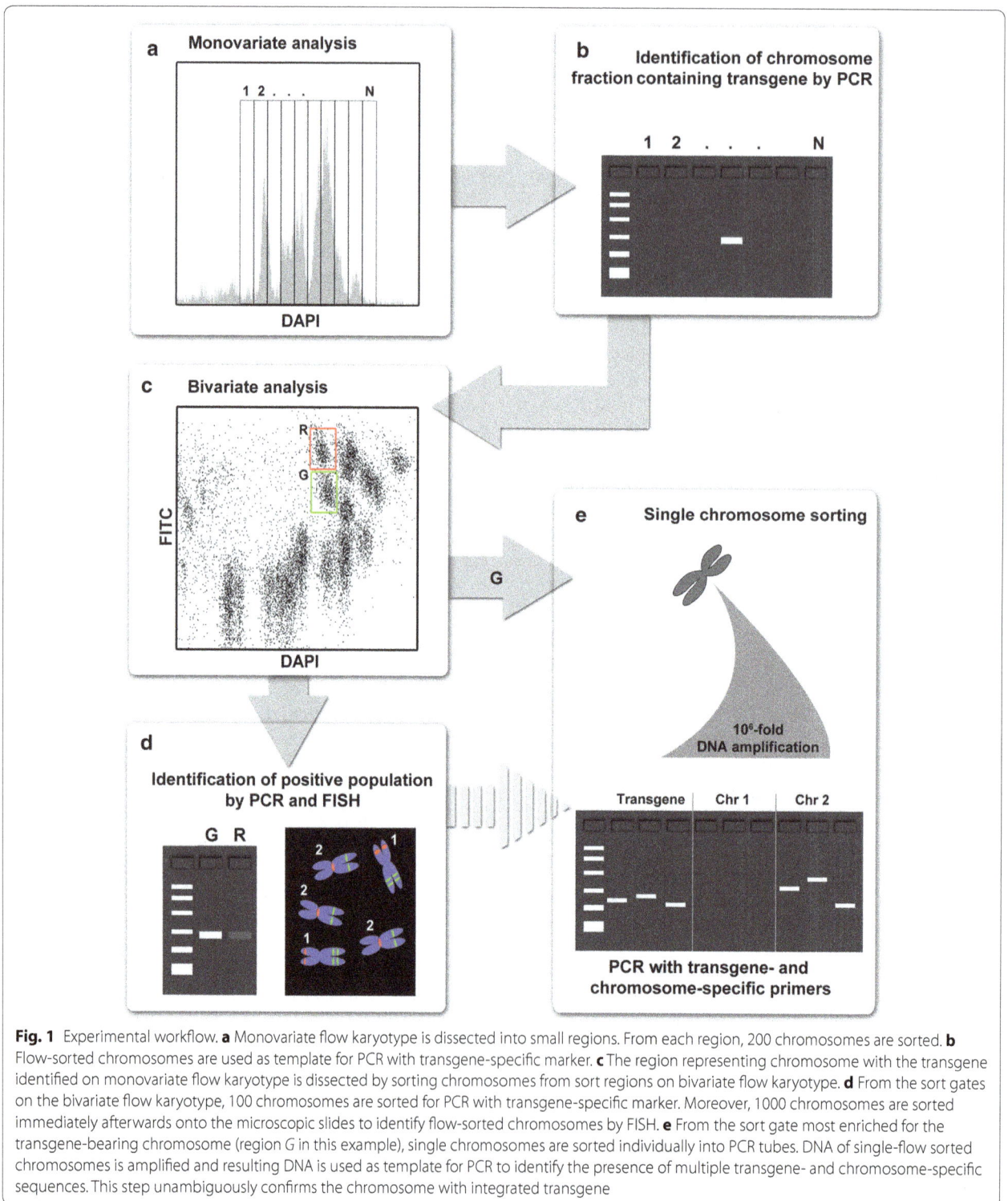

**Fig. 1** Experimental workflow. **a** Monovariate flow karyotype is dissected into small regions. From each region, 200 chromosomes are sorted. **b** Flow-sorted chromosomes are used as template for PCR with transgene-specific marker. **c** The region representing chromosome with the transgene identified on monovariate flow karyotype is dissected by sorting chromosomes from sort regions on bivariate flow karyotype. **d** From the sort gates on the bivariate flow karyotype, 100 chromosomes are sorted for PCR with transgene-specific marker. Moreover, 1000 chromosomes are sorted immediately afterwards onto the microscopic slides to identify flow-sorted chromosomes by FISH. **e** From the sort gate most enriched for the transgene-bearing chromosome (region G in this example), single chromosomes are sorted individually into PCR tubes. DNA of single-flow sorted chromosomes is amplified and resulting DNA is used as template for PCR to identify the presence of multiple transgene- and chromosome-specific sequences. This step unambiguously confirms the chromosome with integrated transgene

transgene and one marker for the wheat chromosome. The reason for occasional failure to amplify DNA from single chromosomes, which was observed previously [15] is not clear. One explanation is that a droplet with sorted chromosome lands on side wall of PCR tube and the

chromosome is excluded from the MDA reaction. The amount of chromosomal DNA in successfully amplified samples ranged from 0.3 to 1.7 µg DNA.

Chromosome specificity of sequence tagged site (STS) markers used in this work to identify individual

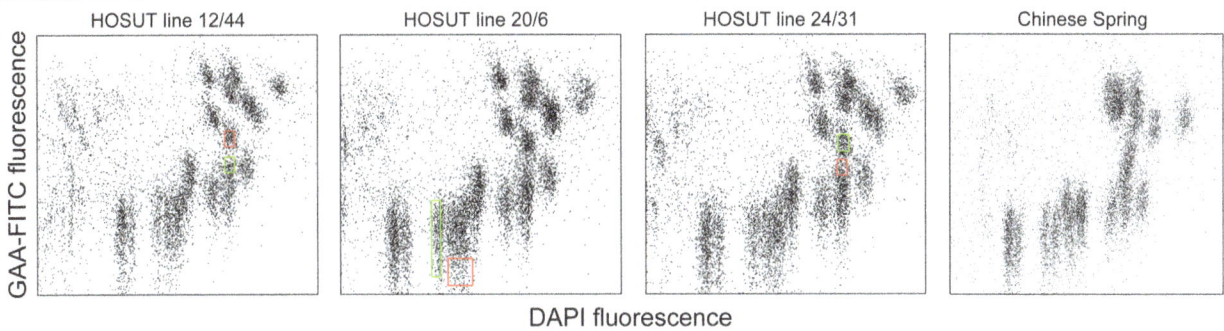

**Fig. 2** Bivariate flow karyotypes of three transgenic HOSUT lines of wheat obtained after the analysis of chromosomes with FITC-labelled (GAA)$_n$ microsatellites and stained by DAPI. The position of *red* and *green* regions used to sort particular chromosomes is indicated. The *green* sort gate was found to represent chromosomes carrying transgene. Chromosomes were flow-sorted also from the neighboring population delineated by *red* gate and were used as a control. Although the transgene-bearing chromosome should not be included in this region, the sorted population could potentially be contaminated with transgene-bearing chromosomes due to similarity in chromosome size and DNA content

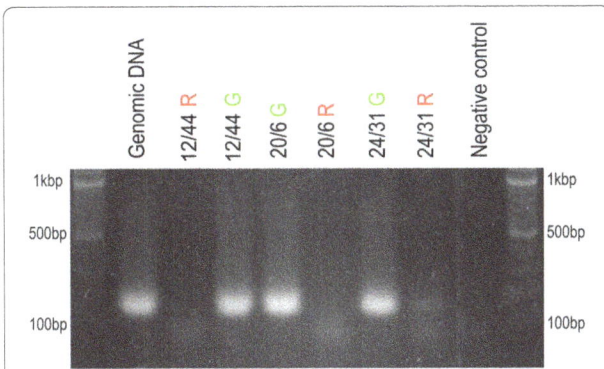

**Fig. 3** Agarose gel electrophoresis of PCR products obtained with primers for the transgene and DNA of chromosomes flow-sorted from three HOSUT lines using the *green* and *red* sort regions as shown in Fig. 1. The amplicon of HvSUT-RT (169 bp) was obtained with chromosomes sorted from the *green* sort region in all three HOSUT lines. When chromosomes were sorted from the *red* sort regions, no PCR amplification occurred for HOSUT 12/44 and HOSUT 20/6. However, a weak band was observed for HOSUT 24/31. Genomic DNA of the transgenic lines served as positive control

chromosomes was first tested using the euploid and corresponding nulli-tetrasomic lines of Chinese Spring (Additional file 2: Figure S2). The results confirmed that the markers were suitable for unambiguous identification of wheat chromosomes 1A, 4A, 5D and 7A. PCR analysis using both transgene- and chromosome-specific markers clearly confirmed chromosome location of transgenes as determined in the first part of this study. In case of HOSUT 24/31, where the location of the transgene was ambiguous, all four transgene markers were detected in DNA amplified from single chromosomes sorted from the green region (Fig. 5), and all four 4A-specific markers were also amplified in the same amplicons. None of the four 7A-specific markers was found in the same amplicons.

## Conclusions

Coupling PCR and FISH mapping using flow-sorted mitotic chromosomes as templates narrowed down the list of candidate chromosomes harboring the transgene to one or two chromosomes. PCR on DNA amplified

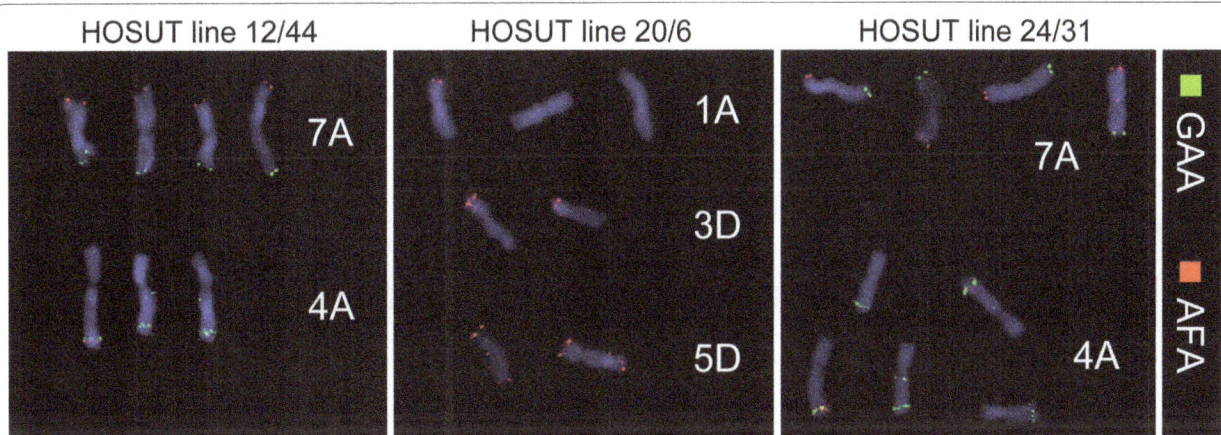

**Fig. 4** Representative images of chromosomes flow-sorted from three HOSUT lines using the *green* and *red* sort regions on bivariate flow karyotypes as shown in Fig. 2. FISH was done using probes for Afa-family (*red* signals) and GAA microsatellites (*green* signals). Chromosomes were counterstained with DAPI (*blue*)

**Table 1 PCR and FISH analysis of chromosomes sorted from each of the sort gates in three HOSUT lines**

| Transgenic line | Sort gate[a] | PCR result | Chromosomes identified FISH[c] |
|---|---|---|---|
| HOSUT 12/44 | Red | Negative | 4A (92.65 %) |
| | Green | Positive | 7A (90.90 %) |
| | | | 4A (4.45 %) |
| HOSUT 20/6 | Red | Negative | 1A (63.75 %) |
| | | | 3D (36.25 %) |
| | Green | Positive | 5D (94.66 %) |
| | | | 1A (5.33 %) |
| HOSUT 24/31 | Red | Semi-positive[b] | 7A (83.19 %) |
| | | | 4A (12.39 %) |
| | | | 2A (4.42 %) |
| | Green | Positive | 4A (97.30 %) |

[a] Sort gates delineated with green and red rectangles in Fig. 2

[b] A faint band was visible after agarose gel electrophoresis of PCR product

[c] More than 1000 chromosomes were examined in each sorted fraction in each line

from single flow-sorted chromosomes then unambiguously identified the chromosomes with the integrated transgene. If chromosome-specific PCR-based markers are available, mapping on single copy chromosomes could be an ultimate approach to assign single copy DNA sequences, including transgenes, to particular chromosomes. Moreover, the sequence assembly of amplicons

from the chromosome could allow detecting the position of transgene insertion, if enough sequence information on the chromosome is available. However the main purpose of this work was to assign a transgene to particular chromosomes. The approach presented here is currently applicable to more than 25 plant species, which include important cereals and legumes [14] where liquid suspensions of mitotic chromosomes suitable for flow cytometric sorting can be prepared.

## Methods

### Plant material

We used German winter wheat cultivar Certo (*Triticum aestivum* L., 2n = 6x = 42, genome formula AABBDD) and its three transgenic lines, HOSUT 12/44, HOSUT 20/6 and HOSUT 24/31. The transgenic lines contain a single copy of the HvSUT1-cDNA (1894 bp) fused to the barley HorB1 promoter (550 bp) and the barley HorB1 terminator (1663 bp) [12]. We also used euploid and nullisomic–tetrasomic (Nt1A1B, Nt7A7B, Nt5D5B) lines of hexaploid wheat cultivar Chinese Spring (obtained from NBRP-wheat) to confirm the specificity of PCR markers to particular wheat chromosomes.

### Flow cytometric chromosome sorting

Cell cycle synchronization and metaphase accumulation of root tip meristem cells was performed as described

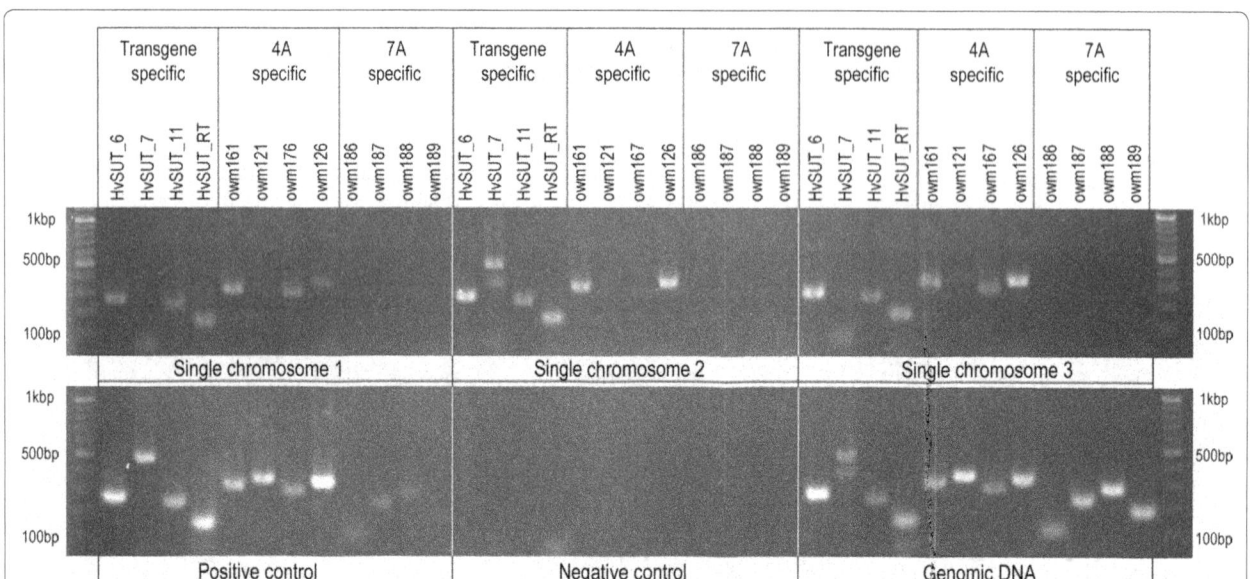

**Fig. 5** Agarose gel electrophoresis of PCR products obtained using DNA produced by multiple displacement amplification of three single chromosomes flow-sorted from the sort region representing chromosome 4A in HOSUT 24/31 line. PCR with primers for the four transgenes resulted in products of expected length. The same was true for the chromosome 4A-specific STS markers. Note that none of the chromosome 7A-specific markers was detected in the samples of single chromosome DNA. PCR with genomic DNA of HOSUT 24/31 as template detected both 4A and 7A chromosome-specific markers. PCR with the positive control (represented by 1000 chromosomes sorted from *green* sorting region and amplified) showed slight PCR bands of chromosome 7A, which reflects a minor contamination of the sorted chromosome 4A by chromosome 7A

previously [17], except for the formaldehyde fixation, which was shortened to 15 min. Isolated chromosomes were labelled by FISHIS (fluorescence in situ hybridization in suspension) using FITC-labeled GAA probe following the protocol of Giorgi et al. [18]. Flow cytometric analysis and sorting was done on BD FACSAria II high speed flow sorter equipped with 390 nm laser for DAPI excitation and 488 nm laser for FITC excitation. Sort gates were initially drawn on monovariate flow karyotypes of DAPI fluorescence (not shown) and subsequently on bivariate flow karyotypes of DAPI fluorescence versus GAA-FITC fluorescence as shown in Fig. 1.

### Fluorescence in situ hybridization (FISH)

For microscopic observations, 1000 chromosomes were sorted onto microscope slides from each of the sort regions. The slides were left to air-dry in the dark overnight. Then the preparations were used for FISH following the protocol of Kubaláková et al. [19] using a Cy5-labeled probe targeting Afa-family repeats, the

chromosomes were already labeled by a GAA microsatellite probe during the FISHIS procedure.

### PCR

PCR was done using primers specific for the HOSUT transgene and for markers specific for candidate wheat chromosomes (Table 2). Of the four HOSUT primers, three were designed in the HvSUT1 region (accession no. AJ272309) and one in the HorB1 terminator region (accession no. FN643080). Wheat chromosome-specific markers were designed by Primer3 based on the chromosome sequences from the International Wheat Genome Sequencing Consortium (IWGSC), while preventing the primers from amplifying the sequence from the homoeologous chromosomes. PCR conditions were set as follows: initial denaturation 95 °C for 3 min, 35 cycles of 30 s denaturation at 95 °C, annealing at 58–62 °C (see Table 2 for $T_a$ of the primer pairs) for 30 s and extension at 72 °C for 30 s, followed by final extension at 72 °C for 5 min. The amount of template DNA was 5 ng for each reaction. PCR

**Table 2  List of PCR primers for the HOSUT transgene construct and PCR primers for wheat STS markers on chromosomes 1A, 4A, 7A and 5D**

| Name | Target | Forward primer sequence | Reverse primer sequence | Amplicon size (bp) | Annealing temperature (°C) |
|------|--------|------------------------|------------------------|--------------------|----------------------------|
| HvSUT_6 | HOSUT1 | AGCGGCGGCGGTCACTGACTG | CCAAAGGACGACACCCCAGCC | 265 | 62 |
| HvSUT_7 | HorB1 terminator | ATTAATTCCTCCCCGACCCTGC | CAATGGAGACGGCGCGTGCAA | 471 | 62 |
| HvSUT_11 | HOSUT1 | GGCGGAACCCGCCGTGCAG | CCTGCGTCTTCCCCATCTGGAAGTA | 241 | 62 |
| HvSUT_RT | HOSUT1 | CGGGCGGTCGCAGCTCGCGTCTATT | CATACAGTGACTCTGACCGGCACACA | 169 | 62 |
| Owm121 | Chromosome 4A | ATTGCCGTCGCGAACTAGA | CGGGACGAGCTTGACGAT | 351 | 60 |
| Owm126 | Chromosome 4A | CCAGTCAGAAATTATTATGAACCTATC | CGCTGTCTCGAGATTGGAGT | 342 | 60 |
| Owm161 | Chromosome 4A | TTTTCAAGCAGGTTTTGTGC | TCACTTCTCTTCTTTGCGTTCA | 324 | 60 |
| Owm167 | Chromosome 4A | TTTTCTTGGTCAGTATAACCTGTTTTT | TGAGCAGAGAAAAATTTCCAAG | 285 | 60 |
| Owm174 | Chromosome 1A | GCATCCTAGTTTCTCTCTCAAGT | AACAAGATCACGAGCGAATTG | 157 | 58 |
| Owm175 | Chromosome 1A | AAACCCCTGATACTCATGCG | GTTTCTTGTCATTCATGTCACTTGT | 530 | 58 |
| Owm176 | Chromosome 1A | TTCCTGTCTGACTCCGCG | AACCACAACCGTCAACCG | 104 | 58 |
| Owm177 | Chromosome 1A | GTAGTCTGCTCCCGAGGAAT | GTCTCTAACCATACATCCATGAAGT | 192 | 58 |
| Owm178 | Chromosome 1A | CAACTTCTTCACATCCCGGAA | ATTTGGCCCTATGAGATATAATTACG | 306 | 58 |
| Owm179 | Chromosome 1A | ACACTGTGATACCTCTAGATGTATG | CACATTGCCTATAAATTCTAAAAGGTC | 425 | 58 |
| Owm180 | Chromosome 5D | CGGACGAGCAGCAGTACC | GCAGATCGGCATAAATTGAATGT | 292 | 58 |
| Owm181 | Chromosome 5D | GGAGGTGTTCTAGGTGTACTTACT | AGAGCAATGTCAGAAGTCATCG | 240 | 58 |
| Owm182 | Chromosome 5D | TCTCCACCTGCAGAGTCG | CATCAGGCCACAGTGTCAAT | 119 | 58 |
| Owm183 | Chromosome 5D | TGTCCACACATTTCCCGTATG | AGTGGTGGATGTGGTTGCT | 196 | 58 |
| Owm184 | Chromosome 5D | AGCATGCTCCCAAAGACTATTAC | GTTATGATGGTGGTAGCAATTTGA | 400 | 58 |
| Owm185 | Chromosome 5D | GTGAACCTATATGACATCTTACCGG | GGGGCAGTTGTCAAGTATTGC | 421 | 58 |
| Owm186 | Chromosome 7A | CTCTCTGTGGCCAATAGTGC | TCTATACCTCAACCCTACATCCA | 112 | 58 |
| Owm187 | Chromosome 7A | GGCCACGAATTCCACAAGTA | CTATCGATCAACCAACCATCCA | 229 | 58 |
| Owm188 | Chromosome 7A | GTACGAGTGCAGACAGTGTG | ACAATTAATTATACGCCCAGTTAAGC | 282 | 58 |
| Owm189 | Chromosome 7A | CGTGCTTTCTTCTTCCTCCG | GCAGGTTAGTTTCTTGTGGTTG | 185 | 58 |
| Owm190 | Chromosome 7A | CGCATGGACATTGTTCTAGTCA | GCACTTAGGCACGCTTGAG | 517 | 58 |
| Owm191 | Chromosome 7A | CGACGACATTAGGAATATGGGAT | TGCGTGTGGGTGTGCTTA | 402 | 58 |

products were run on 1.5 % agarose gels. PCR using 100 sorted chromosomes as template was conducted after a few freeze–thaw cycles to disintegrate the chromosomes and the initial denaturation step was prolonged for 7 min.

## Whole genome amplification of single chromosomes

DNA amplification of single chromosomes was performed by MDA using a GE Healthcare GenomiPhi V2 kit (GE Healthcare Life Sciences, Little Chalfont, UK) according to Cápal et al. [15]. Five individual chromosomes were flow-sorted into five 0.2 ml PCR tubes from green sort gates from each HOSUT line and their DNA amplified. The amplified DNA was evaluated on 1.5 % agarose gel, purified using magnetic beads (AMPure XP system, Beckman Coulter, Inc., Brea, CA, USA) and the concentration was measured by a spectrophotometer (NanoDrop, Thermo Fisher Scientific Inc., Waltham, MA, USA).

## Additional files

**Additional file 1: Figure S1.** Flow karyotypes (histograms of fluorescence intensity) obtained after the analysis of DAPI-stained chromosomes isolated from three transgenic lines and cv. Chinese Spring of common wheat. Flow karyotypes of the transgenic lines are indistinguishable from each other, and slightly differ in profiles of the major composite peaks from those of Chinese Spring.

**Additional file 2: Figure S2.** Verification of marker specificity. PCR with a full set of chromosome-specific wheat STS markers was performed using genomic DNA of cv. Chinese Spring and corresponding nullitetrasomic lines for chromosomes 1A, 4A, 5D and 7A. The markers, which resulted in amplification products only in Chinese Spring and not in the nullitetrasomic lines, were used in this study.

## Authors' contributions
The study was conceived and designed by TE, PC and JD, experiments were performed by PC, TE, M Ka, M Ku, JV, IMR and EK, manuscript was written by PC, TE, JD and WW. All authors read and approved the final manuscript.

## Author details
[1] Institute of Experimental Botany, Centre of the Region Haná for Biotechnological and Agricultural Research, Šlechtitelů 31, 78371 Olomouc, Czech Republic. [2] Faculty of Agriculture, Ryukoku University, 1-5 Yokotani, Seta Oe-cho, Otsu, Shiga 520-2194, Japan. [3] Leibniz Institute of Plant Genetics and Crop Plant Research (IPK) Gatersleben, Corrensstrasse 3, 06466 Stadt Seeland, Germany.

## Acknowledgements
We thank Zdeňka Dubská for technical assistance with flow cytometric chromosome sorting. This work was supported by the National Program of Sustainability (Award No. LO 2014) and the Czech Science Foundation (Award No. P501-12-G090).

## Competing interests
The authors declare that they have no competing interests.

## References
1. Ahmad P, Ashraf M, Younis M, Hu X, Kumar A, Akram NA, Al-Qurainy F. Role of transgenic plants in agriculture and biopharming. Biotechnol Adv. 2012;30:524–40.
2. Fischer RA. The importance of grain or kernel number in wheat: a reply to Sinclair and Jamieson. Field Crops Res. 2008;105:15–21.
3. Svitashev S, Anaiev E, Pawlowski WP, Somers DA. Association of transgene integration sites with chromosome rearrangements in hexaploid oat. Theor Appl Genet. 2000;100:872–80.
4. Kumar S, Fladung M. Gene stability in transgenic aspen (Populus). II. Molecular characterization of variable expression of transgene in wild and hybrid aspen. Planta. 2001;213:731–40.
5. Kooter JM, Matzke A, Meyer P. Listening to the silent genes: transgene silencing, gene regulation and pathogen control. Trends Plant Sci. 1999;4:340–7.
6. Svitashev KS, Somers DA. Characterization of transgene loci in plants using FISH: a picture is worth a thousand words. Plant Cell Tiss Org Cult. 2002;69:205–14.
7. Danilova TV, Friebe B, Gill BS. Development of a wheat single gene FISH map for analyzing homoeologous relationship and chromosomal rearrangements within the Triticeae. Theor Appl Genet. 2014;127(3):715–30.
8. Karafiátová M, Bartoš J, Kopecký D, Ma L, Sato K, Houben A, Doležel J. Mapping nonrecombining regions in barley using multicolor FISH. Chromosome Res. 2013;21:739–51.
9. Pedersen C, Zimny J, Becker D, Jähne-Gärtner A, Lörz H. Localization of introduced genes on the chromosomes of transgenic barley, wheat and triticale by fluorescence in situ hybridization. Theor Appl Genet. 1997;94:749–57.
10. Dong J, Pushpa K, Cervera M, Hall TC. The use of FISH in chromosomal localization of transgenes in rice. Methods Cell Sci. 2001;23:105–13.
11. Khrustaleva LI, Kik C. Localization of single-copy T-DNA insertion in transgenic shallots (Allium cepa) by using ultra-sensitive FISH with tyramide signal amplification. Plant J. 2001;25:699–707.
12. Weichert N, Saalbach I, Weichert H, Kohl S, Erban A, Kopka J, Hause B, Varshney A, Sreenivasulu N, Strickert M, Kumlehn J, Weschke W, Weber H. Increasing sucrose uptake capacity of wheat grains stimulates storage protein synthesis. Plant Physiol. 2010;152:698–710.
13. Saalbach I, Mora-Ramírez I, Weichert N, Andersch F, Guild G, Wieser H, Koehler P, Stangoulis J, Kumlehn J, Weschke W, Weber H. Increased grain yield and micronutrient concentration in transgenic winter wheat by ectopic expression of a barley sucrose transporter. J Cereal Sci. 2014;60:75–81.
14. Doležel J, Vrána J, Cápal P, Kubaláková M, Burešová V, Šimková H. Advances in plant chromosome genomics. Biotechnol Adv. 2014;32:122–36.
15. Cápal P, Blavet N, Vrána J, Kubaláková M, Doležel J. Multiple displacement amplification of the DNA from single flow-sorted plant chromosome. Plant J. 2015. doi:10.1111/tpj.13035.
16. Vrána J, Kubaláková M, Číhalíková J, Valárik M, Doležel J. Preparation of sub-genomic fractions enriched for particular chromosomes in polyploid wheat. Biol Plant. 2015;59:445–55.
17. Vrána J, Kubaláková M, Šimková H, Číhalíková J, Lysák MA, Doležel J. Flow sorting of mitotic chromosomes in common wheat (Triticum aestivum L.). Genetics. 2000;156:2033–41.
18. Giorgi D, Farina A, Grosso V, Gennara A, Ceoloni C, Lucretti S. FISHIS: fluorescence in situ hybridization in suspension and chromosome flow sorting made easy. PLoS One. 2013;8:e57994.
19. Kubaláková M, Kovářová P, Suchánková P, Číhalíková J, Bartoš J, Lucretti S, Watanabe N, Kianian SF, Doležel J. Chromosome sorting in tetraploid wheat and its potential for genome analysis. Genetics. 2005;170:823–9.

# 14

# A rapid phenotyping method for adult plant resistance to leaf rust in wheat

Adnan Riaz[1*], Sambasivam Periyannan[2], Elizabeth Aitken[3] and Lee Hickey[1]

## Abstract

**Background:** Leaf rust (LR), caused by *Puccinia triticina* and is an important disease of wheat (*Triticum aestivum* L.). The most sustainable method for controlling rust diseases is deployment of cultivars incorporating adult plant resistance (APR). However, phenotyping breeding populations or germplasm collections for resistance in the field is dependent on weather conditions and limited to once a year. In this study, we explored the ability to phenotype APR to LR under accelerated growth conditions (AGC; i.e. constant light and controlled temperature) using a method that integrates assessment at both seedling and adult growth stages. A panel of 21 spring wheat genotypes, including disease standards carrying known APR genes (i.e. *Lr34* and *Lr46*) were characterised under AGC and in the field.

**Results:** Disease response displayed by adult wheat plants grown under AGC (i.e. flag-2 leaf) was highly correlated with field-based measures ($R^2 = 0.77$). The integrated method is more efficient—requiring less time, space, and labour compared to traditional approaches that perform seedling and adult plant assays separately. Further, this method enables up to seven consecutive adult plant LR assays compared to one in the field.

**Conclusion:** The integrated seedling and adult plant phenotyping method reported in this study provides a great tool for identifying APR to LR. Assessing plants at early growth stages can enable selection for desirable gene combinations and crossing of the selected plants in the same plant generation. The method has the potential to be scaled-up for screening large numbers of fixed lines and segregating populations. This strategy would reduce the time required for moving APR genes into adapted germplasm or combining traits in top crosses in breeding programs. This method could accelerate selection for resistance factors effective across diverse climates by conducting successive cycles of screening performed at different temperature regimes.

**Keywords:** Wheat, Leaf rust, *Puccinia triticina*, Adult plant resistance, Accelerated growth conditions, Disease screening, Wheat breeding

## Background

Wheat provides more than 20 % of the calorific intake for almost two-thirds of the human population [1]. With an expected global population of 9–10 billion by the year 2050, world food security is paramount. *Puccinia triticina* f. sp. *tritici*, which causes leaf rust (LR), is regarded one of the most geographically widespread disease of wheat and can incur yield losses ranging 10–70 % [2, 3]. It results in reduction of kernels per head, lower kernel weight, degradation in grain quality and increased costs associated with chemical control [4, 5]. In Australia, wheat diseases, including rusts, cause an estimated average annual loss of almost $913 million to the wheat industry [6]. Among the various control methods, the most profitable and sustainable disease minimization strategy is the deployment of genetically resistant cultivars [7].

To date, research around the world has resulted in designation of 73 genes for resistance to LR (i.e. *Lr*), which have been characterised and mapped to chromosomal locations [8]. Genetic resistance is broadly classed into two forms: seedling and adult-plant resistance (APR). Seedling resistance, or 'all stage resistance' (ASR), is typically expressed at all growth stages,

*Correspondence: a.riaz@uq.edu.au
[1] Queensland Alliance for Agriculture and Food Innovation, The University of Queensland, St Lucia, QLD 4072, Australia
Full list of author information is available at the end of the article

conferred by a single 'major effect' gene often associated with a hypersensitive response and is often race specific. On the other hand, APR is typically best expressed in adult plants and often polygenic in nature, controlled by multiple 'minor effect' genes that may influence factors such as pustule size, infection frequency and latent period, thus commonly referred to as 'slow rusting' genes [9, 10]. While APR is often non-race specific, there are exceptions where some genes provide race-specific resistance (e.g. *Lr13* and *Lr37* [10, 11]) and confer a hypersensitive response (e.g. *Lr48* and *Lr49* [12]). Notably, some APR genes have been deployed for almost 100 years, such as *Sr2* and *Lr34*, which continue to provide resistance to stem rust (SR) and LR, respectively. Three well-characterized APR genes are now available to wheat breeders that appear to convey non-race specific resistance to LR (i.e. *Lr34*, *Lr46* and *Lr67*), for which useful DNA markers are also available [13, 14]. However, additional sources of resistance are needed for stacking or pyramiding in new cultivars, which will serve to protect these highly valuable genes against the rapidly evolving nature of *P. triticina*.

APR to LR is typically identified by phenotyping wheat plants at the seedling stage in the glasshouse, then subsequently evaluating adult plants in the field [10]. However, the accuracy of phenotyping in the field can be compromised by environmental factors that influence the expression of APR, such as weather patterns, inoculum pressure, sequential infection, differences in plant maturity and the presence of other diseases [15]. Further, expression of LR resistance in wheat is sensitive to temperature [16], resulting in variability across environments or years of testing [17]. Some studies have successfully evaluated APR to foliar pathogens in cereals grown under glasshouse or controlled environmental conditions (CEC) [15, 18, 19]. A key advantage is that environmental factors, such as temperature and light, can be controlled. Artificial lighting can also be used to impose an extended photoperiod or constant light to accelerate the growth of wheat plants. A plant management system providing accelerated growth conditions (AGC) could be used to speed up disease screening and plant selection.

In this study, we investigated the ability to rapidly phenotype APR to LR in wheat grown under AGC (i.e. constant light and controlled temperature). Using a panel of 21 spring wheat genotypes we compared LR response displayed by adult plants grown under AGC to levels displayed by adult plants grown in the field. We discuss opportunities to exploit this rapid phenotyping method to accelerate research and wheat breeding efforts to develop rust resistant wheat cultivars.

## Methods

### Plant materials

A panel comprising 21 spring wheat genotypes (Table 1) was used to generate a protocol for phenotyping resistance to LR in wheat grown under AGC. The panel comprised a selection of standards, cultivars and breeding lines from Australia, the International Center for Agriculture Research in the Dry Areas (ICARDA), and the International Maize and Wheat Improvement Center (CIMMYT).

### Rust screening: seedling stage

The panel was evaluated for resistance to LR at the seedling stage in a glasshouse at The University of Queensland, St Lucia, Queensland, Australia. Seeds were imbibed for 24 h at room temperature and were placed in a refrigerator (4 °C) for 48 h to encourage synchronous germination across genotypes. Germinated seeds were transplanted into 140 mm ANOVApot® pots filled with a potting media consisting of composted pine bark fines (0–5 mm) (70 %) and coco peat (30 %) with a pH ranging 5.5–6.5. Slow release Osmocote® fertilizer was applied at a rate of 2 g per pot. Each pot contained four different positions (i.e. positions 1–4 clockwise from the pot tag), where each position contained four germinated seeds of the same genotype clumped together. Each genotype was replicated three times in a completely randomized design. Plants were grown at a temperature regime of 22/17 °C (day/night) and a natural 12 h diurnal photoperiod. After 10 days, (i.e. two-leaf stage) plants were inoculated with *P. triticina* pathotype (*pt*) 104-1,2,3,(6),(7),11,13. This pathotype evolved from pathotype 104-1,2,3,(6),(7),11 via a single step mutation on wheat carrying the resistance gene *Lr24* and was first reported in Australia in 2000 [20]. It currently occurs in wheat production regions throughout the east coast of Australia. The rust isolate used in this study was developed using a single spore culture technique and spores increased using susceptible wheat cultivar Morocco. The inoculum was prepared by suspending urediniospores in light mineral oil (Isopar 6) at a rate of 0.005 g/ml. Inoculum at the concentration of $6 \times 10^5$ spores/ml was applied to the leaves of wheat plants using an air brush (IWATA power jet lite®). Plants were then lightly misted with deionized water and placed in a dew chamber maintained at 100 % humidity using an ultrasonic fogger. After 18 h of incubation, plants were removed from the dew chamber and returned to the glasshouse for subsequent disease development. Twelve days post-inoculation seedlings were assessed for infection type (IT) using the 0–4 Stakman scale [21]. Genotypes that displayed an IT of <3 were considered resistant.

**Table 1** Name, pedigree, breeding program and leaf rust resistance genes present in 21 spring wheat genotypes

| Genotypes | Pedigree | Type | Resistance genes | | Breeding program | Source[a] |
|---|---|---|---|---|---|---|
| | | | Seedling | APR | | |
| Thatcher | MARQUIS/IUMILLO DURUM//MARQUIS/KANRED | Cultivar | –[b] | – | North America | [35] |
| Avocet | THATCHER-*AGROPYRON ELONGATUM* TRANSLOCA-TION/3* PINNACLE//WW15/3/EGRET | Cultivar | – | *Lr13* | Australia | [36] |
| Avocet + *Lr34* | AVOCET NEAR ISOGENIC LINE *LR34* | Near isogenic line | – | *Lr34* | Near Isogenic Line | [37] |
| Avocet + *Lr46* | AVOCET NEAR ISOGENIC LINE *LR46* | Near isogenic line | – | *Lr46* | Near Isogenic Line | [37] |
| Dharwar Dry | DWR39/C306//HD2189 | Cultivar | – | – | India | – |
| Drysdale | HARTOG*3/QUARRION | Cultivar | *Lr1* | *Lr13* | Australia | [27] |
| Janz | 3AG3/4*CONDOR//COOK | Cultivar | *Lr24* | *Lr34* | Australia | [27] |
| Lang | QT3765/SUNCO | Cultivar | *Lr24* | *Lr34* | Australia | [27] |
| EGA Gregory | PELSART/2*BATAVIA | Cultivar | *Lr1, Lr3a, Lr23* | *Lr13, Lr34* | Australia | [27] |
| EGA Wylie | QT2327/COOK//QT2804 | Cultivar | *Lr17a* | *Lr34* | Australia | [27] |
| FAC10-16-1 | 10CB-F/W234 | Breeding line | – | – | ICARDA | – |
| Mace | WYALKATCHEM/STYLET//WYALKATCHEM | Cultivar | *Lr23* | *Lr13, Lr37* | Australia | [27] |
| RIL114 | UQ01484/RSY10//H45 | Breeding line | – | – | Australia | – |
| SB062 | SERI M82/BABAX | Breeding line | – | – | Australia | – |
| Scout | SUNSTATE/QH71-6//YITPI | Cultivar | *Lr1* | *Lr37* | Australia | [27] |
| Suntop | SUNCO/2*PASTOR//SUN436E | Cultivar | – | – | Australia | – |
| SeriM82 | KAVKAZ/(SIB)BUHO//KALYANSONA/BLUEBIRD | Breeding line | *Lr23, Lr26* | – | CIMMYT | – |
| Zebu | – | Cultivar | *Lr26* | – | CIMMYT | [27] |
| ZWB10-37 | TACUPETOF2001/BRAMBLING//KIRITATI | Breeding line | – | – | CIMMYT | – |
| ZWW10-128 | ESDA/KKTS | Breeding line | – | – | CIMMYT | – |
| ZWW10-50 | ONIX/4/MILAN/KAUZ//PRINIA/3/BAV92 | Breeding line | – | – | CIMMYT | – |

[a] Study reporting the status of leaf rust resistance genes

[b] A dash (–) indicates data is unavailable or unknown

## Rust screening: adult plant stage

In total, three adult plant experiments were conducted using the panel. Two phenotyping experiments, namely, "adult plant integrated" and "adult plant independent" were conducted under AGC, while phenotyping in the field was conducted in a disease screening nursery.

### Adult plant experiment 1: integrated method under AGC

Following assessment of disease response at the seedling stage (as describe above), the plants were transferred to a fully-enclosed temperature controlled growth facility (dimensions 5 m × 6 m). The growth facility is fitted with 20 low-pressure sodium vapor lamps (400 watt each) generating 400–550 µmol $M^{-2}$ $S^{-1}$ photosynthetically active radiation (PAR) at pot height and 900 µmol $M^{-2}$ $S^{-1}$ at adult plant height (i.e. about 45 cm above pot level). AGC was achieved by adopting constant (i.e. 24 h) light [19] and a 12 h cycling temperature regime of 22/17 °C. Pots were positioned on a bench according to a completely randomized design in a stainless steel tray (240 × 90 × 10 cm). Plants were grown for 2 weeks under AGC, and then re-inoculated with a suspension of LR urediniospores (*pt* 104-1,2,3,(6),(7),11,13), as

described above. Prior to inoculation, the developmental growth stage (GS) was recorded for each plant using the Zadoks decimal code scoring system [22]. Twelve days post-inoculation IT was recorded for different leaves (i.e. flag, flag-1 and flag-2) on the primary/main tiller of each plant using the 0–4 Stakman scale. Genotypes displaying an IT of <3 were considered resistant.

### Adult plant experiment 2: independent method under AGC

As a control, a new batch of plants were sown for the panel and grown from day one under AGC. Environmental conditions and experimental design was consistent with adult plant experiment 1 (above). Three weeks after sowing, the majority of genotypes achieved the adult plant stage and were inoculated with *pt* 104-1,2,3,(6),(7),11,13, as outlined above. Prior to inoculation, the GS for all plants was recorded using the Zadoks scale. Twelve days later, plants were assessed for IT using the Stakman scale.

### Adult plant experiment 3: in the field

The panel of wheat genotypes was evaluated for response to LR in the field at Redlands Research Facility,

Queensland, Australia, from July to October 2014. Six seeds of each genotype was sown as un-replicated hill plots. The susceptible genotype Morocco was used as a disease spreader in the field nursery, where two rows of Morocco were sown between each bay compromising two rows of hill plots. LR epidemics were initiated by transplanting Morocco seedlings infected with *pt* 104-1,2,3,(6),(7),11,13 (as outlined above) into the field among the spreader rows about 5 weeks after sowing. The LR epidemic was promoted with sprinkler irrigation applied in the late evenings when temperatures were favorable for infection and high humidity and low winds at night were expected. Once the epidemic had sufficiently developed on LR standards to allow a clear differentiation between susceptible and resistant genotypes, disease response was assessed on a whole plot basis using the modified Cobb scale [23]. Multiple disease assessments were conducted from late tillering/stem elongation to early grain filling (i.e. 70, 77, 86 and 96 days after sowing; DAS). Host response and disease severity data were used to calculate coefficient of infection (CI), as per Loegring et al. [24]. Genotypes that displayed a LR response from resistant (R) to moderately resistant-moderately susceptible (MRMS) were considered resistant.

## Statistical analysis

For experiments performed under controlled conditions, LR response was evaluated using the 0–4 Stakman scale, which encompasses both numbers (e.g. 0, 1...4) and symbols (e.g. ;, +). This data was converted to the 0–9 scale, where 0 = immune and 9 = very susceptible, using a conversion table [25]. The IT were converted as follows: 0;, ;n, ;, 1−, 1, 1+, 2−, 2, 2+, 2++, 3−, 3 3+, 3++ and 4 were coded as 0, 0.5, 1, 2.5, 3, 3.5, 4, 5, 6, 6.5, 7, 8, 8.5 and 9 respectively. For heterogeneous ITs, each score was converted individually to the 0–9 scale and the average calculated. The converted datasets were then used for further statistical analysis.

Data analysis was performed using GenStat 17.1 © 2000–2015 VSN International Ltd. Analysis of variance (ANOVA) was performed using the converted data for experiments including; seedling, adult plant integrated and adult plant independent. Mean disease response and standard error means (SEM) for each genotype were calculated for comparison of disease reactions.

Regression analyses were performed to investigate the correlation between phenotypes observed for the different experiments and to determine which leaf (i.e. flag, flag-1 and flag-2) under AGC provided the best estimate for LR response in the field for each disease assessment (i.e. 70, 77, 86 and 96 DAS). For the field dataset, CI values obtained from the un-replicated hill plots were used for regression analyses. The CI values were divided by 10

to convert to the 0–9 scale. The converted scores were used in comparison of mean LR response and principle component analysis. To investigate trends in disease response displayed by genotypes across multiple experiments, a principle component analysis (PCA) was performed and results visualized in the form of a biplot. This was performed using the following phenotype datasets: (1) seedling, (2) adult plant integrated, (3) adult plant independent, and (4) adult plant in the field (i.e. fourth assessment at 96 DAS). The disease response for flag-2 was used for both adult plant experiments conducted under AGC.

## Results

### Rust screening: seedling stage

Of the 21 spring wheat genotypes in the panel, 8 displayed susceptibility, while 13 displayed resistance to LR pathotype 104-1,2,3,(6),(7),11,13 at the seedling growth stage (Fig. 1). Thatcher, Avocet, Avocet + *Lr34*, Avocet + *Lr46*, Dharwar dry, Drysdale, Lang and Janz displayed susceptibility with characteristic symptoms of large uredia without chlorosis (i.e. mean disease responses ranging 7–9; Fig. 1). The susceptible standard, Thatcher, lacks effective LR resistance genes and displayed a mean disease response of 9.0. Notably, Avocet carries a race specific APR gene *Lr13* [26] and displayed seedling susceptibility (9.0; Fig. 1). The Indian cultivar Dharwar dry, previously uncharacterized for LR resistance genes, also displayed susceptibility (8.0). Drysdale carries *Lr1* (Table 1), which is ineffective against the pathotype used in this study [27] and displayed a susceptible response (8.0; Fig. 1). Janz and Lang displayed susceptibility at the seedling stage (i.e. 8.0; Fig. 1); both genotypes carry *Lr24* and *Lr34* (Table 1). The seedling gene *Lr24* is ineffective against *pt* 104-1,2,3,(6),(7),11,13 [20], whereas *Lr34* is an APR gene and best expressed at adult plant growth stage [13]. Based on the Stakman scale, the IT of seedling susceptible genotypes range from 3 to 4 (see Additional file 1: Table S1).

EGA Gregory carries *Lr1*, *Lr3a*, *Lr13*, *Lr23* and *Lr34* (Table 1) and displayed a moderately resistant (MR) response (2.3; Fig. 1). The seedling resistance displayed by EGA Gregory was likely due to *Lr13*, as both *Lr1*, *Lr3a* and *Lr23* are ineffective against the pathotype. The MR response displayed by Mace (1.5; Fig. 1) was also likely due to *Lr13* and *Lr37* (Table 1). *Lr13* and *Lr37* are APR genes and are effective against the pathotype used in this study (Table 1). Previous studies have reported early expression of *Lr13* at the seedling stage [28]. Scout carries *Lr1* and *Lr37* (Table 1), where *Lr1* is ineffective against this pathotype, while *Lr37* is effective. Scout displayed a MR response (1.5) at the seedling stage, which could be due to an uncharacterized seedling resistance or early

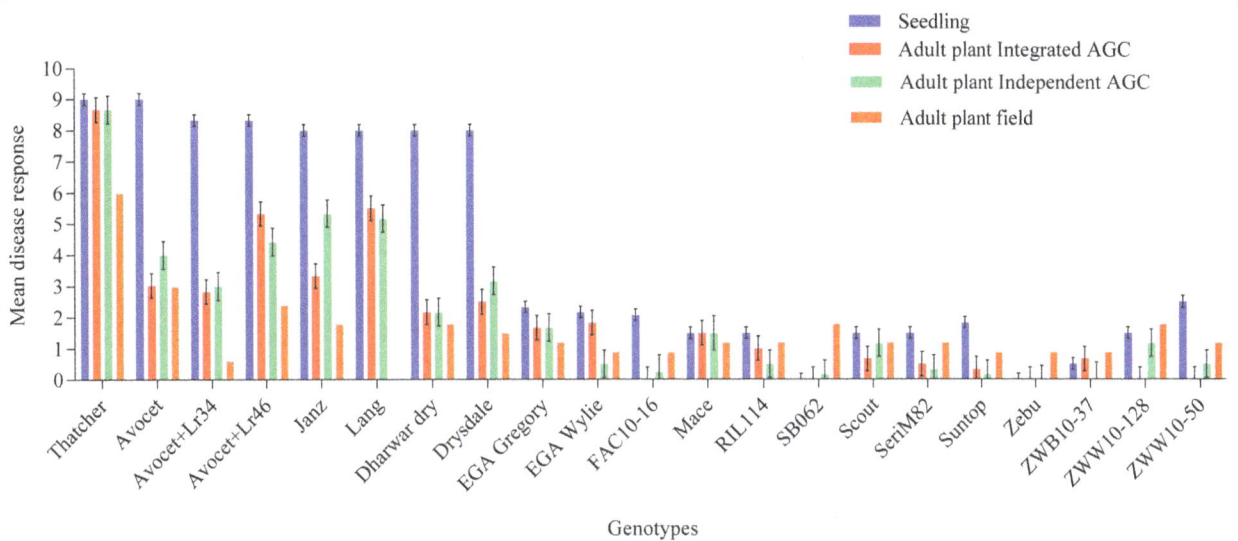

**Fig. 1** Mean leaf rust response for the panel of 21 spring wheat genotypes evaluated in the following experiments: seedling (standard glasshouse), adult plant integrated and adult plant independent under accelerated growth conditions (AGC), and in the field. The disease response for the seedling and adult plant AGC experiments was collected using the 0–4 scale and converted to the 0–9 scale (displayed). Whereas, the disease response in the field was collected using the modified Cobb scale, which was used to calculate coefficient of infection, and was converted to the 0–9 scale (displayed)

expression of *Lr37* at the seedling stage (Fig. 1) [29]. EGA Wyile carries *Lr17a* and *Lr34* (Table 1) and displayed a MR response (2.2; Fig. 1), as the pathotype used in this study is avirulent on *Lr17a*. SeriM82 and Zebu carry *Lr26* and both displayed a highly resistant response (1.5 and 0.0, respectively; Table 1 and Fig. 1). The previously, uncharacterized ICARDA line (FAC10-16-1) displayed a MR response (2.1; Fig. 1). Other genotypes previously uncharacterised for LR resistance genes, including RIL114, Suntop, SB062, ZWB10-37 and ZWW10-128 depicted high levels of resistance with mean disease response ranging 0–1.5 (Fig. 1). Based on the Stakman scale, the IT of the seedling resistant genotypes ranged from 0; to 12+; (see Additional file 1: Table S1).

### Rust screening: adult stage under AGC

In both adult plant experiments performed under AGC (i.e. integrated and independent), 20 of the 21 genotypes in the panel displayed varying levels of resistance (Fig. 1). In both experiments, Thatcher displayed a very susceptible (VS; 9.0) response with urediniospores freely sporulating on leaves (Fig. 1). Avocet displayed a resistant-moderately resistant (RMR) response with a mean disease response ranging 3–4 (Fig. 1). As mentioned earlier, Avocet carries race specific APR gene *Lr13*, which is effective against the pathotype used in this study. In the Avocet background, resistance to LR was slightly enhanced with the addition of *Lr34* and *Lr46* (i.e. *Avocet + Lr34* and *Avocet + Lr46*), which are considered

multi-resistance APR genes (Fig. 1). Avocet + *Lr34* displayed a RMR response with mean disease response ranging 2.8–3.0 and Avocet + *Lr46* displayed a MR response, ranging 4.4–5.3 in the adult plant independent and integrated experiments, respectively. On the Stakman scale, the IT displayed by Avocet + *Lr34* and Avocet + *Lr46* ranged; n12-(independent) to 12-(integrated), where pustules were smaller in comparison to Avocet and some necrosis in case of *Lr34* (see Additional file 1: Table S1). The Indian cultivar Dharwar dry displayed a resistant response in both AGC experiments (Fig. 1). Dharwar dry has not been previously characterized for rust resistance genes, thus the underlying genes are unknown. Drysdale carries *Lr1* along with race specific APR *Lr13* and displayed resistance (Table 1; Fig. 1). Both Janz and Lang carry *Lr24* and *Lr34* in combination (Table 1) however *Lr24* was not effective against the pathotype used in this study. These genotypes displayed a MRMS response, likely due to expression of APR gene *Lr34* (Fig. 1). The mean disease response for Janz and Lang was 3.3 and 5.5 in adult plant integrated experiment, respectively, and displayed similar responses in the adult plant independent experiment (i.e. 5.3 and 5.2, respectively; Fig. 1). EGA Gregory (1.7) and Mace (1.5) displayed a resistant response in both AGC experiments (Fig. 1). EGA Gregory carries *Lr1*, *Lr3a*, *Lr13*, *Lr23* and *Lr34* and Mace carries *Lr1*, *Lr23*, and *Lr37* (Table 1). The LR *pt* 104-1,2,3,(6),(7),11,13 is virulent on both *Lr1*, *Lr3a* and *Lr23*, but avirulent on APR genes *Lr13*, *Lr34* and

*Lr37.* Thus, resistance displayed at adult growth stages by EGA Gregory and Mace is likely a combination of these genes. Scout displayed resistance (1.5) (Fig. 1), most likely attributable to *Lr37* (Table 1). EGA Wylie displayed a highly resistant (HR) response in the integrated (1.8) and independent (0.5) AGC experiments (Fig. 1). This was most likely a result of the combined effect of seedling gene *Lr17a* and APR gene *Lr34* (Table 1). SeriM82 depicted a HR response in AGC experiments (Fig. 1), most likely due to the presence of seedling gene *Lr26* (Table 1). Genotypes previously uncharacterised for LR resistance genes (including SB062, RIL114, Suntop, Zebu, ZWW10-50, ZWW10-37, ZWW10-128 and FAC10-16-1) displayed high levels of resistance in AGC experiments (Fig. 1), indicating effective resistance to the pathotype used in this study. The detailed IT for these genotypes is provided in Additional file 1: Table S1. Overall, comparison of datasets from the integrated and independent experiments performed under AGC revealed only minor differences in infection and response types displayed by the panel of genotypes. Genotypes either displayed the same response or it varied within only one response type across both experiments. For instance, Drysdale displayed a RMR response in the independent experiment, but displayed R response in the integrated experiment (Fig. 1; Additional file 1: Table S1). The GS of plants evaluated under AGC ranged between GS25-45 and GS23-43 (i.e. tillering to booting stage) for the integrated and independent experiments, respectively (Table 2).

### Rust screening: in the field

All genotypes in the panel displayed varying levels of resistance to LR, with the exception of Thatcher, which consistently displayed a susceptible response (60 S; Additional file 1: Table S1). Avocet displayed a MRR response for the first three disease assessments; however on the fourth assessment, Avocet displayed a 50 MRMS response (Additional file 1: Table S1). In the Avocet background, the APR gene *Lr34* (i.e. Avocet + *Lr34*) displayed a 20 MRR response, while Avocet + *Lr46* displayed a MRMS response (40 MRMS; Additional file 1: Table S1). Dharwar dry displayed a MRMS response (30 MRMS), likely due to the presence of uncharacterised APR gene(s) (see Additional file 1: Table S1). Drysdale displayed a MRR response in the field, likely due to race specific APR *Lr13* (50 MRR). Janz carries *Lr24* and *Lr34* in combination and displayed the MRMS response (30 MRMS). As the pathotype used in this study is virulent on *Lr24*, the resistance displayed by Janz is likely due to *Lr34* (see Additional file 1: Table S1). CIMMYT lines (ZWW10-128 and SB062) both displayed a MRR response in the first three disease assessments, however, on the fourth assessment, each was considered MRMS (30 MRMS). ICARDA breeding line FAC10-16-1 was considered RMR (30 RMR)

**Table 2 Zadoks growth stages for the panel of 21 spring wheat genotypes at inoculation under accelerated growth conditions**

| Genotypes | Growth stage at inoculation | |
|---|---|---|
| | Adult plant integrated | Adult plant independent |
| Thatcher | 31 | 37 |
| Avocet | 33 | 43 |
| Avocet + *Lr34* | 34 | 41 |
| Avocet + *Lr46* | 39 | 41 |
| Dharwar dry | 37 | 31 |
| Drysdale | 37 | 25 |
| Janz | 32 | 31 |
| Lang | 31 | 31 |
| EGA Gregory | 30 | 25 |
| EGA Wylie | 32 | 25 |
| FAC10-16-1 | 33 | 25 |
| Mace | 30 | 25 |
| RIL114 | 45 | 41 |
| SB062 | 32 | 26 |
| Scout | 37 | 25 |
| SeriM82 | 33 | 37 |
| Suntop | 39 | 37 |
| Zebu | 28 | 26 |
| ZWB10-37 | 30 | 31 |
| ZWW10-50 | 37 | 26 |
| ZWW10-128 | 37 | 26 |

in the field. Other genotypes, such as EGA Gregory, EGA Wyile, Mace, Scout, RIL114, Suntop, Zebu, ZWW10-50, and ZWW10-37, displayed high levels of resistance (i.e. MRR) in the field with mean disease response ranging 30–40 MRR (see Additional file 1: Table S1). Lang failed to germinate in the field. The detailed host response and disease severity data is provided in Additional file 1: Table S1.

### Adult plant assessment under AGC is predictive of field response

Based on regression analyses, the LR response for different leaves showed very good correspondence across the two adult plant AGC experiments: $R^2 = 0.90$ (flag), 0.88 (flag-1) and 0.96 (flag-2). Despite all leaves showing good correspondence, the flag-2 leaf was considered to provide the most consistent LR response across AGC experiments. Regression analysis was also performed using data from the adult plant integrated AGC experiment and the field. The highest correlation was found for the response displayed by the flag-2 leaf versus the fourth (final) disease assessment in the field ($R^2 = 0.77$; Table 3). Correlations for the other leaves (flag and flag-1) corresponding with the four disease assessments ranged between 0.43–0.57 and 0.63–0.76, respectively (Table 3).

**Table 3** Results from regression analysis (R² values) for the panel of 21 spring wheat genotypes evaluated for leaf rust response in the adult plant integrated experiment versus the field

| Leaf number | Number of observations (n) | Days after sowing (DAS) | | | |
|---|---|---|---|---|---|
| | | 70 | 77 | 86 | 96 |
| Flag | 15 | 0.55 | 0.43 | 0.51 | 0.57 |
| Flag-1 | 19 | 0.76 | 0.63 | 0.71 | 0.73 |
| Flag-2 | 19 | 0.76 | 0.60 | 0.74 | 0.77 |

Regression analysis was performed for the disease response displayed by each leaf under accelerated growth conditions (i.e. Flag, Flag-1 and Flag-2) in comparison to the field response observed for each of the four assessment dates (i.e. 70, 77, 86 and 96 days after sowing, DAS)

Results from PCA displayed in the biplot (Fig. 2) revealed a high correlation between both adult plant experiments conducted under AGC, where the adult plant integrated experiment appeared to be slightly more correlated to the field disease response. The field response was moderately correlated with the adult plant independent experiments performed under AGC (Fig. 2). Notably, only a weak correlation was observed between field and seedling response (Fig. 2).

## Discussion

This study presents a novel method that permits rapid phenotyping for APR to LR in wheat by exploiting AGC to speed up plant development and involves two sequential inoculations to detect APR. Characterization of a panel of 21 wheat genotypes revealed that the LR response displayed under AGC was indicative of levels expressed by adult plants grown in the field. Phenotyping for APR to LR can be completed within just 7 weeks and performed all-year-round, thus provides a useful tool to accelerate breeding and research aiming to develop rust resistant cultivars.

### Detection of APR to LR under AGC
Of the 21 spring wheat genotypes evaluated, 7 were determined to carry APR to LR, including; Avocet, Avocet + *Lr34*, Avocet + *Lr46*, Janz, Lang, Drysdale and Dharwar dry. These genotypes were considered susceptible in the seedling experiment, but displayed resistance in adult plant experiments. Genotypes known to carry APR genes, in particular *Lr13*, *Lr34* and *Lr46*, consistently displayed resistance at the adult pant stage under AGC—similar to levels displayed in the field. For instance, both Janz and Lang carry seedling gene *Lr24* and APR gene *Lr34* in combination; however *Lr24* is not effective against the pathotype used in this study. Therefore, these genotypes displayed a susceptible response in the seedling experiment, but a MRMS response under

AGC at the adult plant stage, likely due to expression of *Lr34*. In some genotypes, the expression of *Lr34* was likely masked by the presence of effective seedling resistance genes, such as *Lr13* in EGA Gregory and *Lr17a* in EGA Wylie. Another good example of APR expression under AGC was observed for Avocet and the Avocet near-isogenic lines for *Lr34* (i.e. Avocet + *Lr34*) and *Lr46* (i.e. Avocet + *Lr46*). Notably, Avocet carries race specific APR gene *Lr13*, which is effective against the pathotype used in this study. The RMR response displayed by Avocet indicated that *Lr13* was successfully detected in the adult plant AGC experiments. In the Avocet background (*Lr13*), the addition of *Lr34* and *Lr46* enhanced the levels of resistance displayed in the adult plant experiments. This indicates the additive effect of APR genes can be detected under AGC. However, to detect the effectivity of the APR against different races the developed method can also be applied by conducting multiple screens using different pathotypes.

### Disease response under AGC is related to field-based measures
The GS of plants evaluated under AGC ranged between tillering to booting stage at time of inoculation with *P. triticina* and plants displayed adult plant phenotypes. This aligns well with previous studies on wheat that report early expression of APR to YR at mid-tillering growth stages in the field [30] and at the stem elongation stage in plants grown under controlled environment [15]. Regression analyses for the panel revealed that the flag-2 leaf expressed levels of APR most similar to those observed in the field. The upper-most infected leaf (i.e. flag leaf) displayed increased susceptibility to the pathogen in comparison to lower leaves. Thus, it appears APR is best expressed in 'older' leaves (that are more aged) compared to 'younger' leaves.

In the field, the inoculum pressure fluctuates due to infection cycles of rust urediniospores and weather conditions. One of the advantages of phenotyping under AGC is the application of inoculum can be controlled. It might be expected that the inoculum concentration applied under AGC using a single inoculation would correlate better with disease assessment performed early in the season (i.e. low disease pressure) as opposed to late in the season (i.e. high disease pressure). However, our results under AGC correlated well with measurements early in the season (i.e. 70 DAS) and late in the season (i.e. 96 DAS). It is feasible that phenotyping based on IT on a single leaf using a controlled single inoculation is indicative of factors important for reducing overall disease severity in the field under polycyclic conditions; such as pustule size and infection frequency.

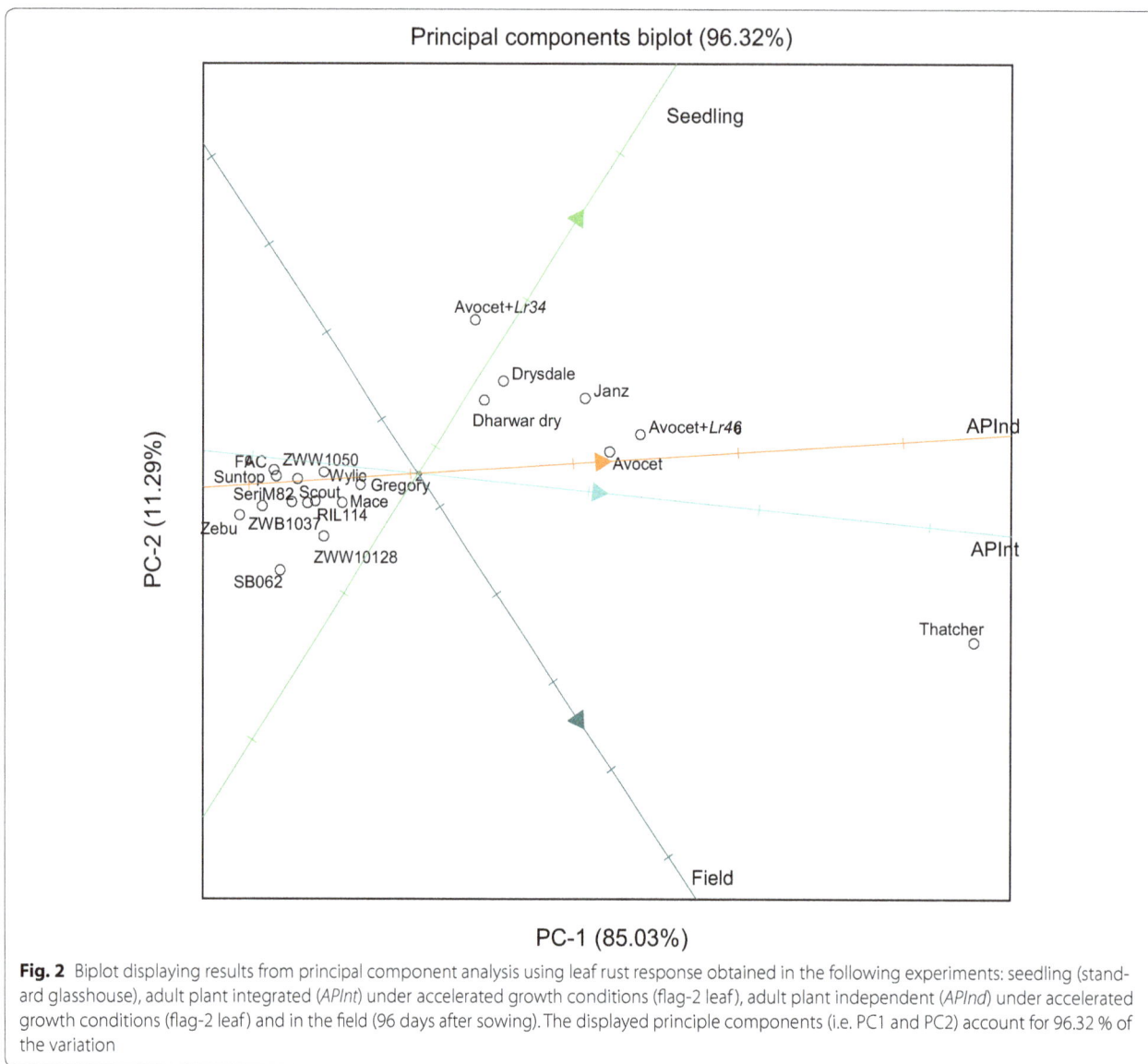

**Fig. 2** Biplot displaying results from principal component analysis using leaf rust response obtained in the following experiments: seedling (standard glasshouse), adult plant integrated (*APInt*) under accelerated growth conditions (flag-2 leaf), adult plant independent (*APInd*) under accelerated growth conditions (flag-2 leaf) and in the field (96 days after sowing). The displayed principle components (i.e. PC1 and PC2) account for 96.32 % of the variation

## Importance of temperature and light to detect APR under AGC

AGC involves constant light and temperature regimes during the early plant growth phase to achieve adult plant stage rapidly. However, to assist a successful infection, diurnal light and temperature regime was implemented post-inoculation until disease assessment. Post-inoculation conditions are important for a successful host-pathogen interaction and become more important when plants are raised and inoculated in an artificial environment, such as the AGC adopted in this study. As discussed above, plant growth stage, along with temperature and light (i.e. quantity and quality) are considered key factors determining disease development [15].

All known *Lr* genes are sensitive to fluctuating post-inoculation temperatures, for instance expression of *Lr13* at the adult growth stage [16]. In the present study, plants were grown under a 12 h cycling temperature regime of 22/17 °C. This temperature enabled rapid plant growth, and importantly, provided healthy plants prior to inoculation. Notably, this falls within the optimal temperature range for LR development (i.e. 10–25 °C) [31]. Under AGC, a warmer growing temperature (e.g. >24 °C) can compromise plant health, which is critical if plants are to be subjected to disease assays. The increase or decrease in temperature can also influence latent period [16, 32]. The fluctuations in latent period are critical in wheat rust infections and AGC could serve as a tool to study the latent period under different temperature regimes.

Light is another key component of the rapid phenotyping method, where it not only affects plant photosynthetic activity, but also plays a role in disease development. Under AGC, wheat plants were grown under constant (24 h) light to quickly obtain adult plants. The importance of light influencing disease development both pre- and post-inoculation has been previously reported for both LR and YR in wheat [33]. We employed a diurnal (12 h) photoperiod post-inoculation until disease assessment. High quality light is important for disease development, particularly for good sporulation [34]. In addition, the diurnal light appears to be important, as constant (24 h) light can impede pathogen development, thus reducing the ability to differentiate between resistant and susceptible genotypes (unpublished data).

## Conclusion

Breeding for rust resistance requires a continuous effort to stay ahead of the rapidly evolving pathogen. This requires robust phenotypic screening and ongoing deployment of new resistance genes. The method reported in this study provides a great tool for detecting APR to LR at levels similar to those observed in the field. It can be scaled-up for screening large numbers of fixed lines and segregating populations, similar to that reported for YR in wheat [15]. Using this technique, it is possible to conduct up to seven consecutive screens annually, compared to just one in the field. It is possible to phenotype APR prior to anthesis under AGC, as genotypes inoculated at or beyond GS30 display resistance representative of adult plants. Assessing plants at early growth stages can enable selection of desirable gene combinations for APR and crossing of the selected plants in the same plant generation. This strategy would reduce time required for moving APR genes into adapted germplasm (from donor sources) or combining traits in top crosses in breeding programs.

## Additional file

**Additional file 1: Table S1.** Mean leaf rust response including infection type and host response for the panel of 21 spring wheat genotypes evaluated in the following experiments: seedling (standard glasshouse), adult plant integrated and adult plant independent under accelerated growth conditions (AGC), and in the field. The disease response for the seedling and adult plant AGC experiments was collected using the 0–4 Stakman scale whereas, the disease response in the field was collected using the modified Cobb scale, which was used to calculate coefficient of infection. A dash (-) indicates data is unavailable or unknown.

**Authors' contributions**
AR, EA, and LH conceived and designed the experiments. AR performed the experiments and analysed the data. AR, EA, SP and LH wrote and reviewed the paper. All authors read and approved the final manuscript.

**Author details**
[1] Queensland Alliance for Agriculture and Food Innovation, The University of Queensland, St Lucia, QLD 4072, Australia. [2] Commonwealth Scientific and Industrial Research Organization (CSIRO) Agriculture, General Post Office Box 1600, Canberra, ACT 2601, Australia. [3] School of Agriculture and Food Science, The University of Queensland, St Lucia, QLD 4072, Australia.

**Acknowledgements**
This research was supported by a Ph.D. scholarship from The University of Queensland, Australia. We would like to acknowledge Dr. Mark Dieters at The University of Queensland for concepts relating to AGC and Dr. Miranda Mortlock at UQ for advice in relation to statistical analyses performed in this study. We also wish to acknowledge contributions from Ph.D. students Mr. Eric Dinglasan and Miss Laura Ziems for assisting with field and AGC experiments.

**Competing interests**
The authors declare that they have no competing interests.

**References**
1. Hawkesford MJ, Araus J-L, Park R, Calderini D, Miralles D, Shen T, Zhang J, Parry MAJ. Prospects of doubling global wheat yields. Food Energy Secur. 2013;2(1):34–48. doi:10.1002/fes3.15.
2. Samborski D. Wheat leaf rust. In: Roelfs AP, Bushnell WR, editors. The cereal rusts-II. 1985. p. 58–76.
3. Huerta-Espino J, Singh RP, Germán S, McCallum BD, Park RF, Chen WQ, Bhardwaj SC, Goyeau H. Global status of wheat leaf rust caused by *Puccinia triticina*. Euphytica. 2011;179(1):143–60. doi:10.1007/s10681-011-0361-x.
4. Everts KL, Leath S, Finney PL. Impact of powdery mildew and leaf rust on milling and baking quality of soft red winter wheat. Plant Dis. 2001;85(4):423–9. doi:10.1094/PDIS.2001.85.4.423.
5. Bolton MD, Kolmer JA, Garvin DF. Wheat leaf rust caused by *Puccinia triticina*. Mol Plant Pathol. 2008;9(5):563–75. doi:10.1111/j.1364-3703.2008.00487.x.
6. Murray G, Brennan J. Estimating disease losses to the Australian wheat industry. Australas Plant Pathol. 2009;38(6):558–70. doi:10.1071/AP09053.
7. Pink DC. Strategies using genes for non-durable disease resistance. Euphytica. 2002;124(2):227–36. doi:10.1023/A:1015638718242.
8. Park RF, Mohler V, Nazari K, Singh D. Characterisation and mapping of gene *Lr73* conferring seedling resistance to *Puccinia triticina* in common wheat. Theor Appl Genet. 2014;127(9):2041–9. doi:10.1007/s00122-014-2359-y.
9. Caldwell RM. Breeding for general and/or specific plant disease resistance. In: Finlay KW, Shepherd KW, editors. Proceedings of the Third International Wheat Genetics Symposium. Australian Academy of Sciences, Canberra: Australia; 1968. p. 263–72.
10. Ellis JG, Lagudah ES, Spielmeyer W, Dodds PN. The past, present and future of breeding rust resistant wheat. Front Plant Sci. 2014;5:641. doi:10.3389/fpls.2014.00641.
11. McIntosh RA, Wellings CR, Park RF. Wheat rusts: an atlas of resistance genes. Melbourne: CSIRO Publishing; 1995. p. 199.
12. Bansal UK, Hayden MJ, Venkata BP, Khanna R, Saini RG, Bariana HS. Genetic mapping of adult plant leaf rust resistance genes *Lr48* and *Lr49* in common wheat. Theor Appl Genet. 2008;117:307–12. doi:10.1007/s00122-008-0775-6.
13. Lagudah ES, McFadden H, Singh RP, Huerta-Espino J, Bariana HS, Spielmeyer W. Molecular genetic characterization of the *Lr34/Yr18* slow rusting resistance gene region in wheat. Theor Appl Genet. 2006;114(1):21–30. doi:10.1007/s00122-006-0406-z.
14. Moore JW, Herrera-Foessel S, Lan C, Schnippenkoetter W, Ayliffe M, Huerta-Espino J, et al. A recently evolved hexose transporter variant confers resistance to multiple pathogens in wheat. Nature genetics. 2015. doi:10.1038/ng.3439.
15. Hickey LT, Wilkinson PM, Knight CR, Godwin ID, Kravchuk OY, Aitken EAB, et al. Rapid phenotyping for adult-plant resistance to stripe rust in wheat. Plant Breed. 2012;131(1):54–61. doi:10.1111/j.1439-0523.2011.01925.x.

16. Kaul K, Shaner G. Effect of temperature on adult-plant resistance to leaf rust in wheat. Phytopathology. 1989;79(4):391–4.

17. Risk JM, Selter LL, Krattinger SG, Viccars LA, Richardson TM, Buesing G, et al. Functional variability of the *Lr34* durable resistance gene in transgenic wheat. Plant Biotechnol J. 2012;10(4):477–87. doi:10.1111/j.1467-7652.2012.00683.x.

18. Singh D, Macaigne N, Park RF. *Rph20*: adult plant resistance gene to barley leaf rust can be detected at early growth stages. Eur J Plant Pathol. 2013;137(4):719–25. doi:10.1007/s10658-013-0282-8.

19. Hickey LT, Dieters MJ, DeLacy IH, Kravchuk OY, Mares DJ, Banks PM. Grain dormancy in fixed lines of white-grained wheat (*Triticum aestivum* L.) grown under controlled environmental conditions. Euphytica. 2009;168(3):303–10.

20. Park RF, Bariana HS, Wellings CR, Wallwork H. Detection and occurrence of a new pathotype of *Puccinia triticina* with virulence for *Lr24* in Australia. Aust J Agric Res. 2002;53(9):1069–76. doi:10.1071/AR02018.

21. Stakman E, Stewart D, Loegering W. Identification of physiologic races of *Puccinia graminis var. tritici*. Washington: USDA; 1962.

22. Zadoks JC, Chang TT, Konzak CF. A decimal code for the growth stages of cereals. Weed Res. 1974;14(6):415–21.

23. Peterson RF, Campbell A, Hannah A. A diagrammatic scale for estimating rust intensity on leaves and stems of cereals. Can J Res. 1948;26(5):496–500.

24. Loegering W. Model of recording cereal rust data. USA: USDA Intern Spring Wheat Rust Nursery; 1959.

25. Ziems LA, Hickey LT, Hunt CH, Mace ES, Platz GJ, Franckowiak JD, et al. Association mapping of resistance to *Puccinia hordei* in Australian barley breeding germplasm. Theor Appl Genet. 2014;127(5):1199–212. doi:10.1007/s00122-014-2291-1.

26. Singh D, Park R, editors. Inheritance of adult plant resistance to leaf rust in four European winter wheat cultivars. In: 11th international wheat genetics symposium 2008 proceedings; 2008: Sydney University Press, Sydney

27. Wellings C, Bariana H, Bansal U, Park R. Expected responses of Australian wheat and triticale varieties to the cereal rust diseases in 2012. Cereal Rust Report, Season. 2012;10(1):1–5.

28. Pretorius ZA, Wilcoxson RD, Long DL, Schaffer JF. Detecting wheat leaf rust resistance gene *Lr13* in seedlings. Plant Dis. 1984;68:585–6.

29. Kloppers F, Pretorius Z. Expression and inheritance of leaf rust resistance gene *Lr37* in wheat seedlings. Cereal Res Commun. 1994;22(1–2):91–7.

30. Park R, Rees R, editors. The epidemiology and control of stripe rust of wheat in the summer rainfall area of Eastern Australia. Second national stripe rust workshop, Wagga Wagga, New South Wales, Australia; 1987

31. Dyck P, Johnson R. Temperature sensitivity of genes for resistance in wheat to *Puccinia recondita*. Can J Plant Pathol. 1983;5(4):229–34.

32. Eversmeyer M, Kramer C, Browder L. Effect of temperature and host: parasite combination on the latent period of *Puccinia recondita* in seedling wheat plants. Phytopathology. 1980;70(10):938–41.

33. De Vallavieille-Pope C, Huber L, Leconte M, Bethenod O. Preinoculation effects of light quantity on infection efficiency of *Puccinia striiformis* and *P. triticina* on wheat seedlings. Phytopathology. 2002;92(12):1308–14.

34. Roelfs A, Huerta-Espino J, Marshall D. Barley stripe rust in Texas. Plant Dis. 1992;76:538.

35. Hayes HK, Ausemus E, Stakman E, Bailey C, Wilson H, Bamberg R et al. Thatcher wheat. Minn Bull. 1936;325:1–39.

36. Fitzsimmons R, Martin R, Wrigley C. Australian wheat varieties: identification according to plant, head and grain characteristics. Clayton: CSIRO Publishing; 1983.

37. Lillemo M, Singh RP, Huerta-Espino J, Chen XM, He ZH, Brown JKM. Leaf rust resistance gene *Lr34* is involved in powdery mildew resistance of cimmyt bread wheat line Saar. In: Buck HT, Nisi JE, Salomón N, editors. Wheat production in stressed environments. Developments in plant breeding. Netherlands: Springer; 2007. p. 97–102.

# 15

# Application of unmanned aerial systems for high throughput phenotyping of large wheat breeding nurseries

Atena Haghighattalab[1], Lorena González Pérez[2], Suchismita Mondal[2], Daljit Singh[3], Dale Schinstock[4], Jessica Rutkoski[2,5], Ivan Ortiz-Monasterio[2], Ravi Prakash Singh[2], Douglas Goodin[1] and Jesse Poland[6*]

## Abstract

**Background:** Low cost unmanned aerial systems (UAS) have great potential for rapid proximal measurements of plants in agriculture. In the context of plant breeding and genetics, current approaches for phenotyping a large number of breeding lines under field conditions require substantial investments in time, cost, and labor. For field-based high-throughput phenotyping (HTP), UAS platforms can provide high-resolution measurements for small plot research, while enabling the rapid assessment of tens-of-thousands of field plots. The objective of this study was to complete a baseline assessment of the utility of UAS in assessment field trials as commonly implemented in wheat breeding programs. We developed a semi-automated image-processing pipeline to extract plot level data from UAS imagery. The image dataset was processed using a photogrammetric pipeline based on image orientation and radiometric calibration to produce orthomosaic images. We also examined the relationships between vegetation indices (VIs) extracted from high spatial resolution multispectral imagery collected with two different UAS systems (eBee Ag carrying MultiSpec 4C camera, and IRIS+ quadcopter carrying modified NIR Canon S100) and ground truth spectral data from hand-held spectroradiometer.

**Results:** We found good correlation between the VIs obtained from UAS platforms and ground-truth measurements and observed high broad-sense heritability for VIs. We determined radiometric calibration methods developed for satellite imagery significantly improved the precision of VIs from the UAS. We observed VIs extracted from calibrated images of Canon S100 had a significantly higher correlation to the spectroradiometer (r = 0.76) than VIs from the MultiSpec 4C camera (r = 0.64). Their correlation to spectroradiometer readings was as high as or higher than repeated measurements with the spectroradiometer per se.

**Conclusion:** The approaches described here for UAS imaging and extraction of proximal sensing data enable collection of HTP measurements on the scale and with the precision needed for powerful selection tools in plant breeding. Low-cost UAS platforms have great potential for use as a selection tool in plant breeding programs. In the scope of tools development, the pipeline developed in this study can be effectively employed for other UAS and also other crops planted in breeding nurseries.

**Keywords:** Unmanned aerial vehicles/systems (UAV/UAS), Wheat, High-throughput phenotyping, Remote sensing, GNDVI, Plot extraction

*Correspondence: jpoland@ksu.edu
[6] Wheat Genetics Resource Center, Department of Plant Pathology and Department of Agronomy, Kansas State University, Manhattan, KS 66506, USA
Full list of author information is available at the end of the article

## Background

In a world of finite resources, climate variability, and increasing populations, food security has become a critical challenge. The rates of genetic improvement are below what is needed to meet projected demand for staple crops such as wheat [1]. The grand challenge remains in connecting genetic variants to observed phenotypes followed by predicting phenotypes in new genetic combinations. Extraordinary advances over the last 5–10 years in sequencing and genotyping technology have driven down the cost and are providing an abundance of genomic data, but this only comprise half of the equation to understand the function of plant genomes and predicting plant phenotypes [2, 3]. High throughput phenotyping (HTP) platforms could provide the keys to connecting the genotype to phenotype by both increasing the capacity and precision and reducing the time to evaluate huge plant populations. To get to the point of predicting the real-world performance of plants, HTP platforms must innovate and advance to the level of quantitatively assessing millions of plant phenotypes. To contribute to this piece of the challenge, we describe here a semi-automated HTP analysis pipeline using a low cost unmanned aerial system (UAS) platform, which will increase the capacity of breeders to assess large numbers of lines in field trials.

A plant phenotype is a set of structural, morphological, physiological, and performance-related traits of a given genotype in a defined environment [4]. The phenotype results from the interactions between a plant's genes and environmental (abiotic and biotic) factors. Plant phenotyping involves a wide range of plant measurements such as growth development, canopy architecture, physiology, disease and pest response and yield. In this context, HTP is an assessment of plant phenotypes on a scale and with a level of speed and precision not attainable with traditional methods [5], many of which include visual scoring and manual measurements. To be useful to breeding programs, HTP methods must be amenable to plot sizes, experimental designs and field conditions in these programs. This entails evaluating a large number of lines within a short time span, methods that are lower cost and less labor intensive than current techniques, and accurately assessing and making selections in large populations consisting of thousands to tens-of-thousands of plots. To rapidly characterize the growth responses of genetically different plants in the field and relate these responses to individual genes, use of information technologies such as proximal or remote sensing and efficient computational tools are necessary.

In recent years, there has been increased interest in ground-based and aerial HTP platforms, particularly for applications in breeding and germplasm evaluation activities [6–8]. Ground-based phenotyping platforms include modified vehicles deploying proximal sensing sensors [9–12]. Measurements made at a short distance with tractors and hand-held sensors that do not necessarily involve measurements of reflected radiation, are classified as proximal sensing. Proximal, or close-range sensing, is expected to provide higher resolution for phenotyping studies as well as allowing collection of data with multiple view-angles, illumination control and known distance from the plants to the sensors [13]. However, these ground-based platforms do have limitations mainly on the scale at which they can be used, limitations on portability and time required to make the measurements in different field locations.

As a complement to ground-based platforms, aerial-based phenotyping platforms enable the rapid characterization of many plots, overcoming one of the limitations associated with ground-based phenotyping platforms. There is a growing body of literature showing how these approaches in remote and proximal sensing enhance the precision and accuracy of automated high-throughput field based phenotyping techniques [14–16]. One of the emerging technologies in aerial based platforms is UAS, which have undergone a remarkable development in recent years and are now powerful sensor-bearing platforms for various agricultural and environmental applications [17–21]. UAS can cover an entire experiment in a very short time, giving a rapid assessment of all of the plots while minimizing the effect of environmental conditions that change rapidly such as wind speed, cloud cover, and solar radiation. UAS enables measuring with high spatial and temporal resolution capable of generating useful information for plant breeding programs.

Different types of imaging systems for remote sensing of crops are being used on UAS platforms. Some of the cameras used are RGB, multispectral, hyperspectral, thermal cameras, and low cost consumer grade cameras modified to capture near infrared (NIR) [2, 9, 21–23]. Consumer grade digital cameras are widely used as the sensor of choice due to their low cost, small size and weight, low power requirements, and their potential to store thousands of images. However, consumer grade cameras often have the challenge of not being radiometrically calibrated. In this study, we evaluated the possibility of using traditional radiometric calibration methods developed for satellite imagery to address this issue.

Radiometric calibration accounts for both variations from photos within an observation day along with changes between different dates of image. The result of radiometric calibration is a more generalized and, most importantly, repeatable, method for different image processing techniques (such as derivation of VIs, change detection, and crop growth mapping) applied to the orthomosaic image instead of each individual image in a dataset. There are well-established radiometric calibration approaches for satellite imagery. However, these approaches are not necessarily applicable in UAS workflows due to several factors such as conditions of data acquisition during the exact time of image capture using these platforms.

Empirical line method is a technique often applied to perform atmospheric correction to convert at-sensor radiance measurements to surface reflectance for satellite imagery. This technique can be modified to apply radiometric calibration and convert digital numbers (DNs) to reflectance values [24]. The method is based on the relationship between DNs and surface reflectance values measured from calibration targets located within the image using a field spectroradiometer. The extracted DNs from the imagery are then compared with the field-measured reflectance values to calculate a prediction equation that can be used to convert DNs to reflectance values for each band [25].

Advances in platform design, production, standardization of image geo-referencing, mosaicking, and information extraction workflow are needed to implement HTP with UAS as a routine tool in plant breeding and genetics. Additionally, developing efficient and easy-to-use pipelines to process HTP data and disseminate associated algorithms are necessary when dealing with big data. The primary objective of this research was to develop and validate a pipeline for processing data captured by consumer grade digital cameras using a low cost UAS to evaluate small plot field-based research typical of plant breeding programs.

## Methods

### Study area

The study was conducted at Norman E Borlaug Experiment Station at Ciudad Obregon, Sonora, Mexico (Fig. 1). The experiment consisted of 1092 advanced wheat lines distributed in 39 trials, each trial consisting of 30 entries. Each trial was arranged as an alpha lattice with three replications. The experimental units were 2.8 m × 1.6 m in size consisting of paired rows at 0.15 m spacing, planted on two raised beds spaced 0.8 m apart. The trials were planted later than optimal on February 24, 2015 to simulate heat stress with temperatures expected to be above optimum for the entire growing season (the local recommended planting date is November 15 to December 15). Irrigation and nutrient levels in the heat trials were maintained at optimal levels, as well as weed, insect and disease control. Grain yield was obtained by using a plot combine harvester.

### Platforms and cameras

We evaluated two unmanned aerial vehicles (UAV); the IRIS+ (3D Robotics, Inc, Berkeley, CA 94710, USA)

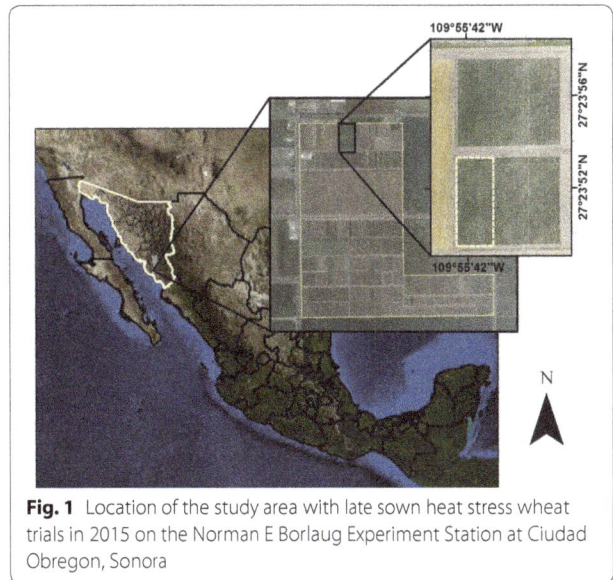

Fig. 1 Location of the study area with late sown heat stress wheat trials in 2015 on the Norman E Borlaug Experiment Station at Ciudad Obregon, Sonora

and eBee Ag (senseFly Ltd., 1033 Cheseaux-Lausanne, Switzerland).

The IRIS+ is a low cost quadcopter UAV with a maximum payload of 400 g. The open-source Pixhawk autopilot system was programmed for autonomous navigation for the IRIS+ based on ground coordinates and the 'survey' option of Mission Planner (Pixhawk sponsored by 3D Robotics, www.planner.ardupilot.com). The IRIS+ is equipped with an uBlox GPS with integrated magnetometer. A Canon S100 modified by MaxMax (LDP LLC, Carlstadt, NJ 07072, USA, www.maxmax.com) to Blue–Green–NIR (400–760 nm) was mounted to the UAV with a gimbal designed by Kansas State University. The gimbal compensates for the UAV movement (pitch and roll) during the flight to allow for nadir image collection.

The eBee Ag, designed as a fixed wing UAV for application in precision agriculture has a payload of 150 g. This UAV was equipped with MultiSpec 4C camera developed by Airinov (Airinov, 75018 Paris, France, www.airinov.fr/en/uav-sensor-agrosensor/) and customized for the eBee Ag. It contains four distinct bands with no spectral overlap (530–810 nm): green, red, red-edge, and near infrared bands, and is controlled by the eBee Ag autopilot during the flight.

For 'ground-truth' validation, spectral measurements were taken using ASD VNIR handheld point-based spectroradiometer (ASD Inc., Boulder, CO, http://www.asdi.com/) with a wavelength range of 325–1075 nm, a wavelength accuracy of ±1 nm and a spectral resolution of <3 nm at 700 nm with a fiber optics of 25° (aperture) full conical angle. The instrument digitizes spectral values to 16 bits. Table 1 summarizes the specification for the cameras and sensor used in this study.

**Table 1  Sensor specification**

| Sensor | Platform (UAV) | Sensor resolution (MP) | Focal length (mm) | Full width at half maximum (FWHM) | Peak wavelength |
|---|---|---|---|---|---|
| Canon S100 | IRIS+ | 12.1 | 5.2 | Blue: 400–495<br>Green: 490–550<br>NIR: 680–760 | Blue: 460<br>Green: 525<br>NIR: 710 |
| MultiSpec 4C | eBee Ag | 1.2 (four sensors) | 3.6 | Green: 530–570<br>Red: 640–680<br>Rededge: 730–740<br>NIR: 770–810 | Green: 550<br>Red: 660<br>Rededge: 735<br>NIR: 790 |
| | | **Spectral resolution** | | **Wavelength accuracy** | **Wavelength range** |
| ASD spectroradiometer | | <3 nm at 700 nm | | ±1 nm | 325–1075 nm |

## Ground control points

In order to geo-reference the aerial images, at the beginning of the season ten ground control points (GCPs) were distributed across the field. The GCPs were 25 cm × 25 cm square white metal sheets mounted on a 50 cm post. The GCPs were white to provide easy identification for the image processing software. The GCP coordinates were measured with a Trimble R4 RTK GPS, with a horizontal accuracy of 0.025 m and a vertical accuracy of 0.035 m. These targets were maintained throughout the crop season to enable more accurate geo-referencing of UAS aerial imagery and overlay of measurements from multiple dates.

## Reflectance calibration panel

For radiometric calibration, spectra of easily recognizable objects (e.g. gray scale calibration board) are needed. A black–gray–white grayscale board with known reflectance values was built and placed in the field during flights for further image calibration. This grayscale calibration panel met the requirements for further radiometric calibration including (1) the panel was spectrally homogenous, (2) it was near Lambertian and horizontal, (3) it covered an area many times larger than the pixel size of the Canon S100, and (4) covered a range of reflectance values [25].

The calibration panel used in this study had 6 levels of gray from 0 % being white and 100 % being black, printed on matte vinyl. This made it possible to choose several dark and bright regions in the images to provide a more accurate regression for further radiometric calibration analysis. Photos of the calibration panel were taken in the field before a UAS flight using the same Canon S100 NIR camera mounted on the IRIS+. Using an ASD spectroradiometer on a sunny day, the spectral reflectance was measured of the panel at a fixed altitude of 0.50 m from different angles.

## Field data collection

At the study site, image time series were acquired using IRIS+ and Canon S100 camera during the growing season as evenly spaced as possible, depending on the weather condition; April 10, April 22, April 28, May 6, and May 18, 2015.

Field data measurements for the experiment in this study included (1) IRIS+ carrying modified NIR Canon S100, and (2) eBee Ag carrying MultiSpec 4C camera, and (3) spectral reflectance measurements using handheld ASD spectroradiometer collected on May 6, 2015 when the trials were at mid grain filling. We took all field measurements within 1 h around solar noon to minimize variation in illumination and solar zenith angle [26]. This limited the number of plots we could measure using handheld spectroradiometer to 280 plots. Four spectra per plot were taken with the beam of the fiber optics placed at 0.50 m over the top of the canopy with a sample area of 0.04 m$^2$. Due to this very small sampling area, great care was taken in data collection to make sure the measured spectral responses were only from the crops.

Both the eBee and IRIS+ were able to image a much larger area in the same time frame. Each system was flown using a generated autopilot path that covered the 2 ha experiment area with an average of 75 % overlap between images (Table 2).

For the IRIS+ with Canon S100, all images were taken in RAW format (.CR2) to avoid loss of image information. The Canon S100 settings were TV mode, which allowed setting of a constant shutter speed. The aperture was set to be auto-controlled by the camera to maintain a good exposure level, and ISO speed was set to Auto with a maximum value of 400 to minimize noise in the images. Focus range was manually set to infinity. We used CHDK (www.chdk.wikia.com) to automate Canon S100's functionality by running intervalometer scripts off an SD card in order to take pictures automatically at intervals during flight. The CHDK script allowed the UAS autopilot system to send electronic pulses to trigger the camera shutter. The camera trigger was set at the corresponding distance interval during the mission planning for the desired image overlap.

**Table 2 Flight information using IRIS+ (multirotor) and eBee Ag (fixed wing) UASs, on May 6, 2015, at CIMMYT, Cd Obregon, Mexico**

| Camera | Platform (UAS) | Flight speed (m/s) | Altitude (m) | Percent overlap | | No. of images | Image format | Spatial resolution (cm) |
|---|---|---|---|---|---|---|---|---|
| | | | | Side (%) | Forward (%) | | | |
| Modified Canon S100 | IRIS+ | 2 | 30 | 75 | 75 | 144 | RAW (16 bits TIFF)[a] | 0.8 |
| MultiSPEC 4C | eBee Ag | 10 | 48 | 75 | 75 | 40 | 10 bits TIFF | 5 |

[a] RAW images of Canon S100 were converted to 16 bits TIFF imagery after pre-processing in Canon Digital Photo Professional software (DPP)

The eBee MultiSpec 4C camera had a predefined setting by Sensefly; ISO and shutter speed was set to automatic, maximum aperture was set to f/1.8 and focal distance was fixed at 4 mm.

### Developing an image processing pipeline for HTP

In this research, a UAS-based semi-automated data analyses pipeline was developed to enable HTP analysis of large breeding nurseries. Data analysis was primarily conducted using Python scripts [27, 28]. The dataset used to develop the image processing pipeline was collected on May 6, 2015 with the IRIS+ flights carrying the modified NIR Canon S100 camera. In order to analyze hundreds of images taken by a UAS that represent the entire experiment area, we developed a semi-automated data analysis pipeline. The developed pipeline completed the following steps: (1) image pre-processing, (2) georeferencing and orthomosaic generation, (3) image radiometric calibration, (4) calculation of different VIs and (5) plot-level data extraction from the VI maps. The overall workflow of the

developed pipeline is presented in Fig. 2, and a detailed description of each segment is provided.

The RAW images of Canon S100 (.CR2 format) were pre-processed using Digital Photo Professional (DPP) software developed by Canon (http://www.canon.co.uk/support/camera_software/). This software to convert RAW to Tiff also included lens distortion correction, chromatic aberration, and gamma correction. For white balance adjustment, we used the pictures of grayscale calibration panel taken during the flights. After pre-processing the images were exported as a tri-band 16 bit linear TIFF image.

### Image stitching and orthomosaic generation

We then generated a geo-referenced orthomosaic image using the TIFF images. There are several software packages for mosaicking UAS aerial imagery, all based around the scale-invariant feature transform algorithm (SIFT) algorithm for feature matching between images [29]. While there are slight differences between packages each

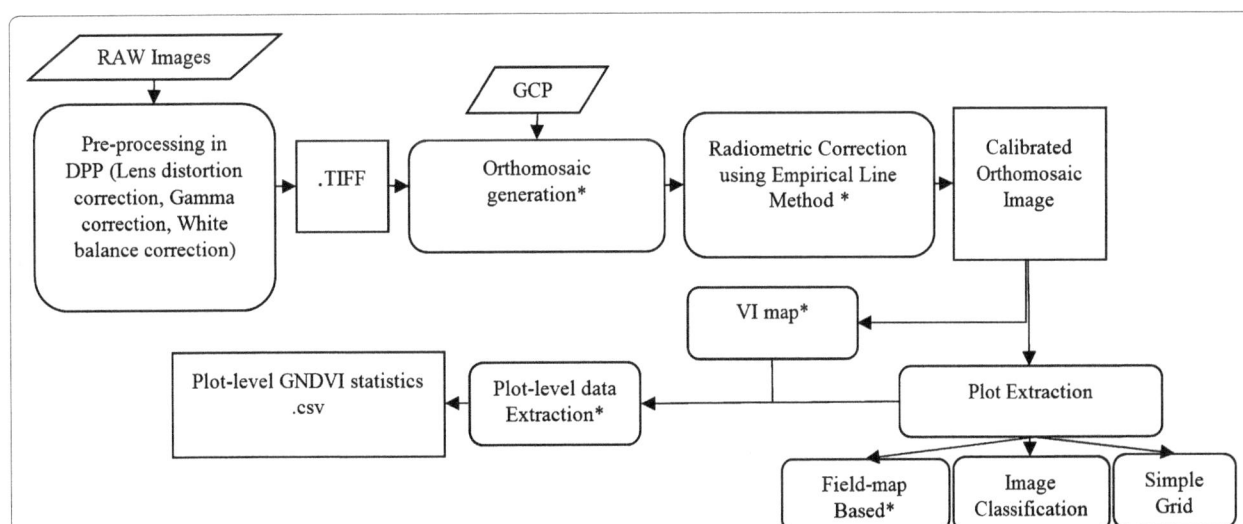

**Fig. 2** Image processing workflow for the low cost UAS imaging system. The developed pipeline steps are as follows: image pre-processing using DPP software, Orthomosaic Generation using Python scripting in PhotoScan, image radiometric correction using empirical line method commonly used for satellite imagery, extraction of wheat plot boundary, and calculation of different VI's. *Asterisk* designates steps done using Python script

performs the same operation namely loading photos; aligning photos; importing GCP positions; building dense point cloud; building digital elevation model (DEM); and finally generating the orthomosaic image. We used the commercial Agisoft PhotoScan package (Agisoft LLC, St. Petersburg, Russia, 191144) to automate the orthomosaic generation within a custom Python script. The procedure of generating the orthomosaic using PhotoScan comprises five main stages (1) camera alignment, (2) importing GCPs and geo-referencing, (3) building dense point cloud, (4) building DEM and (5) generating orthomosaic [30].

For alignment, PhotoScan finds the camera position and orientation for each photo and builds a sparse point cloud model; the software uses the SIFT algorithm to find matching points of detected features across the photos. PhotoScan offers two options for pair selection: generic and reference. In generic pre-selection mode the overlapping pairs of photos are selected by matching photos using the lower accuracy setting. On the other hand, in reference mode the overlapping pairs of photos are selected based on the measured camera locations. We examined both methods to generate the orthomosaic and compared the results.

With aerial photogrammetry the location information delivered with cameras is inadequate and the data does not align properly with other geo-referenced data. In order to do further spatial and temporal analysis with aerial imagery, we performed geo-referencing using GCPs. We used ten GCPs evenly distributed around the orthomosaic image. Geo-referencing with GCPs is the step that still requires manual input.

The main source of error in georeferencing is performing the linear transformation matrix on the model. In order to remove the possible non-linear deformation components of the model, and minimizing the sum of re-projection error and reference coordinate misalignment, PhotoScan offers an optimization of the estimated point cloud and camera parameters based on the known reference coordinates [30].

After alignment optimization, the next step is generating dense point clouds. Based on the estimated camera positions PhotoScan calculates depth information for each camera to be combined into a single dense point cloud [30]. PhotoScan allows creating a raster DEM file from a dense point cloud, a sparse point cloud or a mesh to represent the model surface as a regular grid of height values. Although most accurate results are calculated based on dense point cloud data.

The final step is orthomosaic generation. An orthomosaic can be created based on either mesh or DEM data. Mesh surface type can be chosen for models that are not referenced. The blending mode has two options (mosaic and average) to select how pixel values from different photos will be combined in the final texture layer. Average blending computes the average brightness values from all overlapping images to help reduce the effect of the bidirectional reflectance distribution function that is strong within low altitude wide-angle photography [31]. Blending pixel values using mosaic mode does not mix image details of overlapping photos, but uses only images where the pixel in question is located within the shortest distance from the image center. We compared these two methods by creating orthomosaic images using both mosaic and average mode.

### Radiometric calibration

After the orthomosaic was completed, we then performed radiometric calibration to improve the accuracy of surface spectral reflectance obtained using digital cameras. We applied empirical line correction to each orthomosaic within the data set using field measurements taken with the ASD. The relationship between image raw DNs and their corresponding reflectance values is not completely linear for all the bands for Canon S100 camera and most digital cameras (Fig. 3). To relate remote sensing data to field-based measurements we used a modified empirical line method [32]. We used the grayscale calibration board combined with reflectance measurements from the ASD spectroradiometer to find the relationship between DNs extracted form Canon S100 and surface reflectance values measured from grayscale calibration board using a field spectroradiometer. We then calculated the prediction equations for each band separately to convert DNs to reflectance values.

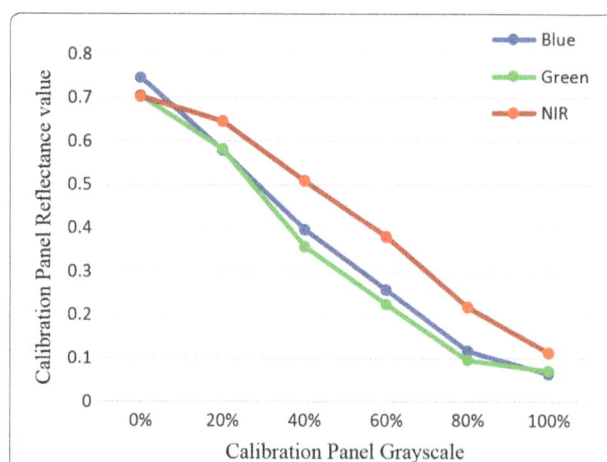

**Fig. 3** Empirical observation of non-linear relationship between percent gray values of the calibration panel and their corresponding mean reflectance values at the spectral region of Canon S100 wavebands (B: 400–495, G: 490–550, NIR: 680–760)

We calculated the mean pixel value of the calibration panel extracted from images for each band separately, and plotted against the mean band equivalent reflectance (BER) of field spectra [33, 34]. We found a linear relation between the natural logarithm of image raw DN values and their corresponding BER, although the relation between image raw DNs and their corresponding BER was not completely linear for all bands of Canon S100 (Fig. 3). We derived the correction coefficient needed to fit uncalibrated UAS multispectral imagery to field-measured reflectance spectra. Using the correction coefficient for each band, we then performed this correction on the entire image.

With the corrected orthomosaic, a map showing the vegetation indices can be generated. The most used VIs derivable from a three-band multispectral sensor are: normalized difference vegetation index (NDVI), and green normalized difference vegetation index (GNDVI) (Table 4). NDVI is calculated from reflectance measurements in the red and NIR portion of the spectrum and its values range from $-1.0$ to $1.0$. GNDVI is computed similarly to the NDVI, where the green band is used instead of the red band and its value ranges from 0 to 1.0. Like NDVI, this index is also related to the proportion of photosynthetically absorbed radiation and is linearly correlated with leaf area index (LAI) and biomass [22, 35, 36].

*Field plot extraction*
In order to get useful information about each wheat plot in the field, we need to extract plot level data from the orthomosaic VI image. Individual wheat plot boundaries need to be extracted and defined separately from images with an assigned plot ID that defines their genomic type. We examined three approaches for extracting plot boundaries from the orthomosaic image: a simple grid based plot extraction, field-map based plot extraction, and image classification/segmentation.

To apply a simple grid superimposed on top of the orthomosaic image we used the Fishnet function using Arcpy scripting in ArcGIS. This function creates a net of adjacent rectangular cells. The output is polygon features defining plot boundaries. Creating a fishnet requires two basic pieces of information: the spatial extent of the desired net, and the number of rows and columns. The number of rows and ranges (columns) and height and width of each cell in the fishnet can be defined by the user. The user can also define the extent of the grid by supplying both the minimum and maximum x- and y-coordinates.

The next method we evaluated in this study is a field-map based method. Experimental or field data are usually stored in a spreadsheet. To accomplish this, we reformatted the field map in order to represent the plots in the field; the first row of the excel sheet is the length (or width, depending on how plots are located in the field) of the plots, and the first column is the width (or length). Each cell in the excel sheet is the plot ID and other information regarding that particular plot available in the field map.

In this approach, first we created a KML file from the field map (.CSV) using Python [27], we generated polygon shapefiles with known size for each plot from KML file, and assigned plot ID to each plot using the field map. To define the geographic extent of the field, the python script has inputs for the coordinate of two points in the field: the start point of the first plot on the top right and the end point of the last plot on the bottom left. The script starts from the top right and builds the first polygon using the defined plot size, and skips the gap between plots and generates the next one until it gets to the last plot on the bottom left. In this approach the plot IDs are assigned automatically and simultaneously from the field map excel sheet. The most important advantage of this approach is that it can be generalized to other crop types as long as the field map is provided and the plots are planted in regular distance and have a consistent size within a trial.

The third method is to extract wheat plot boundaries from the orthomosaic image directly. To examine this method, we used the April 10 data set instead of May 6. The reason for choosing this set of data was that in this dataset we had the most distinguished features (e.g. vegetation vs. bare soil) in the image compared to the image data taken on May 6. We classified the GNDVI image created from the orthomosaic image into four classes: "Wheat", "Shadow", "Soil", and a class named "Others" for all the GCP targets in the field. The class "shadow" stands for the shadow projected on the soil surface. In this supervised classification, we defined training samples and signature files for each defined class. After classifying the image, we merged the "Shadow", "Soil" and "Others" classes together and ended up with two classes: "Wheat", and "Non-wheat". The classified image often needs further processing to clean up the random noise and small isolated regions to improve the quality of the output. With a simple segmentation, we extracted wheat plot polygons from the image.

This method works best with high-resolution imagery taken at the beginning of the growing season when all the classes are clearly distinguishable from each other on the image. Otherwise time consuming post-processing is needed in order to separate the mixed classes.

To analyze the result of VI extraction from the orthomosaic, we used the zonal statistics plugin in the open source software QGIS. We calculated several values of the pixels of orthomosaic (such as the average VI values,

Min, Max, standard deviation, majority, minority and also the median of VI values for each plot and the total count of the pixels that are within a plot boundary), using the polygonal vector layer of plot boundaries generated by one of the methods above. We then merged the data from the field map spreadsheet (plot ID, entry numbers, trial name, number of rep, number of block, row and column, and other information on planting) with the generated plots statistics.

### Accuracy evaluation of different types of aerial imagery sensors and platforms

To evaluate the accuracy and reliability of our low cost UAS and consumer grade camera, we calculated correlations between calibrated DN values from the orthomosaic images of Canon S100 and MultiSpec 4C camera and band equivalent reflectance (BER) of field spectra. For ground-truth evaluation we used the ASD Fieldspectro 2 spectroradiometer to measure spectral reference of 280 sample plots after running the UAS within the time window of 1 h around noon. The ASD spectroradiometer measures a wavelength range of 325–1075 nm, a wavelength accuracy of ±1 nm and a spectral resolution of <3 nm at 700 nm (Table 1). Using the ASD, the top of canopy reflectance was measured approximately 0.5 m above the plants. To account for the very small field of view of this sensor, we constantly monitored the spectral response curves to avoid measuring mixed pixels in our samplings and we averaged four reflectance readings within each sample plots. Reflectance measurements were calibrated at every 15 plots using a white (99 % reflectance) Spectralon calibration panel.

To evaluate the accuracy of each platform, we calculated the correlations between different VIs extracted from Canon S100 and VIs extracted from MultiSpec 4C compared to the BER of field spectra. We calculated BER for each camera separately by taking the average of all the spectroradiometer bands that are within the full width at half maximum (FWHM) of each instrument (Table 1). The Pearson correlation between the average VI values of the sample plots and field spectra for each platform was calculated. The resulting correlation coefficients were tested using a two-tail test of significance at the $p \leq 0.05$ level. Accuracy assessment was based on computing the root mean square error (RMSE) for GNDVI image by comparing the pixel values of the reflectance image to the corresponding BER of field spectra at the sample sites not selected to generate the regression equations.

### Broad-sense heritability

Broad-sense heritability, commonly known as repeatability, is an index used by plant breeders and geneticists to measure precision of a trial [37–39]. The ratio of the genetic variance to the total phenotypic variance is broad-sense heritability ($H^2$), which varies between 0 and 1. A heritability of 0 indicates there is no genetic variation contributing to phenotypic variation, and $H^2$ of 1 if the entire phenotypic variation is due to genetic variation and no environmental noise. For the purpose of comparing phenotyping tools and approaches, broad-sense heritability calculated for measurements on a given set of plots under field conditions at the same point in time, is a direct assessment of measurement error as the genetics and environmental conditions are constant.

In order to assess the accuracy of the different HTP platforms, we calculated broad-sense heritability on an entry-mean basis for each individual trial and phenotypic sampling date on May 6, 2015. To calculate heritability we performed a two steps process where we first calculated the plot average and then fit a mixed model for replication, block and entry as random effects. Heritability was calculated from Eq. 1 [40].

$$H^2 = \frac{\sigma^2_{genotypic}}{\sigma^2_{phenotypic}} = \frac{\sigma^2_{genotypic}}{\sigma^2_{genotypic} + \frac{\sigma^2_{error}}{replication}} \quad (1)$$

where $\sigma^2_{genotypic}$ is genotypic variance and $\sigma^2_{error}$ is error variance.

## Results and discussion

### Developing an image processing pipeline for HTP

To evaluate the potential of UAS for plant breeding programs, we deployed a low cost, consumer grade UAS and imaging system at multiple times throughout the growing season. Using the aerial imagery collected on May 6, we built a robust pipeline scripted in Python that facilitated image analysis and allowed a semi-automated image analysis algorithm to run with minimum user support. The only requirement for manual input in the processing pipeline is to validate the GCP positions. This pipeline was tested on eight temporal datasets collected from IRIS+ flights over heat trials during the growing season.

The individual parameters selected during orthomosaic generation and plot extraction can have a large impact on the quality of the orthomosaic and subsequent data analysis. Because of the multiple parameter settings that users could choose, we investigated how changing various parameters would affect the data quality. To obtain more accurate camera position estimates, we selected higher accuracy at the alignment step. For pair selection mode, we chose reference since in this mode the overlapping pairs of photos are selected based on the estimated camera locations (i.e. from the GPS log of the control system). For the upper limit of feature points on every image to be taken into account during alignment stage, we accepted the default value of 40,000. We imported positions for all of the ten GCPs available on the field to geo-reference the aerial images. The GCPs were also used in the self-calibrating bundle adjustment to ensure correct geo-location and to avoid any systematic error or block deformation. To improve georeferencing accuracy, we performed alignment optimization.

Finally, to generate orthomosaic, we tested the two methods of blending: mosaic and average. We found the orthomosaic generated by the mosaic method gives more quality for the orthomosaic and texture atlas than the average mode. On the other hand, mosaic blending mode generated artifacts in the image since it uses fewer images by selecting pixels with the shortest distance from the image center (Fig. 4). We examined the correlation between orthomosaic generated from these two approaches and the ground truth data and found the averaging method has slightly higher correlation of 0.68 compared to mosaic method with the correlation of 0.65.

The 16 bit orthomosaic TIFF image was generated semi-automatically using python scripting in PhotoScan software with a resolution of 0.8 cm, matching the ground sampling distance (GSD) of the camera. Due to the short flight times (less than 20 min) and low flying altitude (30 meters above ground level), we assume there is no need for atmospheric corrections [28].

To convert DNs to reflectance values, a modified empirical approach was implemented using pixel values of IRIS+ aerial imagery and field-based reflectance measurements from handheld spectroradiometer. Calibration Eqs. (2, 3, and 4) calculated from non-linear relationship between average field spectra and relative spectra response of each band separately were determined as:

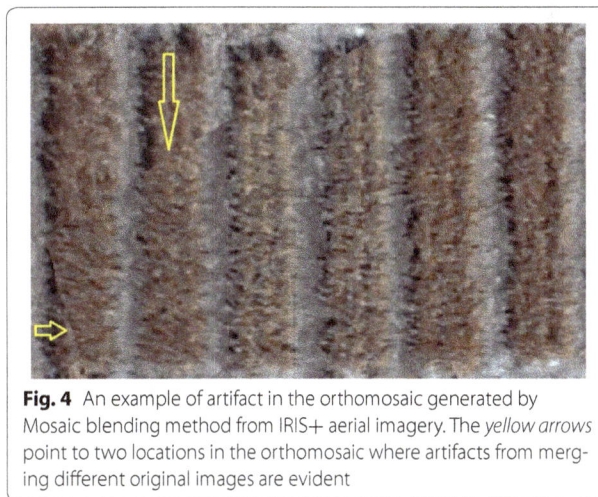

**Fig. 4** An example of artifact in the orthomosaic generated by Mosaic blending method from IRIS+ aerial imagery. The *yellow arrows* point to two locations in the orthomosaic where artifacts from merging different original images are evident

As breeding trials are regularly containing thousands or tens of thousands of plots, an automated or semi-automated approach to generating and overlaying plot boundaries is needed.

### Simple-grid based method

We first implemented a simple overlay grid (Fig. 5a) using QGIS open source software. This method, although fast and easy, is not accurate for plot assigning as it does not account for within plot gaps (for plots planted in beds), gaps between plots and also gaps between each range of wheat plots. It simply generates polygons attached to each other and not spaced as needed to correctly capture the actual plot

$$Reflectance_{Blue} = -\exp(-\log(4.495970e^{-01} + (3.191189e^{-05}) * DN_{Blue}))$$
$$(r^2 = 0.98, \ P < 0.001)$$
(2)

$$Reflectance_{Green} = -\exp(-\log(4.705485e^{-01} + (3.735676e^{-05}) * DN_{Green}))$$
$$(r^2 = 0.97, \ P < 0.001)$$
(3)

$$Reflectance_{NIR} = -\exp(-\log(1.197440e^{+00} + (4.712299e^{-05}) * DN_{NIR}))$$
$$(r^2 = 0.94, \ P < 0.001)$$
(4)

We applied each calibration equation to their corresponding band image and converted the image raw DNs into reflectance values.

### Wheat plot boundary extraction results

We evaluated multiple approaches for defining and overlaying plot coordinate boundaries for the large field trials.

boundaries (Fig. 5a). In the example below, it can also be noticed that a boundary polygon may cover part of the neighbor wheat plot, which is due to the assumption of a fixed plot boundary size in this method. To account for the gap between plots, beds and also between ranges of plots, we developed a field map based method.

*Field-map based method*

To accurately reflect the actual field planting and plot size, we developed a script to overlay defined plot sizes with known spacing. We reformatted the field map using the information in the spreadsheet such as: plot ID, and plot's location based on the block number, column and row's number, and also knowing the plots were planted on beds with a plot size of 2.8 m length and 1.6 m wide. This method allows us to eliminate border effect by changing the plot size to 2.3 m by 1.5 m. In the reformatted field map spreadsheet, the first row of the excel sheet is the length of the plots (1.5 m) and the length of the gap between plots in beds (0.15 m) and gaps between each bed (0.8 m). The first column is the width of wheat plots (2.3 m) and the width of the gaps between each range (0.5 m). The cells in the reformatted spreadsheet are filled with the plot's information including plot ID, trial name, and planting date. From this plot level information, we overlay plots of the defined size with plot-to-plot spacing that accurately reflects the field configuration (Fig. 5b).

*Image classification plot extraction*

To examine an image classification based plot extraction method, we first generated GNDVI from the orthomosaic image generated from the April 10 aerial imagery rather than the May 6 dataset analyzed in the rest of this study. The reason for choosing this set of data was that in this dataset from earlier in the season had the sharpest contrast between plots and the surrounding soil. We then applied a Maximum Likelihood Classification to the GNDVI image. We converted the classified image into polygons to obtain plot shape polygons (Fig. 5c). This approach is more reliable if the early season aerial imagery is provided. The drawback with this approach is that it requires post-processing and cleanup of the final result if the classes are not quite distinguished and separated from each other in the orthomosaic image. Frequently, this approach did not separate plots properly and manual editing was required (Fig. 5c). As it can be seen in the Fig. 5c, the two plots slightly merged with each other and it caused the classification to classify them as one single plot. To have more accurate results from image classification method, the presence of early season imagery is needed, since this process can have more post-processing and cleanup if the wheat plots and the gaps between plots are not visually discrete from each other in the images.

**Plot-level GNDVI extraction results**

After generating VI maps, we compared these three plot-level extraction techniques using GNDVI values extracted form canon S100 aerial imagery and the

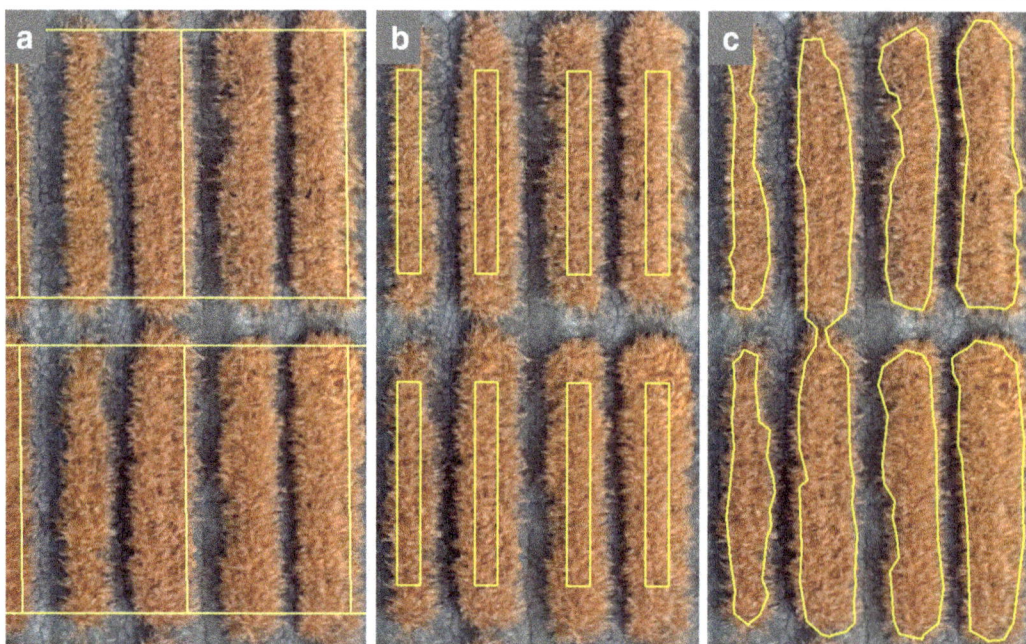

**Fig. 5** Results of different plot extraction method overlay on a subset of plots from IRIS+ calibrated orthomosaic. **a** Simple-grid plots, **b** field-map based method, **c** image classification approach and an example of misclassification; this example confirms the need of post processing for image classification method

corresponding values extracted from spectroradiometer in the field. The field-map based approach for plot polygons generated the best correlation to spectroradiometer measurements, while the classification technique had the lowest correlation with ground-truth data (Table 3). The lower reliability of classification technique compared to field-map based, could be due to the absence of early season imagery. In generating plot boundaries using field-map based, we also considered a buffer around the polygons to avoid any possible mixed plants or border effects for a plot which likely improved performance. We then correlated plot level GNDVI calculated from the spectral reflectance for those same plots using different boundary plot extraction methods (Table 3). From the comparison of different plot extraction methods with correlation to spectral reflectance we found that the boundaries defined using the map-based algorithm were superior to both the simple grid and the classification-based approach (Table 3).

### Empirical line correction

Using Python scripting in QGIS, we generated different VI maps from the calibrated orthomosaic image and extracted plot boundaries using the field-map based method. Using the zonal statistics plugin, we then calculated statistic values of each plot of wheat in the orthomosaic. For each individual plot, the average VI value was used for further analysis and correlation. We tested the utility of applying an empirical line correction to improve the converting of digital numbers to reflectance values for the cameras. Using the same field-map based plot extraction, the correlation between raw GNDVI values extracted from the orthomosaic image and the BER increased from 0.68 to 0.76 after performing empirical line method as the radiometric calibration method.

### Comparison of different vegetation indices

To test the accuracy of VI from the digital cameras by comparison to the ASD spectroradiometer, we calculated BER for Canon S100 by taking the average of all the spectroradiometer bands that are within the FWHM of this camera (Table 1). For example, to calculate BER GNDVI for Canon S100, we averaged the reflectance values between 490 and 550 nm as green band, and averaged all the values between 680 and 760 nm as NIR band.

Among all the calculated VIs for the Canon S100, GNDVI had the highest correlation with the spectroradiometer (Table 4; Fig. 6). Since the MultiSpec 4C camera has 4 bands, NIR, red-edge, red and green, we were able to generate a red-edge normalized difference vegetation index (RENDVI) as well as other VIs. Using the developed pipeline, we extracted wheat plots from the generated orthomosaic for RENDVI by means of the field-map based method, which had also proved to be the best method of plot extraction for the MultiSpec 4C camera. The RENDVI plot values had higher correlation with spectroradiometer when compared to other VIs extracted from MultiSpec 4C camera (Fig. 7).

We found the correlation between VIs extracted from spectroradiometer and calibrated Canon S100 was significantly higher than the one with MultiSpec 4C camera. This could be due to lower flight altitude, and slower movements of multi-rotor IRIS+ compare to fixed wing eBee Ag, which results in higher resolution imagery captured by IRIS+/Canon S100. The MultiSpec 4C camera is a narrow band multispectral camera that does not have any spectral overlap in its band response. The better spectral performance could be negated by the lower

**Table 3 Correlation analysis between mean GNDVI values extracted from raw Canon S100 digital number values, Calibrated Canon S100 digital number values and corresponding band equivalent reflectance (BER) GNDVI values from the spectroradiometer using different plot boundary extraction method**

| | Raw digital numbers | Calibrated digital numbers | | |
| --- | --- | --- | --- | --- |
| | Field-map based | Field-map based | Classification | Simple-grid |
| Spectroradiometer BER | 0.68 | 0.76 | 0.58 | 0.68 |

**Table 4 Calculated vegetation indices for each cameras, based on their spectral bands and correlation between vegetation index values extracted from UAS imagery and band equivalent reflectance values from Spectroradiometer**

| Index | Formula | Correlation | |
| --- | --- | --- | --- |
| | | Canon S100 | MultiSpec 4C |
| NDVI | $\frac{NIR-Red}{NIR+Red}$ | – | 0.33 |
| GNDVI | $\frac{NIR-Green}{NIR+Green}$ | 0.68 | 0.63 |
| RENDVI | $\frac{NIR-Red\_edge}{NIR+Red\_edge}$ | – | 0.64 |
| ENDVI | $\frac{NIR+Green-2*Blue}{NIR+Green+2*Blue}$ | 0.54 | – |
| GIPVI | $\frac{NIR}{NIR+Green}$ | 0.59 | – |

The values from Canon S100 values are before applying modified empirical line correction method

GNDVI values for Canon S100, and RENDVI for MultiSpec 4C have higher correlation with spectroradiometer compoare to other vegetation indices (shown in italics font)

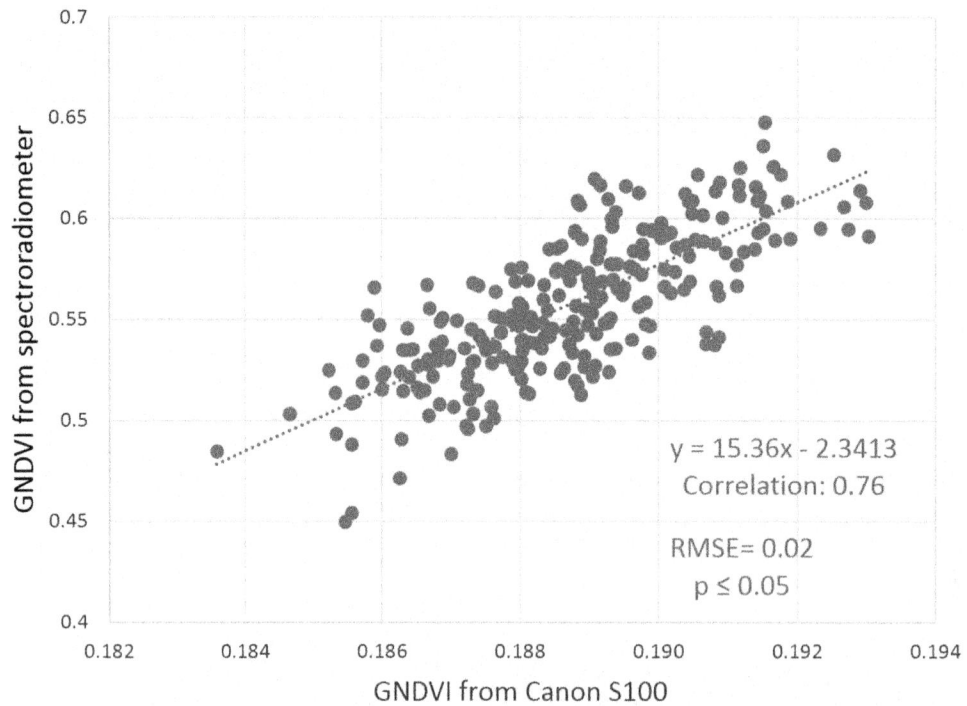

**Fig. 6** Linear relation between GNDVI from Canon S100 and BER GNDVI from spectroradiometer. Sample plot GNDVI values were extracted from calibrated orthomosaic of Canon S100 imagery, IRIS+ and band equivalent GNDVI for Canon S100 camera calculated from ASD spectroradiometer readings for 280 sample plots

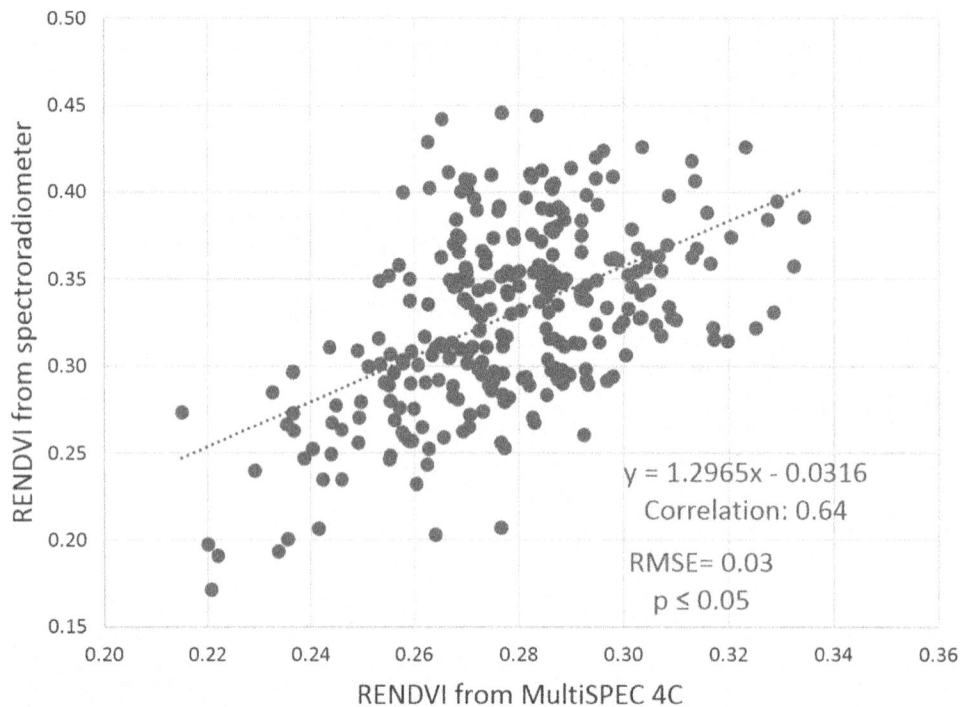

**Fig. 7** Scatterplot of the plot-level RENDVI from MultiSpec 4C imagery versus spectroradiometer. Plot-level RENDVI were extracted from orthomosaic generated from MultiSPEC 4C camera imagery using field-map based plot extraction method, and Band equivalent RENDVI for Spec 4C calculated from spectroradiometer reading for 280 sample plots

resolution of the sensor, higher flight altitude, and faster travel speed of the fixed wing UAV. It was also clear that the modified empirical line correction also played a role in higher correlation of Canon S100 values with ground truth spectral measurements.

We also calculated the correlation between the spectra readings using the handheld spectroradiometer. As we had four readings per plot, we were able to determine the repeatability within the spectroradiometer measurements. We averaged the first two measurements of and last two measurements for each plot independently and calculated an overall correlation for 280 sample plots (Table 5). Based on this assessment, the UAS imaging had repeatability as good as, or better than the spectroradiometer.

### Broad sense heritability

As the Canon S100 and MultiSpec 4C measurements were taken at the same time on the same day and on the same plots, the environmental variance and genetic variance can be considered as negligible. Therefore, the broad sense heritability values ($H^2$) can be interpreted as a level of precision of sensor measurement, which is the only remaining significant source of error variance. Overall the heritability for VIs was high. We found VIs derived from the Canon S100 imagery had a higher heritability than the MultiSpec 4C data for nine out of eleven trials (Table 6). The trials had lower, but still high, heritability for grain yield. There were some unexpected rains during the crop cycle followed by irrigation issues in the borders of the field, however, the heritability for yield was still above 0.60 which is a moderate repeatability for breading trials under heat (Table 6).

### Conclusion

We developed a semi-automated pipeline for data analysis of a low cost UAS imagery. The raw images of a consumer grade digital camera were pre-processed and used as the input of the image-processing pipeline. During the mosaicking step, we found that the averaging method had marginally higher correlation of 0.68 compared to mosaic

**Table 5 Correlation between repeated measurements from Handheld ASD for two different vegetation indices**

|  | Correlation |
| --- | --- |
| Green normalized difference vegetation index; GNDVI (Canon S100) | 0.62 |
| Red-edge normalized difference vegetation index; RENDVI (MultiSpec 4C) | 0.88 |

Four measurements were taken on each of 280 plots. To determine repeatability, the first two measurements were averaged and correlated to the average of the second two measurements on a per plot basis across all plots. The VIs presented are band equivalent indices for the Canon S100 (green normalized difference vegetation index; GNDVI) and MultiSpec 4C (red-edge normalized difference vegetation index: RENDVI) used in this study

**Table 6 Broad-sense heritability in 11 trials for vegetation indices and grain yield**

| Trial name | Broad-sense heritability ($H^2$) | | Grain yield |
| --- | --- | --- | --- |
|  | Vegetation indices | |  |
|  | Canon S100 (GNDVI) | MultiSpec 4C (RENDVI) |  |
| EYTBWBLHT_01 | 0.89 | 0.84 | 0.66 |
| EYTBWBLHT_02 | 0.90 | 0.90 | 0.91 |
| EYTBWBLHT_03 | 0.84 | 0.90 | 0.78 |
| EYTBWBLHT_04 | 0.87 | 0.70 | 0.75 |
| EYTBWBLHT_05 | 0.81 | 0.80 | 0.82 |
| EYTBWBLHT_06 | 0.73 | 0.66 | 0.80 |
| EYTBWBLHT_07 | 0.83 | 0.70 | 0.85 |
| EYTBWBLHT_08 | 0.64 | 0.61 | 0.86 |
| EYTBWBLHT_09 | 0.78 | 0.64 | 0.87 |
| EYTBWBLHT_10 | 0.66 | 0.57 | 0.65 |
| EYTBWBLHT_11 | 0.72 | 0.57 | 0.74 |

The VIs derived from the Canon S100 had a higher heritability than the MultiSpec 4C for nine out of eleven trials (shown in italics font)

method with the correlation of 0.65. Using a modified empirical line method, we radiometrically calibrated the DN values extracted from canon S100 and compared the calibrated digital values with the reflectance values of the spectroradiometer to evaluate the comparability of our data. Our results confirm that radiometric calibration is important for consumer grade cameras to convert the DN values to reflectance measurements and can improve the value of the image data.

We examined three different ways of wheat plot extraction: simple grid, field-map based, and image classification. We found that the field-map based technique is more accurate and fastest compared to other techniques, and had a higher correlation to ground truth data. The advantage of this method is that it is applicable to any crop types as long as the field map is provided, and also this method is fully automated in Python.

We evaluated two types of UAS; (1) low cost: IRIS+ and (2) commercial grade agriculture-use: eBee Ag, each carrying different sensors with (1) low cost consumer grade camera: modified NIR Canon S100 on the IRIS+ and (2) a more specialized multispectral camera: MultiSpec 4C camera on the eBee. We compared their performance with ground-truth reflectance data and found overall good performance to spectroradiometer measurements. Based on correlation to the spectral readings and assessment of heritability, the Canon S100 had better performance than the MultiSpec 4C mounted on the fixed wing, which was likely a result of higher resolution of the sensor, lower altitude and slower travel speed for the Canon S100 carried on the quadcopter IRIS+.

Our study provides strong evidence of the value of UAS for HTP applied to large wheat breeding nurseries. Further work is needed to investigate the strength of the relationship between remotely sensed derived plant phenological traits and the wheat biophysical properties collected in the field such as plant height, biomass, and yield. With the overall vision of integrating multiple measurements extracted from UAS (plant height, ground cover, etc.) with plant growth simulations to maximize the biological utility of the estimated phenotypes new avenues will be opened to breeders for predicting yield. Moreover, the development of new sensors and imaging systems undoubtedly will continue to improve our ability to phenotype very large experiments or breeding nurseries. When combined with genomic and physiological modeling, the rapid, low-cost evaluation of large field trials in plant breeding with UAS platforms has the potential to greatly accelerate the breeding process through more accurate selections on larger populations.

## Abbreviations
BER: band equivalent reflectance; CIMMYT: International Maize and Wheat Improvement Center; DEM: digital elevation model; DN: digital number; DPP: digital photo professional; ENDVI: enhanced normalized difference vegetation index; FWHM: full width at half maximum; GCP: ground control points; GIPVI: infrared percentage vegetation index; GNDVI: green normalized difference vegetation index; GSD: ground sampling distance; HTP: high-throughput phenotyping; LAI: leaf area index; NDVI: normalized difference vegetation index; NIR: near infrared; RENDVI: red-edge normalized difference vegetation index; RMSE: root mean square error; SIFT: scale-invariant feature transform; UAS: unmanned aerial system; UAV: unmanned aerial vehicle; VI: vegetation index.

## Authors' contributions
Conceived and designed the study: DS, RPS, DG and JP; Developed and Validated Methods: AH, LGP, DS, and IOM; Collected Data: AH, LGP, SM, JR, and RPS; Analyzed Data and Interpreted Results: AH, SM and JP. All authors read and approved the final manuscript.

## Author details
[1] Department of Geography, Kansas State University, Manhattan, KS 66506, USA. [2] International Maize and Wheat Improvement Center (CIMMYT), Int. Apdo., Postal 6-641, 06600 Mexico, D.F., Mexico. [3] Interdepartmental Genetics Program, Kansas State University, Manhattan, KS 66506, USA. [4] Department of Mechanical and Nuclear Engineering, Kansas State University, Manhattan, KS 66506, USA. [5] International Programs of the College of Agriculture and Life Science, Cornell University, Ithaca, NY 14853, USA. [6] Wheat Genetics Resource Center, Department of Plant Pathology and Department of Agronomy, Kansas State University, Manhattan, KS 66506, USA.

## Acknowledgements
This work was done through the International Maize and Wheat Improvement Center (CIMMYT), Mexico and supported through the National Science Foundation—Plant Genome Research Program (IOS-1238187) and the US Agency for International Development Feed the Future Innovation Lab for Applied Wheat Genomics (Cooperative Agreement No. AID-OAA-A-13-00051).

## Competing interests
The authors declare that they have no competing interests.

## References
1. Ray DK, Mueller ND, West PC, Foley JA. Yield trends are insufficient to double global crop production by 2050. PLoS One. 2013;8:e66428.
2. White JW, Andrade-Sanchez P, Gore MA, Bronson KF, Coffelt TA, Conley MM, Feldmann KA, French AN, Heun JT, Hunsaker DJ, Jenks MA, Kimball BA, Roth RL, Strand RJ, Thorp KR, Wall GW, Wang G. Field-based phenomics for plant genetics research. Field Crop Res. 2012;133:101–12.
3. Cobb JN, DeClerck G, Greenberg A, Clark R, McCouch S. Next-generation phenotyping: requirements and strategies for enhancing our understanding of genotype–phenotype relationships and its relevance to crop improvement. Theor Appl Genet. 2013;126:867–87.
4. Granier C, Vile D. Phenotyping and beyond: modelling the relationships between traits. Curr Opin Plant Biol. 2014;18:96–102.
5. Dhondt S, Wuyts N, Inzé D. Cell to whole-plant phenotyping: the best is yet to come. Trends Plant Sci. 2013;18:428–39.
6. Furbank R, Tester M. Phenomics–technologies to relieve the phenotyping bottleneck. Trends Plant Sci. 2011;16:635–44.
7. Fiorani F, Schurr U. Future scenarios for plant phenotyping. Annu Rev Plant Biol. 2013;64:267–91.
8. Walter A, Liebisch F, Hund A. Plant phenotyping: from bean weighing to image analysis. Plant Methods. 2015;11:14.
9. Busemeyer L, Mentrup D, Möller K, Wunder E, Alheit K, Hahn V, Maurer H, Reif J, Würschum T, Müller J, Rahe F, Ruckelshausen A. BreedVision—a multi-sensor platform for non-destructive field-based phenotyping in plant breeding. Sensors. 2013;13:2830–47.
10. Andrade-Sanchez P, Gore MA, Heun JT, Thorp KR, Carmo-Silva AE, French AN, Salvucci ME, White JW. Development and evaluation of a field-based high-throughput phenotyping platform. Funct Plant Biol. 2014;41:68–79.
11. Crain J, Wei Y, Barker J, Sean T, Alderman P, Reynolds M, Zhang N, Poland J. Development and deployment of a portable field phenotyping platform. Crop Sci. 2015;56(3):965–75.
12. Deery D, Jimenez-Berni J, Jones H, Sirault X, Furbank R. Proximal remote sensing buggies and potential applications for field-based phenotyping. Agronomy. 2014;4:349–79.
13. White J, Andrade-Sanchez P, Gore MA, Bronson KF, Coffelt TA, Conley MM, Feldmann KA, French AN, Heun JT, Hunsaker DJ, Jenks MA, Kimball BA, Roth RL, Strand RJ, Thorp KR, Wall GW, Wang G. Field-based phenomics for plant genetics research. Field Crop Res. 2012;133:101–12.
14. Berger B, Parent B, Tester M. High-throughput shoot imaging to study drought responses. J Exp Bot. 2010;61:3519–28.
15. Munns R, James RA, Sirault XRR, Furbank RT, Jones HG. New phenotyping methods for screening wheat and barley for beneficial responses to water deficit. J Exp Bot. 2010;61:3499–507.
16. Araus JL, Cairns JE. Field high-throughput phenotyping: the new crop breeding frontier. Trends Plant Sci. 2014;19:52–61.
17. Baluja J, Diago MP, Balda P, Zorer R, Meggio F, Morales F, Tardaguila J. Assessment of vineyard water status variability by thermal and multispectral imagery using an unmanned aerial vehicle (UAV). Irrig Sci. 2012;30:511–22.
18. Chao H, Cao Y, Chen Y. Autopilots for small unmanned aerial vehicles: a survey. Int J Control Autom Syst. 2010;8:36–44.
19. Dunford R, Michel K, Gagnage M, Piégay H, Trémelo M-L. Potential and constraints of unmanned aerial vehicle technology for the characterization of Mediterranean riparian forest. Int J Remote Sens. 2009;30:4915–35.
20. Gonzalez-Dugo V, Zarco-Tejada P, Nicolás E, Nortes PA, Alarcón JJ, Intrigliolo DS, Fereres E. Using high resolution UAV thermal imagery to assess the variability in the water status of five fruit tree species within a commercial orchard. Precis Agric. 2013;14:660–78.
21. Chapman S, Merz T, Chan A, Jackway P, Hrabar S, Dreccer M, Holland E, Zheng B, Ling T, Jimenez-Berni J. Pheno-Copter: a low-altitude, autonomous remote-sensing robotic helicopter for high-throughput field-based phenotyping. Agronomy. 2014;4:279–301.
22. Hunt ER Jr, Hively WD, Fujikawa SJ, Linden DS, Daughtry CST, McCarty GW. Acquisition of NIR-green–blue digital photographs from unmanned aircraft for crop monitoring. Remote Sens. 2010;2:290–305.
23. Liebisch F, Kirchgessner N, Schneider D, Walter A, Hund A. Remote, aerial phenotyping of maize traits with a mobile multi-sensor approach. Plant Methods. 2015;11:9.

24. Haest B, Biesemans J, Horsten W, Development S, Everaerts J, Manager P, Van Camp N, Van Valckenborgh J. Radiometric calibration of digital photogrammetric. In: Proceedings of ASPRS annual meeting 2009.

25. Smith GM, Milton EJ. The use of the empirical line method to calibrate remotely sensed data to reflectance. Int J Remote Sens. 1999;20:2653–62.

26. Gu XF, Guyot G, Verbrugghe M. Evaluation of measurement errors in ground surface reflectance for satellite calibration. Int J Remote Sens. 1992;13:2531–46.

27. Haghighattalab A. Plot boundary extraction. 10.5281/zenodo.46732. 2015.

28. Haghighattalab A, Agisoft LLC. Orthomosaic generation. 10.5281/zenodo.46734. 2014.

29. Lowe DG. Distinctive image features from scale-invariant keypoints. Int J Comput Vis. 2004;60:91–110.

30. Agisoft LLC: Agisoft PhotoScan User Manual: Professional Edition, Version 1.2. 2016.

31. Wang C, Price KP, Van Der Merwe D, An N, Wang H. Modeling above ground biomass in tallgrass prairie using ultra high spatial resolution sUAS imagery. Photogramm Eng Remote Sens. 2014;80:1151–9.

32. Staben GW, Pfitzner K, Bartolo R, Lucieer A. Empirical line calibration of WorldView-2 satellite imagery to reflectance data: using quadratic prediction equations. Remote Sens Lett. 2012;3:521–30.

33. Staben G, Pfitzner K. Calibration of WorldView-2 satellite imagery to reflectance data using an empirical line method. In: Proceedings of 34th international symposium on remote sensing environment GEOSS Era Towar Oper Environ Monit Sydney, Aust 2011.

34. Wang C, Myint SW. A Simplified empirical line method of radiometric calibration for small unmanned aircraft systems-based remote sensing. IEEE J Sel Top Appl Earth Obs Remote Sens. 2015;8:1876–85.

35. Kross A, McNairn H, Lapen D, Sunohara M, Champagne C. Assessment of RapidEye vegetation indices for estimation of leaf area index and biomass in corn and soybean crops. Int J Appl Earth Obs Geoinf. 2015;34:235–48.

36. Gitelson A, Merzlyak M. Remote sensing of chlorophyll concentration in higher plant leaves. Adv Spaces Res. 1998;22:689–92.

37. Xu NW, Xu S, Ehlers J. Estimating the broad-sense heritability of early growth of cowpea. Int J Plant Genom. 2009;2009:984521.

38. Piepho H-P, Möhring J. Computing heritability and selection response from unbalanced plant breeding trials. Genetics. 2007;177:1881–8.

39. Visscher PM, Hill WG, Wray NR. Heritability in the genomics era—concepts and misconceptions. Nat Rev Genet. 2008;9:255–66.

40. Holland J, Nyquist W, Cervantes-Martinez C. Estimating and interpreting heritability for plant breeding: an update. Plant Breed Rev. 2003;22:9–112.

# 16

# A simple test for the cleavage activity of customized endonucleases in plants

Nagaveni Budhagatapalli[1†], Sindy Schedel[1†], Maia Gurushidze[1], Stefanie Pencs[1,2], Stefan Hiekel[1], Twan Rutten[3], Stefan Kusch[4], Robert Morbitzer[5], Thomas Lahaye[5], Ralph Panstruga[4], Jochen Kumlehn[1] and Goetz Hensel[1*]

## Abstract

**Background:** Although customized endonucleases [transcription activator-like effector nucleases (TALENs) and RNA-guided endonucleases (RGENs)] are known to be effective agents of mutagenesis in various host plants, newly designed endonuclease constructs require some pre-validation with respect to functionality before investing in the creation of stable transgenic plants.

**Results:** A simple, biolistics-based leaf epidermis transient expression test has been developed, based on reconstituting the translational reading frame of a mutated, non-functional *yfp* reporter gene. Quantification of mutation efficacy was made possible by co-bombarding the explant with a constitutive *mCherry* expression cassette, thereby allowing the ratio between the number of red and yellow fluorescing cells to serve as a metric for mutation efficiency. Challenging either stable mutant alleles of a compromised version of *gfp* in tobacco and barley or the barley *MLO* gene with TALENs/RGENs confirmed the capacity to induce site-directed mutations.

**Conclusions:** A convenient procedure to assay the cleavage activity of customized endonucleases has been established. The system is independent of the endonuclease platform and operates in both di- and monocotyledonous hosts. It not only enables the validation of a TALEN/RGEN's functionality prior to the creation of stable mutants, but also serves as a suitable tool to optimize the design of endonuclease constructs.

**Keywords:** Transcription activator-like effector nucleases, RNA-guided endonucleases, Biolistic gene transfer, Site-directed mutagenesis, Transient expression

## Background

The exploitation of either customizable transcription activator-like effector nucleases (TALENs) [1, 2] or RNA-guided endonucleases (RGENs) [3, 4] has opened up numerous possibilities for site-directed mutagenesis and precise genome editing in plants. The gaps induced in the host's DNA are repaired by either non-homologous end joining or by homology-directed repair. The former process is error-prone and so randomly introduces insertions and deletions (indels), while the latter, which exploits the respective locus of the sister chromatid or an artificially

provided DNA that combines homology to the target site with an alteration of choice as repair template, is highly precise [5, 6]. The documented use of customized endonucleases in plants to date has largely involved *Arabidopsis thaliana, Nicotiana benthamiana*, maize or rice [7]. A few inherited TALEN- or RGEN-induced mutations have been reported in both barley [8, 9] and wheat [10]. Budhagatapalli et al. [11] have demonstrated the feasibility of homology-directed editing of barley at the cellular level. In mammalian cells, mutation frequencies of up to, respectively, 60 and 80 % have been achieved following the application of TALENs and RGENs [12].

Customizing endonucleases remains a somewhat empirical process, which would benefit from the development of a simple validation assay able to be carried out prior to transformation. The purpose of the present research was to establish a chimeric expression construct

---

*Correspondence: hensel@ipk-gatersleben.de

†Nagaveni Budhagatapalli and Sindy Schedel contributed equally to this work

[1] Plant Reproductive Biology, Leibniz Institute of Plant Genetics and Crop Plant Research (IPK), Corrensstr. 3, 06466 Stadt Seeland/OT Gatersleben, Germany

Full list of author information is available at the end of the article

comprising a target sequence positioned upstream of a mutated, non-functional copy of *yfp* (encoding the cellular marker yellow fluorescent protein). The compromised *yfp* sequence features a frameshift mutation which can be reversed by the activity of a sequence-specific endonuclease; thus the frequency of cells producing YFP was expected to reflect the cleavage activity of the endonuclease. The assay has been deliberately designed to be usable with any customizable endonuclease platform in both di- and monocotyledonous plant species. One of the chosen targets in barley was the *MLO* gene, the product of which is associated with susceptibility to infection by *Blumeria graminis*, the causative fungal pathogen of powdery mildew; *mlo* alleles, which fail to produce a functional MLO protein, typically imbue the host with a durable, broad-spectrum resistance against the disease [13, 14].

## Results

### The test system in the dicotyledonous species tobacco

The principle of the assay is that the error-prone repair of endonuclease-induced double-strand breaks would generate indels at the target site. Consequently, a proportion of induced mutations were expected to restore the functionality of the compromised *yfp* sequence positioned downstream of the target sequence (Fig. 1). When tobacco (*Nicotiana tabacum* cv. SR1) leaf segments were co-bombarded with a nuclease-specific vector construct along with a TALEN or an RGEN construct, a number of epidermal cells began to accumulate YFP (Fig. 2a). The mutation frequency was quantified by co-bombarding a constitutive *mCherry* expression cassette; the ratio between the number of red and yellow fluorescing cells was then used to estimate the efficiency of the endonuclease construct (Fig. 2b). The co-bombardment of pTARGET-gfp1 with the pair of *gfp*-specific TALEN constructs (Table 1) generated $100 \pm 71$ yellow and $363 \pm 148$ red fluorescent cells, corresponding to a ratio of 27 %; meanwhile, the co-bombardment of pTARGET-gfp1 with the *gfp*-specific RGEN construct yielded a ratio of 75 % (Table 2). Bombardment with the pTARGET-gfp1 vector on its own failed to induce any cells to synthesize YFP.

To estimate the frequency of mutations achievable in a stably transformed plant, a transgenic tobacco plant harboring a single copy of *gfp* [15] was co-transformed with the pair of *gfp*-specific TALEN constructs to produce plants carrying only one of the TALEN units; no mutations were detectable in the *gfp* sequence in either transgenic. The two TALEN units were then brought together by intercrossing the primary transgenics. Of the 35 progeny found to harbor both *gfp*-specific TALEN units (Table 3), one carried a mutated form of *gfp*. In a similar experiment based on the *gfp*-specific RGEN construct, 17 transgenics were obtained; of these, 15 carried *Cas9* and guide-RNA (gRNA), and of these, 12 contained mutations (Table 3). The mutations obtained were diverse in nature [15].

### The test system in the monocotyledonous species barley

Co-bombardment of barley (*Hordeum vulgare* cv. 'Golden Promise') leaves with pTARGET-gfp1 and the pair of *gfp*-specific TALEN constructs (Table 1) produced $77 \pm 25$ epidermal cells showing yellow (YFP) fluorescence and $250 \pm 59$ exhibiting red (mCherry) fluorescence (Table 2), yielding a ratio of 30.7 %. Bombardment with the *gfp*-specific RGEN construct and pTARGET-gfp2 led to a relative cleavage activity of 31.4 %. To validate the functionality of TALENs targeting an endogenous barley gene, a further co-bombardment of the same barley material was conducted using the *MLO*-specific pTARGET-MLO reporter plasmid and the appropriate TALEN pair #3: this resulted in a relative cleavage activity of 42.0 % (Table 2). The *MLO*-specific TALEN constructs were also tested in combination with the *GUS* reporter gene in barley (cv. 'Ingrid'; using bombardment) and in *N. benthamiana* (using *Agrobacterium tumefaciens*-mediated transient expression) using five different TALEN pairs targeting two different exons of the *MLO* gene. Out of the five tested TALEN pairs (Additional file 1), only TALEN pairs #2 and #3 showed detectable activity and restored *GUS* expression when the TALEN constructs were co-transformed with the respective reporter construct (Additional file 2). While TALEN pair #2 yielded only very few GUS-positive cells in barley and *N. benthamiana*, TALEN pair #3 was more active and reproducibly led to GUS-positive cells in both plant systems (Additional file 3). Note that in the case of this assay, there was no means of normalizing the transformation efficiency due to the lack of a second reporter.

To assess the efficacy of the TALEN constructs in a stably transformed barley line, a transgenic version of cv. 'Igri' harboring *gfp* [8] was re-transformed. TALEN-induced mutations in *gfp* were detected in four out of 66 $T_0$ plants (Table 3; Fig. 3). The induced mutations included various deletions in the size range of 15–172 nt. Therefore, one can conclude mono-allelic mutations for the *gfp* gene. When the *MLO*-specific TALEN construct was transformed into cv. 'Golden Promise', three of the six regenerants included a deletion in *MLO*: one harbored a single 15 nt deletion, while clones derived from the other two harbored a range of 4–8 nt deletions (Fig. 4). While two plants were considered bi-allelic mutants, in one *MLO* mutant plant, also wild-type alleles were detectable leading to the conclusion of a mono-allelic alteration.

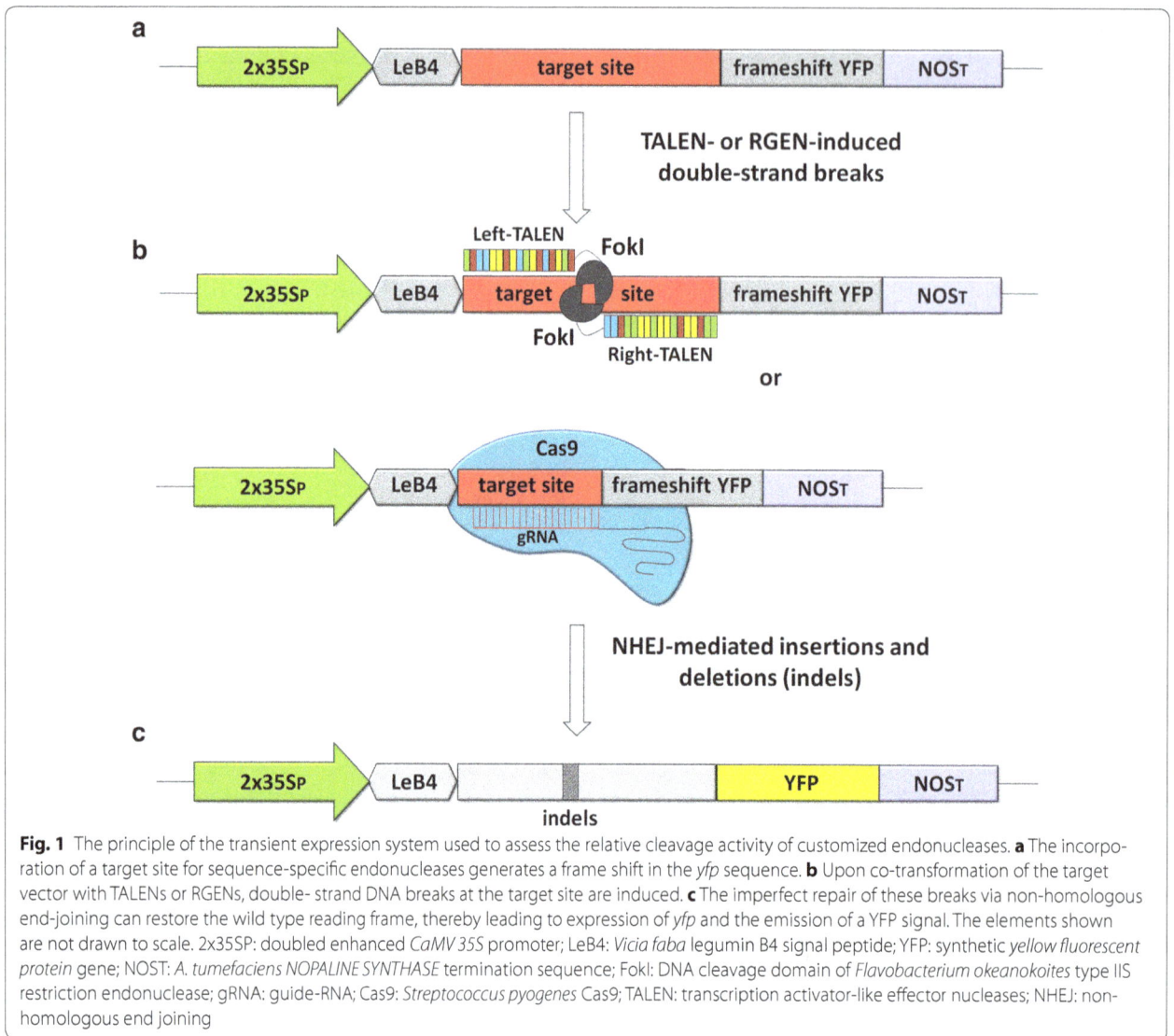

**Fig. 1** The principle of the transient expression system used to assess the relative cleavage activity of customized endonucleases. **a** The incorporation of a target site for sequence-specific endonucleases generates a frame shift in the *yfp* sequence. **b** Upon co-transformation of the target vector with TALENs or RGENs, double- strand DNA breaks at the target site are induced. **c** The imperfect repair of these breaks via non-homologous end-joining can restore the wild type reading frame, thereby leading to expression of *yfp* and the emission of a YFP signal. The elements shown are not drawn to scale. 2x35SP: doubled enhanced *CaMV 35S* promoter; LeB4: *Vicia faba* legumin B4 signal peptide; YFP: synthetic *yellow fluorescent protein* gene; NOST: *A. tumefaciens NOPALINE SYNTHASE* termination sequence; FokI: DNA cleavage domain of *Flavobacterium okeanokoites* type IIS restriction endonuclease; gRNA: guide-RNA; Cas9: *Streptococcus pyogenes* Cas9; TALEN: transcription activator-like effector nucleases; NHEJ: non-homologous end joining

## Discussion

Although endonuclease-enabled site-directed mutagenesis is known to be effective in a range of plant species [7], the tools presently available to aid the in silico design of binding modules (Target Finder [16], Talvez and Storyteller [17, 18], and TALgetter [19]) or gRNAs (CRISPR design [20], CRISPRer [21] and Deskgen [22]) do not consistently produce the desired outcome [23]. There is thus clearly potential for optimization based on reshuffling the current endonuclease systems, while entirely novel systems are also emerging (such as the Cpf1, see [4]). Here, a convenient platform for detecting the cleavage activity of an endonuclease construct was elaborated, based on the restoration of function to a compromised version of *yfp*, a gene which encodes the readily assayable yellow fluorescent protein. The principle of customized

endonuclease-induced restoration of reporter gene function has been used previously in the context of transient expression via infiltration of leaves using Agrobacterium, which is amenable only for a limited number of plant species [24, 25]. By contrast, the method presented here relies on particle bombardment and can thus be readily adopted in any plant species. An additional novel feature which was included was co-transformation with *mCherry*, in order to allow for the comparative quantification of mutation frequency induced by the TALEN/ RGEN construct. The concept of using two fluorescent reporters is related to a previously established assay system designed to assess the efficiency of gene silencing constructs [26].

The tobacco TALEN transient expression experiment established that around one third of the cells showing red

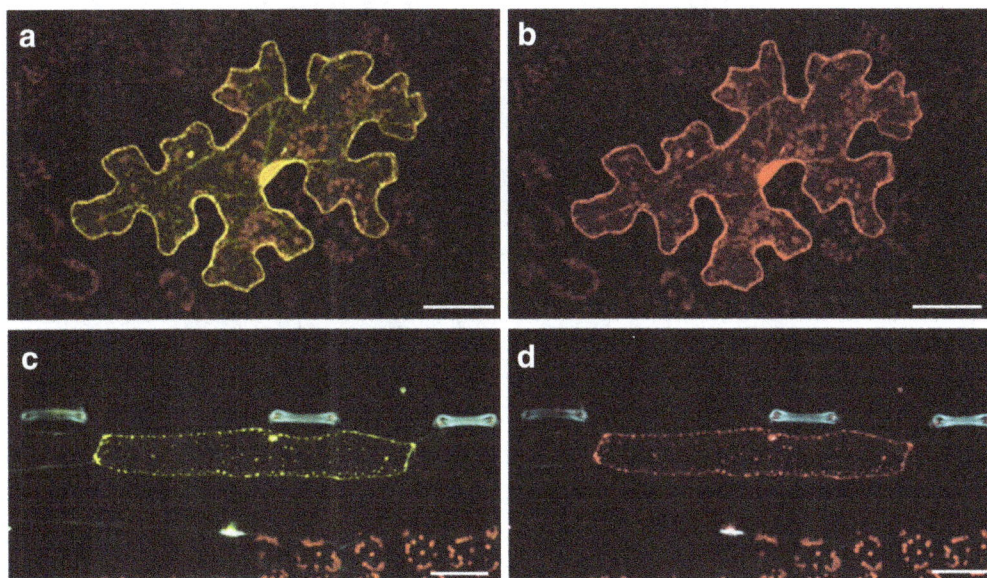

**Fig. 2** Induced mutations as detected by *yfp* expression. Representative epifluorescing transgenic **a**, **b** tobacco and **c**, **d** barley cells visualized 1 day after bombardment with a nuclease-specific vector together with a TALEN or RGEN; *mCherry* was co-transformed to allow quantification. *Bar* 50 μm

**Table 1  List of TALEN and RGEN target site sequences and experiments used**

| Construct | Target sequence | Species | Approach (constructs) |
|---|---|---|---|
| pTARGET-gfp1 | TGGTGAACCGCATCGAGCTGAAGGGCATCGACTTCAAGGAGGACGGCAAGT | Tobacco | RGEN[t, st] (pSI24) |
| | | | TALEN[t, st] (pSP10, pSP11) |
| | | Barley | TALEN[t, st] (pGH297, pGH400) |
| pTARGET-gfp2 | GTCTTTGCTCAGGGCGGACTGGG | Barley | RGEN[t] (pSH92) |
| MLO pair #1 | TGGTGCTCGTGTCCGTCCTCATGGAACACGGCCTCCACAAGCTCGGCCATGTA | Tobacco/barley | TALEN[t] (p110/111) |
| MLO pair #2 | TCCTCATGGAACACGGCCTCCACAAGCTCGGCCATGTAAGTCCCGTTACCCTA | | TALEN[t] (p112/113) |
| MLO pair #3 | TGGAACACGGCCTCCACAAGCTCGGCCATGTAAGTCCCGTTACCCTAGCTCAA | | TALEN[t] (p114/115) |
| MLO pair #4 | TGCTGGCTTTGTATGCAGATGGGATCAAACATGAAGAGGTCCATCTTCGACGA | | TALEN[t] (p124/125) |
| MLO pair #5 | TGGCTTTGTATGCAGATGGGATCAAACATGAAGAGGTCCATCTTCGACGAGCA | | TALEN[t] (p126/127) |
| pTARGET-MLO | GCTGGAACACGGCCTCCACAAGCTCGGCCATGTAAGTCCCGTTACCCTAGCTCA | Barley | TALEN[t, st] (p114/115) |

*t* transient, *st* stable transgenic plants

fluorescence (*mCherry* expressing) also generated a YFP signal (Fig. 2; Table 2), while this frequency was raised to about 75 % when the RGEN construct was assayed (Table 2). The higher success rate achieved with the RGEN construct accords with the literature, and has been at least partly explained by the prediction that in contrast to TALENs, which exhibit a broad range of mutational events, most of the mutations induced by RGEN activity are 1 nt insertions [27, 28]. Consequently, the three possible reading frames of the reporter gene are unlikely to occur in balanced proportions in the latter platform. In the case of the *gfp*-RGEN approach in tobacco, the preferentially occurring insertion of one nucleotide restores the *yfp* open-reading-frame and therefore, the

comparatively high score achieved here is in accordance with the literature. By contrast, the *gfp*-RGEN approach in barley uses a different target site and in this case frame shifts by 1 nt back or 2 nt forward cause a functional *yfp*, which is not expected to occur as frequently as the shift by 1 nt forward. In general, it is clear that comparisons of the efficacy of multiple nucleases based on a single target gene are only possible if the same frame-shift is used for the *yfp* coding sequence succeeding the respective target motifs.

The induction of mutations to *gfp* induced by stably incorporated TALEN and RGEN constructs corroborated the behavior shown in the transient expression assays. While only a low mutation frequency (2.9 %) was

**Table 2** Relative cleavage activity of RGEN and TALEN constructs in transiently transformed barley and tobacco leaf explants

| Plant species | Target gene | Type of customized endonuclease | Constructs used | Experiment | YFP cells | mCherry cells | Ratio YFP/mCherry cells (%) |
|---|---|---|---|---|---|---|---|
| Tobacco | *gfp* | RGEN | pTARGET-gfp1 | 1 | 273 | 371 | 73.6 |
| | | | + RGEN-gfp | 2 | 342 | 389 | 87.9 |
| | | | | 3 | 324 | 499 | 64.9 |
| | | | | Average | 313 ± 29 | 420 ± 57 | 74.6 |
| | | | pTARGET-gfp1 | 1 | 0 | 106 | 0 |
| | | | | 2 | 0 | 204 | 0 |
| | | | | 3 | 0 | 81 | 0 |
| | | | | Average | 0 | 130 ± 65 | 0 |
| | | TALEN | pTARGET-gfp1 | 1 | 195 | 361 | 54.0 |
| | | | + TALEN-gfp | 2 | 25 | 183 | 13.7 |
| | | | | 3 | 80 | 545 | 14.7 |
| | | | | Average | 100 ± 71 | 363 ± 148 | 27.5 |
| Barley | *gfp* | RGEN | pTARGET-gfp2 | 1 | 72 | 207 | 34.8 |
| | | | + RGEN-gfp | 2 | 55 | 206 | 26.7 |
| | | | | 3 | 83 | 255 | 32.5 |
| | | | | Average | 70 ± 12 | 223 ± 23 | 31.4 |
| | | | pTARGET-gfp2 | 1 | 0 | 46 | 0 |
| | | TALEN | pTARGET-gfp1 | 1 | 71 | 223 | 31.8 |
| | | | + TALEN-gfp | 2 | 110 | 331 | 33.2 |
| | | | | 3 | 49 | 195 | 25.1 |
| | | | | Average | 77 ± 25 | 250 ± 59 | 30.7 |
| | | | pTARGET-gfp1 | 1 | 0 | 44 | 0 |
| | *MLO* | TALEN | pTARGET-MLO | 1 | 134 | 350 | 38.3 |
| | | | + TALEN-MLO | 2 | 107 | 254 | 42.1 |
| | | | | 3 | 112 | 237 | 47.3 |
| | | | | Average | 118 ± 12 | 280 ± 50 | 42.0 |
| | | | pTARGET-MLO | 1 | 0 | 52 | 0 |

**Table 3** Stable transgenic plants expressing RGEN or TALEN constructs

| Plant species | Target gene | Type of customized endonuclease | PCR positive | | Mutant plants | Ratio PCR+/mutants (%) |
|---|---|---|---|---|---|---|
| | | | *Cas9/FokI* | gRNA | | |
| Tobacco | *gfp* | RGEN | 17 | 15 | 12 | 80.0 |
| | | TALEN | 35 | n.a. | 1 | 2.9 |
| Barley | *gfp* | TALEN | 66 | n.a. | 4 | 6.1 |
| | *MLO* | TALEN | 6 | n.a. | 3 | 50.0 |

*n.a.* not analysed

achieved using the *gfp*-specific TALENs, 80 % of the primary transgenic tobacco plants harboring the *gfp*-specific RGEN-encoding sequence contained indels in the target sequence (Table 3). Nekrasov et al. [29] have reported a mutation frequency of 6.6 % in the *N. benthamiana PHYTOENE DESATURASE* (*PDS*) sequence by co-expressing *Cas9* and gRNA. At the same time, Li et al. [30] have shown that both the expression level of the gRNA and the size of the ratio of Cas9 to gRNA are important determinants of mutagenic potential. A very high efficiency (>84 %) has been claimed in RGEN experiments involving *N. tabacum PDS* or *PLEIOTROPIC DRUG RESISTANCE6* (*PDR6*), in which Cas9 and gRNA were presented within a single expression vector [31]. The efficiency of

```
GFP-WT      AGGGCGACACCCTGGTGAACCGCATCGAGCTGAAGGGCATCGACTTCAAGGAGGACGGCAACATCCTGGGG

BM33_639    AGGGCGACACCCTGGTGAACCGCATCGAGCTGA--------------------CGGCAACATCCTGGGG      d22    3/7
BM35_697    AGGGCGACACCCTGGTGAACCGCATCGAGCTGAA--------------GGAGGCGGCAACATCCTGGGG      d15    5/6
BM36_757    AGGGCGACACCCTGGTGAACCGCATCGAGCTGAA--------------GGAGGACGGCAACATCCTGGGG     d15    5/6
BM37_760    ---------------------------------------ACTTCAAGGAGGACGGCAACATCCTGGGG      d172   5/5
```

**Fig. 3** Alignment of mutated *gfp* sequences recovered from four barley transformants (BM33_639, BM35_697, BM36_757, and BM37_760) induced by the presence of a TALEN pair. *gfp*-specific PCR products were subcloned and up to ten clones were sequenced. The sequences shown in *red* and *blue* represent, respectively the *left* and *right* TALEN binding sites; the number of nucleotides deleted (*dashes*) and the number of mutants/number of clones analysed is shown to the right of each sequence

```
WT          GTCCGTCCTCATGGAACACGGCCTCCACAAGCTCGGCCATGTAAGTCCCGTTACCCTAGCTCAATCAGCA

1E01-1      GTCCGTCCTCATGGAACACGGCCTCCACAAGCTCG------TAAGTCCCGTTACCCTAGCTCAATCAGCA      d6    7/10
1E01-2      GTCCGTCCTCATGGAACACGGCCTCCACAAGCTC--------AAGTCCCGTTACCCTAGCTCAATCAGCA      d8    3/10
1E03-7      GTCCGTCCTCATGGAACACGGCC---------------ATGTAAGTCCCGTTACCCTAGCTCAATCAGCA      d15   4/11
2E02-3      GTCCGTCCTCATGGAACACGGCCTCCACAAGCTCGGCCA-----GTCCCGTTACCCTAGCTCAATCAGCA      d5    6/9
2E02-10     GTCCGTCCTCATGGAACACGGCCTCCACAAGCTCG----TGTAAGTCCCGTTACCCTAGCTCAATCAGCA      d4    3/9
```

**Fig. 4** Alignment of mutated *MLO* sequences recovered from three barley transformants (1E01, 1E03, and 2E02) induced by the presence of a TALEN construct. *MLO*-specific PCR products were subcloned and up to ten clones were sequenced. The sequences shown in *red* and *blue* represent, respectively the *left* and *right* TALEN binding sites; the number of nucleotides deleted (*dashes*) and the number of mutants/number of clones analysed is shown to the right of each sequence

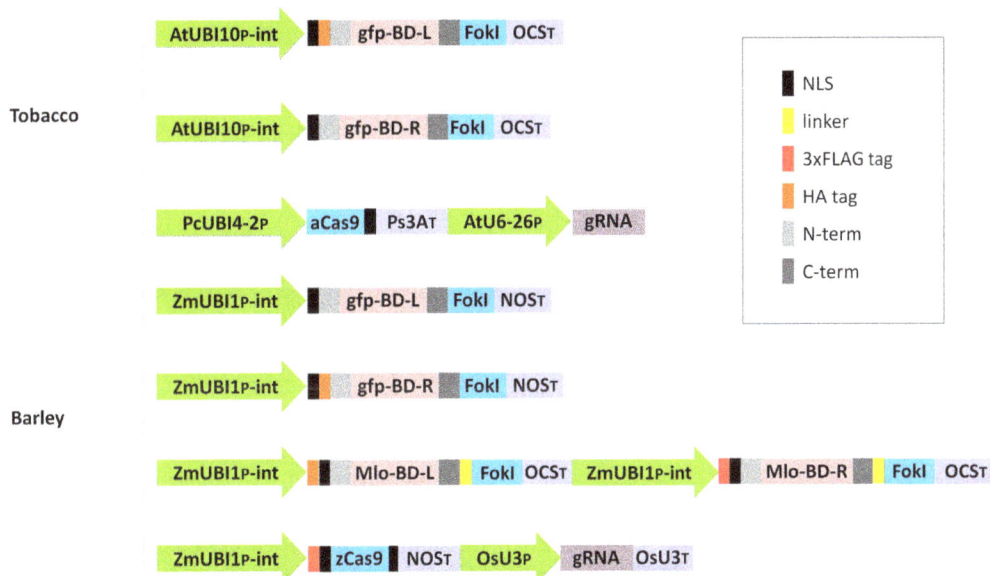

**Fig. 5** The structure of the customized endonuclease constructs. AtUBI10P-int: *A. thaliana UBIQUITIN-10* promoter with intron; gfp-BD-L/R: synthetic S65T green fluorescent protein left/right TALEN binding domain; FokI: FokI cleavage domain; OCST: *A. tumefaciens OCS* terminator; PcUBI4-2P: *Petroselinum crispum UBIQUITIN4-2* promoter; aCas9: *A. thaliana* codon-optimzed *Cas9*; Ps3AT: pea *Pea3A* terminator; AtU6-26P: *A. thaliana U6-26* promoter; gRNA: guide-RNA; ZmUBI1P-int: maize *UBIQUITIN1* promoter with first intron; NOST: *NOS* terminator; NLS: SV40 Simian virus 40 nuclear localization signal; 3xFLAG tag: multimeric synthetic FLAG octapeptide; HA tag: hemagglutinin tag; N-term: N-terminus of a modified version of *Xanthomonas campestris* pv. *vesicatoria AvrBs3*; C-term: C-terminus of a modified version of *AvrBs3*. The elements shown are not drawn to scale

RGEN-mediated mutagenesis is also thought to depend on the accumulation of Cas9 protein. In tobacco, the mutation frequency was reduced when the human codon-optimized variant of *Cas9* was employed [29], while Gao et al. [31] used a tobacco codon-optimized version. Here, a single expression vector was designed on the basis of the *A. thaliana* codon-optimized *Cas9* [32], and this induced a high mutation frequency.

When the system was applied to the more refractory monocotyledonous species barley, the same choice of target gene (*gfp*) was made, since it has been established that *gfp*-specific TALENs are effective inducers of heritable mutations in cultured barley cells [8]. As in tobacco, co-bombardment of pTARGET-gfp1 with the two TALEN units led to the induction of *yfp* expression in about a third of the red fluorescing cells (Table 2). In contrast to tobacco though, the cleavage activities of the *gfp*-specific TALEN and RGEN constructs were identical (Table 2). The comparatively poor performance of the latter in barley indicates that different target sites may have different accessibility for the endonucleases. The generation of stable *gfp* mutants via the transformation of immature embryos proved to be less efficient than that achievable from embryogenic pollen cultures [8], possibly because the effectiveness of the maize *UBI1* promoter is cell-type dependent. The stable generation of *mlo* mutants corroborated the high TALEN activity detected in the transient expression assay. The *MLO*-specific TALEN pair was almost as efficient as the *gfp*-specific RGEN construct in tobacco (Table 3), clearly showing that a pre-validation of TALEN constructs on the basis of their transient expression is an effective procedure, because only one out of five tested *MLO*-specific TALEN constructs showed activity in a reproducible manner (Additional file 3).

The differences between transient expression and stable transgenic plants may be assigned to the fact that expression strength strongly depends on the number of expression units present within a given cell (gene dosage), the specificity and strength of the promoter driving the transgene, the cell type and developmental stage as well as further conditions, which differ in approaches involving different gene transfer methods and target tissues such as barley and tobacco epidermal cell layers and immature embryos, which were used in the present study. In addition, the *gfp*-TALENs were driven by another *ZmUBI1* promoter version than the one used in the *MLO*-TALEN and *gfp*-RGEN constructs.

## Conclusions

In summary, a convenient *in planta* test system for the cleavage activity of sequence-specifically customized endonucleases was established and exemplified using TALENs and RGENs. It is applicable in both dicot and monocot plant species, which makes it a universal tool for the plant science community. This system may not only facilitate the validation of endonuclease functionality prior to the generation of stable mutant plants but also enable researchers to study general principles of endonuclease activity and to optimize construct design.

## Methods
### Plant material
Four week old seedlings of wild type tobacco (*N. tabacum* cv. SR1) were used for the transient transgene expression

experiments. In the stable transformation experiment, seeds of TSP20L1-1, an established transgenic line harboring a single copy of *gfp* [15], were surface-sterilized and germinated on [33] solid medium for 2 weeks, after which the seedlings were transferred to culture boxes (107 × 94 × 96 cm) for a further 4 weeks. Barley (*H. vulgare*) cvs. 'Golden Promise' and 'Ingrid' were used for the transient transgene expression experiments and cv. 'Golden Promise' for the mutation of *MLO*, while a transgenic line of cv. 'Igri' was used for the *gfp*-specific TALEN experiment; the latter's seedlings required vernalization (8 weeks at 4 °C under a 9 h photoperiod).

### Transient expression test vector construction
Details of the cloning steps, based on standard procedures, plasmid maps and sequences (Additional file 6), primers and functional elements, are provided in Additional files 4 and 5, and Fig. 5. The generic vector pNB1 (GenBank: KU705395) carries a modified *yfp* reporter gene [11] driven by a doubled enhanced *CaMV 35S* promoter [34]; it includes a *Bam*HI and an *Eco*RI cloning site between the sequences encoding the pea *Legumin B4* signal peptide, *yfp* and a C-terminal KDEL motif for protein retention in the endoplasmic reticulum. Nuclease-specific vectors were developed by inserting 20–50 bp target motifs (annealed oligos, Additional file 4) into the *Bam*HI and *Eco*RI sites to generate the constructs pTARGET-MLO, pTARGET-gfp1 and pTARGET-gfp2 (Additional file 6).

### TALEN vector construction
The *gfp*-specific TALEN vectors [8] comprise a left and a right TALEN unit, each driven by the maize *UBIQUITIN1* promoter [35], along with the bialaphos resistance-conferring *BAR* gene driven by a doubled enhanced *CaMV 35S* promoter. The *gfp*-specific TALEN sequences used in the tobacco constructs were identical to those used in barley [8]. However, unlike in the barley constructs, in tobacco the left and right TALEN units were introduced into pUbiAt-OCS (DNA-Cloning-Service, Hamburg, Germany), allowing them to be driven by the *A. thaliana UBIQUITIN-10* promoter; their terminal sequence was from the *A. tumefaciens OCTOPINE SYNTHASE* (*OCS*) gene. Each TALEN expression cassette was introduced into the p6N vector via its *Sfi*I cloning sites, which harbors the *HYGROMYCIN PHOSPHOTRANSFERASE* (*HPT*) gene driven by the *A. tumefaciens NOPALINE SYNTHASE* promoter. The pSP10 (left TALEN unit) and pSP11 (right TALEN unit) vectors (Additional file 6) were introduced into *A. tumefaciens* strain GV2260 using a heat shock protocol.

The *MLO*-specific TALEN effector binding elements were preceded by a T, 18 bp long, and separated by a 15 nt spacer sequence (Additional file 1). Repeat lacking TALEN units (pICH47732 TALENΔRep and pICH47742 TALENΔRep) were assembled with *Bsa*I site-flanked modules encoding a truncated *CaMV 35S* promoter (pICH51277, see [36]), HA-NLS [37], a truncated TALE N- and C-terminus, *Fok*I [38] and *OCS* terminator (pICH41432, [36]) into pICH47732 and pICH47742 [36], respectively. The repeat domains of TALEN 114 and TALEN 115 were created following Morbitzer et al. [39] and were cloned into the *Bpi*I site of pICH47732 TALENΔRep and pICH47742 TALENΔRep. TALEN modules with repeats (pICH47732 TALEN 114 and pICH47742 TALEN115) were assembled together with pICH47744 [36] into pUC57_BpiI/KpnI_shuttle via *Bpi*I cut-ligation and thereby flanked by *Kpn*I. Both of the TALEN *Kpn*I fragments were introduced into the *Kpn*I site of the p6int vector (DNA-Cloning-Service, Hamburg, Germany). The TALEN encoding T-DNA vector used for the *GUS* reporter system was assembled via *Bpi*I cut-ligation from pICH47732 TALEN 114, pICH47742 TALEN 115, pICH47751 Kanamycin, a vector which confers resistance to kanamycin, and pICH47766 [36] into pICH50505 [36] (details and sequences given in Additional file 5).

## RGEN vector construction
To generate a monocotyledonous species-specific generic RGEN vector, two *Sfi*I restriction sites were first inserted into pBUN411 [40]. In addition the *BAR* gene was removed to produce pSH91. The *gfp*-specific RGEN vector pSH92 was generated by replacing the spectinomycin resistance gene via *Bsa*I digestion with a synthetic DNA fragment containing a *gfp*-specific protospacer, formed by annealing the partially complementary oligonucleotides GFP_PP1_f and GFP_PP1_r (Additional file 4). For the tobacco *gfp*-specific RGEN, the Gateway®-compatible *Cas9* expression system [32] was used. The above-mentioned *gfp*-specific protospacer sequence was introduced into pEN-Chimera via the pair of *Bbs*I sites. The resulting construct, driven by the *A. thaliana U6-26* promoter, was then transferred into pDe-CAS9 via a single site Gateway® LR reaction [32], ensuring that *Cas9* lay under the control of the *Petroselinum crispum PcUbi 4-2* promoter and the pea *Pea3A* terminator sequence. The resulting pSI24 vector (Additional file 6) was introduced into *A. tumefaciens* strain GV2260 using a heat shock protocol.

## Transient transgenesis via particle bombardment
Barley and *N. tabacum* leaf explants were transiently transformed using a PDS-1000/He Hepta™ device equipped with a 1100 psi rupture disc (Bio-Rad, Munich, Germany). For barley, six primary leaves harvested from 7 to 8 day old seedlings were placed adaxial side up on

1 % agar containing 20 µg/mL benzimidazol and 20 µg/mL chloramphenicol. For *N. tabacum*, a single leaf harvested from a 4 week old plant was placed on solidified (0.8 % agar) Murashige and Skoog [33] medium containing 2 % sucrose and 400 mg/L ticarcillin. A 7 µg aliquot of plasmid DNA was mixed with 3 mg gold micro-carriers by vortexing in the presence of 25 µL 25 mM $CaCl_2$ and 10 µL 0.1 M spermidine. After centrifugation, the pellet was washed with 75 and 100 % ethanol, followed by suspension in 60 µL 100 % ethanol. A total of 4 µL of coated micro-carrier suspension was loaded onto each of the seven macro-carriers, as recommended by the PDS-1000/He manual. Each set of explants were bombarded twice with a total amount of 16–19 µg plasmid DNA (7 µg nuclease test vector, 7 µg endonuclease vector and 2–5 µg mCherry vector), then incubated at room temperature for 1 day before assaying for fluorescence. Each experiment was carried out three times.

## Transient transgenesis via *Agrobacterium tumefaciens* infiltration
The *A. tumefaciens* strain GV3101 pMP90RK [41] was used for transient expression assays in *N. benthamiana*. Liquid culture-grown (28 °C, 180 rpm for 1 day) *Agrobacterium* was set to an $OD_{600}$ of 1.0 with infiltration solution (10 mM $MgCl_2$, 10 mM MES (pH 5.7), 200 µM acetosyringone). The bacteria were delivered into 3–4 week-old *N. benthamiana* leaves using a needleless syringe. After 2 days, infiltrated areas were cut out and de-stained in 80 % (v/v) ethanol for a few days prior to GUS staining (see below).

## Stable transformants of barley and tobacco
Barley was transformed according to Hensel et al. [42], except that the immature embryos harvested from transgenic single-copy, *gfp* expressing cv. 'Igri' were initially cultured for 5 days on BPCM (solid BCIM, 5 mg/L dicamba) before the introduction of *A. tumefaciens*. Tobacco (*N. tabacum* wild type or line TSP20L1-1) plants were transformed with pGH292 or pSI24, respectively. This vector harbors *gfp* controlled by the *A. thaliana UBIQUITIN-10* promoter and the *A. tumefaciens OCS* terminator. The *NEOMYCIN PHOSPHOTRANSFERASE* gene (*NptII*) for kanamycin resistance *in planta* is driven by the *CaMV 35S* promoter and termination sequence. The transgene was introduced into *A. tumefaciens* strain GV2260 using a heat shock protocol. Leaf sections (~1 cm²) excised from sterile-grown plants were laid on Murashige and Skoog [33] medium containing 3 % w/v sucrose, 1 mg/L 6-benzylaminopurine, 0.1 mg/L 1-naphthalene acetic acid and 2 % agar for 1–2 days, before inoculation with the transgenic *A. tumefaciens* for 30 min. The explants were blotted with sterile filter paper

and kept for 3 days at 19 °C in the dark, on medium supplemented with 400 mg/L ticarcillin and either 100 mg/L kanamycin or 5 mg/L bialaphos at 22 °C. Developing calli were sub-cultured every 10 days. After emergence of first shoots, the plates were transferred to light (16 h photoperiod) until shoots had reached 1 cm in length, at which point these were excised and placed on Murashige and Skoog [33] medium containing 2 % w/v sucrose, 0.8 % agar to stimulate root initiation. Plantlets which had developed a viable root system were transferred to the greenhouse.

### Genomic DNA isolation and PCR

Genomic DNA was isolated from snap-frozen leaves following Palotta et al. [43]. Subsequent 20 µL PCRs were formulated with 50–100 ng template DNA, and primed as listed in Additional file 4. The reaction products were purified using a QIAquick PCR Purification kit (Qiagen, Hilden, Germany) to allow for amplicon sequencing. Target-specific PCR products amplified from transgenic individuals were cloned into pGEM-T Easy (Promega, Mannheim, Germany). After blue-white selection, plasmid DNA was isolated from ten positive clones and sequenced.

### Confocal microscopy

Frequency of TALEN or RGEN construct induced mutations was determined from the ratio between the number of yellow-fluorescent (YFP) and red-fluorescent (mCherry) cells. For this a total of six leaves of barley and one leaf of *N. tabacum* were analyzed with a Zeiss LSM780 confocal laser microscope (Carl Zeiss, Jena, Germany). YFP fluorescence was visualized using a 514 nm laser line in combination with a 517–560 nm bandpass; mCherry fluorescence was visualized with a 561 nm laser line in combination with a 570–620 nm bandpass.

### GUS staining

Barley and *N. benthamiana* leaves were harvested for GUS staining three and 2 days, respectively, after bombardment/or *Agrobacterium* infiltration. The leaves were submerged in 5-bromo-4-chloro-3-indolyl-ß-D-glucoronide cyclohexylammonium (X-Gluc) staining solution (42.3 mM $NaH_2PO_4$, 57.7 mM $Na_2HPO_4$, 10 mM EDTA, 20 % methanol, 5 mM $K_3Fe(CN)_6$, 5 mM $K_4Fe(CN)_6 \times 3$ $H_2O$, 1 mg/mL X-Gluc, 0.1 % Triton X-100) and vacuum was applied three times for 10 min. Then, the material was incubated at 37 °C overnight. Afterwards, the leaves were bleached in 80 % EtOH at room temperature for at least 2 days. The leaves were screened for blue-stained cells by bright field microscopy; photographs were taken with the Keyence Biorevo BZ9000 microscope (Keyence Corporation, Neu-Isenburg, Germany).

### Additional files

> **Additional file 1.** *MLO*-specific TALEN target sequences.
>
> **Additional file 2.** (a) Step-wise functional principle of transient expression vector system for assessing the relative cleavage activity of customized endonucleases. Incorporation of a target site for sequence-specific endonucleases deliberately generates a frame shift in the codon sequence of *GUS*. Upon co-transformation of target vector along with respective customized TALENs, double-strand breaks at the target site are induced. The repair of double-strand breaks at target site via non-homologous end-joining, which often introduces indels, render *GUS* back in frame and GUS protein can be detected by X-Gluc staining. (b) Example for successful employment of the reporter system in a transient assay in barley. From left to right: negative control (reporter construct only) and two examples of successful induction of indels indicated by blue-green GUS staining of the cell. Upper panel: barley (*Hordeum vulgare* cv. 'Ingrid', *Hv*) after bombardment; lower panel: *N. benthamiana* (*Nb*) after *Agrobacterium* infiltration. Arrows highlight some GUS-stained cells. Bars: 50 µm (upper panel); 100 µm (lower panel).
>
> **Additional file 3.** Transient expression test of *MLO*-specific TALEN constructs in barley (using bombardment) and *N. benthamiana* (using agroinfection).
>
> **Additional file 4.** Oligomers used to clone pTARGET and RGEN plasmids, and the PCR-based analysis of putative transgenic regenerants.
>
> **Additional file 5.** Details and sequences for the *MLO*-specific TALEN constructs.
>
> **Additional file 6.** Plasmid maps and sequences of the constructs.

### Authors' contributions

JK, GH, TL and RP conceived the study. NB, SS, SP, MG, SH, SK, RM, TR and GH generated the necessary plasmids, performed the experiments and analysed the data. NB, SS, JK and GH wrote the manuscript. All authors read and approved the final manuscript.

### Author details

[1] Plant Reproductive Biology, Leibniz Institute of Plant Genetics and Crop Plant Research (IPK), Corrensstr. 3, 06466 Stadt Seeland/OT Gatersleben, Germany. [2] Present Address: Chair of Plant Breeding, Martin Luther University, Betty-Heimann-Str. 3, 06120 Halle (Saale), Germany. [3] Structural Cell Biology, Leibniz Institute of Plant Genetics and Crop Plant Research (IPK), Corrensstr. 3, 06466 Stadt Seeland/OT Gatersleben, Germany. [4] Unit of Plant Molecular Cell Biology, Institute for Biology I, RWTH Aachen University, Worringerweg 1, 52056 Aachen, Germany. [5] ZMBP-General Genetics, University of Tübingen, Auf der Morgenstelle 32, 72076 Tübingen, Germany.

### Acknowledgements

We appreciate the excellent technical assistance given by Sabine Sommerfeld, Sibylle Freist and Petra Hoffmeister. The research aimed at establishing TALENs as a tool for genome editing in plants is supported by a DFG Grants to TL (LA 1338/5-1). The Gateway®-compatible Cas9 expression system was kindly provided by H. Puchta (KIT, Karlsruhe, Germany).

### Competing interests

The authors declare that they have no competing interests.

## References

1. Cermak T, Doyle EL, Christian M, Wang L, Zhang Y, Schmidt C, Baller JA, Somia NV, Bogdanove AJ, Voytas DF. Efficient design and assembly of custom TALEN and other TAL effector-based constructs for DNA targeting. Nucleic Acids Res. 2011;39:e82.
2. Li T, Liu B, Spalding MH, Weeks DP, Yang B. High-efficiency TALEN based gene editing produces disease-resistant rice. Nat Biotechnol. 2012;30:390–2.
3. Shan Q, Wang Y, Li J, Zhang Y, Chen K, Liang Z, Zhang K, Liu J, Xi JJ, Qiu J-L, Gao C. Targeted genome modification of crop plants using a CRISPR-Cas system. Nat Biotechnol. 2013;31:686–8.
4. Zetsche B, Gootenberg JS, Abudayyeh OO, Slaymaker IM, Makarova KS, Essletzbichler P, Volz SE, Joung J, van der Oost J, Regev A, Koonin EV, Zhang F. Cpf1 Is a single RNA-guided endonuclease of a class 2 CRISPR-Cas system. Cell. 2015;163:759–71.
5. Puchta H, Dujon B, Hohn B. Two different but related mechanisms are used in plants for the repair of genomic double-strand breaks by homologous recombination. Proc Natl Acad Sci USA. 1996;93:5055–60.
6. Jasin M, Rothstein R. Repair of strand breaks by homologous recombination. Cold Spring Harb Perspect Biol. 2013;5:a012740.
7. Baltes NJ, Voytas DF. Enabling plant synthetic biology through genome engineering. Trends Biotechnol. 2015;33:120–31.
8. Gurushidze M, Hensel G, Hiekel S, Schedel S, Valkov V, Kumlehn J. True-breeding targeted gene knock-out in barley using designer TALE-nuclease in haploid cells. PLoS ONE. 2014;9:e92046.
9. Lawrenson T, Shorinola O, Stacey N, Li C, Ostergaard L, Patron N, Uauy C, Harwood W. Induction of targeted, heritable mutations in barley and Brassica oleracea using RNA-guided Cas9 nuclease. Genome Biol. 2015;16:258.
10. Wang YP, Cheng X, Shan QW, Zhang Y, Liu JX, Gao CX, Qiu JL. Simultaneous editing of three homoeoalleles in hexaploid bread wheat confers heritable resistance to powdery mildew. Nat Biotechnol. 2014;32:947–51.
11. Budhagatapalli N, Rutten T, Gurushidze M, Kumlehn J, Hensel G. Targeted modification of gene function exploiting homology-directed repair of TALEN-mediated double-strand breaks in barley. G3 Genes Genomes Genet. 2015;5:1857–63.
12. Kim H, Kim JS. A guide to genome engineering with programmable nucleases. Nat Rev Genet. 2014;15:321–34.
13. Buschges R, Hollricher K, Panstruga R, Simons G, Wolter M, Frijters A, van Daelen R, van der Lee T, Diergaarde P, Groenendijk J, Toepsch S, Vos R, Salamini F, Schulze-Lefert P. The barley MLO gene: a novel control element of plant pathogen resistance. Cell. 1997;88:695–705.
14. Piffanelli P, Ramsay L, Waugh R, Benabdelmouna A, D'Hont A, Hollricher K, Jorgensen JH, Schulze-Lefert P, Panstruga R. A barley cultivation-associated polymorphism conveys resistance to powdery mildew. Nature. 2004;430:887–91.
15. Schedel S, Pencs S, Hensel G, Mueller A, Kumlehn J. RNA-guided endonuclease-driven mutagenesis in tobacco followed by efficient fixation of mutated sequences in doubled haploid plants. doi:10.1101/042291
16. Target Finder. https://tale-nt.cac.cornell.edu. Accessed 7 Jan 2016.
17. Perez-Quintero A, Rodriguez-R LM, Dereeper A, Lopez C, Koebnik R, Szurek B, Cunnac S. An improved method for TAL effectors DNA-binding sites prediction reveals functional convergence in TAL repertoires of Xanthomonas oryzae strains. PLoS ONE. 2013;8:e68464.
18. Talvez. http://bioinfo.mpl.ird.fr/cgi-bin/talvez/talvez.cgi. Accessed 7 Jan 2016.
19. TALgetter. http://galaxy2.informatik.uni-halle.de:8976/tool_runner?tool_id=TALgetter. Accessed 7 Jan 2016.
20. CRISPR design. http://crispr.mit.edu/. Accessed 7 Jan 2016.
21. CRISPRer. http://galaxy2.informatik.uni-halle.de:8976/. Accessed 7 Jan 2016.
22. Deskgen. https://www.deskgen.com/landing/. Accessed 7 Jan 2016.
23. Noel ES, Verhoeven M, Lagendijk AK, Tessadori F, Smith K, Choorapoikayil S, den Hertog J, Bakkers J. A Nodal-independent and tissue-intrinsic mechanism controls heart-looping chirality. Nat Commun. 2013;4:2754.
24. Jiang W, Zhou H, Bi H, Fromm M, Yang B, Weeks DP. Demonstration of CRISPR/Cas9/sgRNA-mediated targeted gene modification in Arabidopsis, tobacco, sorghum and rice. Nucleic Acid Res. 2013;41(20):e188.
25. Yin K, Han T, Liu G, Chen T, Wang Y, Yu AYL, Liu Y. A geminivirus-based guide RNA delivery system for CRISPR/Cas9 mediated plant genome editing. Sci Rep. 2015;5:14926.
26. Panstruga R, Kim MC, Cho MJ, Schulze-Lefert P. Testing the efficiency of dsRNAi constructs in vivo: a transient expression assay based on two fluorescent proteins. Mol Biol Rep. 2003;30:135–40.
27. Zhang H, Zhang JS, Wei PL, Zhang BT, Gou F, Feng ZY, Mao YF, Yang L, Zhang H, Xu NF, Zhu JK. The CRISPR/Cas9 system produces specific and homozygous targeted gene editing in rice in one generation. Plant Biotechnol J. 2014;12:797–807.
28. Feng Z, Mao Y, Xu N, Zhang B, Wei P, Yang DL, Wang Z, Zhang Z, Zheng R, Yang L, Zeng L, Liu X, Zhu JK. Multigeneration analysis reveals the inheritance, specificity, and patterns of CRISPR/Cas-induced gene modifications in Arabidopsis. Proc Natl Acad Sci USA. 2014;111:4632–7.
29. Nekrasov V, Staskawicz B, Weigel D, Jones JDG, Kamoun S. Targeted mutagenesis in the model plant Nicotiana benthamiana using Cas9 RNA-guided endonuclease. Nat Biotechnol. 2013;31:691–3.
30. Li W, Teng F, Li T, Zhou Q. Simultaneous generation and germline transmission of multiple gene mutations in rat using CRISPR-Cas systems. Nat Biotechnol. 2013;31:684–6.
31. Gao J, Wang G, Ma S, Xie X, Wu X, Zhang X, Wu Y, Zhao P, Xia Q. CRISPR/Cas9-mediated targeted mutagenesis in Nicotiana tabacum. Plant Mol Biol. 2015;87:99–110.
32. Fauser F, Schiml S, Puchta H. Both CRISPR/Cas-based nucleases and nickases can be used efficiently for genome engineering in Arabidopsis thaliana. Plant J. 2014;79:348–59.
33. Murashige T, Skoog F. A revised medium for rapid growth and bio assays with tobacco tissue cultures. Physiol Plant. 1962;15:473–97.
34. Odell JT, Nagy F, Chua NH. Identification of DNA-sequences required for activity of the cauliflower mosaic virus-35S promoter. Nature. 1985;313:810–2.
35. Christensen AH, Quail PH. Ubiquitin promoter-based vectors for high-level expression of selectable and/or screenable marker genes in monocotyledonous plants. Transgenic Res. 1996;5:213–8.
36. Weber E, Engler C, Gruetzner R, Werner S, Marillonnet S. A modular cloning system for standardized assembly of multigene constructs. PLoS ONE. 2011;6:e16765.
37. de Lange O, Wolf C, Dietze J, Elsaesser J, Morbitzer R, Lahaye T. Programmable DNA-binding proteins from Burkholderia provide a fresh perspective on the TALE-like repeat domain. Nucleic Acids Res. 2014;42:7436–49.
38. Mussolino C, Morbitzer R, Lutge F, Dannemann N, Lahaye T, Cathomen T. A novel TALE nuclease scaffold enables high genome editing activity in combination with low toxicity. Nucleic Acids Res. 2011;39:9283–93.
39. Morbitzer R, Elsaesser J, Hausner J, Lahaye T. Assembly of custom TALE-type DNA binding domains by modular cloning. Nucleic Acids Res. 2011;39:5790–9.
40. Xing HL, Dong L, Wang ZP, Zhang HY, Han CY, Liu B, Wang XC, Chen QJ. A CRISPR/Cas9 toolkit for multiplex genome editing in plants. BMC Plant Biol. 2014;14:327.
41. Koncz C, Schell J. The promoter of T-DNA gene 5 controls the tissue-specific expression of chimaeric genes carried by a novel type of Agrobacterium binary vector. Mol Gen Genet. 1986;204:383–96.
42. Hensel G, Kastner C, Oleszczuk S, Riechen J, Kumlehn J. Agrobacterium-mediated gene transfer to cereal crop plants: current protocols for barley, wheat, triticale, and maize. Int J Plant Genomics. 2009;2009:835608.
43. Pallotta MA, Graham RD, Langridge P, Sparrow DHB, Barker SJ. RFLP mapping of manganese efficiency in barley. Theor Appl Genet. 2000;101:1100–8.

# 17

# UV crosslinked mRNA-binding proteins captured from leaf mesophyll protoplasts

Zhicheng Zhang[1], Kurt Boonen[1], Piero Ferrari[2], Liliane Schoofs[1], Ewald Janssens[2], Vera van Noort[3], Filip Rolland[1] and Koen Geuten[1*]

## Abstract

**Background:** The complexity of RNA regulation is one of the current frontiers in animal and plant molecular biology research. RNA-binding proteins (RBPs) are characteristically involved in post-transcriptional gene regulation through interaction with RNA. Recently, the mRNA-bound proteome of mammalian cell lines has been successfully cataloged using a new method called interactome capture. This method relies on UV crosslinking of proteins to RNA, purifying the mRNA using complementary oligo-dT beads and identifying the crosslinked proteins using mass spectrometry. We describe here an optimized system of mRNA interactome capture for *Arabidopsis thaliana* leaf mesophyll protoplasts, a cell type often used in functional cellular assays.

**Results:** We established the conditions for optimal protein yield, namely the amount of starting tissue, the duration of UV irradiation and the effect of UV intensity. We demonstrated high efficiency mRNA-protein pull-down by oligo-d(T)$_{25}$ bead capture. Proteins annotated to have RNA-binding capacity were overrepresented in the obtained medium scale mRNA-bound proteome, indicating the specificity of the method and providing in vivo UV crosslinking experimental evidence for several candidate RBPs from leaf mesophyll protoplasts.

**Conclusions:** The described method, applied to plant cells, allows identifying proteins as having the capacity to bind mRNA directly. The method can now be scaled and applied to other plant cell types and species to contribute to the comprehensive description of the RBP proteome of plants.

**Keywords:** Messenger RNA-binding proteins, Messenger ribonucleoprotein complexes, *Arabidopsis thaliana* leaf mesophyll protoplasts, In vivo UV crosslinking, mRNA-bound proteome

## Background

Eukaryotic cells use post-transcriptional gene regulation (PTGR) to determine the fates of RNAs, including RNA processing, transportation, localization, translation and degradation [1]. These processes are controlled by various RNA-binding proteins (RBPs), which interact with RNAs and form ribonucleoprotein complexes (RNPs). Identifying and characterizing RNPs is therefore critical to understand the regulation of cellular RNA metabolism [2]. When considering different RNA metabolic regulation pathways, post-transcriptional regulation of pre-mature mRNAs is particularly important because of the complexity of the pool of mRNAs, their abundance and the additional complexity of translating one or more different protein isoforms from a single gene locus [3].

RBP binding specificities from mainly mammalian cells have been experimentally studied by use of common in vitro methods such as RNA electrophoretic mobility shift assay (REMSA), protein affinity purification, systematic evolution of ligands by exponential enrichment (SELEX), fluorescence methods and nuclear magnetic resonance spectroscopy (NMR) [4–8]. These results have been assembled in an RNA-binding Protein DataBase (RBPDB), which provides us with a comprehensive view of the functions of RNPs, the specificities of RNA-binding domains (RBDs) and the RNA motifs they target [9]. More recently, the first genome-wide mRNA-bound proteome has been characterized for HEK293 and HeLa

*Correspondence: koen.geuten@kuleuven.be
[1] Department of Biology, KU Leuven, Kasteelpark Arenberg 31, 3001 Louvain, Belgium
Full list of author information is available at the end of the article

human cell lines, embryonic stem cells (ESCs) and yeast cells by use of a new experimental strategy called mRNA interactome capture [10–13]. The method entails in vivo UV nucleic acid-protein crosslinking followed by poly(A) tailed mRNA pull-down and protein mass spectrometry (MS). The advantage of UV crosslinking over other types of crosslinking based on chemical fixatives is that it generates covalent bonds specifically between physically interacting proteins and nucleic acids [14, 15]. This allows isolating messenger ribonucleoprotein complexes (mRNPs) from a physiological cellular environment. A recent study has investigated the conservation of the mRNA interactome between yeast and human cells [16]. Interestingly, these authors identified previously unknown but conserved RBPs, suggesting that more proteins have RNA-binding capacities than previously considered. Complementary experimental efforts have been pursued to identify the RNA motifs with which RNA-binding proteins interact through methods such as CLIP or crosslinking and immunoprecipitation. This involves in vivo UV crosslinking, immunoprecipitation and RNA sequencing [10, 13, 16, 17]. Also the RNA-binding sites of UV irradiated RNPs can be detected by a novel approach which combines photo-induced crosslinking, MS and statistical automated analysis [18]. Causal functions of RBPs in plant growth and development have already been clearly established, such as in the regulation of flowering time, in transcriptional regulation of the circadian clock and in the regulation of gene expression in chloroplasts and mitochondria [19–23]. Plant endogenous developmental processes can be tightly integrated with responses to environmental stress, especially to abiotic stress [24]. It is notable that many recent studies have focused on the causal roles of plant RBPs in abiotic stress response, such as salinity, cold, drought or abscisic acid (ABA) signaling [25–28]. In the *Arabidopsis* genome, more than 200 RBP genes have thus far been predicted based on well-defined sequence motifs, such as the RNA recognition motif (RRM) or K homology (KH) domain in the encoded proteins while the number of predicted RBP genes in *Oryza sativa* is approximately 250 [29, 30]. When compared to recent studies of mammalian RBPs, experimental evidence for most of these predicted plant RBPs is mostly missing. Furthermore, many studies used in vitro methods to predict the binding specificities of RBPs and focused on specific RBPs, rather than the entire RBP proteome. The specific RBP association with pre-mRNA in plant cell nuclei by use of in vivo UV crosslinking has been previously reported in Lambermon et al. [31]. Here, we identified in vivo UV crosslinking as a major tool missing from the toolbox to discover RBP proteomes coordinating RNA physiology in plants. Interactome capture is a method that allows the straightforward visual confirmation of the

success of UV crosslinking through the observation of a "halo" produced by the captured proteins on the oligo-dT beads and therefore appeared to be a good method to optimize the important parameters for UV crosslinking in plant cells, such as light intensity, irradiation duration and the amount of starting plant material required. We used *Arabidopsis* leaf mesophyll protoplasts (i.e. cells from which the cell wall is removed) as a source material to provide optimal access of UV light to the cells. This cell type has been extensively used to study other cellular processes and is also amenable to transient gene expression protocols to allow rapid functional characterization [32, 33]. Protoplasting is also applicable to other cell types and other plant species (e.g. [34–36]).

## Results and discussion
### mRNA interactome capture from leaf mesophyll cells
In this study we focus on the mRNA-bound proteome of plant cells, applying the interactome capture method, which was developed for yeast and human cells to plant mesophyll cells, the major type of ground tissue in plant leaves. As illustrated in Fig. 1, the method encompasses ten steps. The first four steps include *Arabidopsis* leaf mesophyll protoplast isolation (1), in vivo mRNA-protein crosslinking by UV irradiation (2), protoplast lysis under denaturing conditions (3) and mRNP pull-down and purification by oligo-d$(T)_{25}$ beads (4). The resulting samples were further analyzed in three ways. RNA quality was checked by proteinase K treatment and mRNA purification (5) followed by qRT-PCR (6). Protein quality control entails RNase treatment and mRBP concentration (7) followed by SDS-PAGE and silver-staining (8). The protein band patterns in the gel are directly compared between a CL sample (in vivo crosslinked mRBPs from UV irradiated protoplasts) and a control sample that was not UV irradiated (non-CL protoplasts as negative control). The final identification of proteins in the CL sample was performed through trypsin digestion of protein bands and peptide purification (9) and Nano-LC–MS (Nano reverse phase liquid chromatography coupled to mass spectrometry assay) (10). Bioinformatic analysis allows identifying mRBPs only present in the CL sample. While the overall procedure is similar to previously reported interactome capture methods for yeast and human cells, some steps had to be modified to be compatible with plant cells.

### Efficiency of mRNA-protein pull-down by oligo-d$(T)_{25}$ beads
We started by verifying the efficiency of UV crosslinking in plant cells by oligo-d$(T)_{25}$ bead capture. A characteristic halo that surrounds the beads pellet after crosslinking was consistently observed in the CL sample during washing with wash buffer 2 (Fig. 2a). Possibly, this halo is a consequence

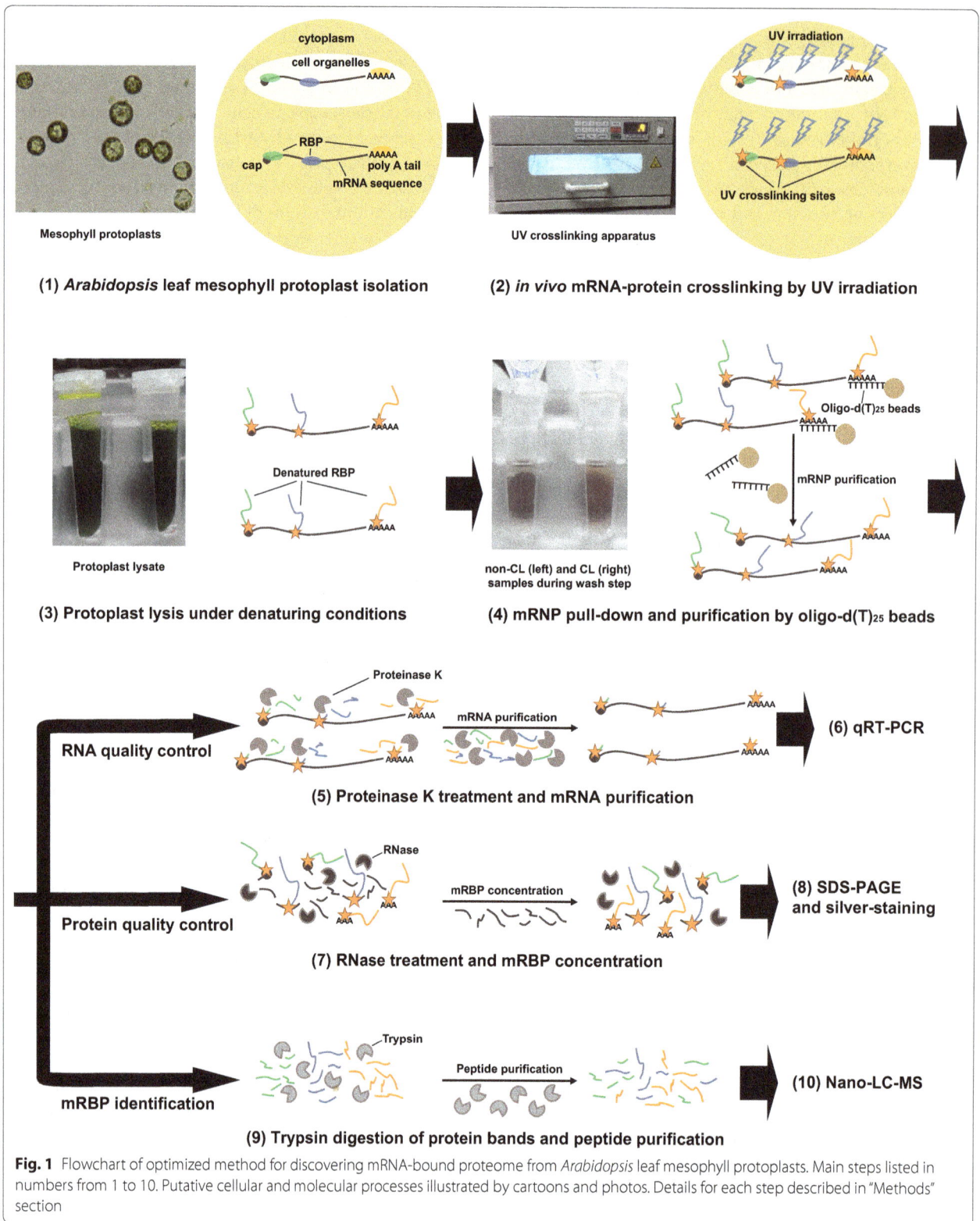

**(1)** *Arabidopsis* **leaf mesophyll protoplast isolation**          **(2)** *in vivo* **mRNA-protein crosslinking by UV irradiation**

**(3) Protoplast lysis under denaturing conditions**          **(4) mRNP pull-down and purification by oligo-d(T)₂₅ beads**

**(5) Proteinase K treatment and mRNA purification**

**(7) RNase treatment and mRBP concentration**

**(9) Trypsin digestion of protein bands and peptide purification**

**Fig. 1** Flowchart of optimized method for discovering mRNA-bound proteome from *Arabidopsis* leaf mesophyll protoplasts. Main steps listed in numbers from 1 to 10. Putative cellular and molecular processes illustrated by cartoons and photos. Details for each step described in "Methods" section

of bound RNPs that inhibit the dense aggregation of beads through the magnetic field, resulting in a more diffuse aggregation on the magnet. The observation of the halo in the CL sample indicated that pull-down of crosslinked mRNPs by oligo-d(T)₂₅ beads was effective [37]. In eukaryotic cells, rRNAs have a higher abundance compared to mature mRNAs [3]. Since oligo-d(T)$_{25}$ beads can only bind poly(A) tailed mRNA, the mRNAs should be enriched in the eluent. In the non-CL control sample, the *UBQ10* reference mRNA is significantly more abundant than 18S rRNA (Fig. 2b). rRNA levels are also low in the CL sample, while mRNA is again significantly enriched. Analysis of the protein samples by SDS-PAGE and silver-staining shows a protein band pattern only present in the CL sample lanes but no specific bands observed in the non-CL control sample that could not be explained by the presence of RNase (Fig. 2c). We conclude that the oligo-d(T)$_{25}$ bead capture is efficient and specific for mRNAs and isolation of mRNPs.

## Optimization of UV crosslinking

We observed that captured proteins in CL samples could only be detected by SDS-PAGE and silver-staining when a minimum of $10^7$ protoplasts is used. Lower concentrations did not yield an observable mRNP pattern on SDS-PAGE and should probably not be used for mass spectrometry because silver-staining and MS detection have similar sensitivity. The duration of UV irradiation and the applied UV light dose is a second critical aspect that determines the efficiency of crosslinking. It is preferable to minimize the duration of irradiation to avoid protoplast damage but sufficient crosslinking still needs to occur. When comparing different UV irradiation times (1–5 min) and UV doses, we obtained protein band patterns in all samples (Fig. 2c). Most optimal was a 1 min UV dose of 0.13 J/cm$^2$ as band intensities were indistinguishable between 1 min and 3 min conditions and lower rather than higher staining intensities were observed with a longer crosslinking duration of 5 min of 0.65 J/cm$^2$. We finally tested the effect of light intensity and continuous versus pulsed irradiation by replacing the conventional UV lamp with a UV laser source [38]. A pulsed UV laser delivers photons for UV crosslinking in nano-second pulse lengths of 10 Hz and could be more efficient in fixing protein-nucleic acid complexes. We compared samples from 1 min UV lamp irradiation with samples from 3 min and 5 min pulsed UV laser irradiation with UV dose 0.94 and 1.56 J/cm$^2$ respectively (Fig. 2d). Again, a

similar band pattern appeared when using UV laser irradiation and but the protein yield from the same number of cells appeared lower. As laser irradiation requires a much more complicated experimental setup and does not appear to provide a specific advantage, we propose that a continuous UV source seems to be more optimal for use in standard biological laboratories.

## High abundance of annotated RBPs in the mRNA-bound proteome

Using these optimized conditions, we then set out to analyze the isolated proteins. Identification of proteins was achieved by qualitative and quantitative proteomics ("Methods" section). In qualitative analysis, we identified a total of 341 proteins in CL samples whereas only 8 proteins were detected in the non-CL control samples and 36 proteins were detected in both non-CL and CL samples (Additional file 1: Fig. S1a right). Such enormous difference in the number of identified proteins between non-CL and CL samples is consistent with the previously observed protein band pattern on SDS-PAGE gel (Fig. 2c). For quantitative analysis, protein fold changes (CL/non-CL) based on peptide fold changes from all qualitatively identified proteins were calculated and the results were illustrated in a volcano plot in which all proteins possessing log2-fold changes greater than 2 were considered as positive hits (Additional file 1: Fig. S1a left). From these proteins, there are 225 proteins with log2-fold changes greater than 2, but below the significance level due to data sparsity (only a few peptides present for per protein and the high variability of peptide intensities of low abundant peptides). Because most of them (210 proteins) were qualitatively identified only in CL samples, they were considered as positive hits as well.

In total, we identified 325 proteins in the mesophyll protoplast mRNA-bound proteome (Additional file 1: Fig. S1a). We further classified the proteins into three categories, namely ribosomal proteins (category I), main RBPs (category II) and candidate RBPs (category III) (Fig. 2e), which we annotated within each category by use of Gene Ontology (GO) ("Methods" section). In category I, a high

(See figure on next page.)
**Fig. 2** mRNA-bound proteome from *Arabidopsis* leaf mesophyll protoplasts. Observed halo surrounding beads pellet in CL sample and not in non-CL sample during wash step (**a**). 18S rRNA and *UBQ10* mRNA expression levels in non-CL and CL samples by qRT-PCR (values were mean ± SD (n = 3); *single asterisk* and *double asterisk* significant differences with $p < 0.05$ and <0.01) (**b**). Separated mRBPs in protein eluent by SDS-PAGE gels and visualized by silver-staining (**c, d**). Protein eluent of non-CL sample compared with CL samples irradiated by continuous UV for 1, 3 and 5 min (**c**). CL sample irradiated by 1 min continuous UV compared with CL samples irradiated by a pulsed UV laser source for 3 and 5 min (**d**). Classification of three categories from mRNA-bound proteome (quantity of identified proteins listed in numbers and the false discovery rate (FDR) at the peptide and protein levels below 5%) (**e**). List of proteins from category I and II according to the annotated RNA-binding domains (**f**). Detection of plant orthologous core RBPs to yeast and human through comparison between our mRNA-bound proteome and the core mRNA-bound proteome of yeast and human from literature Beckmann et al. [16] (**g**). Pie chart of classification of category III. Quantity of identified metabolic enzymes and other candidate proteins listed in numbers (**h**)

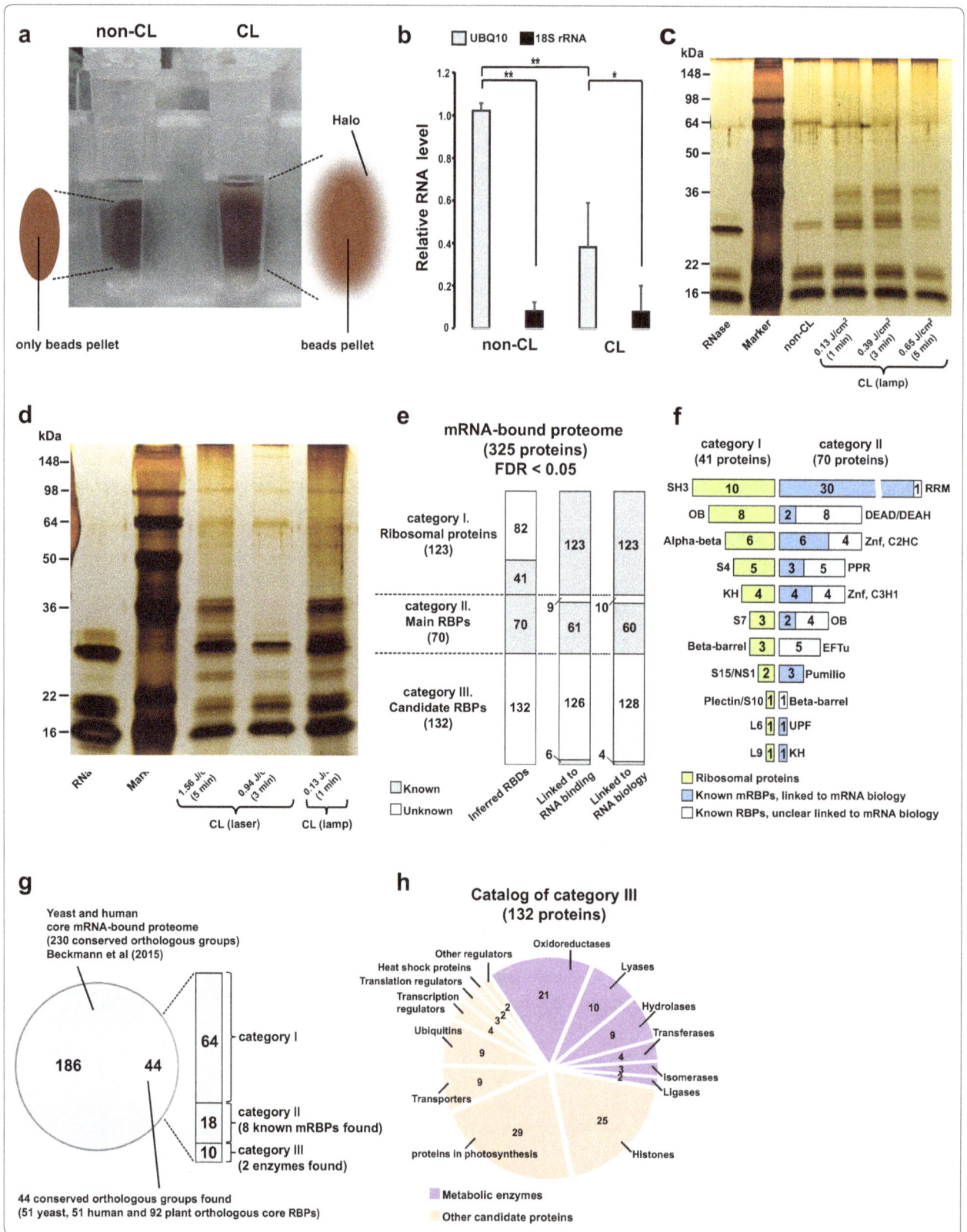

**a**

non-CL    CL

Halo

only beads pellet          beads pellet

**b**

☐ UBQ10    ■ 18S rRNA

Relative RNA level

1.2
1.0
0.8
0.6
0.4
0.2
0

**       *

non-CL          CL

**c**

kDa
148
98
64
50
36
22
16

RNase  Marker  non-CL  0.13 J/cm² (1 min)  0.39 J/cm² (3 min)  0.65 J/cm² (5 min)

CL (lamp)

**d**

kDa
148—
98—
64—
50—
36—
22—
16—

RNase  Marker  1.56 J/... (5 min)  0.94 J/... (3 min)  0.13 J/... (1 min)

CL (laser)    CL (lamp)

**e**

**mRNA-bound proteome
(325 proteins)
FDR < 0.05**

category I.
Ribosomal proteins
(123)

82
41

category II.
Main RBPs
(70)

70

category III.
Candidate RBPs
(132)

132

9        10

123      123

61       60

126      128

6        4

☐ Known
☐ Unknown

Inferred RBDs   Linked to RNA binding   Linked to RNA biology

**f**

category I
(41 proteins)        category II
(70 proteins)

SH3            10        30              1  RRM
OB             8      2    8             DEAD/DEAH
Alpha-beta     6      6      4           Znf, C2HC
S4             5      3      5           PPR
KH             4      4      4           Znf, C3H1
S7             3      2      4           OB
Beta-barrel    3           5            EFTu
S15/NS1        2      3                  Pumilio
Plectin/S10    1    1                    Beta-barrel
L6             1    1                    UPF
L9             1    1                    KH

☐ Ribosomal proteins
☐ Known mRBPs, linked to mRNA biology
☐ Known RBPs, unclear linked to mRNA biology

**g**

Yeast and human
core mRNA-bound proteome
(230 conserved orthologous groups)
Beckmann et al (2015)

186    44

64    category I

18    category II
(8 known mRBPs found)

10    category III
(2 enzymes found)

44 conserved orthologous groups found
(51 yeast, 51 human and 92 plant orthologous core RBPs)

**h**

**Catalog of category III
(132 proteins)**

Other regulators
Heat shock proteins
Translation regulators
Transcription regulators
Ubiquitins
Transporters
proteins in photosynthesis

Oxidoreductases
Lyases
Hydrolases
Transferases
Isomerases
Ligases
Histones

21    10    9    4    3    2
3  2  2
4
9
9
29    25

☐ Metabolic enzymes
☐ Other candidate proteins

number of ribosomal proteins (123 proteins) was revealed and in category II, a moderate number of classical RBPs (70 proteins) was identified. These two categories indeed represent approximately 38 and 22% of the whole mRNA-bound proteome respectively. The last 40% (132 proteins) were placed into category III since these proteins lack conventional RNA-binding domains and most of their roles in RNA binding or RNA biological processes have not been clarified yet (Fig. 2e, Additional file 2: Table S1). Therefore proteins in this category could reveal novel functions in RNA metabolism. In summary, the interactome capture approach successfully pulled down diverse classes of RBPs from mesophyll protoplasts.

## Most conserved orthologous core RBPs are found as ribosomal proteins from category I

Category I is composed of 101 cytosolic small and large ribosomal proteins (40 and 60S) with a smaller number of 22 chloroplast proteins (30 and 50S). Approximately 33% of ribosomal proteins (41 proteins) possess ribosomal RBDs (Fig. 2e, f, Additional file 2: Table S1). When we mapped the proteins to the core mRNA-bound proteome of yeast and human cells [16], a large number of conserved ribosomal orthologous core RBPs (64 proteins, occupying approximate 52% of category I) was found (Fig. 2g). GO enrichment analysis demonstrated that almost all of ribosomal proteins participate in "gene expression" and "translation" of biological process and they possess a molecular function of "structural constituent of ribosome" (Additional file 3: Table S2), indicating the evolutionary conservation of the putative roles in translation across very distant species.

## Multiple and specific roles of main RBPs in category II

The main RBPs in category II were further classified based on their annotated protein domains known to interact with RNAs. We noticed that a very large number of main RBPs is annotated as linked to RNA binding and RNA biology respectively (Fig. 2e). Furthermore, this category includes 41 proteins considered as "known messenger RNA-binding proteins (known mRBPs)" for which roles in mRNA binding and biology have already been annotated using the GO database ("Methods" section). Another 29 proteins are considered more generally as "known RBPs", because their role in mRNA processing is not clear yet. When all inferred classical RBDs are listed in Fig. 2f and Additional file 2: Table S1, we noticed that diverse classes of RBDs were discovered. The RNA Recognition Motif domain (RRM) is most abundant in both "known mRBPs" and "known RBPs" groups. Furthermore, it is noteworthy that domain organization is highly diverse. For example, a single RRM domain with repeated copies was identified in series of polyadenylate-binding proteins (AT4G34110,

AT1G22760, AT2G23350, AT1G71770, AT1G49760) and multiple RBDs were detected in cold shock protein 2 (AT4G38680), containing OB-fold like domain and zinc fingers. This suggests that different RNA targets could be regulated and RBPs may possess multiple roles in RNA biology. Another example that illustrates this is a group of RBPs which has been experimentally discovered as responding to different abiotic stresses. Schmidt et al. [39] investigated a small *Arabidopsis* mRNA-bound proteome involved in response to reactive oxygen species (ROS) such as hydrogen peroxide. In this study, mRNP pull-down was achieved by oligo(dT) chains on cellulose, somewhat similar to our approach. After comprehensive mapping of mRNA-bound proteomes between our study and Schmidt et al., it is notable that the overlap was significant and included a total of 12 RBPs found in both proteomes (Additional file 1: Fig. S1b, Additional file 2: Table S1). Interestingly, in our category II, 5 RBPs were significantly associated with the specific biological process "response to cold" (AT4G13850, AT2G21660, AT4G39260, AT2G37220, AT4G38680, Additional file 3: Table S2). Because our protoplasts were not under oxidative stress but treated with ice-cold cell culture solution ("Methods" section) and GO annotations of these RBPs refers to "response to cold" or "cold acclimation", this suggests that the same RBPs were expressed under different abiotic stresses. In contrast to the large number of conserved plant ribosomal core RBPs, we found only 18 orthologs (8 "known mRBPs") from category II (Fig. 2g). Most of them were significantly enriched in GO annotated "gene expression", "RNA metabolic process" and "response to cadmium ions" and none of them was related to cold shock stress (Additional file 3: Table S2). This small number suggests distinct roles for RBPs involved in response to environmental abiotic stimuli in plants.

## Diverse biological processes associated with candidate RBPs in category III

A final category including 132 proteins is classified as "candidate RBPs", similar to the "enigmRBPs" identified in a recent study of mRNA interactomes from yeast and human cells [16]. Notably, most of these yeast and human enigmRBPs are enzymes involved in diverse biological processes and molecular functions, such as glycolysis, protein folding, cell redox homeostasis, ubiquitination or as having kinase activity. In our category III, we found 49 metabolic enzymes, occupying 37% of candidate RBPs while the rest has no annotated enzymatic functions (Fig. 2h, Additional file 2: Table S1). GO enrichment analysis discovered diverse biological processes for these metabolic enzymes, mainly "photosynthesis", "glycolysis", "oxidation reduction" and response to environmental stimuli, such as "response to cold", "response to light stimuli" and "defense response to bacterium" while another 83 candidate RBPs were also involved in other processes, such as "response to heat", "transmembrane transport" and "nucleosome assembly" (Additional file 3: Table S2). Interestingly, there were 13 metabolic enzymes significantly enriched in "response to cold" (Additional file 3: Table S2), possibly associated with previously discovered RBPs related to "response to cold" from category II. Furthermore, one of these 13 enzymes is the plant ortholog of yeast phosphoglycerate kinase (AT3G12780, Additional file 2: Table S1). RNA-binding capacity of phosphoglycerate kinase has been detected in both yeast and human cells [16], suggesting that plant enzymes could act in RNA metabolism under stress although they lack a conventional RBDs. In the coming years, the role of these metabolic enzymes in RNA biology should be further characterized. Notably, the C-terminal end of ethylene-insensitive protein 2 (EIN2) was recently reported to be cut off in response to ethylene detection and to function in the repression of EIN3-BINDING F-BOX1/2 (EBF1/2) translation through binding of their 3′UTRs in Arabidopsis [40, 41]. Our study provides support for EIN2 as a candidate RBP (AT5G03280, Additional file 2: Table S1) with no yeast or human orthologs, suggesting its specific role in direct post-translational regulation of mRNAs in ethylene signaling. Furthermore, mapping to the yeast and human core mRNA-bound proteome indicates that only 10 proteins (2 chloroplast 2-Cys peroxiredoxin enzymes and 8 other candidate proteins) belong to the core RBPs (Fig. 2g, Additional file 2: Table S1). These core RBPs were enriched only in biological processes "response to cadmium ion", "response to biotic stimulus" and "response to heat" (Additional file 3: Table S2). The small number of orthologs detected in this category indicates that most plant candidate RBPs may serve plant specific functions in RNA metabolism.

## Conclusions

In this study we have successfully developed an efficient mRNA interactome capture protocol that allows inventorying the RNA-binding proteins from plant cells. The advantage of this method is that it specifically identifies proteins with the capacity to physically interact with mRNA in vivo. We have optimized experimental conditions, such as the minimum concentration of cells required for sample preparation, UV irradiation time and source. In addition, we demonstrated the efficiency of mRNP pull-down by oligo-d(T)$_{25}$ bead capture. MS identification of captured proteins confirmed the specificity of our method, as the majority of identified proteins were RBPs that were previously annotated as such in silico. We also present the first experimental evidence

in plants for previously unknown RNA-binding activity of protein, with ortholog conserved in yeast cells. Exploring the binding specificities of these candidate RBPs must be continued through other methods, such as CLIP. One example for investigating the binding specificities of a certain RBP to regulate its target mRNA transcript in Arabidopsis by use of CLIP, has been demonstrated by literature Zhang et al. [42]. Recently a new article reported the mRNA-bound interactomes from Arabidopsis cell cultures and leaf tissue by use of a similar interactome capture approach [43], which highlights the likely importance of this method in the future. Our study differs from that study in that we provide more detailed optimized conditions of the interactome capture approach and focus on leaf mesophyll protoplasts, a single plant cell type. Furthermore, an alternative method, called photoactivatable-ribonucleoside-enhanced crosslinking or PAR-CL (UV-A 365 nm) has also been recommended for investigating mRNA interactomes from yeast and human cells [16, 37]. Our protocol is based on conventional UV crosslinking (cCL), denoting as UV-C 254 nm [44]. PAR-CL needs the incorporation of the photoactivatable nucleotide 4-thiouridine (4sU) into nascent RNAs during RNA metabolisms without toxicity. 4sU is stable when UV-light is absent and has similar base pairing properties as natural uridine. Under UV-A (365 nm) irradiation, 4sU is highly reactive towards to other nucleotides to form covalent bonds with amino acids [15]. Although the efficient uptake of exogenous 4sU into mesophyll protoplasts needs to be later detected by other method which has been previously developed for yeast cell lines [45], future experiments will allow to compare the utility and complementarity of both cCL and PAR-CL approaches.

## Methods
### Arabidopsis leaf mesophyll protoplast isolation
Leaf mesophyll protoplasts were isolated essentially as described by Yoo et al. [32] with some modifications. Arabidopsis thaliana Col-0 ecotype seeds were soaked in deionized water for 2 days at 4 °C in darkness. Stratified seeds were then sown on a mixture of soil (Peltracom) and vermiculite (Sibli AS) in 50% (v/v). Plant growth conditions were 12 h light/12 h dark cycle at 23 °C with a light intensity of 100 µmol m$^{-2}$ s$^{-1}$ for 4 to 5 weeks. For one (non-CL or CL) sample around 150 fully expanded 2nd or 3th pair true leaves (3–4 per rosette) were cut into 0.5-1 mm strips using a sharp razorblade and immediately transferred and submerged into the enzyme solution in a large Petri dish (150 × 20 mm, SARSTEDT). The 40 mL isotonic enzyme solution contained 400 mM Mannitol, 20 mM KCl, 20 mM MES buffer (pH 5.7), 0.6 g Cellulase R10 (Yakult, Japan) and 0.16 g Macerozyme R10 (Yakult, Japan), supplemented with 10 mM CaCl$_2$

and 0.1% (w/v) BSA and was filter sterilized. The Petri dish was covered with aluminum foil and leaf strips were vacuum infiltrated for 30 min and then incubated at room temperature for an additional 2.5 h. From this step, the protoplasts are always kept in darkness. Protoplasts were then released into the enzyme solution by gentle horizontal shaking and the cell suspension was filtered through a 35–75 μm nylon mesh (SEFAR NITEX®) using W5 solution (154 mM NaCl, 125 mM $CaCl_2$, 5 mM KCl, and 2 mM MES buffer, pH 5.7) to rinse the Petri dish and recover the rest of the cells. Protoplasts were then washed with W5 solution, using centrifugation for 5 min at 100$g$, and gently resuspended in 10 mL W5 buffer yielding approximately $1 \times 10^7$ cells from 150 leaves. Protoplasts were then kept on ice for 30 min for recovery and resuspended in 20 mL ice-cold MMg solution (400 mM Mannitol, 15 mM $MgCl_2$, and 4 mM MES buffer, pH 5.7).

### In vivo mRNA-protein crosslinking by UV irradiation

Protoplasts of the non-CL sample were kept in MMg solution on ice, while protoplasts of the CL sample were immediately subjected to UV irradiation. For irradiation by the continuous wave 254 nm UV source, the protoplast suspension was transferred into a large Petri dish (150 × 20 mm, SARSTEDT) with addition of an extra 30 mL of ice-cold MMg solution to cover the plate surface. Protoplasts were irradiated at 0.13 $J/cm^2$ for 1 min. For irradiation by the pulsed 254 nm UV laser source (Nd:YAG pumped optical parametric oscillator, Quanta-Ray MOPO710, equipped with a BBO crystal based frequency doubling unit), the protoplast suspension was first divided over 6 wells of a multiwell culture plate (35 × 10 mm, Greiner CELLSTAR®). The volume in each well was adjusted to 4 mL by adding ice-cold MMg solution. Protoplasts in each well were irradiated by a 35 mm diameter laser beam at 5 mJ/pulse (repetition rate of 10 Hz), giving an average fluence of 0.94 $J/cm^2$ for 3 min and 1.56 $J/cm^2$ for 5 min. Protoplasts of both samples were collected (combining the cells from the 6 wells), washed an additional one time with 10 mL MMg buffer to remove any remaining digestive enzymes and harvested by centrifugation for 5 min at 100$g$.

### Protoplast lysis under denaturing conditions

The protoplasts of each sample ($10^7$ cells) were lysed by adding 9 mL lysis/binding buffer (500 mM LiCl, 0.5% (w/v) Lithium Dodecyl Sulphate (LiDS), 5 mM DTT, 20 mM Tris–HCl, pH 7.5, and 1 mM EDTA, pH 8.0) to the cell pellet resulting in a clear green solution. After homogenization by passing twice through a glass syringe (50 mL, FORTUNA® Optima®) with a narrow needle (0.9 × 25 mm, Becton–Dickinson microlance$^{Tm}$ 3) and

incubation on ice for 10 min, the lysates were flash-frozen in liquid nitrogen and stored at −80 °C. Samples can be stored for up to 3 weeks.

### mRNP pull-down and purification by oligo-d(T)$_{25}$ beads

All described materials and reagents here are for one non-CL or CL sample. 1.8 mL oligo-d(T)$_{25}$ magnetic beads stock (5 mg/mL, New England BioLabs, cat no. S1419S) was aliquoted into 6 round bottom microcentrifuge tubes (2 mL, SARSTEDT) on ice. In each tube, the beads suspension was mixed with 600 μL lysis/binding buffer using rotation at 4 °C for 2 min. The oligo-d(T)$_{25}$ bead capture involves the following three steps: In the binding step, tubes were first put into a magnetic rack at 4 °C for 3 min resulting in magnetic capture of the beads and clearing of the suspension. After the supernatant was discarded and tubes were removed from the magnetic rack, 9 mL protoplast lysate was aliquoted into these 6 tubes. The whole suspension was then mixed by pipetting followed by gentle rotation at 4 °C for 1 h. In the wash step, the tubes were put back into the magnetic rack at 4 °C for 3 min. The protoplast lysate must be removed by pipetting and kept at 4 °C for an extra two rounds of oligo-d(T)$_{25}$ bead capture. 1.5 mL ice-cold wash buffer 1 (500 mM LiCl, 0.1% (w/v) LiDS, 5 mM DTT, 20 mM Tris–HCl, pH 7.5, and 1 mM EDTA, pH 8.0) was added to the beads in each tube. The beads were resuspended followed by gentle rotation for 1 min. Tubes were then put back into the magnetic rack at 4 °C for 3 min and the supernatant was discarded. This wash step must be repeated once. Afterwards, the same procedure of washing was repeated twice using 1.5 mL ice-cold wash buffer 2 (500 mM LiCl, 5 mM DTT, 20 mM Tris–HCl, pH 7.5, and 1 mM EDTA, pH 8.0) and one time using 1.5 mL ice-cold low salt buffer (200 mM LiCl, 20 mM Tris–HCl, pH 7.5, and 1 mM EDTA, pH 8.0). In the elution step, finally, 500 μL elution buffer (20 mM Tris–HCl, pH 7.5, and 1 mM EDTA, pH 8.0) was added to the beads in each tube. The beads were resuspended and incubated at 50 °C for 3 min to release the poly(A) tailed RNAs. After gently resuspending the beads, tubes were put back into the magnetic rack at 4 °C for 5 min. All eluents (total 3 mL) were be combined into a clean, sterile RNase-free 15 mL conical bottom tube on ice. The quality and quantity of RNAs can be immediately determined. Samples can be frozen in liquid nitrogen and kept at −80 °C for long term storage. The whole procedure was then repeated twice with the stored protoplast lysate (from the first binding step) to deplete poly(A) tailed RNAs, re-using the oligo-d(T)$_{25}$ beads after washing twice with 1 mL ice-cold elution buffer and once with 1 mL ice-cold lysis/binding buffer to adjust the salt LiCl concentration back to 500 mM.

## Proteinase K treatment and mRNA purification

Each non-CL or CL sample yielded a total of 9 mL eluent after three rounds of oligo-d(T)$_{25}$ bead capture step. The RNA concentration of each sample was approximately 10 ng/μL with an $A_{260}/A_{280}$ ratio around 1.9. 1 mL eluent of each sample was taken for RNA quality control. 16 μg Proteinase K (Invitrogen) was added to the eluent to digest the UV crosslinked proteins. After brief vortex mixing, the eluent was incubated at 37 °C for 1 h. RNA was then purified using the InviTrap® Spin Plant RNA Mini Kit (Stratec Molecular).

## qRT-PCR

Efficient synthesis of first-strand cDNA using 1 μg RNA as template was achieved by use of the GoScript™ reverse transcription system (Promega). The sample was diluted to 5 ng/μL with nuclease-free water. cDNA was then amplified and quantified using the GoTaq® qPCR master mix (Promega) and StepOnePlus™ Real-Time PCR cycler (Thermo Fisher) using 10 ng as template. To quantify RNA levels, the comparative Ct method, namely the $2^{-\Delta\Delta Ct}$ method was used [46]. The reference gene here was an endogenous internal control gene *UBQ10* (AT4G05320). qRT-PCR primers for *UBQ10* and 18S rRNA were described in Li et al. [47] and Durut et al. [48].

## RNase treatment and mRBP concentration

Approximately 100 U RNase Cocktail containing RNase A and RNase T1 was added to the remaining 8 mL eluent. One control sample with RNase Cocktail, in which the eluent was replaced by nuclease-free water, was included. After brief vortexing, all samples were incubated at 37 °C for 1 h. After RNase digestion, the eluent was concentrated using Amicon® Ultra-4 centrifugal filter units (EMD Millipore). After concentration, the end volume of each sample was 100 μL with a total protein yield of approximately 2 μg. Samples can be kept at −80 °C for long term storage.

## SDS-PAGE and silver-staining

25 μL concentrated eluent and a control sample were mixed with 15 μL 2X loading dye and loaded on an SDS-PAGE gel containing 5% stacking gel and 12% resolving gel including a protein marker (SeeBlue® Plus2 Pre-Stained Standard, Invitrogen). Proteins were condensed at 60 V for 40 min and separated at 160 V for approximately 1 h until the loading dye reached to the end of the resolving gel. Silver-staining of the proteins was performed using the Pierce® Silver Stain Kit (Thermo Scientific).

## Trypsin digestion of protein bands and peptide purification

Gel lanes were hydrated with 50 μL 100 mM NH$_4$HCO$_3$ for 10 min and dehydrated afterwards with CH$_3$CN for 10 min. This was repeated two times and spots were dried afterwards. For enzymatic digestion, gel pieces were covered with 25 μL of a digestion buffer [50 mM NH$_4$HCO$_3$, 5 mM CaCl$_2$, and 6 ng/μL trypsin (Promega)] and incubated on ice for 45 min. The enzymatic digestion was done overnight at 37 °C. The tryptic peptides were extracted in three steps of each 30 min: once with 80 μL of 50 mM NH$_4$HCO$_3$ and twice with 80 μL of 50% (w/v) CH$_3$CN and 5% (v/v) formic acid (FA). The samples were dried and redissolved in 25 μL solution containing 2% (w/v) CH$_3$CN and 0.1% (v/v) aqueous trifluoroacetic acid (TFA) and afterwards desalted by use of Millipore Zip Tip μ-C18 columns. The final eluent containing purified peptides was dissolved in 4 μL 60% (w/v) CH$_3$CN and 0.1% (v/v) FA and dried again.

## Nano-LC–MS (Nano reverse phase liquid chromatography coupled to mass spectrometry assay)
### Nano reverse phase liquid chromatography

The LC–MS analysis was performed on a Q Exactive™ Hybrid Quadrupole-Orbitrap™ Mass Spectrometer (Thermo Scientific, San Jose, CA), coupled online to an Ultimate 3000 ultra-high performance liquid chromatography (UHPLC) instrument (Thermo Scientific, San Jose, CA). The UHPLC system was equipped with an Easy Spray Pepmap RSLC C18 column (2 μm particle, 100 Å pore size, and dimensions: 50 μm × 15 cm, Thermo Scientific, San Jose, CA). Before sample separation on the analytical column, the lyophilized sample was resuspended in 16 μL solution containing 2% (v/v) CH$_3$CN and 0.1% (v/v) FA solution. Next, 5 μL sample was injected and loaded on an Acclaim Pepmap 100 C18 precolumn (3 μm particle size, 100 Å pore size, nanoviper, and dimensions: 75 μm × 2 cm, Thermo Scientific, San Jose, CA) at a flow rate of 5 μL/min. Sample separation was performed using a 95 min gradient. Mobile phase A consisted of 99.9% H$_2$O and 0.1% (v/v) FA and mobile phase B of 19.92% H$_2$O, 80% (w/v) CH$_3$CN and 0.08% (v/v) FA. Mobile phase B increased from 4 to 10% in 5 min, 10–25% in 50 min, 25–45% in 18 min followed by a steep increase to 95% in 1 min. A flow rate of 300 nL/min was used. An inherent rinse step (10 min gradient, from 4–95% in 5 min) was applied after every 95 min separation gradient.

### Mass spectrometry assay

The Q Exactive™ Hybrid Quadrupole-Orbitrap™ Mass Spectrometer was operated in data dependent mode. All mass spectra were acquired in the positive ionization mode with an m/z scan range of 400–1600 m/z. For each precursor spectrum, up to the ten most intense ions were selected for the generation of fragmentation spectra. For precursor spectra, a resolving power of 70,000 full width at half maximum (FWHM) was used with an automatic gain control (AGC) target of 3,000,000 ions and a maximum ion injection time (IT) of 256 ms. For fragmentation spectra, a resolving power of 17,000 FWHM was used with an AGC target of 1,000,000 ions and a maximum IT of 64 ms. Dynamic exclusion of 10 s was applied in order to avoid repeated fragmentation of the most abundant ions. Concerning ion selection, a charge exclusion of $1^+$, $6^+$–$8^+$ was applied. The raw data from Q Exactive mass spectrometer (.RAW) are available on request.

### Qualitative proteomics: peptide and protein identification

The Peaks studio software (Version 7, Bioinformatics solutions Inc., Waterloo, ON, Canada) workflow was used to analyze the fragmentation spectra. This software contains four modules: a module for *de novo* sequencing of MS/MS spectra, a Peaks DB search module for database driven peptide identification, a Peaks PTM search module for detection of post-translational modifications and a Peaks Spider search module designed to detect peptide mutations and perform homology search [49–52]. Spectra with the same mass were merged and a default quality threshold of 0.65 was applied. All spectra were searched against the Swiss-Prot database (version December 2013), with the taxonomy set to *Arabidopsis thaliana*. The following search parameters were used: a precursor mass tolerance of 10 ppm using monoisotopic mass and a fragment mass tolerance of 20 mmu. Trypsin was specified as digestion enzyme and maximum 2 missed cleavages were tolerated. Cysteine carbamidomethylation was set as fixed modification, methionine oxidation was set as variable modification. A maximum of 3 variable post-translational modifications was allowed per peptide. Peptide and Protein score thresholds for reliable peptide and protein identification was set such that both had a FDR of <5%.

### Quantitative proteomics: statistical analysis of mass spectrometry data

Progenesis LC–MS (Nonlinear Dynamics, version 4.1) was used for the label-free quantitative analysis of proteomics data. MS1 peak areas of peptides with 2–8 charges were exported and linked to peptides identified by Peaks studio software by their mass (tolerated error of max. 10 ppm). Afterwards, average log2-fold changes (CL/non-CL) were calculated for each peptide. The fold changes of peptides were grouped by the original protein and evaluated for statistical significance by calculation of $p$ values through student $t$ test. $p$ values were corrected for FDR by the Benjamini-Hochberg method was

achieved by use of R language (version 3.3.0). The fold change of a protein was the average of the fold changes of its peptides. The volcano plot was drawn by function package "calibrate" (version 1.7.2) in R language (version 3.3.0) in which the -log10 transformed adjusted $p$ values [−log10 (adj. $p$ value)] was in function of average log2-fold changes. At last, only proteins were considered as positive hits in our mRNA-bound proteome when they possess the average log2-fold changes greater than 2 with or without significance.

### Venn diagrams and hypergeometric tests

Venn diagrams to illustrate overlap of mRNA-bound proteomes among three biological replicates or overlap of mRNA-bound proteomes and core mRNA interactomes between our data and the literature were drawn by function package "venneuler" (version 1.1-0) in R language (version 3.3.0). Hypergeometric tests were used to test the significance of overlap by function "phyper" in R language (version 3.3.0). The overlap is significant when the calculated $p$ value is lower than 0.05. The *Arabidopsis* proteome was based on "Ara Proteome TAIR10_pep_20110103_representative_gene_model" from TAIR database containing total 27416 proteins, as background for the hypergeometric tests.

### Catalog of mRNA-bound proteome

A total of 325 identified proteins was classified into three categories based on the items of "molecular functions and biological process" via the Gene Ontology (GO) database and "family and domain" via the InterPro database. Category I or "Ribosomal proteins" contains all detected ribosomal proteins. Proteins from category II or "Main RBPs" were defined as containing annotated protein domains that interact with RNAs or link to RNA binding with known or unknown functions in RNA biology. Furthermore, subgroup "known mRBPs" contains all mRBPs which was defined if they have "mRNA binding [GO:0003729]", "transcription antitermination factor activity, RNA binding [GO:0001072]", "mRNA 5′-UTR binding [GO:0048027]", "mRNA processing [GO:0006397]", "alternative mRNA splicing, via spliceosome [GO:0000380]", "mRNA splicing, via spliceosome [GO:0000398]", "mRNA modification [GO:0016556]", "mitochondrial mRNA modification [GO:0080156]", "regulation of translation [GO:0006417]", "translational initiation [GO:0006413]", "chloroplast RNA processing [GO:0031425]" in molecular function and/or biological process. Other RBPs belong to subgroup "known RBPs". Proteins demonstrating known or unknown functions in RNA biology without annotated RNA-binding domains were placed into category III or "Candidate RBPs". Enzymes from category III were defined based on annotations from the IntEnz database.

### Definition of plant orthologous core RBPs

Plant core RBPs orthologous to yeast and human were defined via orthologous groups from InParanoid8 dataset [53]. There were total 1933 groups of orthologs containing 5196 *Arabidopsis* (*A. thaliana*) in-paralogs and 2330 yeast (*S. cerevisiae*) in-paralogs. For *Arabidopsis* to human, there were 3119 groups of orthologs containing 7533 *Arabidopsis* (*A. thaliana*) in-paralogs and 5570 human (*H. sapiens*) in-paralogs. The yeast and human core mRNA-bound proteome containing 230 conserved orthologous groups for comparison with our plant proteome was utilized from Beckmann et al. [16].

### Gene ontology (GO) enrichment analysis

GO enrichment analysis for proteins in each category or orthologous groups was achieved through agriGO (http://bioinfo.cau.edu.cn/agriGO/analysis.php). Database "Arabidopsis genemodel (TAIR9)" was set as a reference. As statistical tests we chose "Fisher" and the Multi-test adjustment method was "Yekutieli (FDR under dependency)" with 0.05 as a significance level.

## Additional files

**Additional file 1: Figure S1.** Identification of mesophyll protoplast mRNA-bound proteome by proteomic analyses. In quantitative analysis (left), the volcano plot displaying the average log2-fold changes (CL/non-CL) and related adjusted p values (−log10 (adj. p values)) of all proteins. These proteins identified by qualitative analysis and illustrated in venn diagrams (right). Quantity of proteins listed in numbers. Numbers of proteins in the grey frames based on quantitative and qualitative proteomic results considered as positive hits (**a**). Comparison between our mesophyll protoplast mRNA-bound proteome and the small mRNA-bound proteome from literature Schmidt et al. [39]. The hypergeometric test showing the overlapping significance (**b**).

**Additional file 2: Table S1.** Details of mRNA-bound proteome of *Arabidopsis* leaf mesophyll protoplasts.

**Additional file 3: Table S2.** GO enrichment analysis for mesophyll protoplast mRNA-bound proteome.

### Authors' contributions

KG conceived the study, ZZ, KB, FR, VvN and KG designed the optimized interactome capture method suitable to *Arabidopsis thaliana* leaf mesophyll protoplasts. PF and EJ designed the pulsed UV laser irradiation approach. PF gave assistance to ZZ with irradiating protoplasts by pulsed UV laser. KB provided statistical analysis of raw mRNA-bound proteomic data to ZZ and performed the analysis of proteome by mass spectrometry. VvN performed and guided the bioinformatic data analysis and ZZ performed the experimental results by qRT-PCR and silver-staining, demonstrated the catalog of mRNA-bound proteome and wrote the article with input from all other authors. LS, EJ, VvN, FR and KG supervised the writing.

### Author details
[1] Department of Biology, KU Leuven, Kasteelpark Arenberg 31, 3001 Louvain, Belgium. [2] Department of Physics and Astronomy, KU Leuven, Celestijnenlaan 200d, 3001 Louvain, Belgium. [3] Department of Microbial and Molecular Systems, KU Leuven, Kasteelpark Arenberg 22, 3001 Louvain, Belgium.

### Acknowledgements
We would like to thank the lab of Prof. Joris Winderickx who provided the UV crosslinking apparatus equipped with the conventional UV lamp. PF acknowledges CONICyT for Becas Chile scholarship. VvN and KG are supported by the KU Leuven research fund and KG acknowledges support from FWO grant G065713 N. We thank bachelor students Freek Vanneste, Stijn Vereecke and Thomas-Wolf Verdonckt for help in optimizing the method.

### Competing interests
The authors declare that they have no competing interests.

### Funding
PF is supported by CONICyT for Becas Chile scholarship. VvN and KG are supported by the KU Leuven research fund. KG is supported by FWO Grant G065713N.

### References
1. Gerstberger S, Hafner M, Tuschl T. A census of human RNA-binding proteins. Nat Rev Genet. 2014;15:829–45.
2. Glisovic T, Bachorik JL, Yong J, Dreyfuss G. RNA-binding proteins and post-transcriptional gene regulation. FEBS Lett. 2008;582:1977–86.
3. Jankowsky E, Harris ME. Specificity and nonspecificity in RNA-protein interactions. Nat Rev Mol Cell Biol. 2015;16:533–44.
4. Patel GP, Ma S, Bag J. The autoregulatory translational control element of poly(A)-binding protein mRNA forms a heteromeric ribonucleoprotein complex. Nucleic Acids Res. 2005;33:7074–89.
5. Song JK, McGivern JV, Nichols KW, Markley JL, Sheets MD. Structural basis for RNA recognition by a type II poly(A)-binding protein. Proc Natl Acad Sci USA. 2008;105:15317–22.
6. Lin Q, Taylor SJ, Shalloway D. Specificity and determinants of Sam68 RNA-binding. Implications for the biological function of K homology domains. J Biol Chem. 1997;272:27274–80.
7. Kattapuram T, Yang S, Maki JL, Stone JR. Protein kinase CK1 alpha regulates mRNA-binding by heterogeneous nuclear ribonucleoprotein c in response to physiologic levels of hydrogen peroxide. J Biol Chem. 2005;280:15340–7.
8. Deo RC, Bonanno JB, Sonenberg N, Burley SK. Recognition of polyadenylate RNA by the poly(A)-binding protein. Cell. 1999;98:835–45.
9. Cook KB, Kazan H, Zuberi K, Morris Q, Hughes TR. RBPDB: a database of RNA-binding specificities. Nucleic Acids Res. 2011;39:D301–8.
10. Baltz AG, Munschauer M, Schwanhausser B, Vasile A, Murakawa Y, Schueler M, Youngs N, Penfold-Brown D, Drew K, Milek M, Wyler E, Bonneau R, Selbach M, Dieterich C, Landthaler M. The mRNA-bound proteome and its global occupancy profile on protein-coding transcripts. Mol Cell. 2012;46:674–90.
11. Castello A, Fischer B, Eichelbaum K, Horos R, Beckmann BM, Strein C, Davey NE, Humphreys DT, Preiss T, Steinmetz LM, Krijgsveld J, Hentze MW. Insights into RNA biology from an atlas of mammalian mRNA-binding proteins. Cell. 2012;149:1393–406.
12. Kwon SC, Yi H, Eichelbaum K, Fohr S, Fischer B, You KT, Castello A, Krijgsveld J, Hentze MW, Kim VN. The RNA-binding protein repertoire of embryonic stem cells. Nat Struct Mol Biol. 2013;20:1122–30.
13. Mitchell SF, Jain S, She M, Parker R. Global analysis of yeast mRNPs. Nat Struct Mol Biol. 2013;20:127–33.
14. Pashev IG, Dimitrov SI, Angelov D. Crosslinking proteins to nucleic acids by ultraviolet laser irradiation. Trends Biochem Sci. 1991;16:323–6.
15. Steen H, Jensen ON. Analysis of protein-nucleic acid interactions by photochemical cross-linking and mass spectrometry. Mass Spectrom Rev. 2002;21:163–82.

16. Beckmann BM, Horos R, Fischer B, Castello A, Eichelbaum K, Alleaume AM, Schwarzl T, Curk T, Foehr S, Huber W, Krijgsveld J, Hentze MW. The RNA-binding proteomes from yeast to man harbour conserved enigmRBPs. Nat Commun. 2015;6:10127.

17. Huppertz I, Attig J, D'Ambrogio A, Easton LE, Sibley CR, Sugimoto Y, Tajnik M, Konig J, Ule J. iCLIP: protein–RNA interactions at nucleotide resolution. Methods. 2014;65:274–87.

18. Kramer K, Sachsenberg T, Beckmann BM, Qamar S, Boon KL, Hentze MW, Kohlbacher O, Urlaub H. Photo-cross-linking and high-resolution mass spectrometry for assignment of RNA-binding sites in RNA-binding proteins. Nat Methods. 2014;11:1064–70.

19. Quesada V, Macknight R, Dean C, Simpson GG. Autoregulation of FCA pre-mRNA processing controls Arabidopsis flowering time. EMBO J. 2003;22:3142–52.

20. Lim MH, Kim J, Kim YS, Chung KS, Seo YH, Lee I, Kim J, Hong CB, Kim HJ, Park CM. A new Arabidopsis gene, FLK, encodes an RNA binding protein with K homology motifs and regulates flowering time via FLOWERING LOCUS C. Plant Cell. 2004;16:731–40.

21. Hornyik C, Terzi LC, Simpson GG. The spen family protein FPA controls alternative cleavage and polyadenylation of RNA. Dev Cell. 2010;18(2):203–13.

22. Schmal C, Reimann P, Staiger D. A circadian clock-regulated toggle switch explains AtGRP7 and AtGRP8 oscillations in Arabidopsis thaliana. PLoS Comput Biol. 2013;9:e1002986.

23. Stern DB, Goldschmidt-Clermont M, Hanson MR. Chloroplast RNA metabolism. Annu Rev Plant Biol. 2010;61:125–55.

24. Kumar AA, Mishra P, Kumari K, Panigrahi KC. Environmental stress influencing plant development and flowering. Front Biosci (Schol Ed). 2012;4:1315–24.

25. Kwak KJ, Kim YO, Kang H. Characterization of transgenic Arabidopsis plants overexpressing GR-RBP4 under high salinity, dehydration, or cold stress. J Exp Bot. 2005;56:3007–16.

26. Liu HH, Liu J, Fan SL, Song MZ, Han XL, Liu F, Shen FF. Molecular cloning and characterization of a salinity stress-induced gene encoding DEAD-box helicase from the halophyte Apocynum venetum. J Exp Bot. 2008;59:633–44.

27. Wang SC, Liang D, Shi SG, Ma FW, Shu HR, Wang RC. Isolation and characterization of a novel drought responsive gene encoding a glycine-rich RNA-binding protein in Malus prunifolia (Willd.) Borkh. Plant Mol Biol Report. 2011;29:125–34.

28. Raab S, Toth Z, de Groot C, Stamminger T, Hoth S. ABA-responsive RNA-binding proteins are involved in chloroplast and stromule function in Arabidopsis seedlings. Planta. 2006;224:900–14.

29. Ambrosone A, Costa A, Leone A, Grillo S. Beyond transcription: RNA-binding proteins as emerging regulators of plant response to environmental constraints. Plant Sci. 2012;182:12–8.

30. Lorkovic ZJ, Barta A. Genome analysis: RNA recognition motif (RRM) and K homology (KH) domain RNA-binding proteins from the flowering plant Arabidopsis thaliana. Nucleic Acids Res. 2002;30:623–35.

31. Lambermon MHL, Simpson GG, Kirk DAW, Hemmings-Mieszczak M, Klahre U, Filipowicz W. UBP1, a novel hnRNP-like protein that functions at multiple steps of higher plant nuclear pre-mRNA maturation. EMBO J. 2000;19:1638–49.

32. Yoo SD, Cho YH, Sheen J. Arabidopsis mesophyll protoplasts: a versatile cell system for transient gene expression analysis. Nat Protoc. 2007;2:1565–72.

33. Niu Y, Sheen J. Transient expression assays for quantifying signaling output. Methods Mol Biol. 2012;876:195–206.

34. Petersson SV, Johansson AI, Kowalczyk M, Makoveychuk A, Wang JY, Moritz T, Grebe M, Benfey PN, Sandberg G, Ljung KSV. An auxin gradient and maximum in the Arabidopsis root apex shown by high-resolution cell-specific analysis of IAA distribution and synthesis. Plant Cell. 2009;21:1659–68.

35. Bargmann BO, Birnbaum KD. Fluorescence activated cell sorting of plant protoplasts. J Vis Exp. (2010); 36. doi:10.3791/1673.

36. Hong SY, Seo PJ, Cho SH, Park CM. Preparation of leaf mesophyll protoplasts for transient gene expression in brachypodium distachyon. J Plant Biol. 2012;55:390–7.

37. Castello A, Horos R, Strein C, Fischer B, Eichelbaum K, Steinmetz LM, Krijgsveld J, Hentze MW. System-wide identification of RNA-binding proteins by interactome capture. Nat Protoc. 2013;8:491–500.

38. Zhang L, Eggers-Schumacher G, Schoffl F, Prandl R. Analysis of heat-shock transcription factor-DNA binding in Arabidopsis suspension cultures by UV laser crosslinking. Plant J. 2001;28:217–23.

39. Schmidt F, Marnef A, Cheung MK, Wilson I, Hancock J, Staiger D, Ladomery M. A proteomic analysis of oligo(dT)-bound mRNP containing oxidative stress-induced Arabidopsis thaliana RNA-binding proteins ATGRP7 and ATGRP8. Mol Biol Rep. 2010;37:839–45.

40. Li W, Ma M, Feng Y, Li H, Wang Y, Ma Y, Li M, An F, Guo H. EIN2-directed translational regulation of ethylene signaling in Arabidopsis. Cell. 2015;163:670–83.

41. Merchante C, Brumos J, Yun J, Hu Q, Spencer KR, Enriquez P, Binder BM, Heber S, Stepanova AN, Alonso JM. Gene-specific translation regulation mediated by the hormone-signaling molecule EIN2. Cell. 2015;163:684–97.

42. Zhang Y, Gu L, Hou Y, Wang L, Deng X, Hang R, Chen D, Zhang X, Zhang Y, Liu C, Cao X. Integrative genome-wide analysis reveals HLP1, a novel RNA-binding protein, regulates plant flowering by targeting alternative polyadenylation. Cell Res. 2015;25:864–76.

43. Marondedze C, Thomas L, Serrano NL, Lilley KS, Gehring C. The RNA-binding protein repertoire of Arabidopsis thaliana. Sci Rep. 2016;6:29766.

44. Kovacs E, Keresztes A. Effect of gamma and UV-B/C radiation on plant cells. Micron. 2002;33:199–210.

45. Miller C, Schwalb B, Maier K, Schulz D, Dümcke S, Zacher B, Mayer A, Sydow J, Marcinowski L, Dölken L, Martin DE, Tresch A, Cramer P. Dynamic transcriptome analysis measures rates of mRNA synthesis and decay in yeast. Mol Syst Biol. 2011;7:458.

46. Livak KJ, Schmittgen TD. Analysis of relative gene expression data using real-time quantitative PCR and the 2(T)(-Delta Delta C) method. Methods. 2001;25:402–8.

47. Li Y, Van den Ende W, Rolland F. Sucrose induction of anthocyanin biosynthesis is mediated by DELLA. Mol Plant. 2014;7:570–2.

48. Durut N, Abou-Ellail M, Pontvianne F, Das S, Kojima H, Ukai S, de Bures A, Comella P, Nidelet S, Rialle S, Merret R, Echeverria M, Bouvet P, Nakamura K, Saez-Vasquez J. A duplicated NUCLEOLIN gene with antagonistic activity is required for chromatin organization of silent 45S rDNA in Arabidopsis. Plant Cell. 2014;26:1330–44.

49. Han X, He L, Xin L, Shan B, Ma B. PeaksPTM: mass spectrometry-based identification of peptides with unspecified modifications. J Proteome Res. 2011;10:2930–6.

50. Han Y, Ma B, Zhang K. SPIDER: software for protein identification from sequence tags with de novo sequencing error. J Bioinform Comput Biol. 2005;3:697–716.

51. Ma B, Lajoie G. De novo interpretation of tandem mass spectra. Curr Protoc Bioinform Chapt. 13, Unit 13–10 (2009).

52. Zhang J, Xin L, Shan B, Chen W, Xie M, Yuen D, Zhang W, Zhang Z, Lajoie GA, Ma B. PEAKS DB: de novo sequencing assisted database search for sensitive and accurate peptide identification. Mol Cell Proteomics. 2012;11(M111):010587.

53. Sonnhammer ELL, Ostlund G. InParanoid 8: orthology analysis between 273 proteomes, mostly eukaryotic. Nucleic Acids Res. 2015;43:D234–9.

# 18

# Measures for interoperability of phenotypic data: minimum information requirements and formatting

Hanna Ćwiek-Kupczyńska[1], Thomas Altmann[2], Daniel Arend[2], Elizabeth Arnaud[3], Dijun Chen[4], Guillaume Cornut[5], Fabio Fiorani[6], Wojciech Frohmberg[1,7], Astrid Junker[2], Christian Klukas[8], Matthias Lange[2], Cezary Mazurek[9], Anahita Nafissi[6], Pascal Neveu[10], Jan van Oeveren[11], Cyril Pommier[5], Hendrik Poorter[6], Philippe Rocca-Serra[12], Susanna-Assunta Sansone[12], Uwe Scholz[2], Marco van Schriek[11], Ümit Seren[13], Björn Usadel[6,14], Stephan Weise[2], Paul Kersey[15] and Paweł Krajewski[1*]

## Abstract

**Background:** Plant phenotypic data shrouds a wealth of information which, when accurately analysed and linked to other data types, brings to light the knowledge about the mechanisms of life. As phenotyping is a field of research comprising manifold, diverse and time-consuming experiments, the findings can be fostered by reusing and combining existing datasets. Their correct interpretation, and thus replicability, comparability and interoperability, is possible provided that the collected observations are equipped with an adequate set of metadata. So far there have been no common standards governing phenotypic data description, which hampered data exchange and reuse.

**Results:** In this paper we propose the guidelines for proper handling of the information about plant phenotyping experiments, in terms of both the recommended content of the description and its formatting. We provide a document called "Minimum Information About a Plant Phenotyping Experiment", which specifies what information about each experiment should be given, and a Phenotyping Configuration for the ISA-Tab format, which allows to practically organise this information within a dataset. We provide examples of ISA-Tab-formatted phenotypic data, and a general description of a few systems where the recommendations have been implemented.

**Conclusions:** Acceptance of the rules described in this paper by the plant phenotyping community will help to achieve findable, accessible, interoperable and reusable data.

**Keywords:** Data standardisation and formatting, Experimental metadata, Minimum information recommendations, Plant phenotyping, Experiment description

## Background

Plant research routinely uses a multitude of techniques and increasingly advanced types of analyses. Scientists delve into a wide range of characteristics manifesting themselves at all levels of plant structure and over their life cycles. The resulting data encompassing genome, epigenome, transcriptome, proteome, and metabolome, and the expression of all other traits (economically or otherwise important) should be integrated to provide a better understanding of the plant systems. The quality and cost of such integration is, however, critically conditioned by the interoperability of the underlying data, i.e., by the availability of adequate metadata describing datasets, and the compatibility of the metadata and data contributed by different scientists, both in terms of the content and the structure. Meanwhile, some plant research fields, especially phenotyping, still lack proper standardization policies to facilitate effective data exchange and integration [1].

*Correspondence: pkra@igr.poznan.pl
[1] Institute of Plant Genetics, Polish Academy of Sciences, ul. Strzeszyńska 34, 60-479 Poznań, Poland
Full list of author information is available at the end of the article

Phenotyping is a very wide and heterogeneous research field. It analyses both static quantities and dynamic processes. Sensitivity of the phenotypic observations to environmental conditions (in the sense of the genotype-by-environment interaction, G × E) requires scrupulous data handling for the acquired signal to be optimally preserved and persisted in databases to deliver most substantial scientific value. Meanwhile, differing amounts of metadata about experiment set-ups, lots of different trait names and their synonyms, and diverse rating scales are used (e.g. [2, 3]), leading to ambiguity and inconsistency of phenotypic data description. Hence, both correct integration and interpretation of phenotyping experiments is hampered. Actions undertaken so far for phenotypic data have either been project-specific (DROPS [4]), platform-specific (PODD [5, 6]; Phenome FPPN [7]), or database-specific (MaizeGDB [8], Triticeae Toolbox [9], Phenopsis DB [10], GnpIS-Ephesis [11]). The lack of common standards of plant phenotyping experiments' description, both in terms of its content and the format, hampers the correct usage and re-usage of phenotypic data.

A proper description of experimental metadata is a key to the correct interpretation of the outcome. In many research domains there have been initiatives aiming at provisioning of recommendations for the set of metadata needed to describe experimental results of particular biological assays. Most of them have resulted in a formulation of a "Minimum Information" or a similar "checklist" document, containing assay-specific recommendations. For example, the Genomic Standards Initiative formulated requirements for reporting sequences of nucleotides (MIxS [12]). The Microarray Gene Expression Database Group suggested the requirements for the description of transcriptomic data (MIAME/Plant [13]). The Proteomics Standards Initiative published a corresponding set of recommendations for protein data (MIAPE [14]). Finally, the Metabolomics Standards Initiative provided rules concerning metabolomic observations (CIMR [15–17]) that were recently considered as a basis for more formal standardization by Rocca-Serra et al. [18]. These documents agree—in principle—on how to describe the experimental material and the treatments applied to it. A similar approach seems advisable to provide metadata recommendations for plant phenotypic data.

As far as data formatting is concerned, for most data types the existing policies are database-specific. Formats that gained wider acceptance are MAGE-TAB [19], a text, tabular format required by the ArrayExpress database [20], storing gene expression data, and PRIDE XML or mzIdentML, required by the PRIDE database [21] for proteomics data. The ISA-Tab format [22] has been developed to address descriptions for many types of experiments and assays. Its flexibility and focus on the experimental metadata, clearly separated from the data itself, make ISA-Tab a generic solution, now used by a number of research communities [23], with a potential to constitute a general experimental metadata description standard, also for phenotypes.

In this paper, we report the measures taken to standardize the description of plant phenotypic data. We present solutions that are a concrete implementation of the opinions expressed recently by many partners of two European infrastructural projects, transPLANT (Transnational Infrastructure for Plant Genomic Science [24]) and EPPN (European Plant Phenotyping Network [25]) in [1]. The solutions are generic and intended to systematize the way of describing all types of phenotypic data independently of the particular local requirements of a project or database, and thus aim for a better interoperability. At the same time, our propositions take into account the achievements of other omics- and phenotype-oriented initiatives, including the above mentioned.

We provide a document called "Minimum Information About a Plant Phenotyping Experiment" (MIAPPE). It constitutes a list of attributes that, based on our experience, are necessary for a useful description of a plant phenotyping experiment and understanding of the data obtained in it. In particular, it comprises recommendations given by Poorter et al. [26] and Hannemann et al. [27] about the documentation of environmental parameters during the experiment, which is a crucial aspect in a G × E-aware phenotype analysis.

As to the way of formatting the metadata, we propose using the above-mentioned ISA-Tab structure for experimental metadata collection and exchange. We show that ISA-Tab, thanks to its generality and flexibility, can handle multitude of phenotyping experiment types and designs. Also, due to its application by several projects and platforms (see [23]), it promotes compatibility of our propositions with those concerning other data types.

Interoperability cannot take place without semantic annotation of the data with respect to the publicly available, controlled vocabularies and ontologies, which provide a community vetted language. This must be done at least for properly identified pivot objects, or key resources, i.e. the elements of a given dataset that allow its integration with other datasets. While the use of particular ontologies is not our main topic, we provide some recommendations in this area. Importantly, all annotations can be conveyed by the ISA-Tab formatted files.

Finally, we present example datasets constructed according to the methods described. Technical aspects of dataset construction and data annotation using recommended ontologies are not covered in this paper; we give some general remarks and refer to existing tools designed for these tasks. We present a few examples of systems where the recommendations have been (or are being) implemented and tested. Some of them are based on own tools and databases, others make use of publicly available utilities provided by the developers of ISA-Tab format [28]. They demonstrate some use cases where the approach described in this paper proved suitable.

## Results

### Minimum Information About a Plant Phenotyping Experiment (MIAPPE)

The Minimum Information About a Plant Phenotyping Experiment is a list of attributes that we recommend for the description of phenotypic observations. It contains the properties that should be provided (by a person or system depositing the data) alongside experimental results to ensure easy and correct interpretation, assessment, review and reproducibility.

To create the recommendations contained in MIAPPE, we took into account previously created Minimum Information documents for various branches of biological research: MIxS for sequences, MIAME/Plant for transcriptomics, MIAPE for proteomics, and CIMR for metabolomics, and have re-used their attribute definition where appropriate. In many cases, where several standards touch upon the same data type (e.g. general metadata, timing and location, treatments), they do so in a compatible fashion, making it straightforward to adopt existing recommendations. Yet, for a number of data types we had to make a choice which approach to adopt. Finally, some information had not been described in the existing documents, which called for provision of such a description from scratch.

The MIAPPE checklist consists of attributes that can be classified within the following sections:

- General metadata,
- Timing and location,
- Biosource,
- Environment,
- Treatments,
- Experimental design,
- Sample collection, processing, management,
- Observed variables.

Each section aggregates attributes detailing specific aspects of an experiment that are important to note, where applicable. The full list of MIAPPE attributes, their origins, and the reasons behind their selection, are given in Table 1. Below, we justify the presence of particular MIAPPE sections.

The attributes from the "General metadata" section should allow to identify the research by providing some basic formal facts. First of all, an identifier of the dataset should be given, possibly a unified and permanent one. Additional important characteristics include a list of the contacts and other people involved, institutions, related projects and publications, data use policy, etc.

Another important aspect of research is to take note of the location and timing of an experiment. Depending on the nature of the study and scientific objectives, different initial time points might be crucial—sowing date or transfer date, treatment application time, etc. Duration of particular stages is also important. As regards location, certain amount of information about the experimental site should be provided for most types of research, in the form of a geographical identifier.

Plant material identification is a critical interoperability pivot and should receive careful attention when building a dataset. In the MI documents, a name "Biosource" has been coined for it. We recommend to define the biosource, i.e. biological object under study, by at least two attributes (as suggested by MIxS): one describing the organism's species name, and the other the infraspecific name—either in the strict sense of McNeill et al. [29], or otherwise simply in the sense of the name of the plant accession, line, or variety, preferably included in a public collection of names, or in a namespace of an experimental station or a genebank (see also similar recommendations on the FAO/Bioversity Multi-Crop Passport Descriptors [30]). We also recommend indicating the source of the seeds for the experiment. Any additional descriptors, further specifying the biosource are optional, yet appreciated.

Owing to the central influence of environmental conditions on the phenotypic expression, accurate reporting on the conditions in which an experiment is performed is critical and warrants the level of details of the section "Environment" of the MIAPPE recommendations. It is our proposition to follow here Poorter et al. [26], who provided a table of attributes recommended to characterise the environment in which plant experiments are conducted. These recommendations encompass environmental descriptors for plants grown in growth chambers, greenhouses, and experimental fields and gardens. Collectively, they constitute a list of descriptors that should be used to describe basic properties of the experimental environment: aerial conditions, light, rooting conditions, fertilizing regimes, watering, and salinity.

Treatments are an inherent element of most phenotyping experiments. While it is impossible to list the types or names of all possible interventions that are used to test the reactions of plants, in MIAPPE's section "Treatment" we provide some suggestions of experimental factors that should be added to the description, if applicable. Some of them are related to the environmental properties,

**Table 1  Minimum Information About a Plant Phenotyping Experiment (MIAPPE)**

| Checklist section | Attributes | Source list/biosharing ID/reference | Recommended ontologies |
|---|---|---|---|
| General metadata | Unique identifier*<br>Title*<br>Description*<br>Submission date<br>Public release date<br>Publications<br>Laboratory address and contact details | ISA reporting standard [38] | OBI, Ontology for Biomedical Investigations [66]<br>CRO, Crop Research Ontology [35] |
| Timing and location | Timing:<br>Start of experiment (date/hour)*<br>Duration (hours/days/months/years)*<br>Experiment location:<br>Geographic location*<br>Latitude and longitude<br>Altitude<br>Inclination and aspect<br>Habitat | Poorter et al. [26]<br>Morrison et al. [17]<br>CIMR [67]: Environmental Analysis Context | OBI, Ontology for Biomedical Investigations [66]<br>GAZ, Gazetteer [68] |
| Biosource | Organism (taxon)*<br>Infraspecific_name*<br>Infraspecific_rank<br>Common name<br>Genotype<br>Organism age<br>Life stage<br>Seed preparation:<br>Seed source*<br>Pretreatments<br>Conservation conditions | MIxS Plant-associated environmetal package [69]<br>Yilmaz et al. [12]<br>FAO/Bioversity Multi-Crop Passport Descriptors V.2 (MCPD V.2) [30] | UNIPROT Taxonomy [70]<br>NCBI Taxonomy [71] |
| Environment | Growth facility* (growth chamber, GC/greenhouse, GH/open top chamber, OTC/experimental garden/experimental field)<br><br>Aerial conditions*<br>$CO_2$<br>For GC and GH:<br>Controlled/uncontrolled<br>Average $CO_2$ during the light and dark period ($\mu$mol mol$^{-1}$)<br>Air humidity (moisture)*<br>Average VPDair during the light period (kPa) or average humidity during the light period (%)<br>Average VPDair during the night (kPa) or average humidity during the night (%)<br>Daily photon flux (light intensity)*<br>Average daily integrated PPFD measured at plant or canopy level (mol m$^{-2}$ day$^{-1}$)<br>Average length of the light period (h)<br>For GC:<br>Light intensity ($\mu$mol m$^{-2}$ s$^{-1}$)<br>Range in peak light intensity ($\mu$mol m$^{-2}$ s$^{-1}$)<br>For GH and OTC:<br>Fraction of outside light intercepted by growth facility components and surrounding structures<br>Light quality:<br>For GC and GH:<br>Type of lamps used<br>R/FR ratio (mol mol$^{-1}$)<br>Temperature (°C)* | Poorter et al. [26]<br>Hanneman et al. [27] | XEO, XEML Environment Ontology [36]<br>ENVO, Ontology of environmental features and habitats [72]<br>Crop Research Ontology [35] |

**Table 1** continued

| Checklist section | Attributes | Source list/biosharing ID/reference | Recommended ontologies |
| --- | --- | --- | --- |
| | Average day temperature<br>Average night temperature<br>Change over the course of experiment | | |
| | Rooting conditions*<br>Rooting medium*: aeroponics/hydroponics (water-based, solid-media based)/soil type (sand, peat, clay, mixed, …)<br>For greenhouse:<br>Container type*<br>Volume (L)*<br>Height<br>Other dimensions*<br>Number of plants per container*<br>For field:<br>Plot size*<br>Sowing density*<br>pH*<br>Frequency and volume of replenishment or addition<br>Soil parameters:<br>Soil penetration strength (Pa m$^{-2}$)<br>Water retention capacity (g g$^{-1}$ dry weight)<br>Organic matter content (%)<br>Porosity (%)<br>Rooting medium temperature | | |
| | Nutrients<br>For hydroponics:<br>Composition*<br>Concentration<br>For soil:<br>Extractable N content per unit ground area before fertiliser added*<br>Type and amount of fertiliser added per container or m$^2$*<br>Concentration of P and other nutrients before start of the experiment<br>Extractable N content per unit ground area at the end of the experiment | | |
| | Watering<br>Irrigation type: irrigation from top/bottom/drip irrigation*<br>Volume (L) and frequency of water added per container or m$^2$*<br>For soil:<br>Range in water potential (MPa) | | |
| | Salinity<br>Concentration of Na, Cl and Mg in the water used for irrigation<br>For soils and hydroponics:<br>Electrical conductivity (dS m$^{-1}$) | | |
| | Aquatic environment<br>If sample was submerged and emerged<br>Depth<br>Time<br>Water temperature<br>Tidal phase | | |
| | Biotic environment<br>Description of interacting organism (pathogens, mutualists, herbivores, endophytes, etc.) | | |

**Table 1 continued**

| Checklist section | Attributes | Source list/biosharing ID/reference | Recommended ontologies |
|---|---|---|---|
| Treatments | Seasonal environment<br>Air temperature regime<br>Soil temperature regime<br>Antibiotic regime<br>Chemical administration<br>Disease status<br>Fertilizer regime<br>Fungicide regime<br>Gaseous regime<br>Gravity<br>Growth hormone regime<br>Herbicide regime<br>Mechanical treatment<br>Mineral nutrient regime<br>Humidity regime<br>Non-mineral nutrient regime<br>Radiation (light, UV-B, X-ray) regime<br>Rainfall regime<br>Salt regime<br>Watering regime<br>Water temperature regime<br>Standing water regime<br>Pesticide regime<br>pH regime<br>Other perturbation | MIxS Plant-associated environ-metal package [69]<br>Yilmaz et al. [12] | XEO, XEML Environment Ontology [36]<br>CRO, Crop Research Ontology [35] |
| Experimental design | Spatial coordinates<br>Plant ID<br>Plot ID<br>Plot (x, y) coordinates<br>Blocking<br>Block ID<br>Sub-block ID<br>Sub-sub-block ID<br>Superblock ID<br>Row ID<br>Column ID<br>Other ID<br>Replication<br>Biological replication<br>Technical replication<br>Experimental unit | | OBI, Ontology for Biomedical Investigations [66]<br>STATO, Statistics Ontology [37]<br>CRO, Crop Research Ontology [35] |
| Sample collection, processing, management | Plant body of interest (organ)*<br>Plant product<br>Organism count<br>Sample temperature<br>Oxygenation status of sample<br>Sample salinity<br>Sample storage duration<br>Sample storage location<br>Sample storage temperature<br>Sampling time | CIMR [67]: Plant Biology Context<br>Fiehn et al. [16] | |
| Observed variables | Phenotypic variables<br>Trait*<br>Method*<br>Scale*<br>Environmental variables<br>Trait*<br>Method*<br>Scale*<br>Data processing protocols | "Trait/Method/Scale" triplet approach applied by Genera-tion Challenge Program, Crop Ontology [32]<br>Shrestha et al. [33]<br>Poorter et al. [26]<br>Hanneman et al. [27] | PTO, Plant Trait Ontology [73]<br>PO, Plant Ontology [74]<br>CO, Crop Ontology [32]<br>PATO, Phenotypic Quality Ontology [75]<br>XEO, XEML Environment Ontology [36] |

Attributes (concepts, subconcepts—in terms of ontology) marked by asterisk (*) are essential for a description of experiment (e.g. by Poorter et al. [26]); the rest forms an extended description. For some attributes possible values are listed (after colon) whereas others are of artificial nature (e.g. mechanical treatment). With the help of this general list of treatments provided in MIAPPE, the description of the experiment should be completed with the details of all of the perturbations that appeared during the trial.

Plant phenotyping experiments are performed in a wide range of experimental designs. To obey the basic rules of replication and local control defined by Ronald A. Fisher, the (incomplete) block, row and column, or other layouts are used, both in the field and in greenhouse experiments. The description of the experimental design is an important part of metadata because any data analysis unaware of it cannot be valid. Especially, experimental units should be defined, i.e. "the groups of material to which a treatment is applied in a single trial" [31]; examples of the entities that play the role of experimental units in plants experiments are: single plant, a plot, or a pot (understood not as containers, but groups of plants).

Sample collection and processing information should include metadata related to phenotyping procedures, in particular sample collection protocol, sample preparation and treatments. If sampling is repeated in time, the time points must be specified.

A specific feature of phenotyping assays is the wide spectrum of observed variables and protocols (methods) used for measurements. This is reflected in MIAPPE in the section "Observed Variables". Following the approach of the Crop Ontology platform [32, 33], we propose to describe the observed variables by three basic attributes: trait name, method, and scale. In this section, in addition to phenotypic variables (any plant characteristics that are measured in a phenotyping experiment), we also consider environmental variables, i.e. any attributes of the environment in which the phenotypic variables are recorded. Such variables are defined here because it is frequently necessary to measure various characteristics influencing the phenotype (potential covariates), possibly (or even usually) not just once, but periodically during the course of the experiment. Indeed, in the limiting situation one can imagine an assay in which the only variables measured are of the environmental type.

We are fully aware that MIAPPE suggests a description of the experiment that is rather extended in comparison to current practices. Hence, although we think that all of the attributes in Table 1 are needed to adequately describe each dataset, we accept that, in practice, the full complement of information may not be possible to collect, or might be unavailable to the person building the dataset. Therefore, we have selected and marked those descriptors deemed absolutely essential. These are also the attributes that we have used as defaults for constructing practical configurations and templates for data formatting (see "Metadata formatting" below). The rest of the attributes form an extended description.

## Annotation

Without proper semantic annotation, the wording used to name particular metadata elements might remain obscure. Referencing publicly available dictionaries and ontologies clarifies the concepts involved in the description, and should be done wherever possible. Ideally, the semantic layer present in an experiment's description should also enable its use by automatic analysis and reasoning tools. In Table 1 we recommend ontologies for use in metadata annotation.

The selection of ontologies is based on [1] and on recent developments in this area. In addition to the reference ontologies for plants recommended by the Planteome project [34], e.g. Plant Trait Ontology (PTO), Plant Ontology (PO), ontology of phenotypic qualities (PATO), widely recognized and already frequently used vocabularies like Ontology for Biomedical Investigations (OBI), Gazetteer (GAZ), Environment Ontology (ENVO), NCBI Taxonomy, EURISCO catalogue, and species-specific ontologies developed as part of the Crop Ontology project, we recommend using the recently constructed:

- Crop Research Ontology [35]—especially for the MIAPPE sections General metadata, Environment, Treatments, and Experimental Design,
- XEO, XEML Environment Ontology [36]—especially for the section Environment and for environmental variables,
- STATO, Statistics Ontology [37]—for the section Experimental design and for unambiguously describing key statistical measures, such as p value, mean, standard deviation.

## Metadata formatting

As a sustainable exchange format for describing phenotyping experiments, we use the ISA-Tab, "Investigation-Study-Assay" format [22]. To facilitate formatting of MIAPPE-compliant datasets, we designed a novel ISA-Tab Phenotyping Configuration that satisfies the recommendations of the Minimum Information document.

ISA-Tab is a general-purpose format to handle experimental metadata description. It consists of a set of tab-delimited text files, namely Investigation, Study, and Assay files, that are linked to each other to form a hierarchy, and describe different properties of a scientific undertaking (Fig. 1). In each dataset a sole Investigation file contains formal general information, e.g. the title, goals, methods, participants, etc. It also lists and formally describes one or more studies performed as parts of that undertaking. Each Study file represents a practical experiment, i.e. it describes the biosources (biological objects), experimental design, environmental conditions and treatments. An Assay file accommodates information about

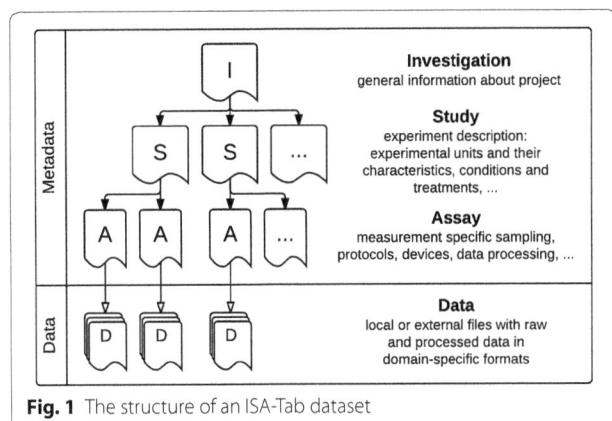

**Fig. 1** The structure of an ISA-Tab dataset

measurements, including description of samples collected from an experiment described in the Study for specific type of analysis, in particular their characteristics, processing and measuring procedures. The actual results of the measurements (or quantities derived from them—statistics) are contained in separate data files and linked to the corresponding metadata through a reference in the Assay file. There can be multiple Assay files per Study, each of them dedicated to a different assay type.

The Study and Assay files consist of columns describing properties of the objects under study; the objects are defined in rows. The allowed types of objects' properties and the rules of their arrangements are defined in the ISA-Tab format specification [38]. Among the columns in Study and Assay files the main ones are so called "data nodes" (identifiers of groups of objects, objects, their parts, or samples taken from them; e.g. *Source Name, Sample Name, Extract Name, Assay Name*) that represent consecutive stages of the experiment. They are described by *Characteristics* (providing detailed object characterisation), *Factors* (naming experimental factors and their levels applied to each object), *Protocols* with *Parameters* (describing conditions and handling of the objects between particular stages), and *Comments* (any other unclassified content). All properties can be accompanied by their semantic annotation in dedicated fields (*Term Source REF* and *Term Accession Number* columns following the property column). *Raw Data File* and *Derived Data File* columns contain references to files in which raw and processed results of measurements are stored.

ISA-Tab configurations are extensions of the general specification, and provide additional requirements for types and arrangement of properties for particular purposes. Configurations can also be used to convey formatting to tools and services dealing with ISA-Tab files.

We propose a Phenotyping Configuration which facilitates formatting of MIAPPE-compliant ISA-Tab datasets. Within the configuration we define a dedicated Study

file which provides a backbone for detailed description of field and greenhouse plant experiments, and a new type of Assay, a Phenotyping Assay, which deals with the information about phenotypic trait measuring procedures. The phenotyping Study files are compatible with other ISA-Tab Assays, so they can be useful for describing any plant experiment in which the environmental conditions are worth recording, irrespective of the types of measurements performed. The Phenotyping Assay can also be used with the default ISA-Tab configuration, and thus integrated in complex, multi-assay datasets that combine ISA-Tab-formatted results of diverse aspects of the analysed phenomena.

## MIAPPE to ISA-Tab mapping

The application of the format for phenotypic datasets consists in defining an ISA-Tab structure that serves as a container for MIAPPE concepts. This structure is defined in an XML file called ISA-Tab configuration. When preparing an ISA-Tab configuration for plant phenotyping, we had to allow for differences that occur between particular types of plant experiments, e.g. performed in different growth facilities. This is reflected in a varying set of attributes recommended in MIAPPE. Therefore, we propose an ISA-Tab Phenotyping Configuration that consists of a standard Investigation file, a Phenotyping Assay (described later) and three versions of a Study file:

- Basic Study—a general ordering of plant experiment specific metadata. It is a default initial description of all plant experiments, and needs to be extended by adding recommended MIAPPE attributes that are applicable in particular cases. In practice, it can be also used when very little is known about the origin of observations, e.g. for simple, external or legacy phenotypic datasets that should be formatted as ISA-Tab, without the ambition to satisfy the MIAPPE recommendations.
- Field Study/Greenhouse Study—extensions of the basic plant Study, featuring specific attributes for growth facilities and environmental information. They satisfy the MIAPPE requirements in terms of the most essential experiment attributes, yet should be further extended to include specific experimental factors present in the trial, and all of the other applicable recommended attributes that can be captured.

The three versions of plant Study use one common Phenotyping Assay file that describes phenotyping procedures and observed variables.

In Table 2 we describe the proposed ISA-Tab Phenotyping Configuration by showing how the MIAPPE

**Table 2 Mapping of essential MIAPPE attributes to ISA-Tab structures in Phenotyping Configuration: Basic, Field and Greenhouse**

| Checklist section | ISA-Tab level | Checklist attribute | ISA-Tab structure | Basic | Field | Greenhouse |
|---|---|---|---|---|---|---|
| General metadata | Investigation | Unique identifier | | ● | ● | ● |
| | | Title | | ● | ● | ● |
| | | Description | | ● | ● | ● |
| Timing and location | Study | Timing<br>Start of experiment (date)<br>Duration (days/months/year) | Characteristics | ● | ● | ● |
| | | Experiment location<br>Geographic location | | ● | ● | ● |
| Biosource | Study | Organism (taxon) | Characteristics | ● | ● | ● |
| | | Infraspecific name | | ● | ● | ● |
| | | Seed origin | | ● | ● | ● |
| Environment | Study | Growth facility (growth chamber, GC/greenhouse, GH/open top chamber, OTC/experimental garden/field) | Characteristics | ● | ● | ● |
| | | Aerial conditions<br>Air humidity (moisture)<br>Daily photon flux (light intensity)<br>Temperature (°C):<br>Average day temperature<br>Average night temperature | Protocol "Aerial conditions" with parameters | | ● | ● |
| | | Rooting conditions<br>Rooting medium: aeroponics/hydroponics (water-based, solid-media based)/soil type (sand, peat, clay, mixed, etc.)<br>pH<br>For field:<br>Plot size<br>Sowing density<br>For greenhouse:<br>Container type<br>Container volume<br>Container dimensions<br>Number of plants per container | Protocol "Rooting" with parameters | | ● | ● |
| | | Nutrients<br>For soil:<br>Extractable N content per unit ground area before fertiliser added<br>Type and amount of fertiliser added, | Protocol "Nutrition" with parameters | | ● | ● |
| | | Watering<br>For soil:<br>Range in water potential (MPa)<br>Irrigation from top/bottom/drip irrigation | Protocol "Watering" with parameters | | ● | ● |
| Treatments | Study or Assay | All interventions being part of the experiment | Factor or Protocol with parameters | | □ | □ |
| Experimental design | Study | Experimental units and their grouping (into blocks, super-blocks etc.) | Characteristics, Factor, Protocol "Sampling" with parameters | | ● | ● |
| Sample collection, processing, management | Assay | Plant body of interest (organ) | Characteristics | ● | ● | ● |
| Observational variables | Assay | Phenotypic variables<br>Trait<br>Method<br>Scale | | ● | ● | ● |

**Table 2 continued**

| Checklist section | ISA-Tab level | Checklist attribute | ISA-Tab structure | Basic | Field | Greenhouse |
|---|---|---|---|:---:|:---:|:---:|
| | | Environmental variables<br>Trait .<br>Method<br>Scale | | ☐ | ☐ | ☐ |
| Observations | Assay | Raw data | Raw data file | ☐ | ☐ | ☐ |
| | | Derived data | Derived data file | ● | ● | ● |

●—included in the ISA-Tab configuration; ☐—not included in the configuration, specific per experiment

attributes are mapped to the ISA-Tab elements in different plant Studies and in the Phenotyping Assay, and demonstrate how the description of the environment is included in field and greenhouse extensions through adding a number of protocols. A comparison of those protocols is shown in Table 3.

**Table 3 Comparison of default fields in the Study file in Basic, Field and Greenhouse ISA-Tab configurations**

| Basic | Field | Greenhouse |
|---|---|---|
| Source Name | Source Name | Source Name |
| Characteristics[Organism] | Characteristics[Organism] | Characteristics[Organism] |
| Characteristics[Infraspecific name] | Characteristics[Infraspecific name] | Characteristics[Infraspecific name] |
| Characteristics[Seed origin] | Characteristics[Seed origin] | Characteristics[Seed origin] |
| Characteristics[Study start] | Characteristics[Study start] | Characteristics[Study start] |
| Characteristics[Study duration] | Characteristics[Study duration] | Characteristics[Study duration] |
| Characteristics[Growth facility] | Characteristics[Growth facility] | Characteristics[Growth facility] |
| Characteristics[Geographic location] | Characteristics[Geographic location] | Characteristics[Geographic location] |
| | Protocol REF[Rooting] | Protocol REF[Rooting] |
| | Parameter Value[Rooting medium] | Parameter Value[Rooting medium] |
| | | Parameter Value[Container type] |
| | | Parameter Value[Container volume] |
| | Parameter Value[Plot size] | Parameter Value[Container dimension] |
| | Unit | Unit |
| | Parameter Value[Sowing density] | Parameter Value[Number of plants per container] |
| | Parameter Value[pH] | Parameter Value[pH] |
| | Protocol REF[Aerial conditions] | Protocol REF[Aerial conditions] |
| | Parameter Value[Air humidity] | Parameter Value[Air humidity] |
| | Parameter Value[Daily photon flux] | Parameter Value[Daily photon flux] |
| | Parameter Value[Length of light period] | Parameter Value[Length of light period] |
| | Parameter Value[Day temperature] | Parameter Value[Day temperature] |
| | Parameter Value[Night temperature] | Parameter Value[Night temperature] |
| | Protocol REF[Nutrition] | Protocol REF[Nutrition] |
| | Parameter Value[N before fertilisation] | Parameter Value[N before fertilisation] |
| | Parameter Value[Type of fertiliser] | Parameter Value[Type of fertiliser] |
| | Parameter Value[Amount of fertiliser] | Parameter Value[Amount of fertiliser] |
| | Protocol REF[Watering] | Protocol REF[Watering] |
| | Parameter Value[Irrigation type] | Parameter Value[Irrigation type] |
| | Parameter Value[Volume] | Parameter Value[Volume] |
| | Parameter Value[Frequency] | Parameter Value[Frequency] |
| | Protocol REF[Sampling] | Protocol REF[Sampling] |
| | Parameter Value[Experimental unit] | Parameter Value[Experimental unit] |
| Sample Name | Sample Name | Sample Name |

The ISA-Tab Phenotyping Configuration is available online via our record registered with the BioSharing community [39].

## Observed variables

The specificity of the Phenotyping Assay (among other ISA-Tab assays, see [40]) lies in the fact that it collects information about different phenotypic and environmental variables that can be measured using different methods. The description of those variables is contained in a separate dedicated file, so-called Trait Definition File, referenced in the Phenotyping Assay as a parameter *Trait Definition File* of "Data transformation" *Protocol*. This file is an extension of the ISA-Tab specification, similar to the one that has been used in the ISA-Tab metabolomic configuration (see [41]) to describe metabolites.

The Trait Definition File contains a table with rows corresponding to variables and columns corresponding to the appropriate MIAPPE attributes, describing the trait, method and scale. In particular, it consists of the following columns:

- *Variable ID*—a local unique identifier of a variable, e.g. a short name, that is a key linking the definitions of variables with observations in Derived Data File,
- *Trait*—a name of the trait mapped to an external ontology; if there is no exact mapping, an informative description of the trait,
- *Method*—a name of the measurement method mapped to an external ontology; if there is no exact mapping, an informative description of the measurement procedure,
- *Scale*—units of the measurement or a scale in which the observations are expressed; if possible, standard units and scales should be used and mapped to existing ontologies; in case of a non-standard scale a full explanation should be given.

## Data

The data (observations or their functions) are represented in ISA-Tab in separate files, contained within the dataset or external, and are referenced in the Assay file as *Raw Data File* or *Derived Data File* properties. Formatting of the data file is not governed by the ISA-Tab specification, yet some recommendations usually exist within particular communities. In our implementation of MIAPPE, we do not restrict the format of the raw data in any way; it can be any custom, platform- or device-specific format, including texts, images, binary data, etc. Similarly, we do not restrict the format of any file referred to as *Derived Data File*; however, we require that the format be fully described in the corresponding *Protocol* "Data transformation" (a field that should precede the

data reference, and explain how it was derived from the raw data, or from the previous derived data). If there is no description, the *Derived Data File* should be a standard, plain tab-separated sample-by-variable matrix. Its first column should contain (in the simplest situation) values from the *Assay Name* column in the Assay file, and the rest of the columns provide values for all variables. The names of those columns should correspond to the values in the *Variable ID* column in the Trait Definition File (see above). So, a default derived data format is an "Assay Name × Variable" matrix of observations, that can be quantitative or qualitative.

An extension of the above rule governing the format of the *Derived Data File* is possible by using values from another "data node" column (e.g. *Source Name, Sample Name, Extract Name*, etc.) as unique identifiers of the rows in the table with the associated observations. Thus, we can provide separate data files with measurements taken for different observational units, e.g., morphological traits like 'height' and 'number of leaves' can be assigned to the whole plant, whereas physiological traits can be restricted to samples taken from particular leaf of a plant. Also conveying data aggregated over "data nodes" is possible in this way.

## Implementations

The developed standard as well as the solutions proposed in this paper were first applied by the project partners dealing with phenotypic data. The main implementations, demonstrating possible approaches to follow the specification, are described below.

### BII database at IPG PAS

At the Institute of Plant Genetics PAS, a BII database serving as an ISA-Tab-compliant storage for phenotypic data compatible with the MIAPPE standards has been launched. The BII software is part of the ISA Software Suite [28]. It consists of BII-Manager application which is used to validate ISA-Tab formatted datasets and store information to the database backend, and of BII Web application that provides a database front-end accessible via an Internet browser. The installation runs on a server at Poznań Supercomputing and Networking Center and is publicly available [42]. The system serves as a proof of concept and an illustration of the application of a generic, out-of-the-box tool for the basic needs of plant phenotypic data management.

Upon submission of the ISA-Tab archive to the administrator, the software is used to validate the files against a suitable configuration. If the validation is successful, the files get stored, and selected metadata are parsed into the internal structures for indexing and search. The content of the database is accessible via the web interface.

Datasets can be browsed online, searched for by selected metadata terms, filtered according to the organism name and assay properties, and downloaded as ISA-Tab archives. It is also possible to declare a dataset as private, so that it is stored in private sections of the database and is inaccessible for unauthorized users. In its present version, the BII software cannot be used to retrieve data filtered by all metadata, so it does not use the full potential of the ISA-Tab format.

### GnpIS-Ephesis at INRA

GnpIS [43] is an information system that allows data discovery and mining of genomic, genetic and phenomic data for plants and their bioagressors. GnpIS-Ephesis [11] allows experimental phenomics data mining, additionally including extended phenotype, genotype and environmental data and metadata integration. It offers users the possibility of creating multi-trial datasets suitable for various analyses (G × E meta-analysis, GWAS, etc.). GnpIS can be used, for example, to retrieve all data for a given diversity panel across several years or locations, all observations of a given phenological variable over several years, or all the data of a specific scientific study or project. Furthermore, all GWAS and genetic data integrated in GnpIS can be linked to a GnpIS-Ephesis experiment, allowing a better traceability and data exploration.

GnpIS-Ephesis allows to dynamically build and export ISA-Tab datasets, which demonstrates the capability of the format to handle results of diverse experiments, and to serve as a dataset exchange format. In the exported dataset the Investigation file represents the whole search results, and it integrates all the metadata, including the search parameters. There is one Study per trial. The Study contains only the subset of data corresponding to the user query with all the metadata necessary to ensure the reusability and the traceability of the data. The advantage of this implementation is that many public datasets are available through GnpIS, which allows to demonstrate the ISA-Tab format features.

An example the reader may look at is a dataset [44] that covers the winter wheat phenotypic observations from a French experimental network. It includes different traits (agronomic, quality, disease, phenology, etc.) measured at 10 experimental locations during 15 years (more than 700 trials) and for more than 1700 winter wheat genotypes [45], in the experimental network that allows to produce new varieties which can be registered to the French catalogue of varieties (CTPS) after their eight's generation. Their identification is centralized by the French Wheat Genebank at Clermont Ferrand and is available through GnpIS. Several treatments were applied, like low fertilization, high nitrogen, etc. Each trial is stored as a single Study in ISA-Tab. Each Study lists the varieties used in a specific trial. The observation variables are collected in a dedicated ontology which is referenced in the ISA-Tab archive. Only derived data files are available.

### Research data at IPK

IPK's research data infrastructure comprises four layers [46]:

1. Primary research data: data generated manually or automatically in the course of experiments, derived data after post-processing of primary research data. Those data files are stored in IPK's storage backend.
2. An in-house Laboratory Information Management System (LIMS), used for documentation of experimental metadata (experimental setup, used protocols etc.), based on primary data from layer 1.
3. Dedicated web-based information systems and databases, which provide access to curated and relationally structured data from layer 1, and which optionally link to the information from the LIMS (layer 2) [47].
4. The e!DAL data publication infrastructure [48], which provides DOIs for layer 1 data (especially datasets which are not covered by databases of layer 3), and which enables the public download of these datasets and registration of related technical metadata in the DataCite repository.

The ISA-Tab-based exchange format for plant phenotyping data was discussed among the collaborators from the German Plant Phenotyping Network (DPPN), the German Network for Bioinformatics Infrastructure (de. NBI), and partners from the European transPLANT project. Its application for future exchange of phenotypic data was agreed among partners from DPPN (especially IPK, *German Research Center for Environmental Health*, HMGU, Munich and *Research Center Jülich GmbH*, FZJ). It will serve as an exchange format for the semantic description of published data.

As an initial step, a reference experiment comprising multiple data domains was described using ISA-Tab structure and published [49] as a part of a research article of Junker et al. [50]. This dataset combines results and metadata from metabolite profiling, high throughput automated imaging and image analysis, as well as manual phenotypic measurements. All semantic and technical documentations, measured parameters, protocols and references to ontologies were manually described using ISA-Tab format. All raw files of such ISA-Tab formatted data publications are stored in the Plant Genomics and Phenomics Data Repository (PGP [51]), hosted at IPK using *e!DAL* as software infrastructure [52]. Recently IPK has published the first MIAPPE-compliant ISA-Tab container describing a high throughput plant phenotyping experiment including metadata, raw and processed images, extracted phenotypic features and manual validation data ([53], also stored in the PGP repository) as a data descriptor accepted at Nature's Scientific Data journal [54]. The ISA-Tab files were manually filled and will be used as templates for the automated export of respective standardized metadata files describing all future high throughput plant phenotyping experiments. This dataset is shortly described as Dataset III in Discussion below.

### GWA-Portal at GMI

GWA-Portal [55] is a web-application that allows researchers to upload their phenotypes and easily carry out Genome Wide Association Studies (GWAS) without installing any software. The GWAS results as well as the phenotypes can be shared with collaborators. By storing information ranging from phenotypes, germplasm to GWAS results in a single database, a comprehensive genotype-phenotype map can be constructed and thus allows researchers to do meta-analysis of pleiotropy. The development of GWA-Portal started before the MIAPPE was formulated and relies on the Genomic Diversity and Phenotype Data Model (GDPDM [56]) that was originally developed by Terry Casstevens from Ed Buckler's lab. Although GDPDM was primarily designed for maize, it is not plant specific. In fact, the GWA-Portal instance that is hosted at the GMI, is used by the Arabidopsis community for storing phenotypes of the model plant *A. thaliana*. Initially GWA-Portal allowed the user to upload and download phenotypes as simple comma separated files. In the course of the transPLANT project the functionality was extended to support the ISA-Tab format. As GDPDM stores less information about phenotypes than what is defined in MIAPPE, we use the basic phenotyping configuration. Phenotypes in GDPDM are always stored as part of a study. This hierarchical structure maps quite well to the Investigation-Study-Assay set of the ISA-Tab format, with a study in GWA-Portal being equivalent to an investigation in ISA-Tab. As a result the mapping is quite straightforward.

The export functionality was implemented first. In order to avoid re-inventing the wheel, we tried to leverage the ISA-Tab toolchain and libraries as much of as possible. Specifically we used the ISAcreator library [57]. The import functionality was implemented shortly after. The ISAcreator library that we used for the export and import functionality is a GUI application and because we only use a small part of it, we suggested to the ISA-Tools team to create a dedicated lightweight library for parsing and creating ISA-Tab files.

## Discussion: best practices
### MIAPPE

MIAPPE recommendations provide a list of attributes that might be necessary to sufficiently describe a phenotypic dataset. One of its goals is to raise awareness of the researchers about the need to record a rich set of experimental metadata, especially environmental qualities which constitute a factor determining the phenotype in interaction with the genotype. Therefore, the MIAPPE requirements should serve as a checklist for the researchers recording the data to make them consider all aspects that might influence the experimental process and take note of those aspects. We suggest that the MIAPPE recommendations should be used in phenotyping projects already at the data management planning stage and be implemented according to the plan at all later stages of data collection.

We have selected a subset of MIAPPE attributes that seem common to the basic plant phenotyping cases, and marked them as obligatory ones. They should always be provided by the data producers to ensure some minimum standardisation in terms of data content. Inclusion of other attributes depends on the type of particular research, and it is up to the data owner to collect and describe all the factors in a responsible way, so that the dataset is correctly interpretable.

Selection of obligatory attributes raises the question of acceptance of the datasets by repositories. This is a community-wide issue. Repositories may wish to first flag submissions which are syntactically valid (a bare minimum for interoperation). Then, repositories may wish to insist on compliance with MIAPPE guidelines because there is an obvious long term benefit in terms of reuse, related to the notion of making data FAIR, i.e. Findable, Accessible, Interoperable and Reusable [58].

### ISA-Tab

Application of ISA-Tab format for plant phenotyping can be seen as a reference implementation of MIAPPE requirements. The textual and tabular nature of this format makes it usable for everyone without any dedicated tools or skills. We recommend using ISA-Tab as a format for experimental metadata collection and exchange. Whether to use the format to also store the datasets internally is a matter of individual decisions, based on existing solutions and needs.

The ISA-Tab Phenotyping Configuration contains the basic common subset of attributes that are necessary to describe a phenotyping experiment according to MIAPPE requirements. We propose using the configuration to ensure consistency of the phenotyping datasets formatted as ISA-Tab. Preparation of each dataset should involve providing all of the attributes named in the configuration, as well as identifying and adding to the dataset all other qualities present in the experiment (e.g. experimental factors and treatments, or supplementary protocols) as additional columns (e.g. *Factor, Characteristics* or *Protocol REF* and *Protocol Parameter*). Preparation of the ISA-Tab files can be done in three ways:

- manually in a text editor, adhering to the rules of the ISA-Tab format specification and Phenotyping Configuration; practically, the easiest way for the researchers recording the data might be to fill in a template (an empty dataset) provided by a data manager who prepares it based on the suitable Phenotyping Configuration through extending it by all adequate MIAPPE attributes and distributes it among the researchers;
- partly-manually, by using the ISA-Creator tool from official ISA software suite distribution with the Phenotyping Configuration to fill in and annotate experimental metadata;
- automatically, by preparing own scripts, possibly using the existing APIs, to construct ISA-Tab datasets based on manual data input (e.g. in GUI) or export from phenotypic databases.

Validation of the completed datasets against the rules provided in the ISA-Tab format specification and in the configuration can be done automatically by dedicated tools, e.g. ISA Validator.

In individual cases where adding the same new qualities for a number of experiments is necessary, we suggest creating a new local configuration based on the Phenotyping Configuration through extending it by the missing attributes, which will ensure the same structure for all of the experiments. It is important that the names of fields inherited from the original Phenotyping Configuration should not be changed in such derived configurations, and no fields should be removed, even if not used.

Similarly, the definition of Phenotyping Assay that we propose can be used as a starting point for building more specific extensions to the Phenotyping Configuration that would be appropriate for other common phenotyping measurements. For example, a high-throughput phenotyping protocol could be handled by an extension to the Phenotyping Assay, which should involve additional attributes defining phenotyping-facility-specific settings. Such extensions for the popular phenotyping platforms could be published, and included in the Phenotyping Configuration.

ISA-Tab is a very general format, suitable for a structured description of different kinds of experiments. The Investigation-Study-Assay model may look complicated at first; however, this very structure makes the format

adjustable to various types of studies, and serves as a method of normalizing the metadata. Accepting a standard universal structure should remove the burden of learning new metadata arrangement formats every time a different dataset is produced. In the Phenotyping Configuration, we propose a data arrangement that should be applicable to the vast majority of plant experiments and phenotyping procedures, and which permits a straightforward integration with different assay types.

How to use ISA-Tab? Imagine a situation in which a collection of seeds of a number of crop varieties is given to researchers at different sites to compare the influence of the local environment on yield. They perform separate trials on, assumingly, the same set of objects, in similar—but not exactly the same—experimental designs. All general information about Biosource, Environment, Treatments and Experimental Design is to be given in separate Study files for each site. Data can later be aggregated across locations according to the obligatory attribute "Geographic location". Imagine another situation, where an experiment is performed in one location, and many different researchers take samples from it, taking note of the identifier of the plant they analyse. In such a case, there is just one Study file, and a number of Assays for the individual researchers to record detailed description of handling of the samples and measurements.

We discuss the application of the presented approach by three examples of formatted datasets.

### Dataset I

Data contained in 'dataset_basic_GMI_Atwell' (Additional file 1) comes from the investigation described by Atwell et al. [59], and concerns *Arabidopsis* accessions. The data was downloaded in the ISA-Tab format from GWA-Portal at GMI. It has been formatted according to the basic phenotyping configuration. The Study file "s_Study1.txt" lists all the Biosources, i.e. *Arabidopsis* accessions, which are annotated by their identifiers in the GMI's accession list. There are multiple replications of each accession; each one is assigned a unique Sample Name. The Sample Names are repeated in the Assay file "a_study1.txt" which links them to the rows of the Derived Data File "d_data.txt" through *Assay Name* column. The columns of the Derived Data File correspond to the 107 phenotypic variables stored on the GWAS platform and defined in the Trait Definition File named "tdf.txt".

This example illustrates a situation in which the structure of the ISA-Tab archive does not reflect any actual experiment; the data, exported from an intermediary database, are in fact detached from most of their original metadata. Therefore, the information that is to be conveyed is very simple. One may say that in this situation the ISA-Tab structure, even in its basic configuration, is too complicated. However, obeying the rules even for simple datasets enhances greatly their interoperability.

## Dataset II

Data contained in 'dataset_field_IPGPAS_Polapgen' (Additional file 2) were obtained in a project aimed at studying reaction to drought in populations of barley recombinant inbred lines (RIL) [60]. The GeH population, obtained from a cross between Georgie and Harmal, consisting of 100 lines, was observed in a two-year field experiment in 2012 and 2013. The RILs and their parental forms constitute 102 biosources defined at the Study level in two files "s_study1.txt" and "s_study2.txt", corresponding to the two years. The most important environmental data concerning soil type, field size, sowing density, and day temperature are provided as values of *Parameters* of the appropriate *Protocols*. Some information required by MIAPPE was not available, therefore a few columns in both Study files are empty. The phenotyping done on samples taken from field experiments is described in two Phenotyping Assay files, "a_study1_phenotyping_field2012.txt" and "a_study1_phenotyping_field2013.txt". In the experiment eight phenotypic traits were measured; these are named and annotated in the Trait Definition File "tdf_polapgen_field.txt". Additionally, two environmental variables were recorded: "water vapor pressure deficit" and "total precipitation"; they are also described in the Trait Definition File. The observations of phenotypic traits and of environmental variables are contained in data files "d_polapgen_field2012.txt" and "d_polapgen_field2013.txt", corresponding to the two assays.

The GeH RIL dataset represents a very common case of a multi-environment study made with the same set of plant accessions. We decided to take the two environments—years—as two separate Studies; data are distinguishable upon processing by the value of the *Characteristics[Study start]* attribute. Another approach to handle different environments would consist in describing them within one Study file. In our case, however, the separation based on time-depended attribute seemed more convenient for data collection and the management of a whole series of experiments. In general, time points of sampling or data collection can be specified as a *Factor* or *Characteristic*.

Values of the environmental variables are constant over assays, as they represent the mean for the whole experimentation period and the whole experimental field. The same structure would hold single per-plot measurements. An environmental variable measured many times per experimental unit can be handled by splitting into a number of separate variables for each time point. Another approach would be to define a *Factor* "Time" and use it to

define individual *Assay Name* for combinations of experimental units and time. Yet another solution would be to define a separate Assay to keep measurements of environmental variables.

## Dataset III

The experiment described and data contained in [53] have been acquired in the frame of a series of validation experiments in IPK's high throughput plant phenotyping system for small plants. It assessed the effect of plant rotation during imaging (*Factor* "rotating"/"stationary") as well as of soil covers (*Factor* "covered"/"uncovered") on growth and development of 484 Arabidopsis plants. The dataset contains raw and processed images, extracted phenotypic features relevant for quantification of biomass (growth) and manual validation data. Detailed information about the experimental procedures and results can be found in [50]. The study has been described according to a MIAPPE-compliant ISA-Tab phenotyping configuration (Greenhouse Study) and was a part of data descriptor article [54]. The raw image files can be found in the "1135FA_images" folder. The subfolders are ordered and categorized into "camera_sensor" (vis/fluo/nir), "camera_view" (top/side) and "das" (day after sowing). The corresponding ISA-Tab files (Investigation, Study, and Assay files) for the semantical description are located in the "metadata" folder.

This dataset demonstrates the application of the ISA-Tab configuration (and MIAPPE) for a high throughput phenotyping experiment comprising time series measurements with different camera sensors. On the basis of this example the integration and representation of further related data (novel sensors, and importantly, environmental data) will be done at IPK.

## Conclusions

The results of research funded from public resources are expected to be publicly available, not only as a proof that the research was done, but also as a source of knowledge, or even input for further analyses. Open access to data is usually provided through open repositories (e.g. Dryad or Zenodo). They implement different policies of data formatting and description. Some accept objects (including datasets) of any type, assigning them simply an ID; others require adding a set of general attributes describing an object; some more ask for a specific data format. Repositories and databases of particular institutions and projects provide their own way for storage and access to data, most suitable for their needs, with an increasing policy toward Open Access. Future usability of datasets dispersed across all those repositories relies upon numerous factors: possibility to extract a specific dataset together with its metadata, comprehensible dataset formatting, completeness of its description, and clear meaning of individual elements of this description. Compatibility with other experimental results, also those of different types, is also not to be neglected in the context of data interoperability. Our work has been aimed at moving phenotyping data towards these objectives:

- The MIAPPE document, defining recommendations for phenotypic dataset description elements, helps to provide the right metadata in the dataset.
- The ISA-Tab format allows experimental metadata formatting, and thus inclusion of all important information within the dataset, making it exchangeable and independent of a data repository's metadata policy. Flexibility of the format allows to export databases' internal structures as ISA-Tab, while the definite rules for element arrangement make the experimental process traceable.
- The ISA-Tab Phenotyping Configuration provides mapping of MIAPPE requirements to ISA-Tab structures for the basic phenotyping situations, and thus facilitates dataset construction. Thanks to holding information on ontologies for particular attributes, it supports data annotation. A list of recommended ontologies for annotation of particular elements of experiment description assists in choosing formal terminology to clarify the wording, and thus avoiding ambiguity of the description. Ontological annotation is accommodated in ISA-Tab datasets.

The Minimum Information About a Plant Phenotyping Experiment document has been constructed as a result of consultations with a number of research groups within the transPLANT project and beyond, especially EPPN and DPPN. Although it is focused on classical phenotyping experiments, some attention in MIAPPE is also given to less frequently performed, but nonetheless important, experiments in aquatic and biotic conditions. Yet, a real application of MIAPPE in such situations would require more discussion with relevant practitioners. The same remark applies to observational studies.

Based on experimental data from high throughput plant phenotyping experiments at IPK using the LemnaTec platform, a first version of a high throughput phenotyping configuration has been prepared. This work builds the basis for a comprehensive plant phenomics experiment documentation and data publication pipeline. Indeed this kind of experiments comprising automated multisensor-imaging-based procedures can produce terabytes of data for each experiment. Handling such Big Data needs dedicated technologies and the level of resolution of related experimental metadata to

be represented and published using ISA-Tab archives is still a matter of discussion. The selection of an adequate level of detail (geographical location of every single pot vs. location of the greenhouse), data volume (whether to remove low quality images or not) and processing stage (raw images vs. compressed/processed images) for data publication is linked to the technical capability of publication servers as well as institutional or journal policies. Nevertheless, the continuous documentation of the data lifecycle is a basic requirement for a consistent and seamless creation of ISA-Tab archives. We hope that the discussion with interested parties dealing with this type of experiments will allow a general or platform-specific High-Throughput Phenotyping Assay to be developed.

The textual nature of the ISA-Tab format makes it directly readable for everyone, without the need for any special software and support from computer scientists or bioinformaticians. Similarly, the construction of a dataset is possible manually, in a text or spreadsheet editor, by filling in a prepared template. A more advanced option is the preparation of an own implementation of data export/import as ISA-Tab based on the format specification to combine ISA-Tab with existing databases and tools. ISA-Tab is also supported by a free software suite, ISA-Tools, developed by ISA group [61] and members of the community. There are a number of tools and APIs for dataset construction, validation, analysis, management, and export to other formats. Certain functionalities of this official tools distribution are not yet provided, but the implementation of new user-friendly environments for dataset management is in progress [62, 63]. Further development of tools supporting formatting of data according to the given rules is an important step to promote adoption of the metadata standards.

Since the textual nature of ISA-Tab makes it not particularly convenient for automatic processing, the possibility to export ISA-Tab dataset structure to other formats is a useful feature. The existing tools provide, among others, JSON and RDF representations, as well as OWL for compatibility with the Linked Data. ISA-API [64] is going to further simplify programmatic approach to data formatting and management.

The ISA-Tab format has been accepted by the Nature Publishing Group for dataset publication, which additionally popularizes the format and encourages new users. More work is needed to achieve a widespread acceptance of the policy of data publication in the form of open resources. The FAIR Data Principles [58] that define the properties of a good dataset are a convenient reminder of the targets that are to be aimed at. Acceptance of the rules described in this paper will help to achieve these targets by the plant phenotyping community.

## Abbreviations

MIAPPE: Minimum Information About a Plant Phenotyping Experiment specification; ISA-Tab: investigation-study-assay tabular format for experimental metadata description; G × E: genotype by environment interaction; GWAS: genome-wide association study.

## Authors' contributions

PKr and PKe planned and coordinated the research. HĆK and PKr proposed the solutions, coordinated the exchange of ideas of other contributors and their incorporation in the final solution, drafted the first version of the manuscript, and prepared its final version after corrections. All authors participated in shaping the standard by bringing the theoretical contribution to Minimum Information and formatting requirements, or implementing and testing the solutions described in the paper at their institutions. All of the authors worked together to edit the manuscript. All authors read and approved the final manuscript.

## Author details

[1] Institute of Plant Genetics, Polish Academy of Sciences, ul. Strzeszyńska 34, 60-479 Poznań, Poland. [2] Leibniz Institute of Plant Genetics and Crop Plant Research (IPK), OT Gatersleben, Corrensstraße 3, 06466 Stadt Seeland, Germany. [3] Bioversity International, parc Scientifique Agropolis II, 34397 Montpellier Cedex 5, France. [4] Institute for Biochemistry and Biology, University of Potsdam, 14476 Potsdam, Germany. [5] INRA, UR1164 URGI - Research Unit in Genomics-Info, INRA de Versailles-Grignon, Route de Saint-Cyr, 78026 Versailles, France. [6] Forschungszentrum Jülich GmbH, IBG-2 Plant Sciences, 52425 Jülich, Germany. [7] Institute of Computing Science, Poznań University of Technology, ul. Piotrowo 3a, 60-479 Poznań, Poland. [8] LemnaTec GmbH, Pascalstraße 59, 52076 Aachen, Germany. [9] Poznań Supercomputing and Networking Center Affiliated to the Institute of Bioorganic Chemistry, Polish Academy of Sciences, ul. Jana Pawła II 10, 61-139 Poznań, Poland. [10] UMR MISTEA, INRA SupAgro, Place Pierre Viala 2, 34060 Montpellier, France. [11] Keygene N.V., Agro Business Park 90, 6708 PW Wageningen, The Netherlands. [12] Oxford e-Research Centre, University of Oxford, 7 Keble Road, Oxford OX1 3QG, UK. [13] Gregor Mendel Institute, Austrian Academy of Sciences, 1030 Vienna, Austria. [14] Institute of Biology I, BioSC, RWTH Aachen, Worringer Weg 3, Aachen, Germany. [15] The European Molecular Biology Laboratory–The European Bioinformatics Institute, Wellcome Trust Genome Campus, Hinxton, Cambridgeshire CB10 1SD, UK.

## Acknowledgements

We are grateful to all partners of the supporting projects for useful discussions, and to anonymous reviewers for helpful comments. Thanks to the URGI engineers for their help in implementing GnpIS ISA-Tab export, and in particular Thomas Letellier.

## Competing interests

The authors declare that they have no competing interests.

## Funding

The work was supported by the European Union under the FP7 thematic priority Infrastructures, Project No. 283496 transPLANT and Project No. 28443 EPPN—European Plant Phenotyping Network, by the German Federal Ministry of Education and Research (BMBF) Grant No. 031A053A/B/C DPPN—German Plant Phenotyping Network, and by the Infrastructure Biologie Santé 'Phenome-FPPN' supported by the National Research Agency and the "Programme d'Investissements d'Avenir" (PIA) as ANR-11-INBS-0012.

## References

1. Krajewski P, Chen D, Ćwiek H, et al. Towards recommendations for metadata and data handling in plant phenotyping. J Exp Bot. 2015;66:5417–27.
2. Pérez-Harguindeguy N, Díaz S, Garnier E, et al. New handbook for standardised measurement of plant functional traits worldwide. Aust J Bot. 2013;61:167–234.
3. PrometheusWiki. http://prometheuswiki.publish.csiro.au. Accessed 30 Mar 2016.
4. DROPS project. http://drops-project.eu. Accessed 30 Mar 2016.
5. PODD repository. http://plantphenomics.org.au/projects/podd. Accessed 30 Mar 2016.
6. Li YF, Kennedy G, Davies F, Hunter J. PODD: an ontology-driven data repository for collaborative phenomics research. The role of digital libraries in a time of global change. Berlin: Springer; 2010. p. 179–88.
7. Phenome network. https://www.phenome-fppn.fr/phenome_eng. Accessed 30 Mar 2016.
8. MaizeGDB portal. http://www.maizegdb.org. Accessed 30 Mar 2016.
9. Triticeae Toolbox portal. https://triticeaetoolbox.org. Accessed 30 Mar 2016.
10. Phenopsis DB database. http://bioweb.supagro.inra.fr/phenopsis. Accessed 30 Mar 2016.
11. GnpIS-Ephesis database. https://urgi.versailles.inra.fr/ephesis. Accessed 30 Mar 2016.
12. Yilmaz P, et al. The genomic standards consortium: bringing standards to life for microbial ecology. ISME J. 2011;5:1565–7.
13. Zimmerman P, et al. MIAME/Plant—adding value to plant microarrray experiments. Plant Methods. 2006;2:1.
14. Taylor CF, et al. The minimum information about a proteomics experiment (MIAPE). Nat Biotechnol. 2007;25:887–93.
15. Fiehn O, et al. The metabolomics standards initiative (MSI). Metabolomics. 2007;3:175–8.
16. Fiehn O, et al. Minimum reporting standards for plant biology context information in metabolomic studies. Metabolomics. 2007;3:195–201.
17. Morrison N, et al. Standard reporting requirements for biological samples in metabolomics experiments: environmental context. Metabolomics. 2007;3:203–10.
18. Rocca-Serra P, Salek RM, Arita M, et al. Data standards can boost metabolomics research, and if there is a will, there is a way. Metabolomics. 2016;12:14.
19. Rayner TF, Rocca-Serra P, Spellman PT, et al. A simple spreadsheet-based, MIAME-supportive format for microarray data: MAGE-TAB. BMC Bioinform. 2006;7:489.
20. ArrayExpress database. https://www.ebi.ac.uk/arrayexpress. Accessed 30 Mar 2016.
21. PRIDE database. http://www.ebi.ac.uk/pride. Accessed 30 Mar 2016.
22. Rocca-Serra P, Brandizi M, Maguire E, et al. ISA software suite: supporting standards-compliant experimental annotation and enabling curation at the community level. Bioinformatics. 2010;26:2354–6.
23. ISA commons. http://isacommons.org. Accessed 30 Mar 2016.
24. Trans-national Infrastructure for Plant Genomic Science. http://transplantdb.eu. Accessed 30 Mar 2016.
25. European Plant Phenotyping Network. http://plant-phenotyping-network.eu. Accessed 30 Mar 2016.
26. Poorter H, Fiorani F, Stitt M, et al. The art of growing plants for experimental purposes: a practical guide for the plant biologist. Funct Plant Biol. 2012;39:821–38.
27. Hannemann J, Poorter H, Usadel B, et al. Xeml Lab: a software suite for a standardised description of the growth environment of plants. Plant, Cell Environ. 2009;32:1185–200.
28. ISA-Tab software suite. http://www.isa-tools.org/software-suite. Accessed 30 Mar 2016.

29. McNeill J, et al., editors. International code of botanical nomenclature (Vienna Code) adopted by the seventeenth International Botanical Congress, Vienna, Austria, July 2005. 2006. http://web.archive.org/web/20121006231936/, http://ibot.sav.sk/icbn/main.htm. Accessed 30 Mar 2016.

30. FAO/Bioversity Multi-Crop Passport Descriptors, http://www.bioversity-international.org/e-library/publications/detail/faobioversity-multi-crop-passport-descriptors-v2-mcpd-v2. Accessed 30 Mar 2016.

31. Cochran WG, Cox GM. Experimental designs. New York: Wiley; 1957.

32. Crop Ontology Platform. http://www.cropontology.org. Accessed 30 March 2016.

33. Shrestha R, et al. Bridging the phenotypic and genetic data useful for integrated breeding through a data annotation using the CropOntology developed by the crop communities of practice. Front Physiol. 2012;3:326.

34. Planteome project. http://planteome.org. Accessed 30 Mar 2016.

35. Crop Research Ontology. http://cropontology.org/ontology/CO_715/Crop%20Research. Accessed 30 Mar 2016.

36. XEML Environment Ontology. http://purl.bioontology.org/ontology/XEO. Accessed 30 Mar 2016.

37. Statistics Ontology. http://bioportal.bioontology.org/ontologies/STATO. Accessed 30 Mar 2016.

38. ISA-Tab format specification. http://www.isa-tools.org/format/specification. Accessed 30 Mar 2016.

39. MIAPPE at BioSharing portal. https://www.biosharing.org/bsg-000543. Accessed 30 Mar 2016.

40. ISA-Tab configurations. http://www.isa-tools.org/format/configurations. Accessed 30 Mar 2016.

41. MetaboLights database. http://www.ebi.ac.uk/metabolights. Accessed 30 Mar 2016.

42. BII database for phenotypic data. http://www.igr.poznan.pl/bb/bii. Accessed 30 Mar 2016.

43. Steinbach D, Alaux M, Amselem J, et al. GnpIS: an information system to integrate genetic and genomic data from plants and fungi. Database. 2013;. doi:10.1093/database/bat058.

44. An exemplary dataset stored at Ephesis database. https://urgi.versailles.inra.fr/ephesis/ephesis/viewer.do#dataResults/trialSetIds=5,6,7. Accessed 30 Mar 2016.

45. Oury FX, Heumez E, Rolland B, et al. A dataset: winter wheat (*Triticum aestivum* L.) phenotypic data from the multiannual, multilocal field trials of the INRA small grain cereals. Network. 2015;. doi:10.15454/1.4489662216568333E12.

46. Arend D, Colmsee C, Knüpffer H, et al. Data management experiences and best practices from the perspective of a plant research institute. In: Galhardas H, Rahm E, editors. Data integration in life sciences. Lecture notes in bioinformatics. Springer 2014; 8574:41–9. doi:10.1007/978-3-319-08590-6_4.

47. Laboratory Information Management System at IPK. http://www.ipk-gatersleben.de/en/databases. Accessed 30 Mar 2016.

48. Arend D, Lange M, Chen J, et al. e!DAL—a framework to store, share and publish research data. BMC Bioinform. 2014;15:214.

49. Junker A. A dataset "Optimizing experimental procedures for quantitative evaluation of crop plant performance in high throughput phenotyping systems". 2014. doi:10.5447/IPK/2014/4.

50. Junker A, Muraya MM, Weigelt-Fischer K, et al. Optimizing experimental procedures for quantitative evaluation of crop plant performance in high throughput phenotyping systems. Front Plant Sci. 2015;5:770.

51. Plant Genomics and Phenomics Data Repository. http://edal.ipk-gatersle-ben.de/repos/pgp. Accessed 30 Mar 2016.

52. Arend D, Junker A, Scholz U, et al. PGP repository: a plant phenomics and genomics data publication infrastructure. Database. 2016;2016:1–11. doi:10.1093/database/baw033.

53. Junker A. A dataset "Raw images files from quantitative monitoring of 484 Arabidopsis thaliana plants using high-throughput plant phenotyping". 2016. doi:10.5447/IPK/2016/7.

54. Arend D, et al. Quantitative monitoring of Arabidopsis thaliana growth and development using high-throughput plant phenotyping. Sci Data. 2016;. doi:10.1038/sdata.2016.55.

55. GWA-Portal. http://gwas.gmi.oeaw.ac.at. Accessed 30 Mar 2016.

56. Genomic Diversity and Phenotype Data Model. http://tassel.bitbucket.org/gdpdm. Accessed 30 Mar 2016.

57. ISAcreator library. https://github.com/ISA-tools/ISAcreator. Accessed 30 March 2016.

58. Wilkinson M, et al. The FAIR guiding principles for scientific data management and stewardship. Scientific Data. 2016.

59. Atwell S, et al. Genome-wide association study of 107 phenotypes in Arabidopsis thaliana inbred lines. Nature. 2010;465:627–31.

60. Mikołajczak K, Ogrodowicz P, Gudyś K, et al. Quantitative trait loci for yield and yield-related traits in spring barley populations derived from crosses between European and Syrian cultivars. PLoS ONE. 2016;. doi:10.1371/journal.pone.0155938.

61. The ISA Team. http://www.isa-tools.org/team. Accessed 30 Mar 2016.

62. ISA-explorer. http://scientificdata.isa-explorer.org. Accessed 30 Mar 2016.

63. COPO project. https://documentation.tgac.ac.uk/display/COPO. Accessed 9 June 2016.

64. ISA-API. https://github.com/ISA-tools/isa-api. Accessed 9 June 2016.

65. ISA-Tab reporting guideline. http://www.biosharing.org/bsg-000078. Accessed 30 Mar 2016.

66. Ontology for Biomedical Investigations. http://obi-ontology.org. Accessed 30 Mar 2016.

67. Core Information for Metabolomics Reporting. http://www.biosharing.org/bsg-000175. Accessed 30 Mar 2016.

68. Gazetteer. http://purl.bioontology.org/ontology/GAZ. Accessed 30 Mar 2016.

69. Minimum Information about any (x) Sequence, reporting guideline. http://www.biosharing.org/bsg-000518. Accessed 30 Mar 2016.

70. UNIPROT Taxonomy. http://www.uniprot.org/taxonomy. Accessed 30 Mar 2016.

71. NCBI Taxonomy http://www.ncbi.nlm.nih.gov/taxonomy. Accessed 30 Mar 2016.

72. Ontology of environmental features and habitats http://purl.bioontology.org/ontology/ENVO. Accessed 30 Mar 2016.

73. PTO, Plant Trait Ontology. https://bioportal.bioontology.org/ontologies/PTO. Accessed 30 Mar 2016.

74. PO, Plant Ontology. http://www.plantontology.org. Accessed 30 Mar 2016.

75. PATO, Phenotypic Quality Ontology. http://purl.bioontology.org/ontology/PATO. Accessed 30 Mar 2016.

# Permissions

# List of Contributors

**Marta Vazquez-Vilar, Joan Miquel Bernabé-Orts, Asun Fernandez-del-Carmen, Antonio Granell and Diego Orzaez**
Instituto de Biología Molecular y Celular de Plantas (IBMCP), Consejo Superior de Investigaciones Científicas, Universidad Politécnica de Valencia, Camino de Vera s/n, 46022 Valencia, Spain

**Pello Ziarsolo and Jose Blanca**
Centro de Conservación y Mejora de la Agrodiversidad Valenciana (COMAV), Universidad Politécnica de Valencia, Camino de Vera s/n, 46022 Valencia, Spain

**Mitchell A. Ellison**
Department of Biology, University of Pittsburgh School of Medicine, Pittsburgh, PA, USA

**Michael B. McMahon, Morris R. Bonde and Douglas G. Luster**
USDA-ARS Foreign Disease-Weed Science Research Unit, Ft. Detrick, MD, USA

**Cristi L. Palmer**
IR-4 Project, Rutgers University, Princeton, NJ, USA

**Andrew S. Fister**
The Huck Institutes of the Life Sciences, The Pennsylvania State University, 422 Life Sciences Building, University Park, PA 16802, USA

**Siela N. Maximova and Mark J. Guiltinan**
The Huck Institutes of the Life Sciences, The Pennsylvania State University, 422 Life Sciences Building, University Park, PA 16802, USA
The Department of Plant Science, The Pennsylvania State University, University Park, PA 16802, USA

**Zi Shi**
Center for Applied Genetic Technologies, University of Georgia, Athens, GA 30602, USA

**Yufan Zhang**
Department of Electrical Engineering, Princeton University, Princeton, NJ 08544, USA

**Emily E. Helliwell**
Department of Botany and Plant Pathology, Center for Genome Research and Biocomputing, Oregon State University, Corvallis, OR 97331, USA

**Michal Goralski and Paula Sobieszczanska**
Institute of Bioorganic Chemistry, Polish Academy of Sciences, Noskowskiego 12/14, 61-704 Poznan, Poland

**Aleksandra Swiercz, Agnieszka Zmienko and Marek Figlerowicz**
Institute of Bioorganic Chemistry, Polish Academy of Sciences, Noskowskiego 12/14, 61-704 Poznan, Poland
Institute of Computing Science, Poznan University of Technology, Piotrowo 2, 60-965 Poznan, Poland

**Aleksandra Obrepalska-Steplowska**
Institute of Plant Protection – National Research Institute, Wladyslawa Wegorka 20, 60-318 Poznan, Poland

**Peter Lootens, Tom Ruttink and Isabel Roldán-Ruiz**
Plant Sciences Unit - Growth and Development, ILVO, Caritasstraat 39, 9090 Melle, Belgium

**Antje Rohde**
Plant Sciences Unit - Growth and Development, ILVO, Caritasstraat 39, 9090 Melle, Belgium
Bayer CropScience, Technologiepark 38, 9052 Ghent, Belgium.

**Didier Combes and Philippe Barre**
INRA – UR4 P3F, BP 6, 86600 Lusignan, France

**Anthony C. Bryan, Sara S. Jawdy, Xiaohan Yang, Jin-Gui Chen and Gerald A. Tuskan**
Biosciences Division, Oak Ridge National Laboratory, Oak Ridge, TN 37831, USA

**Olaf Czarnecki**
Biosciences Division, Oak Ridge National Laboratory, Oak Ridge, TN 37831, USA
KWS SAAT SE, Grimsehlstraße 31, 37555 Einbeck, Germany

**Zong-Ming Cheng**
Department of Plant Sciences, University of Tennessee, Knoxville, TN 37996, USA

**Sarah R. Hind and Susana De la Torre Diaz**
Boyce Thompson Institute for Plant Research, Ithaca, NY 14853, USA

**Simon Schwizer, Christine M. Kraus and Gregory B. Martin**
Boyce Thompson Institute for Plant Research, Ithaca, NY 14853, USA
Plant Pathology and Plant–Microbe Biology Section, School of Integrative Plant Science, Cornell University, Ithaca, NY 14853, USA

**Patrick C. Boyle**
Boyce Thompson Institute for Plant Research, Ithaca, NY 14853, USA
Monsanto Company, St. Louis, MO 63141, USA

**Bin He**
Department of Chemistry and Chemical Biology, Cornell University, Ithaca, NY 14853, USA
College of Pharmacy, Guiyang Medical University, Guiyang 550004, Guizhou, China

**David Dobnik, Ana Lazar, Tjaša Stare, Kristina Gruden and Jana Žel**
Department of Biotechnology and Systems Biology, National Institute of Biology, Večna Pot 111, 1000 Ljubljana, Slovenia

**Vivianne G. A. A. Vleeshouwers**
Wageningen UR Plant Breeding, Wageningen University and Research Centre, P.O. Box 386, 6700 AJ Wageningen, The Netherlands

**De-Zhu Li and Zhen-Hua Guo and Xiao-Yan Wang**
Germplasm Bank of Wild Species, Kunming Institute of Botany, Chinese Academy of Sciences, Kunming 650201, China

**Guo-Qian Yang, Yun-Mei Chen, Cen Guo, Lei Zhao and Ying Guo**
Germplasm Bank of Wild Species, Kunming Institute of Botany, Chinese Academy of Sciences, Kunming 650201, China
Kunming College of Life Sciences, University of Chinese Academy of Sciences, Kunming 650201, China

**Jin-Peng Wang and Li Li**
Key Laboratory of Experimental Marine Biology, Institute of Oceanology, Chinese Academy of Sciences, Qingdao 266071, China

**Anna Ostendorp, Steffen Pahlow, Jennifer Deke, Melanie Thieß and Julia Kehr**
Molecular Plant Genetics, University Hamburg, Biocenter Klein Flottbek, Ohnhorststr. 18, 22609 Hamburg, Germany

**Caroline Dean and Stefanie Rosa**
Department of Cell and Developmental Biology, John Innes Centre, Norwich Research Park, Norwich NR4 7UH, UK

**Susan Duncan, Tjelvar S. G. Olsson and Matthew Hartley**
Department of Computational and Systems Biology, John Innes Centre, Norwich Research Park, Norwich NR4 7UH, UK

**Chantal Le Marié, Norbert Kirchgessner, Patrick Flütsch, Johannes Pfeifer, Achim Walter and Andreas Hund**
Institute of Agricultural Sciences, ETH Zurich, Universitätstrasse 2, 8092 Zurich, Switzerland

**Petr Cápal, Jan Vrána, Marie Kubaláková, Miroslava Karafiátová, Eva Komínková and Jaroslav Doležel**
Institute of Experimental Botany, Centre of the Region Haná for Biotechnological and Agricultural Research, Šlechtitelů 31, 78371 Olomouc, Czech Republic

**Takashi R. Endo**
Institute of Experimental Botany, Centre of the Region Haná for Biotechnological and Agricultural Research, Šlechtitelů 31, 78371 Olomouc, Czech Republic
Faculty of Agriculture, Ryukoku University, 1-5 Yokotani, Seta Oe-cho, Otsu, Shiga 520-2194, Japan

**Isabel Mora-Ramírez and Winfriede Weschke**
Leibniz Institute of Plant Genetics and Crop Plant Research (IPK) Gatersleben, Corrensstrasse 3, 06466 Stadt Seeland, Germany

**Adnan Riaz and Lee Hickey**
Queensland Alliance for Agriculture and Food Innovation, The University of Queensland, St Lucia, QLD 4072, Australia

**Sambasivam Periyannan**
Commonwealth Scientific and Industrial Research Organization (CSIRO) Agriculture, General Post Office Box 1600, Canberra, ACT 2601, Australia

**Elizabeth Aitken**
School of Agriculture and Food Science, The University of Queensland, St Lucia, QLD 4072, Australia

**Atena Haghighattalab and Douglas Goodin**
Department of Geography, Kansas State University, Manhattan, KS 66506, USA

**Lorena González Pérez, Suchismita Mondal, Ivan Ortiz-Monasterio and Ravi Prakash Singh**
International Maize and Wheat Improvement Center (CIMMYT), Int. Apdo., Postal 6-641, 06600 Mexico, D.F., Mexico

**Jessica Rutkoski**
International Maize and Wheat Improvement Center (CIMMYT), Int. Apdo., Postal 6-641, 06600 Mexico, D.F., Mexico
International Programs of the College of Agriculture and Life Science, Cornell University, Ithaca, NY 14853, USA

**Daljit Singh**
Interdepartmental Genetics Program, Kansas State University, Manhattan, KS 66506, USA

**Dale Schinstock**
Department of Mechanical and Nuclear Engineering, Kansas State University, Manhattan, KS 66506, USA

**Jesse Poland**
Wheat Genetics Resource Center, Department of Plant Pathology and Department of Agronomy, Kansas State University, Manhattan, KS 66506, USA

**Nagaveni Budhagatapalli, Sindy Schedel, Maia Gurushidze, Stefan Hiekel Jochen Kumlehn and Goetz Hensel**
Plant Reproductive Biology, Leibniz Institute of Plant Genetics and Crop Plant Research (IPK), Corrensstr. 3, 06466 Stadt Seeland/OT Gatersleben, Germany

**Stefanie Pencs**
Plant Reproductive Biology, Leibniz Institute of Plant Genetics and Crop Plant Research (IPK), Corrensstr. 3, 06466 Stadt Seeland/OT Gatersleben, Germany
Chair of Plant Breeding, Martin Luther University, Betty-Heimann-Str. 3, 06120 Halle (Saale), Germany

**Twan Rutten**
Structural Cell Biology, Leibniz Institute of Plant Genetics and Crop Plant Research (IPK), Corrensstr. 3, 06466 Stadt Seeland/OT Gatersleben, Germany

**Stefan Kusch and Ralph Panstruga**
Unit of Plant Molecular Cell Biology, Institute for Biology I, RWTH Aachen University, Worringerweg 1, 52056 Aachen, Germany

**Robert Morbitzer and Thomas Lahaye**
ZMBP-General Genetics, University of Tübingen, Auf der Morgenstelle 32, 72076 Tübingen, Germany

**Zhicheng Zhang, Kurt Boonen, Liliane Schoofs, Filip Rolland and Koen Geuten**
Department of Biology, KU Leuven, Kasteelpark Arenberg 31, 3001 Louvain, Belgium

**Piero Ferrari and Ewald Janssens**
Department of Physics and Astronomy, KU Leuven, Celestijnenlaan 200d, 3001 Louvain, Belgium

**Vera van Noort**
Department of Microbial and Molecular Systems, KU Leuven, Kasteelpark Arenberg 22, 3001 Louvain, Belgium

**Hanna Ćwiek-Kupczyńska and Paweł Krajewski**
Institute of Plant Genetics, Polish Academy of Sciences, ul. Strzeszyńska 34, 60-479 Poznań, Poland

**Wojciech Frohmberg**
Institute of Plant Genetics, Polish Academy of Sciences, ul. Strzeszyńska 34, 60-479 Poznań, Poland
Institute of Computing Science, Poznań University of Technology, ul. Piotrowo 3a, 60-479 Poznań, Poland

**Thomas Altmann, Daniel Arend, Astrid Junker, Matthias Lange, Uwe Scholz and Stephan Weise**
Leibniz Institute of Plant Genetics and Crop Plant Research (IPK), OT Gatersleben, Corrensstraße 3, 06466 Stadt Seeland, Germany

**Elizabeth Arnaud**
Bioversity International, parc Scientifique Agropolis II, 34397 Montpellier Cedex 5, France

**Dijun Chen**
Institute for Biochemistry and Biology, University of Potsdam, 14476 Potsdam, Germany

**Guillaume Cornut and Cyril Pommier**
INRA, UR1164 URGI - Research Unit in Genomics-Info, INRA de Versailles-Grignon, Route de Saint-Cyr, 78026 Versailles, France

**Fabio Fiorani, Anahita Nafissi and Hendrik Poorter**
Forschungszentrum Jülich GmbH, IBG-2 Plant Sciences, 52425 Jülich, Germany

**Björn Usadel**
Forschungszentrum Jülich GmbH, IBG-2 Plant Sciences, 52425 Jülich, Germany
Institute of Biology I, BioSC, RWTH Aachen, Worringer Weg 3, Aachen, Germany

**Christian Klukas**
LemnaTec GmbH, Pascalstraße 59, 52076 Aachen, Germany

**Cezary Mazurek**
Poznań Supercomputing and Networking Center Affiliated to the Institute of Bioorganic Chemistry, Polish Academy of Sciences, ul. Jana Pawła II 10, 61-139 Poznań, Poland

**Pascal Neveu**
UMR MISTEA, INRA SupAgro, Place Pierre Viala 2, 34060 Montpellier, France

**Jan van Oeveren and Marco van Schriek**
Keygene N.V., Agro Business Park 90, 6708 PW Wageningen, The Netherlands

**Philippe Rocca-Serra and Susanna-Assunta Sansone**
Oxford e-Research Centre, University of Oxford, 7 Keble Road, Oxford OX1 3QG, UK

**Ümit Seren**
Gregor Mendel Institute, Austrian Academy of Sciences, 1030 Vienna, Austria

**Paul Kersey**
The European Molecular Biology Laboratory–The European Bioinformatics Institute, Wellcome Trust Genome Campus, Hinxton, Cambridgeshire CB10 1SD, UK

# Index

www.ingramcontent.com/pod-product-compliance
Lightning Source LLC
Chambersburg PA
CBHW082100190326
41458CB00010B/3536